ONCODEVELOPMENTAL MARKERS

Biologic, Diagnostic, and Monitoring Aspects

CONTRIBUTORS

Kenneth D. Bagshawe
Stephen B. Baylin
Richard H. J. Begent
William A. Blattner
Hans Bohn
Glenn D. Braunstein
P. Burtin
Harris Busch
Rose K. Busch
William J. Catalona
Pui-Kwong Chan
Frank Dolbeare
M. J. Escribano
William H. Fishman
Robert F. Gagel
Ronald B. Herberman
David Kelsey

Young S. Kim
Hilary Koprowski
Paul H. Lange
D. J. R. Laurence
Lawrence J. McIntyre
Mani Menon
A. M. Neville
Derek Raghavan
Raymond W. Ruddon
Erkki Ruoslahti
Markku Seppälä
Marjatta Son
William H. Spohn
Zenon Steplewski
Kei Takahashi
Kai-Li Xu
Norman Zamcheck

ONCODEVELOPMENTAL MARKERS
Biologic, Diagnostic, and Monitoring Aspects

EDITED BY

William H. Fishman

Cancer Research Center
La Jolla Cancer Research Foundation
La Jolla, California

RC 268.3
O 53
1983

 1983

ACADEMIC PRESS
A Subsidiary of Harcourt Brace Jovanovich, Publishers
New York   London
Paris   San Diego   San Francisco   São Paulo   Sydney   Tokyo   Toronto

COPYRIGHT © 1983, BY ACADEMIC PRESS, INC.
ALL RIGHTS RESERVED.
NO PART OF THIS PUBLICATION MAY BE REPRODUCED OR
TRANSMITTED IN ANY FORM OR BY ANY MEANS, ELECTRONIC
OR MECHANICAL, INCLUDING PHOTOCOPY, RECORDING, OR ANY
INFORMATION STORAGE AND RETRIEVAL SYSTEM, WITHOUT
PERMISSION IN WRITING FROM THE PUBLISHER.

ACADEMIC PRESS, INC.
111 Fifth Avenue, New York, New York 10003

United Kingdom Edition published by
ACADEMIC PRESS, INC. (LONDON) LTD.
24/28 Oval Road, London NW1 7DX

Library of Congress Cataloging in Publication Data
Main entry under title:

Oncodevelopmental markers.

Includes index.
1. Carcinoembryonic antigens. 2. Tumor antigens.
I. Fishman, William H. [DNLM: 1. Genetic marker.
2. Neoplasms--Diagnosis. 3. Oncogenes. QZ 241 O58]
RC268.3.O53 1983 616.9'940792 83-3826
ISBN 0-12-257701-9

PRINTED IN THE UNITED STATES OF AMERICA

83 84 85 86 9 8 7 6 5 4 3 2 1

Contents

CONTRIBUTORS xiii
PREFACE xvii

Part I
BIOLOGIC ASPECTS

CHAPTER 1
Oncodevelopmental Markers
WILLIAM H. FISHMAN

I.	Introduction	3
II.	Embryologic Basis of Tumor Nomenclature	4
III.	Oncodevelopmental Alkaline Phosphatase Isoenzymes	9
IV.	Extent of Oncodevelopmental Expression	13
V.	Current Perspective	14
	References	16

CHAPTER 2
Cell–Matrix Interactions as Determinants of Differentiation and Tumor Invasion
ERKKI RUOSLAHTI

I.	Introduction	21
II.	Structure of Extracellular Matrices and Their Components	22
III.	Interaction of Cells with Extracellular Matrix	27
	References	32

v

CHAPTER 3
Human Tumor Nucleolar Antigens
HARRIS BUSCH, ROSE K. BUSCH, PUI-KWONG CHAN, DAVID KELSEY, KEI TAKAHASHI, WILLIAM H. SPOHN, and MARJATTA SON

I.	Introduction	38
II.	Nucleolar Antigens	38
III.	Discussion	62
	References	66

CHAPTER 4
Systematic Identification of Specific Oncoplacental Proteins
HANS BOHN

I.	Introduction	69
II.	Detection and Classification of Placental Proteins	70
III.	Purification of the Specific Oncoplacental Proteins	73
IV.	Characterization of the Specific Oncoplacental Proteins	75
V.	Localization in Normal Tissues and Cells	78
VI.	Quantitation in Placental Extracts and Normal Body Fluids	78
VII.	Occurrence in Tumor Patients and Application as Markers in Oncology	81
	References	84

CHAPTER 5
Marker Expression by Cultured Cancer Cells
RAYMOND W. RUDDON

I.	Tumor-Associated Antigens	88
II.	Cell Membrane-Associated Glycoproteins and Glycolipids	91
III.	Hormones	94
IV.	Enzymes	98
V.	Tumor-Derived Growth Factors	102
	References	103

CHAPTER 6
The Study of Oncodevelopmental Markers in Heterotransplanted Tumors
DEREK RAGHAVAN

I.	Introduction	109
II.	Oncodevelopmental Proteins	112
III.	Hormones and Other Substances	123
IV.	Cell Surface Antigens	125
V.	Summary	126
	References	126

CHAPTER 7
Specific Antibodies in Cytopathology and Immunohistology
D. J. R. LAURENCE and A. M. NEVILLE

I.	Introduction	131
II.	The Improved Classification of Human Tumors	133
III.	The Detection of Micrometastases	139
IV.	Endocrine Aspects of Cancer	143
V.	Monoclonal Antibodies and Immunohistology	147
VI.	Conclusions	152
	References	152

CHAPTER 8
Use of Monoclonal Antibodies to Recognize Tumor Antigens
ZENON STEPLEWSKI and HILARY KOPROWSKI

I.	Introduction	155
II.	Hybridomas	156
III.	Monoclonal Antibody-Defined Antigens of Human Melanoma	156
IV.	Monoclonal Antibody-Defined Antigens of Human Gastrointestinal Tumors	159
V.	Oncodevelopmental Character of the Monoclonal Antibody-Defined Antigens	163
	References	164

CHAPTER 9
Radioimmunolocalization of Cancer
RICHARD H. J. BEGENT and KENNETH D. BAGSHAWE

I.	Introduction	167
II.	Antibody Localization in Xenografts of Human Tumors in Experimental Animals	168
III.	Antibody Localization in Patients with Cancer	169
IV.	Potential Improvements	184
V.	Conclusions	186
	References	187

CHAPTER 10
Familial Cancer: An Opportunity to Study Mechanisms of Neoplastic Transformation
WILLIAM A. BLATTNER

I.	Introduction	189
II.	Genetic Syndromes and Cancer	190
III.	Dysplastic Nevi and Risk for Cutaneous Malignant Melanomas	194
IV.	Defective Immune Response to EB Virus and Lymphoproliferation	196

V.	HLA-Associated Defective Immune Response Genes	197
VI.	Diverse Neoplasms in a Family and a Novel Radiation Repair Defect	198
VII.	Chromosomal Rearrangement and Cancer Risk in Families	199
VIII.	Genetic Markers Linked to Familial Cancer Risk	200
IX.	Defects in Cellular Growth Regulation and Risk for Familial Colon Cancer	201
X.	Conclusions	202
	References	202

Part II
DIAGNOSTIC AND MONITORING ASPECTS

CHAPTER 11
Flow Cytoenzymology—An Update
FRANK DOLBEARE

I.	Introduction	207
II.	Flow Cytoenzymology	208
III.	Approaches to Analyzing Cellular Enzymes by Flow Cytometry	209
IV.	Analytical Considerations	212
V.	Enzymes in Oncology	214
VI.	Future State of the Art	216
	References	217

CHAPTER 12
Tumor Markers of Medullary Thyroid Carcinoma
ROBERT F. GAGEL

I.	Calcitonin Secretion by the Normal C Cell	222
II.	Calcitonin as a Tumor Marker	222
III.	The Histologic Evolution of Medullary Thyroid Carcinoma	227
IV.	Other Tumor Markers	229
V.	The Rat Model of Medullary Thyroid Carcinoma	233
VI.	The Clonal Nature of Hereditary Medullary Thyroid Carcinoma	234
VII.	Control of Gene Expression in MTC	235
VIII.	A Model for the Future Study of Hereditary Medullary Thyroid Carcinoma	236
	References	236

CHAPTER 13
Markers for Germ Cell Tumors of the Testis
PAUL H. LANGE

I.	Introduction	241
II.	Clinical Aspects of Testicular Cancer	242
III.	Background of α-Fetoprotein and Human Chorionic Gonadotropin	243
IV.	Sensitivity and Specificity of α-Fetoprotein and Human Chorionic Gonadotropin in Testicular Cancer	245

V.	α-Fetoprotein and Human Chorionic Gonadotropin in the Diagnosis of Testicular Tumors	247
VI.	α-Fetoprotein and Human Chorionic Gonadotropin in Clinical Staging of Nonseminomatous Germ Cell Tumors	247
VII.	α-Fetoprotein and Human Chorionic Gonadotropin in the Clinical Monitoring of Nonseminomatous Germ Cell Tumors	248
VIII.	Tumor Markers in Seminona	249
IX.	Newer Applications of α-Fetoprotein and Human Chorionic Gonadotropin in Monitoring Nonseminomatous Germ Cell Tumors	251
X.	Other Tumor Markers for Nonseminomatous Germ Cell Tumors	252
XI.	Radioimmunolocalization of Testicular Tumors	253
XII.	Immunohistochemistry of Testicular Tumor Markers	254
XIII.	The Future	254
	References	255

CHAPTER 14
Biochemical Markers of Human Small (Oat) Cell Lung Carcinoma—Biological and Clinical Implications
STEPHEN B. BAYLIN

I.	Introduction	259
II.	Biology of Human SCC	260
III.	The Biochemistry of SCC and Implications for Clinical Behavior	270
IV.	Summary	274
	References	275

CHAPTER 15
Tumor Markers in Prostatic Cancer
MANI MENON and WILLIAM J. CATALONA

I.	Introduction	279
II.	Serum Markers	283
III.	Urinary Markers	292
IV.	Prostate Fluid Proteins	295
V.	Leukocyte Adherence Inhibition (LAI) and Leukocyte Migration Inhibition (LMI)	295
VI.	Summary and Conclusions	296
	References	296

CHAPTER 16
Markers of Gastrointestinal Cancer
YOUNG S. KIM and LAWRENCE J. MCINTYRE

I.	Introduction	299
II.	Oncodevelopmental Markers of Colorectal Cancer	300
III.	Oncodevelopmental Markers of Pancreatic Cancer	305
IV.	Oncodevelopmental Markers of Gastric Cancer	308
V.	Conclusions	310
	References	311

CHAPTER 17
The Carcinoembryonic Antigen and Its Cross-Reacting Antigens
P. BURTIN and M. J. ESCRIBANO

I.	Introduction	315
II.	Cell and Tissue Localization	316
III.	Biosynthesis and Catabolism of CEA	318
IV.	Physicochemical Characteristics	319
V.	CEA Evaluation as a Marker of Neoplasia	322
VI.	Monoclonal Antibodies against CEA	324
VII.	Utilization of Anti-CEA Antibodies in Radioimmunodetection	325
VIII.	Use of Anti-CEA Antibodies in Radioimmunotherapy	325
IX.	Antigens That Cross-React with CEA	326
X.	Reviews	331
	References	331

CHAPTER 18
Colorectal Cancer Markers: Clinical Value of CEA
NORMAN ZAMCHECK

I.	CEA-like Antigens	335
II.	Non-CEA-like Antigens	336
III.	Clinical Use of CEA	336
IV.	Pathology	339
V.	Radioimmunodiagnosis	342
VI.	CEA_p	343
VII.	Serum Galactosyltransferase Isoenzyme (GT-II)	343
VIII.	Colon Mucoprotein Antigens (CMA)	344
IX.	The Zinc Glycinate Marker	344
X.	Combinations of Markers	344
XI.	Urine Marker	345
XII.	Comment	346
	References	346

CHAPTER 19
hCG Expression in Trophoblastic and Nontrophoblastic Tumors
GLENN D. BRAUNSTEIN

I.	Historical Overview	351
II.	Chemistry, Biologic Activity, and Sites of Production	352
III.	Methods of Measurement	354
IV.	hCG in Gestational Trophoblastic Disease (GTD)	355
V.	hCG in Germ Cell Tumors of the Testis	358
VI.	hCG in Nontrophoblastic Neoplasms	360
	References	365

CHAPTER 20
Oncodevelopmental Antigens in Gynecologic Cancer
MARKKU SEPPÄLÄ

I.	Introduction	373
II.	Placental Proteins	374
III.	α-Fetoprotein	381
IV.	Carcinoembryonic Antigen	383
V.	Other Antigens	387
VI.	Summary	389
	References	390

CHAPTER 21
Large-Scale AFP Screening for Hepatocellular Carcinoma in China
KAI-LI XU

I.	Introduction	395
II.	A Project for Large-Scale Screening	396
III.	Results of Large-Scale AFP Screening	398
IV.	Study on the Follow-up of Low-Level AFP Subjects	401
V.	The Possibility of Prevention of Liver Cancer	405
VI.	Conclusions	406
	References	407

CHAPTER 22
Uses and Limitations of Tumor Markers
RONALD B. HERBERMAN

I.	Introduction: Experience of the Past Ten Years	409
II.	The Promise of Current Research	411
III.	Future Directions	412
IV.	Markers in Cancer Detection and Diagnosis	415
	References	417

SUBJECT INDEX 419

Contributors

Numbers in parentheses indicate the pages on which the authors' contributions begin.

Kenneth D. Bagshawe (167), Department of Medical Oncology, Charing Cross Hospital, London W6 8RF, England

Stephen B. Baylin (259), Oncology Center, Department of Medicine, The Johns Hopkins University School of Medicine, Baltimore, Maryland 21205

Richard H. J. Begent (167), Department of Medical Oncology, Charing Cross Hospital, London W6 8RF, England

William A. Blattner (189), Family Studies Section, Epidemiology Branch, National Cancer Institute, Bethesda, Maryland 20205

Hans Bohn (69), Research Laboratories of Behringwerke AG, Diagnostika-Forschung, D-3550 Marburg/Lahn, West Germany

Glenn D. Braunstein (351), Division of Endocrinology, Department of Medicine, Cedars-Sinai Medical Center, Los Angeles, California 90048, and UCLA School of Medicine, Los Angeles, California 90048

P. Burtin (315), Laboratoire d'Immunochimie, Institut de Recherches Scientifiques sur le Cancer, 94802 Villejuif, France

Harris Busch (37), Department of Pharmacology, Baylor College of Medicine, Houston, Texas 77030

Rose K. Busch (37), Department of Pharmacology, Baylor College of Medicine, Houston, Texas 77030

William J. Catalona (279), Department of Surgery, Washington University School of Medicine, St. Louis, Missouri 63110

Pui-Kwong Chan (37), Department of Pharmacology, Baylor College of Medicine, Houston, Texas 77030

Frank Dolbeare (207), Lawrence Livermore National Laboratory, Biomedical Sciences Division, University of California, Livermore, California 94550

M. J. Escribano (315), Laboratoire d'Immunochimie, Institut de Recherches Scientifiques sur le Cancer, 94802 Villejuif, France

William H. Fishman (3), Cancer Research Center, La Jolla Cancer Research Foundation, La Jolla, California 92037

Robert F. Gagel (221), Departments of Medicine and Cell Biology, Baylor College of Medicine, Houston, Texas 77211 and Veterans Administration Medical Center, Houston, Texas 77211

Ronald B. Herberman (409), Biological Therapeutics Branch, Biological Response Modifiers Program, Division of Cancer Treatment, National Cancer Institute, Frederick, Maryland 20205

David Kelsey (37), Department of Pharmacology, Baylor College of Medicine, Houston, Texas 77030

Young S. Kim (299), Gastrointestinal Research Laboratory, Veterans Administration Medical Center, San Francisco, California 94121, and Department of Medicine, University of California, San Francisco, San Francisco, California 94121

Hilary Koprowski (155), The Wistar Institute of Anatomy and Biology, Philadelphia, Pennsylvania 19104

Paul H. Lange (241), Department of Urologic Surgery, University of Minnesota College of Health Sciences, Minneapolis, Minnesota, and Veterans Administration Medical Center, Minneapolis, Minnesota 55417

D. J. R. Laurence (131), Ludwig Institute for Cancer Research (London Branch), Royal Marsden Hospital, Sutton, Surrey SM2 5PX, England

Lawrence J. McIntyre (299), Gastrointestinal Research Laboratory, Veterans Administration Medical Center, San Francisco, California 94121, and Department of Medicine, University of California, San Francisco, San Francisco, California 94121

Mani Menon (279), Division of Urology, Washington University School of Medicine, St. Louis, Missouri 63110

A. M. Neville (131), Ludwig Institute for Cancer Research (London Branch), Royal Marsden Hospital, Sutton, Surrey SM2 5PX, England

Derek Raghavan (109), Department of Medical Oncology, Royal Prince Alfred Hospital, Camperdown, New South Wales, Australia, and Ludwig Institute for Cancer Research, Camperdown, New South Wales, Australia

Raymond W. Ruddon (87), Department of Pharmacology, The University of Michigan Medical School, Ann Arbor, Michigan 48109

Erkki Ruoslahti (21), Cancer Research Center, La Jolla Cancer Research Foundation, La Jolla, California 92037

Markku Seppälä (373), Department 1 of Obstetrics and Gynecology, University Central Hospital, Helsinki, Finland

Marjatta Son (37), Department of Pharmacology, Baylor College of Medicine, Houston, Texas 77030

William H. Spohn (37), Department of Pharmacology, Baylor College of Medicine, Houston, Texas 77030

Zenon Steplewski (155), The Wistar Institute of Anatomy and Biology, Philadelphia, Pennsylvania 19104

Kei Takahashi (37), Department of Pharmacology, Baylor College of Medicine, Houston, Texas 77030

Kai-Li Xu (395), Section of Biochemical and Immunological Diagnosis, Shanghai Cancer Institute, Shanghai, China 200032

Norman Zamcheck (333), Mallory Institute of Pathology Foundation, Boston City Hospital, Boston, Massachusetts 02118, Department of Pathology, Harvard Medical School, Boston, Massachusetts 02118, and Boston University School of Medicine, Boston, Massachusetts 02118

Preface

The recognition that in cancer one is dealing primarily with a change in gene expression rather than a substantial change in the gene per se is becoming widespread with the recent interest in cellular oncogenes. The latter, in our opinion, should now be referred to as "oncodevelopmental genes."

The field of oncodevelopmental genes was first defined in 1976 (*Oncodevelopmental Gene Expression*, W. H. Fishman and S. Sell, Academic Press, Inc.). That treatise dealt with areas of developmental biology, carcinogenesis, and the diagnostic implications of oncodevelopmental gene products.

The literature on tumor-associated markers is voluminous and confusing primarily because each marker is treated as a world unto itself without regard to its relationship to other tumor markers. Previous books have usually been compendiums of papers presented at meetings. Consequently, it was felt that there was a need for a treatise that would deal with the subject conceptually rather than empirically, a treatise that would illustrate for the reader examples of biologic, diagnostic, and monitoring aspects of oncodevelopmental markers, that would define clearly the limitations as well as the promise of such markers, and would be sufficiently provocative to be a springboard for meaningful inquiry.

In attempting to meet these needs, the editor has gathered together outstanding scientists who have written authoritative chapters in their fields. The first part of the book forms a conceptual exposition and examination of a number of theoretical and methodological approaches; through these the reader should achieve an understanding of the uses and limitations of various cell lines, xenotransplanted tumors, and polyclonal and monoclonal antibodies.

The volume begins with a conceptual treatment of "Oncodevelopmental Markers" (W. H. Fishman), an examination of "Cell-Matrix Interactions as Determinants of Differentiation and Tumor Invasion" (E. Ruoslahti), and an exploration of "Human Tumor Nucleolar Antigens" (H. Busch *et al.*).

The following three chapters deal with markers and cell model systems for their study. H. Bohn describes a "Systematic Identification of Specific Oncoplacental Proteins." R. W. Ruddon discusses "Marker Expression by Cultured Cancer Cells," and D. Raghavan covers "The Study of Oncodevelopmental Markers in Heterotransplanted Tumors."

The next three chapters are centered on the use of immunochemical and immunologic tools in the study of markers by recognized leaders in these areas. Thus, D. J. R. Laurence and A. M. Neville review "Specific Antibodies in Cytopathology and Immunohistology"; Z. Steplewski and H. Koprowski evaluate the "Use of Monoclonal Antibodies to Recognize Tumor Antigens"; R. H. J. Begent and K. D. Bagshawe update "Radioimmunolocalization of Cancer."

Finally, W. Blattner presents "Familial Cancer: An Opportunity to Study Mechanisms of Neoplastic Transformation."

In the second part of the book F. A. Dolbeare updates flow cytoenzymology in the first chapter, and the next three chapters provide insights into the origin of cancer from ancestral migratory embryonic cells. Thus, R. F. Gagel's "Tumor Markers of Medullary Thyroid Carcinoma" illustrates the C-cells (calcitonin producing cells) as progenitors of C-cell hyperplasia and neoplasia. P. H. Lange develops a critical view of germ cell tumors of the testis. S. B. Baylin examines carefully the biochemical markers of small (Oat) cell lung carcinoma. M. Menon and W. J. Catalona evaluate "Tumor Markers in Prostatic Cancer."

Gastrointestinal cancer markers have held center stage among clinical oncologists and, in the opinion of some, CEA should be the standard against which to measure the performance of any other tumor marker. Hence, three chapters are devoted to gastrointestinal cancer markers. One by Y. S. Kim and L. J. McIntyre covers the general area, while P. Burtin and M. J. Escribano report on "The Carcinoembryonic Antigen and Its Cross-Reacting Antigens," and N. Zamcheck reviews the clinical value of CEA as a colorectal cancer marker.

Trophoblastic differentiation markers have aroused considerable interest. G. D. Braunstein deals specifically with "hCG Expression in Trophoblastic and Nontrophoblastic Tumors," and M. Seppälä covers "Oncodevelopmental Antigens in Gynecologic Cancer."

Kai-Li Xu discusses "Large-Scale AFP Screening for Hepatocellular Carcinoma in China," and R. B. Herberman writes of "Uses and Limitations of Tumor Markers." Xu's chapter is the first comprehensive review in English of the most extensive controlled evaluation of a tumor marker.

Oncologists, physicians, pathologists, and clinical chemists should find this treatise informative and interesting as a coherent account of biologic, diagnostic, and monitoring aspects of oncodevelopmental markers.

ONCODEVELOPMENTAL MARKERS
Biologic, Diagnostic, and Monitoring Aspects

Part I
BIOLOGIC ASPECTS

CHAPTER 1
Oncodevelopmental Markers

WILLIAM H. FISHMAN
Cancer Research Center
La Jolla Cancer Research Foundation
La Jolla, California

I.	Introduction	3
II.	Embryologic Basis of Tumor Nomenclature	4
III.	Oncodevelopmental Alkaline Phosphatase Isoenzymes	9
	A. Three Gene Loci for Alkaline Phosphatase	10
	B. Immunologic Determinants	11
	C. Neoplastic Expression of Placental Phenotype	11
	D. Tissue-Unspecific Isoenzyme	12
	E. Regulation of Alkaline Phosphatase in Cell Lines	12
	F. Role of Alkaline Phosphatase	13
IV.	Extent of Oncodevelopmental Expression	13
V.	Current Perspective	14
	A. Questions	14
	B. Clinical Aspects	15
	References	16

I. INTRODUCTION

The field of oncodevelopmental biology and medicine promises to substitute gradually for the tumor marker empiricism of the past, a conceptual framework on which to build logical explorations of tumor markers into the future. Specifically, from an era that was populated entirely by tumor antigens of unknown molecules with unclear biological interest, one can now point to a number of characterized tumor proteins, the study of which is giving new insights into the nature of gene expression in cancer cells. More importantly the first oncodevelopmental proteins reported 20 years ago, such as α-fetoprotein, are not only the subjects of ever-increasing investigative interest, but also their utility in oncology as tumor markers is valuable and being defined with greater accuracy.

What is an oncodevelopmental marker? "Oncodevelopmental" describes substances which are produced in both developmental and neoplastic tissues. Developmental tissues from this point of view are defined as those which characterize the gametes and zygote up to the infant at parturition and include blastocyst, embryonic, extraembryonic, trophoblastic, and fetal tissues. Neoplastic tissues are generally understood to include both benign and malignant tumors. Classically, such oncodevelopmental proteins are produced at low levels, if at all, in adult tissues. The expression of α-fetoprotein (AFP) in fetal liver and hepatoma cells is an example of an "ectopic" marker, since normal adult liver lacks AFP.

The situation in which neoplastic transformation leads to a heightened expression of product of genes already active in a tissue is termed "eutopic." Examples are the excessive production of human chorionic gonadotropin (hCG) by the placental neoplasm, choriocarcinoma, and of calcitonin by C cells of medullary thyroid carcinoma. "Ectopic" expression refers to the presence of a gene product in a tumor which is a differentiation phenotype of a normal tissue other than the one giving rise to the tumor. In both the ectopic and eutopic tumor marker phenomena, one is witnessing an aberration in the expression of normal genes, which is usually characteristic of earlier developmental phases in the organism.

Conceptually, therefore, cancer may be viewed as a problem in developmental biology at the center of which is a disorder in the expression of normal genes. A manifestation of this derangement is the reexpression of normal developmental genes in neoplastic cells. Do the products of embryonic genes have an influence on the destiny of the cells in which they appear? Can they move such cells further along the path toward malignancy? Or are they phenotypes of cells expected at a particular state on their differentiation which may persist randomly during neoplasia and play no part in the process of neoplastic progression?

To clarify the potential value of the oncodevelopmental concept, attention has been focused on the embryologic basis of tumor nomenclature. The production of placental alkaline phosphatase in tumors is described in some detail to illustrate the suggestions made in this chapter. The discussion also includes consideration of the genealogy (Fig. 1) of the isoenzyme genes of alkaline phosphatase.

II. EMBRYOLOGIC BASIS OF TUMOR NOMENCLATURE

One can construct a scheme of development designed to illustrate its main branches and to relate to each other the main systems which have their neoplastic counterparts. A novel feature also is the attempt to place chronologically the various migratory embryonic cells in the progression from embryoblast, three germ layers, organogenesis, and fetus. Extraembryonic tissues emphasized here are yolk sac and placenta.

1. Oncodevelopmental Markers

The blastocyst is composed of three germ layers: ectoderm, mesoderm, and endoderm, from which all specialized tissues develop. Ectoderm produces skin and nerve tissue; mesoderm is the source of bone, muscle, and cartilage; endoderm is the precursor of the intestinal tract and related organs such as liver and pancreas.

Tumors are named in the context of the three germ layers (Pitot, 1977) and their combinations. Thus, sarcoma is a malignant growth originating from mesodermal tissue (e.g., fibrosarcoma, osteogenic sarcoma, and chondrosarcoma), while a carcinoma refers to a tumor of embryonic ectodermal or endodermal tissue (e.g., skin cancer or adenocarcinoma of breast). Those tumors which have a primitive embryonic appearance carry the suffix "blastoma" (e.g., hepatoblastoma, neuroblastoma). Finally, tumors derived from all three germ layers are termed teratomas if benign or teratocarcinomas if malignant. Both may exhibit a variety of differentiated tissues.

Within this framework one can list the neoplasias and their characteristic gene products which are recognized as eutopic tumor markers. Except for the putative viral antigens and a few unidentified antigens, these markers are all products of developmental phases of the cells of origin.

From scrutiny of Table I, one can identify the tumors of the gametes as germinomas and seminomas which have a high rate of expression of placental alkaline phosphatase, less so AFP (Lange, 1981; Lange *et al.*, 1982) and hCG (Javadpour *et al.*, 1978). Embryonal carcinoma, the putative counterpart to embryoblast, expresses AFP and chorionic gonadotropin as do the conglomerate tumors of the three germ layers (teratocarcinoma) with the addition of placental alkaline phosphatase (Kurman *et al.*, 1979; Kurman and Scardino, 1981; Taylor *et al.*, 1978a).

Of the neoplasms of extraembryonic tissues, most prominent are trophoblastic and yolk sac tumors (Kurman and Scardino, 1981)—the former displaying a

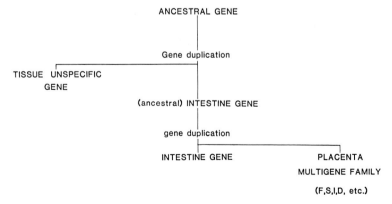

Fig. 1. Hypothetical genealogy of alkaline phosphatase isoenzyme genes.

TABLE I
Neoplasms of Counterpart Developmental Tissues and Their Marker Proteins

A. Gamete neoplasms (seminomas, germinal neoplasms)
    AFP, placental alkaline phosphatase
B. Embryoblast tumors (embryonal carcinoma)
    α-Fetoprotein
    Chorionic gonadotrophin
C. Three germ layer tumors (teratocarcinoma)
    α-Fetoprotein
    Chorionic gonadotropin
    Placental alkaline phosphatase
D. Extraembryonic tumors
    Trophoblastic neoplasms
        Chorionic gonadotropin
        Placental lactogen
        Placental alkaline phosphatase
        Pregnancy-specific glycoprotein (SP_1)
        Placental protein 5 (PP_5)
    Yolk sac neoplasms
        $α_1$Antitrypsin
        α-Fetoprotein
E. One germ layer tumors (ectodermal, endodermal, and mesodermal)
    Ectodermal tumors
        Prostate cancer
            Acid phosphatase
        Breast carcinoma
            α-Lactalbumin
            Casein
            Estrogen receptor
            Placental lactogen
    Endodermal tumors
        Colorectal and other GI cancers
            CEA
    Mesodermal tumors
        Skeletal muscle tumors
            Myoglobin
        Smooth and skeletal muscle tumors
            Myosin
        Osteogenic sarcoma
            Alkaline phosphatase
        Endothelial cell tumors
            Factor VIII related Ag
F. Endocrine tumors
    A variety of pituitary and adrenal tumors
        Adrenocorticotropin
        Growth hormone

(continued)

TABLE I *(Continued)*

        Prolactin, thyroid-stimulating hormone
        Follicle-stimulating hormone
G. Specialized migratory embryonic cell neoplasms
    Neural crest cell tumors
        Medullary thyroid carcinoma
            Calcitonin
            Histaminase
        Melanoma
            Melanin
        Gastrointestinal tract endocrine cell tumors
            Gastrin, somatostatin, secretin, cholecystokinin, 5-hydroxytryptophan, vasoactive intestinal peptide, gastric inhibitory peptide, glicentin, neurotensin, bombesin, pancreatic polypeptide
    Germ cell neoplasms
        Steroid hormones and metabolites
        Placental alkaline phosphatase
        Chorionic gonadotrophin
        α-Fetoprotein
    Hematopoietic cell neoplasms
        Endothelial cell tumors
            Factor VIII related antigen
        B cell lymphomas
            Heavy and light chain immunoglobulins
        Lymphoblastic lymphoma, acute lymphocytic leukemia
            Terminal deoxynucleotidyltransferase
        Histiocytic lymphoma, monocytic leukemia
            Lysozyme

spectrum of placental proteins, as in choriocarcinoma (Braunstein *et al.*, 1973), and the latter, AFP (Abelev, 1971). So far AFP production correlates best with endodermal sinus yolk sac tumor. Also, the expression of placental genes in yolk sac tumors has been conspicuous by its absence and vice versa for yolk sac genes in placenta. In these two extraembryonic tumor tissues, their phenotypic expressions simply do not overlap.

    The category of tumors of single germ layer origin (ectodermal, endodermal, and mesodermal) includes the great majority of neoplasms which afflict man. Ectodermal tumors appear to express late differentiation antigens such as α-lactalbumin (Bussolati and Pick, 1975), casein, and estrogen receptor (Lee, 1981) in breast carcinoma. Also prostatic antigens including prostatic acid phosphatase (Gutman and Gutman, 1938) are evident in cancer of the prostate. Examples of endodermal tumors are colorectal and other gastrointestinal cancers which express carcinoembryonic antigen (Gold and Freedman, 1965). Finally, among the mesodermal tumors, actin, myosin and myoglobin characterize smooth and skeletal muscle tumors (Bussolati et al., 1980; Miller et al., 1976; Mukai et al., 1979; Macartney et al., 1979); alkaline phosphatase characterizes osteogenic sarcoma; and factor VIII related Ag characterizes endothelial cell tumors (Nadji et al., 1980).

A special category of endodermal endocrine tumors includes a variety of pituitary and adrenal neoplasms (Kovacs et al., 1977, 1981; Nakane, 1970; Robert et al., 1978) which are known to overproduce adrenocorticotropin, growth hormone, prolactin, thyroid-stimulating hormone, and follicle-stimulating hormone.

In recent years, more attention has been directed to gene products of tumors of cells derived from specialized migratory embryonic cells. When it is realized that these cell lineages make their appearance in the preembryonic period of development, have no organoid morphology in adult life, and retain their individual cell disposition, it is necessary to treat them as a separate entity in the chronology of development. Many of these have been described as APUD cells or components of the microendocrine system.

Neural crest cell tumors, which include the melanin-producing melanoma and gastrointestinal tract endocrine cell tumors, express a multitude of polypeptide hormones (O'Briain and Dayal, 1981; Dayal and O'Briain, 1981)—gastrin, somatostatin, secretin, cholecystokinin, vasoactive intestinal peptide, gastric inhibitory peptide, glicentin, neurotensin, bombesin, and pancreatic polypeptide. Included also are the calcitonin-producing cells in medullary thyroid carcinoma (DeLellis and Wolfe, 1974; Arnal-Monreal et al., 1977; Mendelsohn et al., 1978; Wolfe et al., 1973), oat cell carcinomas (De Schryver-Keaskemati et al., 1979), and pancreatic islet cell and carcinoid tumors (Creutzfeldt et al., 1975; DeLellis et al., 1976; Larsson, 1978).

Neoplasms of germ cells can occur anywhere along the path traveled by germ cells on their way from the yolk sac to the genital ridge. They may express trophoblast and yolk sac genes to a variable degree (Palmer and Wolfe, 1978)—placental alkaline phosphatase, chorionic gonadropin, and AFP among other markers (Norgaard-Pedersen and Raghavan, 1980).

The hematopoietic system represents a perpetual reenactment of cycles of granulocytic and erythropoietic differentiation which began in the blood islands of embryonic tissue and continued to restricted locales such as the bone marrow and the spleen. In this situation, one notes that the neoplastic cells express products characteristic of early stages in their differentiation. The B cell lymphomas produce immunoglobulins of both heavy and light chains and J chain, lysozyme is increased in histiocytic lymphoma and monocytic leukemia (Mason and Taylor, 1975), and terminal deoxynucleotidyltransferase is produced in histiocytic lymphoma and monocytic leukemia (Banks et al., 1978; Isaacson,

1979a,b; Long et al., 1979; Pengalis and Rappaport, 1977; Taylor et al., 1978a; Warnke et al., 1978).

III. ONCODEVELOPMENTAL ALKALINE PHOSPHATASE ISOENZYMES

The justification for selecting these particular gene products for discussion is that they illustrate some principles which are not a particular topic of the remaining chapters in this volume, and this system is one with which the author is most familiar.

The most unique of the known isozymes of alkaline phosphatase is placental alkaline phosphatase (EC 3.1.3.1.) (PLAP) which is clearly distinguishable from intestinal and tissue-unspecific (liver/bone) alkaline phosphatase by its greater heat stability, differential inhibition by L-phenylalanine, and unique immunological determinants. It is a relatively late evolutionary gene product which is expressed in the last two trimesters in human and certain primate placentas.

Placental alkaline phosphatase is a membrane-bound glycoprotein enzyme (Lin et al., 1976) with many electrophoretically defined allelic forms which exhibit similar biochemical and biological properties. A rare D variant travels very slowly on starch gel electrophoresis and is inhibited by L-leucine.

An alkaline phosphatase isoenzyme similar to PLAP (Regan isoenzyme) is expressed in cancer cells of trophoblastic and nontrophoblastic origin. Its presence in serum of cancer patients is most frequent in ovarian tumors and seminoma. An L-leucine-sensitive form, Nagao isoenzyme, is present in a majority of these tumors (Inglis et al., 1973). An ELISA technique (Millan and Stigbrand, 1981) has been found sensitive and reliable in managing the course of seminoma patients (Lange et al., 1982).

The Kasahara isoenzyme (Higashino et al., 1975) in cancer tissue was found to correspond to fetal intestinal alkaline phosphatase, while the non-Regan isoenzyme was reported to be indistinguishable from first trimester alkaline phosphatase [heat and L-homoarginine sensitive; L-phenylalanine insensitive with liver/bone antigenic determinants (L. Fishman et al., 1976)].

In Table II, the data on four tumor-associated isoenzymes are summarized. Their developmental counterparts are term placenta, first trimester placenta, and fetal intestine.

A number of new insights into the control of specific isoenzyme protein synthesis have been achieved. These are (1) the discovery by rigidly controlled ultrastructural techniques of "cryptic" membrane alkaline phosphatase in the many membrane systems of the cytoplasm (Tokumitsu et al., 1981a,b); (2) the use of specific monoclonal antibodies to neoplastic and placental isoenzymes as probes of structure of isolated molecular forms as well as cellular ones (is the

TABLE II
Tumor-Associated Isoenzymes of Alkaline Phosphatase

Tumor isoenzyme: Developmental Counterpart:	Regan Term placenta	Nagao Term placenta	Non-Regan First trimester placenta	Kasahara Fetal intestine
Molecular weight (subunit)	64,000	—	—	—
NH$_2$-terminal sequence	Ile-Ile-Pro			
pH optimum	10.6	10.6	10.6	10.6
Heat stability (5' at 65°C)	+++	+++	−	−
Amino acid sensitivity (5 mM)				
L-Phenylalanine	+++	+++	+	+++
L-Homoarginine	±	±	+++	±
L-Leucine	+	+++	+	++
Electrophoretic migration (anodal)	2	3	1	1
Neuraminidase sensitivity	+	+	+	+
Reaction with antisera to				
Liver ALP	−	−	+	−
Intestinal ALP	+	+	−	++
Term placental ALP	+++	+++	−	−

"normal" gene being reexpressed in the cancer cell?); (3) the ability to manipulate this expression by adding modulating substances to the cell culture medium and illuminating the mechanisms controlling gene expression at the molecular level; and (4) the expression of oncotrophoblast enzymes and hormones in cancer cell lines, all having clinical relevance because tumor tissues express these gene products in patients. Can they be of value in diagnosis, staging of disease, and monitoring of therapy? Currently, great clinical interest is directed to their potential in immunolocalization of tumors and the development of immunocytotoxic therapy (see Chapter 8 by Steplewski and Koprowski, "Use of Monoclonal Antibodies to Recognize Tumor Antigens"; Chapter 9 by Begent and Bagshawe, "Radioimmunolocalization of Cancer").

A. Three Gene Loci for Alkaline Phosphatase

All available structural data seem to indicate the presence of at least three genes, coding, respectively, for the intestinal, placental, and tissue unspecific forms of the enzyme (McKenna et al., 1979; Seargeant and Stinson, 1979; Badger and Sussman, 1976; Boyer, 1963; Fishman, 1974; Sussman, 1968a,b; Fishman and Sie, 1971; Mulivor et al., 1978). Monospecifc antisera distinguish three different antigenic groups (Sussman et al., 1968b; McKenna et al., 1979; Lehmann, 1975, 1980). The placental and intestinal phosphatase, however, have antigenic determinants in common (Doellgast et al., 1976; Lehmann, 1975).

During the course of evolution successive gene duplications from a common ancestral gene may have occurred, giving rise to the three loci (Fig. 1). The placental form seems to be a late evolutionary product, since it cannot be identified in lower animals (Manning et al., 1969, 1970; Goldstein and Harris, 1979). Doellgast and Benirschke (1979) demonstrated placental type alkaline phosphatase (ALP) in the chimpanzee and orangutan placentas. Thus, an unusually high degree of allelic polymorphism (Harris et al., 1973) at this locus in man might be correlated to its late appearance on the evolutionary scene (Doellgast et al., 1981; Goldstein et al., 1982).

The intestinal type is immunologically unrelated to the tissue-unspecific enzyme (the latter is previously referred to as "liver" or "bone" or liver/bone/kidney), while it shares many characteristics with the placental form. This may indicate a closer evolutionary relation between the intestinal and placental gene products.

The placental type alkaline phosphatase shows the highest degree of polymorphism of any human enzyme so far studied (Harris et al., 1973). Most placentae can be classified into six common phenotypes called SS, FS, FF, II, IS, and IF based on electrophoretic separation on starch gel (S = slow, F = fast, I = intermediate). The three common alleles are responsible for 97.5% of the placental phenotypes, indicating the presence of an unusually high frequency of rare alleles (2.5%).

B. Immunologic Determinants

Wei and Doellgast (1980), by exhaustive absorption with defined variants, obtained a variant-specific polyclonal rabbit anti-human placental alkaline phosphatase antibody, which did not react with the F phenotype phosphatase but did with the products of the other common genes. More recently, we and others have raised monoclonal antibodies to placental alkaline phosphatase and have found among them antibodies that react with determinants present on the product of the S and I, but not that of the F allele (Millan et al., 1982; Slaughter et al., 1981). Thus, altogether, three independent observations indicate that specific reagents to the S and I product can be obtained that do not recognize the F form.

C. Neoplastic Expression of Placental Phenotype

Placental alkaline phosphatase was described as an oncotrophoblast antigen by Fishman et al. (1968a,b). Several groups have recommended placental alkaline phosphatase as a marker for seminomas, where it provides information not available through AFP and hCG levels (Wahren et al., 1979; Lange et al., 1982).

In several tumors studied by both histochemical and immunohistochemical techniques (Miyayama *et al.*, 1976; Nishiyama *et al.*, 1980), placental alkaline phosphatase has been shown to be bound to the plasma membrane. Also, Taylor *et al.* (1981) found that, in the attachment of the enzyme in tumor-derived membrane fragments, some antigenic sites are exposed in the tumor cell membranes which are not detected in the placental cell membranes. Recently, Tokumitsu *et al.* (1981a,b) have demonstrated in HeLa cells, both by enzyme cytochemical and immunocytochemical techniques, alkaline phosphatase in intracytoplasmic structures such as Golgi apparatus, endoplasmic reticulum, and perinuclear membranes. Such technology combined with monoclonal antibody reagents should make possible considerable progress in studies of the membrane attachment of such enzymes in placental and tumor cells.

Some nonmalignant tissues have been shown to synthesize eutopically placental alkaline phosphatase, such as the nonmalignant human cervix (Nozawa *et al.*, 1980), the normal human testis (Chang *et al.*, 1980), and the normal human lung (Goldstein *et al.*, 1982). A clue as to the origin of the low circulating levels of PLAP found in both normal males and females (Anstiss *et al.*, 1971; Benham *et al.*, 1978a,b; Millan and Stigbrand, 1981) is thus evident.

D. Tissue-Unspecific Isoenzyme

Tumor enzymes resembling the tissue-unspecific form were reported by Timperley (1968), Sasaki and Fishman (1973), Timperley *et al.* (1971), Wernes *et al.* (1972), and Tokumitsu *et al.* (1979). In 1976, L. Fishman *et al.* first described isoenzymes from placental tissue during the first weeks of pregnancy (developmental phase specific alkaline phosphatases, early type placental alkaline phosphatases, or chorionic type phosphatases). These isoenzymes were tissue unspecific in type (heat and L-homoarginine sensitive) and occur much more frequently than term placental or intestinal types in tumors.

E. Regulation of Alkaline Phosphatase in Cell Lines

HeLa cell lines produce different forms of alkaline phosphatase (Singer and Fishman, 1974). Benham *et al.* (1978a,b) found that six out of eight HeLa lines contained a placental-like ALP isoenzyme and varying amounts of tissue-unspecific isoenzyme. There was a "line to line" variation in electrophoretic mobility, although the catalytic activity was the placental type. Other enzymes expressed by HeLa cells are phenotypically stable. Cell lines derived from various tissues have been shown to synthesize different forms of phosphatase. These include lung fibroblasts WI-26 and WI-38 (Knaup *et al.*, 1978), choriocarcinoma lines (Hamilton *et al.*, 1979), endometrial cancer cell lines (Suzuki *et al.*, 1980), human osteosarcoma cells (Singh *et al.*, 1978), cultured human term placental cells (Sakiyama et al., 1980), and a human adenocarcinoma of the lower gastrointestinal tract (Singer et al., 1976).

The possibility of studying gene expression in cell lines expressing different forms of alkaline phosphatases and the advantage of being able to modify the cellular microenvironment have opened a new dimension in oncodevelopmental biology. Using the HeLa TCRC-1 cell line, Singer and Fishman (1976) showed that prednisolone induction of the placental type phosphatase (but not the other types) does not require DNA synthesis or cells passing through the S phase in the cell cycle but does take place during G_1. Similar conclusions have been made by Sakiyama et al. (1980). This means that transcription and translation mechanisms can probably explain the induction. Other inducers include sodium butyrate (Griffen et al., 1974) 5-bromodeoxyuridine (BrdUrd) (Edlow et al., 1975) and dibutyryl-cAMP (Chou and Robinson, 1977). Some reports, furthermore, indicate that the induction of different phosphatases within the same cell are under different controls since they can be induced independently. Interestingly, viral transformation of human placental cells causes a significant increase in the synthesis of early placental (tissue unspecific) alkaline phosphatase (Sakiyama et al., 1980).

Nozawa et al. (1983) observed that butyrate induced several developmentally phase-specific trophoblast proteins, early placental alkaline phosphatase, hCG, and SP_1, but not term placental alkaline phosphatase in cervical cancer cell line SKG III.

F. Role of Alkaline Phosphatase

Very recently, Swarup et al. (1981) have reported that phosphotyrosine in histones and membrane proteins is the natural substrate for alkaline phosphatases. That such dephosphorylation may be physiologically significant is suggested by the excellence of membrane proteins as substrates, the physiological pH (7.0 to 8.0) of the phosphotyrosine phosphatase activity, and the relatively low concentrations of enzyme required for dephosphorylation. When it is recalled that the *src* gene produces a protein kinase which phosphorylates tyrosine residues of specific proteins, it is conceivable that the degree of such phosphorylations will depend on the balance between the activity of phosphotyrosine phosphatases and the protein kinases.

IV. EXTENT OF ONCODEVELOPMENTAL EXPRESSION

The finding of placental alkaline phosphatase in human tumors led Fishman (1975) to predict that other embryonic and trophoblastic gene products should be detectable in neoplastic tissue. The first evidence was obtained in ascitic fluid from patients with ovarian cancer. These showed significant levels of human chorionic gonadotropin as well as CEA (Fishman, 1975).

The most complete study of oncotrophoblast gene expression was performed by H. Bohn (see Chapter 4) who examined a spectrum of over a dozen placental proteins. He found several which are expressed in tumors but rarely in normal adult tissues.

V. CURRENT PERSPECTIVE

This chapter has attempted to provide an embryologic framework within which the reader can be expected to place the individual chapters of this treatise.

First, it is an oversimplification to state that tumors represent a complete return to fetal gene expression. Rather one must identify the developmental stage returned to and distinguish it from the gene products of blastocyst, migratory embryonic cells, trophoblast, yolk sac, placenta, amnion, and individual fetal organs.

However, one can be really impressed by the widespread eutopic early and late differentiation gene expression which is chronicled in Table I, listing the gene products of tumors arranged in relation to their embryologic counterparts. The specialized embryonic migratory cells are a remarkably rich source of such gene products. These can all be expected to be clinically useful.

The reexpression of embryonic genes on surfaces of cancer cells may determine their metastatic destination. Thus, Nicolson *et al.* (1982) reported that murine lymphosarcoma cells bear an embryonic liver adhesive glycoprotein which in the embryo presumably accounts for the ability of liver cells to aggregate. Pretreatment with Fab' antibodies of animals inoculated with lymphosarcoma cells completely protects the animals from liver metastases.

Finally, the role of extracellular matrix in controlling differentiation of embryonic cells and explaining metastatic behavior is a whole new intensely interesting field in oncodevelopmental biology (see Chapter 2).

A. Questions

If one adopts the perspective of developmental biology, how should one regard the expression of term placental alkaline phosphatases in lung cancer, as in the case of the Regan isoenzyme (Fishman *et al.*, 1968a,b), when a later report (Chang *et al.*, 1980) found an enzyme with the same properties also being produced by the testis? Should one ascribe this ectopically expressed phosphatase to a testicular gene, or to a placental one, or to a testicular/placental gene?

Certainly the testicular germ cell tumors which consistently overproduce this enzyme would be most reasonably regarded as evidencing eutopic testicular gene expression rather than trophoblast expression.

Would it then be reasonable to assume that trophoblastic neoplasms express testicular genes and that the same genes can be operating in nontrophoblastic neoplasms? If indeed genes active in gametes are significant in the part of the genome being expressed in cancer cells, what meaning, if any, would this have in cancer biology? Could they be developmental "oncogenes?"

On a conceptual basis, does not the Connheim embryonic rest theory updated to 1983 in the oncogene context now acquire greater credibility?

B. Clinical Aspects

With regard to the diagnostic value of oncodevelopmental markers, there are several yardsticks by which these can be measured: screening, histologic diagnosis, differential diagnosis, and monitoring. These are dealt with in detail in other chapters of this treatise.

Although there is often a great disappointment expressed when a tumor marker fails to live up to the original high expectations, this cannot reduce the biological significance of the marker or its potential clinical value. Rather, one should not develop unreasonable expectations in the early stages of exploration of its utility. CEA (carcinoembryonic antigen), for example, has been found useful in staging and monitoring of cancer and has provided the first tumor target for pioneering studies in tumor immunolocalization. Yet, in early evaluation there was a chorus of negative statements regarding its clinical utility. Eventually, clinical utility for each oncodevelopmental marker has become demonstrable.

Expectations should be tempered by an understanding of the biological variability and heterogeneity of tumors. Thus, it is a common clinical observation that oncodevelopmental markers may disappear during the course of the disease. This is understandable in terms of the progressive anaplasticity usually characterizing advanced disease.

On the other hand, there is an instance (Park and Reed, 1980) of unexpected trophoblast differentiation in metastatic adenocarcinoma of the colon whose cells produce hCG: the primary tumor was hCG-negative! No evidence for teratocarcinoma or teratoma could be found. It is thus possible that a well-known marker may appear in metastases and not in the primary tumor as an ectopic phenomenon.

The introduction of monoclonal antibody technology and its promise in the immunolocalization and therapy of tumors has brought renewed attention to oncodevelopmental markers. This is because their high degree of tumor association provides a more specific target for the monoclonal antibodies.

This treatise attempts to provide a conceptual background to this area of contemporary research interest as exemplified in the work of the outstanding scientists who have contributed to this volume. To the extent that the reader finds facts, synthesis, and critical interpretation—to that degree of success—this treatise should be measured.

ACKNOWLEDGMENTS

The author owes much to an excellent source: "Diagnostic Immunohistochemistry" (Masson Publishers, New York, 1981) edited by Ronald A. DeLellis, on which he has drawn heavily. The chapter also has benefited from discussions with E. Ruoslahti, R. M. McIntire, and others. The experimental work referred to from this laboratory was supported by NCI Grants CA-21967, P30-CA-30199, and P01-CA-28896 of the National Institutes of Health, Bethesda, Md. To Bryna Block goes my sincere thanks for her technical help with this manuscript and those of a number of other contributors.

REFERENCES

Abelev, G. I. (1971). *Adv. Cancer Res.* **14,** 295–358.
Anstiss, C. L., Green, S., and Fishman, W. H. (1971). *Clin. Chim. Acta* **33,** 279–286.
Arnal-Monreal, F. M., Goltzman, D., Knaack, J., Wang, N.-S., and Huang, S. N. (1977). *Cancer* **40,** 1060–1070.
Arnold, R., Deuticke, V., Frerichs, H., and Creutzfeldt, W. (1972). *Diabetologia* **8,** 250–259.
Badger, K. S., and Sussman, H. H. (1976). *Proc. Natl. Acad. Sci. U.S.A.* **73,** 2201–2205.
Banks, P. M., Keller, R. H., Ki, C.-Y., and White, W. L. (1978). *Am. J. Med.* **64,** 906–909.
Benham, F., Povey, M. S., and Harris, H. (1978a). *Clin. Chim. Acta* **86,** 201–215.
Benham, F. J., Povey, M. S., and Harris, H. (1978b). *Somatic Cell Genet.* **4,** 13–25.
Boyer, S. H. (1963). *Ann. N.Y. Acad. Sci.* **103,** 938–950.
Braunstein, G. D., McIntire, K. R., and Waldmann, T. A. (1973). *Cancer* **31,** 1065–1068.
Bussolati, G., and Pick, A. (1975). *Am. J. Pathol.* **80,** 117–128.
Bussolati, G., Alfani, V., Weber, K., and Osborn, M. (1980). *J. Histochem. Cytochem.* **28,** 169–173.
Chang, C.-H., Angellis, D., and Fishman, W. H. (1980). *Cancer Res.* **40,** 1506–1510.
Chou, J. Y., and Robinson, J. C. (1977). *J. Cell. Physiol.* **92,** 221–232.
Creutzfeldt, W., Arnold, R., Creutzfeldt, C., and Track, N. S. (1975). *Hum. Pathol.* **6,** 47–76.
Dayal, Y., and O'Briain, D. S. (1981). In "Diagnostic Immunocytochemistry" (R. A. DeLellis, ed.), pp. 111–136. Masson, New York.
DeLellis, R. A. (1981). "Diagnostic Immunochemistry." Masson, New York.
DeLellis, R. A., and Wolfe, H. J. (1976). *Arch. Pathol. Lab. Med.* **100,** 340–350.
DeLellis, R. A., Gagel, R. F., Kaplan, M. M., and Curtis, L. E. (1976). *Cancer* **38,** 201–208.
DeSchryver-Keaskemati, K., Kyriakos, M., Bell, C. E., and Seetharam, S. (1979). *Lab Invest.* **41,** 432–436.
Doellgast, G. J., and Benirschke, K. (1979). *Nature (London)* **280,** 601–602.

1. Oncodevelopmental Markers

Doellgast, G. J., Silver, M. D., Fishman, L., and Guenther, R. A. (1976). *In* "Onco-Developmental Gene Expression" (W. H. Fishman and S. Sell, eds.), pp. 737–741. Academic Press, New York.
Doellgast, G. J., Wei, S. C., Kennedy, M., Stills, H., and Benirschke, K. (1981). *FEBS Lett.* **135**, 61–64.
Edlow, J. B., Ota, T., Relacion, J., Kohler, P. O., and Robinson, J. C. (1975). *Am. J. Obstet. Gynecol.* **121**, 674–681.
Fishman, L., Miyayama, H., Driscoll, S. G., and Fishman, W. H. (1976). *Cancer Res.* **36**, 2268–2273.
Fishman, W. H. (1974). *Am. J. Med.* **56**, 617–650.
Fishman, W. H. (1975). *In* "Isozymes" (C. L. Markert, ed.), Vol. 1, pp. 293–314. Academic Press, New York.
Fishman, W. H., Inglis, N. R., Green, S., Anstiss, C. L., Ghosh, N. K., Reif, A. E., Rustigian, R., Krant, M. J., and Stolbach, L. L. (1968a). *Nature (London)* **219**, 697–699.
Fishman, W. H., Inglis, N. R., Stolbach, L. L., and Krant, M. J. (1968b). *Cancer Res.* **28**, 150–154.
Fishman, W. H., and Sie, H. G. (1971). *Enzymologia* **41**, 141–167.
Gold, P., and Freeman, S. O. (1965). *J. Exp. Med.* **122**, 467–481.
Goldenberg, D. M., Sharkey, R. M., and Primus, F. J. (1978). *Cancer* **42**, 1546–1553.
Goldstein, D. J., and Harris, H. (1979). *Nature (London)* **280**, 602–605.
Goldstein, D. J., Rogers, C., and Harris, H. (1982). *Proc. Natl. Acad. Sci. U.S.A.* **79**, 879–883.
Griffen, M. J., Price, G. H., Bezzell, K. L., Cox, R. P., and Ghosh, N. K. (1974). *Arch. Biochem. Biophys.* **164**, 619–623.
Gutman, A. B., and Gutman, E. B. (1938). *Proc. Soc. Exp. Biol. Med.* **38**, 470.
Hamilton, T. A., Tin, A. W., and Sussman, H. H. (1979). *Proc. Natl. Acad. Sci. U.S.A.* **76**, 323–327.
Harris, H., Hopkinson, D. A., and Robson, E. B. (1973). *Am. Hum. Genet.* **37**, 237.
Higashino, K., Kudo, S., Otani, R., Yamamura, Y., Honda, T., and Sakurai, J. (1975). *Ann. N.Y. Acad. Sci.* **259**, 337–346.
Inglis, N. R., Kirby, S., Stolbach, L. L., and Fishman, W. H. (1973). *Cancer Res.* **33**, 1657–1661.
Isaacson, P. (1979a). *Am. J. Surg. Pathol.* **3**, 431–441.
Isaacson, P. (1979b). *J. Clin. Pathol.* **32**, 802–807.
Javadpour, N., McIntire, K. R., and Waldmann, T. A. (1978). *Cancer* **42**, 2768–2772.
Knaup, G., Pfleiderer, C., and Bayreuther, K. (1978). *Clin. Chim. Acta* **88**, 375–383.
Kovacs, K., Horváth, E., and Ezrin, C. (1977). *Pathol. Annu.* **12**, Part 2, 341–382.
Kovacs, K., Horváth, E., and Ryan, A. R. T. (1981). *In* "Diagnostic Immunochemistry" (R. A. DeLellis, ed.), pp. 277–298. Masson, New York.
Kurman, R. J., Scardino, P. T., McIntire, K. R., Waldman, T. R., and Javadpour, N. (1979). *Cancer*, **40**, 2136–2151.
Kurman, R. J., and Scardino, P. T. (1981). *In* "Diagnostic Immunochemistry" (R. A. DeLellis, ed.), pp. 277–298. Masson, New York.
Lange, P. H. (1981). *In* "Early Detection of Testicular Cancer" (N. E. Skakkeback, J. G. Berthelsen, K. M. Grigor, and J. Visfeldt, eds.), pp. 191–202. Publishers Scriptor.
Lange, P. H., Millan, J. L., Stigbrand, T., Vesala, R. L., Ruoslahti, E., and Fishman, W. H. (1982). *Cancer Res.* **42**, 3244–3247.
Larsson, L.-I. (1978). *Hum. Pathol.* **9**, 401–416.
Lee, S. H. (1981). *In* "Diagnostic Immunocytochemistry" (R. A. DeLellis, ed.), pp. 149–164. Masson, New York.
Lehmann, F.-G. (1975). *Clin. Chim. Acta* **75**, 271–282.
Lehmann, F.-G. (1980). *J. Immunol. Methods* **36**, 137–148.

Lin, C.-W., Sasaki, M., Orcutt, M. L., Miyayama, H., and Singer, R. M. (1976). *J. Histochem. Cytochem.* **24**, 659–667.
Long, J. C., McCaffrey, R. P., Aisenberg, A. C., Marks, S. M., and Kung, P. C. (1979). *Cancer* **44**, 2127–2139.
Macartney, J. C., Trevithick, M. A., Kricka, L., and Curran, R. C. (1979). *Lab. Invest.* **41**, 437–445.
McKenna, M. J., Hamilton, T. A., and Sussman, H. H. (1979). *Biochem. J.* **181**, 67–73.
Manning, J. P., Inglis, N. R., Green, S., and Fishman, W. H. (1969). *Enzymolgia* **37**, 251–261.
Manning, J. P., Inglis, N. R., Green, S., and Fishman, W. H. (1970). *Enzmologia* **39**, 307–318.
Mason, D. Y., and Taylor, C. R. (1975). *J. Clin. Pathol.* **28**, 124–132.
Mendelsohn, J., Eggleston, J. A., Weisburger, W. R., Grann, D. S., and Baylin, S. B. (1978). *Am. J. Pathol.* **92**, 35–52.
Millan, J. L., and Stigbrand, T. (1981). *Clin. Chem. (Winston-Salem, N.C.)* **27**, 2014–2018.
Millan, J. L., Stigbrand, T., Ruoslahti, E., and Fishman, W. H. (1982). *Cancer Res.* **42**, 2444–2449.
Miller, F., Lazarides, E., and Elias, J. (1976). *Clin. Immunol. Immunopathol.* **5**, 416–428.
Miyayama, H., Doellgast, G. J., Memoli, V., Gandbhir, L., and Fishman, W. H. (1976). *Cancer* **38**, 1237–1246.
Mukai, K., Rosai, J., and Hallaway, B. E. (1979). *Am. J. Surg. Pathol.* **3**, 373–376.
Mulivor, R. A., Plothin, L. I., and Harris, H. (1978). *Ann. Hum. Genet.* **42**, 1–13.
Nadji, M., Gonzalez, M. S., Castro, A., and Morales, A. R. (1980). *Lab. Invest.* **42**, 139.
Nakane, P. K. (1970). *J. Histochem. Cytochem.* **18**, 9–20.
Nicolson, G. L., Mascali, J. J., and McGuire, E. J. (1982). *Oncodevelop. Biol. Med.* **4**, 149–160.
Nishiyama, T., Stolbach, L. L., Rule, A. H., DeLellis, R. A., Inglis, N. R., and Fishman, W. H. (1980). *Acta Histochem. Cytochem.* **13**, 245–253.
Norgaard-Pedersen, B., and Raghavan, D. (1980). *Oncodev. Biol. Med.* **1**, 327–358.
Nozawa, S., Ohta, H., Izumi, S., Hayashi, S., Tsuitsui, I., Kurihara, S., and Watanabe, K. (1980). *Acta Histochem. Cytochem.* **13**, 521–530.
Nozawa, S., Udagawa, Y., Ohta, H., Kurihara, S., and Fishman, W. H. (1983). in press.
O'Briain, D. S., and Dayal, Y. (1981). *In* "Diagnostic Immunocytochemistry" (R. A. DeLellis, ed.), pp. 75–110. Masson, New York.
Palmer, P. E., and Wolfe, H. J. (1978). *J. Histochem. Cytochem.* **26**, 523–531.
Pangalis, G. A., and Rappaport, H. (1977). *Lancet* **2**, 880.
Park, C. H., and Reid, J. D. (1980). *Cancer* **46**, 570–575.
Pitot, H. C. (1977). *Chemistry* **50**, 1–6.
Robert, F., Pelletier, G., and Hardy, J. (1978). *Arch. Pathol. Lab. Med.* **102**, 448–455.
Sakiyama, T., Mano, T., and Chou, J. Y. (1980). *J. Biol. Chem.* **225**, 9399–9403.
Sasaki, M., and Fishman, W. H. (1973). *Cancer Res.* **33**, 3008–3018.
Seargeant, L. E., and Stinson, R. A. (1979). *Nature (London)* **281**, 152–154.
Singer, R. M., and Fishman, W. H. (1974). *J. Cell Biol.* **60**, 777–780.
Singer, R. M., and Fishman, W. H. (1976). *Differentiation* **5**, 127–132.
Singh, I., Tsang, K. Y., and Blakemore, W. S. (1978). *Cancer Res.* **38**, 193–198.
Slaughter, C. A., Coseo, M. C., Canero, M. P., and Harris, H. (1981). *Proc. Natl. Acad. Sci. U.S.A.* **78**, 1124–1128.
Sussman, H. H., Bowman, M., and Lewis, J. L. (1968a). *Nature (London)* **218**, 359–360.
Sussman, H. H., Small, P. A., and Cotlove, E. (1968b). *J. Biol. Chem.* **243**, 160–166.
Suzuki, M., Kuramote, H., Hamano, M., Shirane, H., and Watanabe, K. (1980). *Acta Endocrinol. (Copenhagen)* **108**, 113.
Swarup, G., Cohen, S., and Gerbers, D. L. (1981). *J. Biol. Chem.* **256**, 8197–8201.
Taylor, C. R., Kurman, R. J., and Warner, N. E. (1978a). *Hum. Pathol.* **9**, 417–427.

Taylor, C. R., Russell, R., and Chandor, S. (1978b). *Am. J. Clin. Pathol.* **70**, 612–622.
Taylor, D. D., Homesley, H. D., and Doellgast, G. J. (1980). *Cancer Res.* **40**, 4064–4069.
Timperley, W. R. (1968). *Lancet* **2**, 356.
Timperley, W. R., Turner, P., and Davis, S. (1971). *J. Pathol.* **103**, 257–262.
Tokumitsu, S., Tokumitsu, K., Kohnoe, K., and Takeuchi, T. (1979). *Cancer Res.* **39**, 4732–4738.
Tokumitsu, S., Tokumitsu, K., and Fishman, W. H. (1981a). *J. Histochem. Cytochem.* **29**, 1080–1087.
Tokumitsu, S., Tokumitsu, K., and Fishman, W. H. (1981b). *Histochemistry* **73**, 1–13.
Wahren, B., Holmgren, P.-A., and Stigbrand, T. (1979). *Int. J. Cancer* **24**, 749–753.
Warnke, R., Pederson, M., Williams, C., and Levy, R. (1978). *Am. J. Clin. Pathol.* **70**, 867–875.
Wei, S. C., and Doellgast, G. J. (1980). *Biochem. Genet.* **18**, 1097–1107.
Wernes, T. W., Timperley, W. R., Hine, P., and Kay, G. (1972). *Gut* **13**, 513–519.
Wolfe, H. J., Melvin, K. E. W., Cervi-Skinner, S. J., Al Saadi, A. A., Juliar, J. F., Jackson, C. E., and Tashjian, A. H. (1973). *N. Engl. J. Med.* **289**, 437–441.

CHAPTER 2

Cell–Matrix Interactions as Determinants of Differentiation and Tumor Invasion

ERKKI RUOSLAHTI
Cancer Research Center
La Jolla Cancer Research Foundation
La Jolla, California

I.	Introduction	21
II.	Structure of Extracellular Matrices and Their Components	22
	A. Collagens	22
	B. Proteoglycans	22
	C. Glycoproteins	23
	D. Fibronectin—Structural Diversity and Adhesive Properties	24
III.	Interaction of Cells with Extracellular Matrix	27
	A. Attachment of Cells to Extracellular Matrix Components	27
	B. Cell–Matrix Interactions in Development	28
	C. Cell–Matrix Interactions in Cancer	30
	References	32

I. INTRODUCTION

It is commonly thought that cells communicate with the extracellular matrix that surrounds them and that this influences cellular differentiation (Grobstein, 1975). Such communication may be particularly important in cases where epithelial–mesenchymal interactions lead into differentiation of embryonal tissues.

Significant progress has recently been made in the understanding of the chemistry of extracellular and basement membrane macromolecules. Fibronectin (see Ruoslahti *et al.*, 1981a for a review) and laminin (see Timpl *et al.*, 1982) have been identified as major glycoproteins in these structures. It has also become clear that collagen exists as a number of different genetic types of related molecules (see Miller, 1976). For instance, basement membranes have distinct types of collagens, and some of the newly discovered collagens may be predominantly associated with the cell surface (see Bornstein and Sage, 1980). The advances made in the identification and chemical characterization of cell matrix molecules have made it possible to study their influence in the behavior of normal and malignant cells. There is increasing evidence that cells communicate with the surrounding matrix by adhesive interactions and that these interactions are significant for normal cellular differentiation and for the invasive properties of malignant cells.

The purpose of this chapter is to review briefly the chemistry of the known extracellular matrix and basement membrane components and to discuss the current understanding of their role in development and neoplasia.

II. STRUCTURE OF EXTRACELLULAR MATRICES AND THEIR COMPONENTS

The main constituents of extracellular matrices and basement membranes are collagens, proteoglycans and hyaluronic acid, elastin, and various glycoproteins.

A. Collagens

Several genetic types of collagens with different chemical properties and tissue distributions have been identified. Interstitial collagens include types I, II, and III (Miller, 1976). They mainly exist in connective tissues and are assembled in fibrils; type II is specific for cartilage. Basement membranes contain type IV collagen (Kefalides, 1978), while type V seems to be associated with cell surfaces (Gay et al., 1981) and fibrils of interstitial collagens (Martinez-Hernandez et al., 1982). Collagens are synthesized as procollagen polypeptides with molecular weights around 180,000–200,000 and large stretches of repeats of Gly-X-Y sequences. Collagens undergo a unique series of posttranslational modifications which includes hydroxylation of some proline and lysine residues, glycosylation, assembly to triple helices, removal of polypeptides from each end of the polypeptide, and cross-linking (Fessler and Fessler, 1978). Type IV collagen appears to be exceptional in that its polypeptide remains uncleaved in the processing (Kefalides, 1978). The various collagens are involved in cellular attachment and growth in different ways (see below).

B. Proteoglycans

Proteoglycans are composed of glycosaminoglycan side chains attached to a core protein (see Rodén, 1980). The protein core has a characteristically high

2. Cell–Matrix Interactions

content of serine and glycine. The core proteins of different proteoglycans are different (Oldberg et al., 1982). The glycosaminoglycan side chains consist of a repeating disaccharide unit, usually of uronic acid and hexosamine. They exhibit considerable heterogeneity; for instance, they are sulfated to different degrees. Depending on the nature of the glycosaminoglycan, heparin, heparan sulfate, chondroitin sulfate, and keratan sulfate proteoglycans can be distinguished. Dermatan sulfate is often a component of chondroitin sulfate proteoglycan.

C. Glycoproteins

Several noncollagenous glycoproteins are present in extracellular matrices and basement membranes. These appear to form a family of possibly related proteins involved in cellular adhesion. The group so far includes fibronectin (Mosher, 1980; Ruoslahti et al., 1981a), laminin (Chung et al., 1979; Timpl et al., 1979), a recently described sulfated glycoprotein—entactin (Carlin et al., 1981)—which is associated with laminin (Cooper et al., 1981), and the bullous pemphigoid antigen of the skin basement membrane (Stanley et al., 1981).

All of these are large molecular weight glycoproteins. Fibronectin is a disulfide-linked dimer of two polypeptides, each with a molecular weight of about 230,000, while laminin has two disulfide-linked chains with molecular weights of 200,000 and 400,000. They are long, flexible molecules (Fig. 1) with separate binding sites for other macromolecules (see below). The relationship of the bullous pemphigoid antigen to fibronectin is particularly interesting. Its constitu-

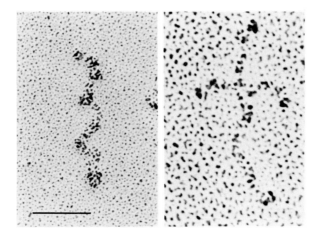

Fig. 1. Electron micrographs of fibronectin (left) and laminin (right). Molecules sprayed on mica were visualized by rotary shadowing. Scale: 50 nm.

ent polypeptide has the same size as that of fibronectin, but it has a restricted distribution suggesting a tissue-specific function.

Fibronectin and laminin mediate attachment of cells to extracellular matrices, and it seems possible that at least the bullous pemphigoid antigen also does this, since patients with bullous pemphigus have antibodies to this protein (Stanley *et al.*, 1981) and their skin blisters at the epidermal basement membrane. The antibodies could be interfering with the attachment of the basal layer of epidermal cells to the basement membrane.

Fibronectin has emerged as a prototype of extracellular matrix and basement membrane proteins active in cellular adhesion. Because of this, its structure and interactions with other macromolecules will be discussed in some detail below.

D. Fibronectin—Structural Diversity and Adhesive Properties

Fibronectin is present at cell surfaces, in connective tissues, and in the blood and other body fluids (for recent reviews, see Yamada and Olden, 1978; Mosher, 1980; Ruoslahti *et al.*, 1981a). It interacts with many other macromolecules, including collagens, glycosaminoglycans, fibrinogen and fibrin, actin, some bacteria, and a structure of unknown nature (receptor?) at the cell surface of most eukaryotic cells. The interaction with cells is manifested through the attachment and spreading of cells to a surface covered with insoluble fibronectin (Klebe, 1974). This phenomenon has led to the hypothesis that fibronectin acts as an anchor for cells in the extracellular matrix. Furthermore, its binding and consequent cross-linking to fibrin direct its incorporation into blood clots which in turn could provide necessary cell attachment sites in wound healing. The involvement of fibronectin in the control of cellular morphology has also been inferred from the observation that whereas fibronectin is abundant in the matrix of normal cells, it is absent from the surface of many transformed cells.

Based on recent structural studies carried out in a number of laboratories, we are beginning to discern the molecular arrangements which underlie this multitude of biological activities of fibronectin.

A main feature of the structure of fibronectin is the occurrence of independent structural domains with different functional activities. Such a domain structure was first proposed based on physical measurements on fibronectin (Alexander *et al.*, 1979). Results obtained by fragmentation of the molecule have confirmed the existence of a domain structure and have shown that some of the various activities of fibronectin reside in different domains and that the domains are structurally and functionally independent.

Proteolytic digestion of fibronectin allows generation of fragments which retain one or more of the activities of the whole molecule. The fibronectin molecule is highly susceptible to proteolytic cleavage and readily yields fragments that have lost the interchain disulfide bond(s) but otherwise retain nearly the entire length of the polypeptide (Chen et al., 1977). Various enzymes also yield smaller fragments that are relatively resistant to further proteolysis. These represent the active domains of the molecule.

The NH_2-terminal domain is 27 kd in size. It contains the binding sites for fibrin(ogen) and staphylococci, and probably also for actin. Intact fibronectin may become covalently cross-linked to fibrin through a glutamine residue in the NH_2-terminal domain in the presence of blood coagulation factor XIII (transglutaminase) (Mosher, 1980). The collagen-binding site is within the 30 kd region immediately adjacent to the COOH-terminal side of the 27 kd, NH_2-terminal domain. Most of the carbohydrate in fibronectin is located in this domain or near it on the COOH-terminal side (Sekiguchi et al., 1981). The cell attachment site is located between the gelatin-binding and heparin-binding domains, 127 to 197 kd from the NH_2-terminus of the fibronectin polypeptide (Ruoslahti et al., 1981b). Peptic digests of fibronectin yield a 15 kd fragment which can be isolated in a homogeneous form by affinity chromatography on a monoclonal antibody that inhibits attachment of cells to fibronectin (Pierschbacher et al., 1981). This 15 kd fragment contains all of the cell attachment-promoting activity of intact fibronectin. This fragment provides a useful probe for the identification of the putative cell surface (receptor?) at the cell surface that interacts with fibronectin. The nature of this interaction is incompletely understood at present.

The glycosaminoglycan-binding domain is located closest to the COOH-terminal end of the polypeptide (Ruoslahti et al., 1981b). It has been isolated as fragments of about 50 kd (Yamada et al., 1980; Pierschbacher et al., 1981). This domain may be structurally similar to the NH_2-terminal domain. Both share affinity to heparin, and there is recent evidence that the COOH-terminal domain may, like the NH_2-terminal domain, have sites for the binding of fibrin and for the cross-linking of fibrin to fibronectin by transglutaminase (Sekiguchi and Hakomori, 1980; Hörmann and Jelinic, 1981). Finally, the interchain disulfide bond(s) that joins the two polypeptides is nearest the COOH terminal of the molecule (Chen et al., 1977; Ruoslahti et al., 1981b), and a free sulfhydryl group has been localized about 170 kd from the NH_2 terminus (Wagner and Hynes, 1980).

Available knowledge about structural arrangements in the fibronectin molecule provides some insight into the function of this protein at the cell surface and in plasma. The collagen- and fibrin-binding activities are likely to be involved in anchoring fibronectin to extracellular matrix and to a fibrin clot. In a wound this might provide scaffolding to which the penetrating cells attach. Heparin and some other glycosaminoglycans enhance the interaction of fibronectin with col-

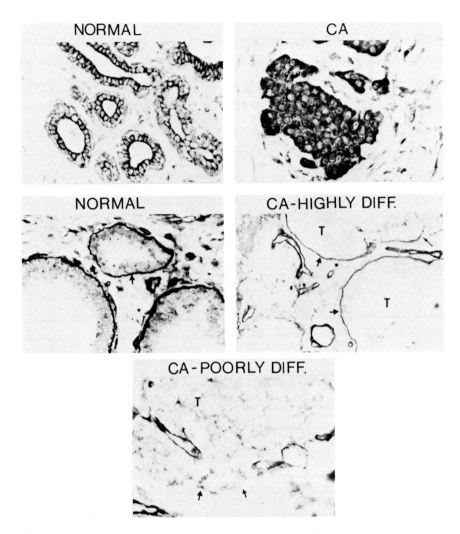

Fig. 2. Immunoperoxidase staining for laminin in normal and malignant human breast tissue. In the upper two panels the formalin-fixed tissue sections representing normal and cancerous (CA) breast tissue were stained with antibodies revealing mostly intracellular laminin staining. In the lower three panels the sections were treated briefly with trypsin prior to the staining. This facilitated the visualization of basement membranes around the normal ducts and in highly differentiated tumors, but revealed no basement membranes in poorly differentiated tumors. (Modified from Albrechtsen *et al.* by permission of *Cancer Research*.)

lagen (see Ruoslahti *et al.*, 1981a). This phenomenon may secure incorporation of fibronectin into the extracellular matrix. These interactions leave the cell attachment site of fibronectin available for binding of cells to the matrix.

The collagen- and heparin-binding activities are not sufficient, however, for the binding of fibronectin to extracellular matrix. Intact human plasma fibronectin injected into the circulation of mice becomes incorporated into the tissues in a manner that makes its distribution indistinguishable from that of the mouse's own tissue fibronectin in immunofluorescence (Oh *et al.*, 1981). In contrast, 200 kd fragments, which bind to both collagen and heparin but are truncated at both ends of the molecule, do not show appreciable tissue incorporation. These fragments lack the NH_2-terminal region as well as the region containing the interchain disulfide bond(s). These missing structures thus seem to play a role in the matrix incorporation of fibronectin.

The capacity of plasma fibronectin to become incorporated into tissue matrix suggests that one of the functions of the circulating fibronectin is to provide a reservoir of tissue fibronectin. Plasma fibronectin has also been found to function as an opsonin. Particles coated with gelatin are taken up more efficiently by liver slices and macrophages in the presence of fibronectin, and fibrin bound to such particles appears to cause macrophage activation (see Saba and Jaffe, 1980). Collagen, fibrin, and actin are the main insoluble proteins at the site of tissue injury. The binding of fibronectin to these proteins may not only facilitate cell attachment and growth but may also enhance phagocytosis of cellular debris.

Much less is known about other matrix glycoproteins, but judging from the fact that laminin mediates cell attachment and binds also to heparin, it seems that molecular arrangements in these proteins may resemble those in fibronectin. The implications of the molecular interactions of the matrix components to differentiation and neoplasia will be discussed below.

III. INTERACTION OF CELLS WITH EXTRACELLULAR MATRIX

Attachment of cells to component macromolecules of extracellular matrices has been used as an indicator of a primary interaction of cells with extracellular matrix. Such interactions may subsequently lead into changes in gene expression of the adhering cells.

A. Attachment of Cells to Extracellular Matrix Components

Cells plated on dishes coated with fibronectin attach and spread rapidly (Grinnell *et al.*, 1977; Hayman *et al.*, 1980). Since fibronectin binds to collagen

(Engvall and Ruoslahti, 1977), the cell attachment-promoting activity of fibronectin can also be demonstrated by constructing an artificial matrix of isolated collagen to which fibronectin is bound. Such a matrix readily supports cell attachment (Klebe, 1974; Pearlstein, 1976).

Fibronectin mediates the attachment of most types of cells. Circulating lymphocytes seem to be an exception (E. Ruoslahti, unpublished results), but attachment of platelets is mediated by fibronectin (Hynes *et al.*, 1978). In spite of its cell attachment-enhancing properties, fibronectin promotes motility of cells (Ali and Hynes, 1978). This property may be important for cell movements during embryonal development (see below).

Laminin also appears to promote cell attachment, but with a more restricted specificity than fibronectin. Laminin promotes the attachment of liver cells (Carlsson *et al.*, 1981; Johansson *et al.*, 1981) and other epithelial cells (Terranova *et al.*, 1980; Vlodavsky and Gospodarowicz, 1981). It remains to be seen how other cell types behave with regard to laminin, but these initial findings suggest that interaction with laminin could form the basis of some cell-type-specific basement membrane interactions.

Fibronectin, by virtue of its interaction with all types of collagens (Dessau *et al.*, 1978; Engvall *et al.*, 1978), apparently can mediate cell attachment to all collagens. In contrast, the attachment of chondrocytes to the cartilage-specific type II collagen is specifically enhanced by a plasma protein distinct from fibronectin (Hewitt *et al.*, 1980). The name chondronectin has been coined for this factor.

Collagens belonging to the different genetic types may also directly serve as attachment and signal proteins without the mediation of other proteins such as fibronectin (reviewed by Adamson, 1982; Kleinman *et al.*, 1980). Epithelial cells seem to have a particular affinity to type IV (basement membrane) collagen (Murray *et al.*, 1980).

B. Cell—Matrix Interactions in Development

In considering the influence of extracellular matrix components on cellular behavior, it is important to note the complexity of the cell surface—matrix interaction. There are two main aspects to this interaction: the production of an extracellular matrix and the ability to interact with this and other matrices. Normally, cells would be likely to do both, whereas some cells may not produce their own matrix, but attach to matrices produced by other cells. The latter situation may be important in cellular migration and invasion.

Several lines of evidence suggest that fibronectin plays a role in directing differentiation and morphogenetic movements of certain cells. Myoblasts, which initially express fibronectin, lose it prior to their fusion into myotubes (Chen, 1977). Addition of fibronectin to cultures of myoblasts inhibits their fusion

(Podleski *et al.*, 1979), and retinal cells cultured on fibronectin-coated plates show increased neurite outgrowth (Akers *et al.*, 1981). A particularly interesting relationship exists between the phenotype of chondrocytes and fibronectin. Differentiated chondrocytes and adult cartilage do not contain fibronectin (Dessau *et al.*, 1978). Addition of exogenous fibronectin to cultures of differentiated chondrocytes makes them assume a fibroblastic phenotype with lowered synthesis of cartilage components, such as type II collagen, proteoglycan, and a chondrocyte-specific cell surface protein (Pennypacker *et al.*, 1979; West *et al.*, 1979). Addition of fibronectin to cultures of differentiated chondrocytes also induces synthesis of endogenous fibronectin in such cultures. Once initiated, the dedifferentiation process may, therefore, become self-perpetuating. Such events could play an important role in degenerative diseases involving cartilage. In another experimental system that appears to have a great deal of potential, it is possible to direct mesenchymal cells to produce cartilage and bone by placing demineralized bone matrix into soft connective tissue (Reddi, 1976; Weiss and Reddi, 1980).

Less direct evidence for the possible involvement of fibronectin in differentiation comes from studies on its expression during differentiation of teratocarcinoma cells *in vitro* and of embryonal tissues *in vivo*. Fibronectin is not present in significant quantities in the undifferentiated stem cells of teratocarcinoma (embryonal carcinoma cells), but when these cells form so-called embryoid bodies, fibronectin appears (Wartiovaara *et al.*, 1978; Zetter *et al.*, 1978). Turning on and turning off fibronectin expression takes place in various differentiative events *in vivo*. For instance, the mesenchyme of the developing tooth becomes devoid of fibronectin in immunofluorescent staining as the cells differentiate into odontoblasts (Thesleff *et al.*, 1979). Such changes in fibronectin may be secondary to the cellular differentiation, but the changes in attachment and motility of cells that must accompany the appearance and disappearance of fibronectin may provide important signals directing differentiation. The expression of fibronectin has also been suggested to be involved in various morphogenetic movements during early embryonal development (Critchley *et al.*, 1979; Mayer *et al.*, 1980).

A particularly interesting set of observations has recently been made using protein-coated latex beads as probes for embryonal cell migration (Bronner-Fraser, 1982). In the embryo, neural crest cells migrate ventrally from the neural crest to give rise to sensory and sympathetic neurons, glial, and chromaffin cells in the gut area. Various types of cells can be injected in the embryo and their migration along the migration pathway observed. When beads coated with albumin and fibronectin were injected to the pathway, the albumin-coated beads (as well as beads left uncoated) migrated along the pathway, but beads coated with fibronectin failed to do so. A general correlation between the ability of various types of cells to become translocated along the neural crest pathway and lack of cell surface fibronectin has also been noted (Erickson *et al.*, 1980; Bronner-Fraser, 1982). The presence or absence of cell surface fibronectin may, therefore, determine whether a given cell participates in migratory movements during development.

Laminin is one of the first known matrix components to be expressed in the development of the mammalian embryo (Leivo *et al.*, 1980). Expression of laminin has been correlated with the development of tubuli in the embryo (Ekblom *et al.*, 1980). Cells of the regenerating liver acquire an ability to attach to laminin. At the same time, increased amounts of laminin, as detected by immunofluorescence, appear in the liver where cell proliferation is taking place (Carlsson *et al.*, 1981). These findings suggest that laminin in the basement membranes may be involved in tissue organization, cell proliferation, and possibly differentiation. A similar role may be played by the various genetic types of collagen which also show changes during development (Adamson *et al.*, 1979).

C. Cell–Matrix Interactions in Cancer

A great deal of the interest in fibronectin has been because the amount of this protein is usually greatly reduced on the surface of malignantly transformed cells as compared with normal cells.

The correlation of loss of surface fibronectin with the transformed phenotype is especially impressive in the experimental systems where mutant viruses, which are temperature-sensitive with respect to transforming capacity, have been used (Gahmberg *et al.*, 1974; Vaheri and Ruoslahti, 1974; Hynes and Wyke, 1975). Cells grown at nonpermissive temperature are phenotypically normal and express fibronectin at their surface. A shift to the permissive temperature brings about the transformation with a concomitant reduction of surface fibronectin. When the cells are returned to nonpermissive temperature, they will regain their normal morphology and cell surface fibronectin. Fibronectin can also be restored at the surface of transformed cells by treatment of cells with glucocorticoids (Furcht *et al.*, 1979), cyclic AMP (Nielson and Puck, 1980), and butyrate (Hayman *et al.*, 1980). Such restoration may provide a useful model for studies on the role of cell surface fibronectin in transformation.

The lack of cell surface fibronectin in malignant cells seems to reflect a generalized absence of extracellular matrix in such cells. Transformed rat kidney cells lack cell surface fibronectin, laminin, and heparan sulfate proteoglycan, which are all present in their normal counterpart cells codistributed in the same structures at the cell surface (Hayman *et al.*, 1982).

Transformed cells that lack surface and pericellular fibronectin nevertheless do produce it (Vaheri and Ruoslahti, 1975; Critchley *et al.*, 1976; Vaheri *et al.*, 1976). This is also true of laminin, at least in the case of transformed rat kidney cells (Hayman *et al.*, 1981). It is not understood why transformed cells do not retain their fibronectin and laminin at the cell surface, but there are reasons to believe that the heparan sulfate proteoglycan may be the key component, the lack of which prevents formation of a matrix. It seems that multiple interactions between the different matrix components could be required for the formation of insoluble extracellular matrix, and that a decrease or alteration of any one of the components, fibronectin, collagen, or proteoglycan, could interfere with this and cause impaired attachment of malignant cells. Since fibronectin and laminin are produced by transformed cells, and these proteins as well as collagens interact with proteoglycans (see Sakashita *et al.*, 1980; Oldberg and Ruoslahti, 1982), the proteoglycan appears to be a likely candidate for the crucial component missing in these cells.

There does not appear to be any straightforward correlation between lack of cell surface fibronectin and tumorigenicity (Der and Stanbridge, 1978; Kahn and Shin, 1979; Neri *et al.*, 1981; Keski-Oja *et al.*, 1982). However, cells isolated from primary tumors have more fibronectin *in vitro* than cells isolated from metastases (Smith *et al.*, 1979; Neri *et al.*, 1981), suggesting that the lack of fibronectin may be one of the factors that influences metastatic capacity of tumor cells. Considering the effects of fibronectin on the migration of embryonal cells discussed above, it stands to reason that the lack of cell surface fibronectin would have an effect on the invasiveness of malignant cells.

The appearance of laminin in normal and malignant cells has not been studied enough to allow any generalizations. Laminin seems to be particularly abundant in various teratocarcinoma-derived endodermal cell lines (Chung *et al.*, 1979). In cultures of mouse teratocarcinoma cells, the differentiation of the embryonal carcinoma stem cells into endodermal cells is accompanied by synthesis of a basement membrane which contains laminin (see Wartiovaara *et al.*, 1980). Laminin is abundant in Reichert's membrane (Hogan, 1980), and it has been shown to be a major constituent of the basement membrane material synthesized by a rat yolk sac tumor (Wewer *et al.*, 1981). Various types of epithelial cells also synthesize laminin *in vitro* (Timpl *et al.*, 1979; Alitalo *et al.*, 1980). As discussed above, at least in the case of the rat kidney cells, the lack of cell surface fibronectin in transformed cells is accompanied by a concomitant absence of cell surface laminin (Hayman *et al.*, 1981).

Studies on the distribution of laminin in breast tumors have revealed a striking absence of laminin-containing basement membranes in some tumors as detected by immunoperoxidase staining with antibodies to laminin (Fig. 2; Albrechtsen *et al.*, 1981). A strong correlation was found between the presence of basement membranes detectable with this method and the degree of morphological differentiation of the tumor: the tumors with the lowest degree of differentiation having no basement membranes. Since basement membranes are thought to form a barrier against tumor spread, staining of tumor tissues for laminin may be useful in evaluating the propensity of a tumor to invade. Whether the lack of basement membranes is due to a lack of deposition or abnormal destruction of the basement membranes is not known. A collagenase that specifically degrades basement membrane collagen may be involved in this phenomenon (Kreiger et al., 1979).

Another possible application of the laminin staining in tumor diagnosis is the detection of micrometastases in lymph nodes. Cells in breast cancer micrometastases show intense intracellular staining with anti-laminin, while the lymph node otherwise contains little laminin. Since laminin is produced by many, if not all, epithelial tissues, these techniques should be applicable to other tumors also.

An important aspect concerning the relationship of malignant cells to basement membranes is the adhesion of tumor cells to basement membranes. To give rise to blood-borne metastasis, tumor cells must adhere to the capillary endothelium and penetrate the capillary walls to the surrounding tissue. Metastatic cells have been found to have a particular affinity to endothelial cell matrix (Kramer et al., 1980) and to basement membrane collagen (Murray et al., 1980), suggesting that such cells attach readily to basement membranes. This and subsequent digestion and penetration of the basement membrane (Liotta et al., 1980) are likely to play a role in extravasation of tumor cells.

In summary, extracellular matrices and basement membranes are composed of a number of different types of collagens, proteoglycans, and noncollagenous glycoproteins. Interaction of cells with these proteins is manifested in enhanced cellular attachment that affects the cellular shape, motility, and state of differentiation. Changes occur in the expression of matrix proteins during embryonal development and in malignant cells, suggesting that cell–matrix interactions may play a role in differentiation and cancer.

ACKNOWLEDGMENTS

I thank Ms. Nancy Beddingfield for assistance with the preparation of the manuscript. Our original work described here, and the preparation of this review were supported by Cancer Center Support Grant P30-CA 30199, and grants CA 28896 and CA 28101 from the National Cancer Institute, DHHS.

REFERENCES

Adamson, E. D. *In* "Collagen in Health and Disease" (M. Jayson and J. Weiss, eds.). Churchill-Livingstone, Edinburgh and London (in press).
Adamson, E. D., Gaunt, S. J., and Graham, C. F. (1979). *Cell* **17,** 469–476.
Akers, R. M., Mosher, D. F., and Lilien, J. E. (1981). *Dev. Biol.* **86,** 179–188.
Albrechtsen, R., Nielsen, M., Wewer, U., Engvall, E., and Ruoslahti, E. (1981). *Cancer Res.* **41,** 5076–5081.
Alexander, S. S., Colonna, G., and Edelhoch, H. J. (1979). *J. Biol. Chem.* **254,** 1501–1505.

Ali, I. U., and Hynes, R. O. (1978). *Cell* **14**, 439–446.
Alitalo, K., Kurkinen, M., Vaheri, A., Krieg, T., and Timpl, R. (1980). *Cell* **19**, 1053–1062.
Bornstein, P., and Sage, H. (1980). *Annu. Rev. Biochem.* **49**, 957–1003.
Bronner-Fraser, M. (1982). *Dev. Biol.* **91**, 50–63.
Carlin, B., Jaffe, R., Bender, B., and Chung, A. E. (1981). *J. Biol. Chem.* **256**, 5209–5214.
Carlsson, R., Engvall, E., Freeman, A., and Ruoslahti, E. (1981). *Proc. Natl. Acad. Sci. U.S.A.* **78**, 2403–2406.
Chen, A. B., Amrani, D. L., and Mosesson, M. W. (1977). *Biochem. Biophys. Acta* **493**, 310–322.
Chen, L. B. (1977). *Cell* **10**, 393–400.
Chung, A. E., Jaffe, R., Freeman, I. L., Vergnes, J.-P., Braginski, J. E., and Carlin, B. (1979). *Cell* **16**, 277–287.
Cooper, A. R., Kurkinen, M., Taylor, A., and Hogan, B. L. M. (1981). *Eur. J. Biochem.* **119**, 189–197.
Critchley, D. R., Wyke, J. A., and Hynes, R. O. (1976). *Biochim. Biophys. Acta* **436**, 335–352.
Critchley, D. R., England, M. A., Wakely, J., and Hynes, R. O. (1979). *Nature (London)* **280**, 498–500.
Der, C. J., and Stanbridge, E. J. (1978). *Cell* **15**, 1241–1251.
Dessau, W., Sasse, J., Timpl, R., Jilek, F., and von der Mark, K. (1978). *J. Cell Biol.* **79**, 342–355.
Ekblom, P., Alitalo, K., Vaheri, A., Timpl, R., and Saxén, L. (1980). *Proc. Natl. Acad. Sci. U.S.A.* **77**, 485–489.
Engvall, E., and Ruoslahti, E. (1977). *Int. J. Cancer* **20**, 1–5.
Engvall, E., Ruoslahti, E., and Miller, E. J. (1978). *J. Exp. Med.* **147**, 1584–1595.
Erickson, C. A., Tosney, K. W., and Weston, J. A. (1980). *Dev. Biol.* **77**, 142–156.
Fessler, J. H., and Fessler, L. I. (1978). *Annu. Rev. Biochem.* **47**, 129–162.
Furcht, L. T., Mosher, D. F., Wendelshafer-Crabb, G., Woodbridge, P. A., and Foidart, J. M. (1979). *Nature (London)* **277**, 393–397.
Gahmberg, C. G., Kiehn, D., and Hakomori, S. I. (1974). *Nature (London)* **248**, 413–415.
Gay, S., Rhodes, R. K., Gay, R. E., and Miller, E. J. (1981). *Collagen Res.* **1**, 53–58.
Grinnell, F., Hays, D. G., and Minter, D. (1977). *Exp. Cell Res.* **110**, 175–190.
Grobstein, C. (1975). *In* "Extracellular Matrix Influences on Gene Expression" (H. C. Slavkin and R. C. Greulich, eds.), pp. 9–16. Academic Press, New York.
Hayman, E. G., Engvall, E., and Ruoslahti, E. (1980). *Exp. Cell Res.* **127**, 478–481.
Hayman, E. G., Engvall, E., and Ruoslahti, E. (1981). *J. Cell Biol.* **88**, 352–357.
Hayman, E. G., Oldberg, Å., Martin, G. R., and Ruoslahti, E. (1982). *J. Cell Biol.* **94**, 28–35.
Hewitt, A. T., Kleinman, H. K., Pennypacker, J. P., and Martin, G. R. (1980). *Proc. Natl. Acad. Sci. U.S.A.* **77**, 385–388.
Hogan, B. L. M. (1980). *Dev. Biol.* **76**, 275–285.
Hörmann, H., and Jelinic, V. (1981). *Hoppe-Seyler's Z. Physiol. Chem.* **362**, 87–94.
Hynes, R. O., and Wyke, J. A. (1975). *Virology* **64**, 492–504.
Hynes, R. O., Ali, I. U., Destree, A. T., Mautner, V., Perkins, M. E., Senger, D. R., Wagner, D. D., and Smith, K. K. (1978). *Ann. N.Y. Acad. Sci.* **312**, 317–342.
Johansson, S., Kjellén, L., Höök, M., and Timpl, R. (1981). *J. Cell Biol.* **90**, 260–264.
Kahn, P., and Shin, S. I. (1979). *J. Cell Biol.* **82**, 1–16.
Kefalides, N. A. (1978). "Biology and Chemistry of Basement Membranes." Academic Press, New York.
Keski-Oja, J., Gahmberg, C. G., and Alitalo, K. (1982). *Cancer Res.* **42**, 1147–1153.
Klebe, R. J. (1974). *Nature (London)* **250**, 248–251.
Kleinman, H. K., Klebe, R. J., and Martin, G. R. (1980). *J. Cell Biol.* **88**, 473–485.
Kramer, R. H., Gonzalez, R., and Nicolson, G. I. (1980). *Int. J. Cancer* **26**, 639–645.
Kreiger, D. T., Liotta, A. S., Nicolson, G. L., and Kizer, J. S. (1979). *Nature (London)* **278**, 562–563.

Leivo, I., Vaheri, A., Timpl, R., and Wartiovaara, J. (1980). *Dev. Biol.* **76,** 100–114.
Liotta, L. A., Tryggvason, K., Garbisa, S., Hart, I., Foltz, C. M., and Shafie, S. (1980). *Nature (London)* **284,** 67–68.
Martinez-Hernandez, A., Gay, S., and Miller, E. J. (1982). *J. Cell Biol.* **92,** 343–349.
Mayer, B. W., Jr., Hay, E. D., and Hynes, R. O. (1980). *Dev. Biol.* **82,** 267–286.
Miller, E. J. (1976). *Mol. Cell. Biochem.* **13,** 165–192.
Mosher, D. F. (1980). *Prog. Hemostasis Thromb.* **5,** 111–151.
Murray, J. C., Liotta, L., Rennard, S. I., and Martin, G. R. (1980). *Cancer Res.* **40,** 347–351.
Neri, A., Ruoslahti, E., and Nicolson, G. L. (1981). *Cancer Res.* **41,** 5082–5095.
Nielson, S. E., and Puck, T. T. (1980). *Proc. Natl. Acad. Sci. U.S.A.* **77,** 985–989.
Oldberg, Å., and Ruoslahti, E. (1982). *J. Biol. Chem.* **257,** 4859–4863.
Oldberg, Å., Schwartz, C. R., and Ruoslahti, E. (1982). *Arch. Biochem. Biophys.* **216,** 400–406.
Oh, E., Pierschbacher, M., and Ruoslahti, E. (1981). *Proc. Natl. Acad. Sci. U.S.A.* **78,** 3218–3221.
Pearlstein, E. (1976). *Nature (London)* **262,** 497–500.
Pennypacker, J. P., Hassell, J. R., Yamada, K. M., and Pratt, R. M. (1979). *Exp. Cell Res.* **121,** 411–415.
Pierschbacher, M. D., Hayman, E. G., and Ruoslahti, E. (1981). *Cell* **26,** 259–267.
Podleski, T. R., Greenberg, I., Schlessinger, J., and Yamada, K. M. (1979). *Exp. Cell Res.* **122,** 317–326.
Reddi, A. H. (1976). *In* "Biochemistry of Collagen" (G. N. Ramachandran and A. H. Reddi, eds.), pp. 449–478. Plenum, New York.
Rodén, L. (1980). *In* "The Biochemistry of Glycoproteins and Proteoglycans" (W. J. Lennarz, ed.), pp. 267–371. Plenum, New York.
Ruoslahti, E., Engvall, E., and Hayman, E. G. (1981a). *Collagen Res.* **1,** 95–128.
Ruoslahti, E., Hayman, E. G., Engvall, E., Cothran, W. C., and Butler, W. T. (1981b). *J. Biol. Chem.* **256,** 7277–7281.
Saba, T. M., and Jaffe, E. (1980). *Am. J. Med.* **68,** 577–594.
Sakashita, S., Engvall, E., and Ruoslahti, E. (1980). *FEBS Lett.* **116,** 243–246.
Sekiguchi, K., and Hakomori, S. I. (1980). *Biochem. Biophys. Res. Commun.* **97,** 709–715.
Sekiguchi, K., Fukuda, M., and Hakomori, S. I. (1981). *J. Biol. Chem.* **256,** 6452–6462.
Smith, H. S., Riggs, J. L., and Mosesson, M. W. (1979). *Cancer Res.* **39,** 4138–4144.
Stanley, J. R., Hawley-Nelson, P., Yuspa, S. H., Shevach, E. M., and Katz, S. I. (1981). *Cell* **24,** 897–903.
Terranova, V. P., Rohrbach, D. H., and Martin, G. R. (1980). *Cell* **22,** 719–726.
Thesleff, I., Stenman, S., Vaheri, A., and Timpl, R. (1979). *Dev. Biol.* **70,** 116–126.
Timpl, R., Rohde, H., Robey, P. G., Rennard, S. I., Foidart, J.-M., and Martin, G. R. (1979). *J. Biol. Chem.* **254,** 9933–9937.
Timpl, R., Rhode, L., Ott, U., Robey, P. G., and Martin, G. R. (1982). *In* "Methods in Enzymology" (L. W. Cunningham and D. W. Frederiksen, eds.), Vol. 82, Pt. A, pp. 831–838. Academic Press, New York.
Vaheri, A., and Ruoslahti, E. (1974). *Int. J. Cancer* **13,** 579–586.
Vaheri, A., and Ruoslahti, E. (1975). *J. Exp. Med.* **142,** 530–535.
Vaheri, A., Ruoslahti, E., Westermark, B., and Pontén, J. (1976). *J. Exp. Med.* **143,** 64–72.
Vlodavsky, I., and Godpodarowicz, D. (1981). *Nature (London)* **289,** 304–306.
Wagner, D. D., and Hynes, R. O. (1980). *J. Biol. Chem.* **255,** 4304–4312.
Wartiovaara, J., Leivo, I., Virtanen, I., Vaheri, A., and Graham, C. F. (1978). *Nature (London)* **272,** 355–356.
Wartiovaara, J., Leivo, I., and Vaheri, A. (1980). *In* "The Cell Surface; Mediator of Developmental Process" (S. Subtleny, ed.), pp. 305–324. Academic Press, New York.
Weiss, R. E., and Reddi, A. H. (1980). *Proc. Natl. Acad. Sci. U.S.A.* **77,** 2074–2078.

West, C. M., Lanza, R., Rosenbloom, J., Lowe, M., and Holtzer, H. (1979). *Cell* **17,** 491–501.
Wewer, U., Albrechtsen, R., and Ruoslahti, E. (1981). *Cancer Res.* **41,** 1518–1524.
Yamada, K. M., and Olden, K. (1978). *Nature (London)* **275,** 179–184.
Yamada, K. M., Kennedy, D. W., Kimata, K., and Pratt, R. M. (1980). *J. Biol. Chem.* **255,** 6055–6063.
Zetter, B. R., Martin, G. R., Birdwell, C. R., and Gospodarowicz, D. (1978). *Ann. N.Y. Acad. Sci.* **312,** 299–316.

CHAPTER 3
Human Tumor Nucleolar Antigens

HARRIS BUSCH ROSE K. BUSCH
PUI-KWONG CHAN DAVID KELSEY
KEI TAKAHASHI WILLIAM H. SPOHN
MARJATTA SON

Department of Pharmacology
Baylor College of Medicine
Houston, Texas

I.	Introduction	38
II.	Nucleolar Antigens	38
	A. Experimental Production of Anti-Nucleolar Antibodies	38
	B. Tissues Employed and Immunofluorescent Microscopy	39
	C. Tumor Specificity	39
	D. Absorption of the Antibodies	43
	E. Purification of Tumor Nucleolar Antigens	44
	F. Liver Nucleolar Antigens	45
	G. mRNA for Nucleolar Antigens	48
	H. Nucleolar Antigens in Human Tumors	50
	I. Antibodies to Other Tumors	51
	J. Tumor Nucleolar Antigens in Nontumor Tissues	51
	K. Breast Cancer	56
	L. Normal Breast Tissue and Benign Tumors	59
	M. "Blind" Study	60
	N. Immune Localization of the Antigen	61
	O. Characterization of the Nucleolar Antigen	61
III.	Discussion	62
	A. Possible Functions of the Nucleolar Antigen	62
	B. The Antigen as Accelerator or Gene Control Element	63
	C. Relation of the Nucleolar Antigens to the "*src* Gene" Product	63
	D. Antigens Migrating into the Nucleolus in Late G_1 Phase	65
	References	66

I. INTRODUCTION

Anti-nucleolar antibodies, like other anti-nuclear antibodies, occur in human autoimmune diseases such as Sjögren's syndrome, lupus erythematosus, mixed connective disease, and rheumatoid arthritis. These mysterious autoimmune ailments differ in the number and types of antibodies elicited, their specificity and binding constants, and their potential destructive effects, which are poorly understood in relation to pathology or to the diseases themselves.

Anti-nucleolar antibodies oppose a variety of macromolecules, including DNA, histones, nonhistone proteins, RNA, and other nuclear elements. Two general types of anti-nucleolar antigens are of interest: those formed in the various disease states mentioned above and those induced by immunization of experimental animals with whole nucleoli or specific nucleolar fractions (Tan, 1979; H. Busch et al., 1979).

The complexity of experimentally produced antibodies increases with the duration of the immunization program and the complexity of the antigens. Specific nucleolar antigens have been purified and found to be at least part protein matter in two tumor studies involving the Novikoff rat hepatoma and the human HeLa and Namalwa tumors (Chan et al., 1980, 1981). These purified protein immunogens probably differ from the DNA-protein complexes utilized by Hnilica and his associates (1978) in their studies on DNP-specific antigens of a variety of tissues and tumors.

Nucleolar antigens present a potential for immunodiagnostic use because they may reflect specific functions in human cancer cells. This chapter presents a general review of the identification, isolation, and functional analysis of the nucleolar antigens of human cancer cells.

II. NUCLEOLAR ANTIGENS

The initial goal of our experimental studies on nucleolar antigens was to define conditions that might permit an analysis of very minute amounts of proteins important to nucleolar functions, some of which could be "promoters" for rDNA. Surprisingly, the results of these studies suggested a possible differentiation between human cancer cells and other cells; this point will be discussed in more detail.

A. Experimental Production of Anti-Nucleolar Antibodies

To determine whether nucleoli or their subfractions could induce immunological responses in rabbits, nucleoli were prepared by standard procedures developed in earlier studies in this laboratory (Busch and Smetana, 1970) and injected into rabbits in Freund's adjuvant (Busch et al., 1974). The sites of injection were initially subcutaneous and intramuscular, but later studies proved injection into footpads to be particularly useful. Initial testing for the immunological response was by the direct immunofluorescence assays of Hilgers et al. (1972). Ouchterlony gel analysis and immunoelectrophoresis were employed to determine whether precipitating antibodies were produced (Busch and Busch, 1977). In the initial studies (Busch et al., 1974) complement fixation was used, more definitive studies were done with other procedures.

B. Tissues Employed and Immunofluorescent Microscopy

In the initial studies, nucleoli from the Novikoff hepatoma cells and normal rat liver cells were employed as the immunogens. In both cases, the rabbits produced antisera which either specifically reacted with the nucleoli and exhibited only comparatively minor extranucleolar reactions (Busch et al., 1974). When the rabbit antisera were not absorbed with other tissues, cross reactions were found. The anti-liver nucleolar antisera produced indirect immunofluorescence in the Novikoff hepatoma cells and the anti-hepatoma nucleolar antisera produced immunofluorescence in the liver cells (Fig. 1). These antisera also produced positive immunofluorescence in normal kidney and in Walker tumor cells (Fig. 2).

Complement fixation studies indicated that, as with most self-antigens, the titers of the antisera were not high; activity was noted in dilutions as low as 1:160.

C. Tumor Specificity

The immunoprecipitin bands between the tumor and liver nucleolar antigens were evaluated by Ouchterlony gel analysis (Busch and Busch, 1977). Three immunoprecipitin bands formed when the liver nucleolar antigens reacted with the anti-liver antibodies, while a single, dense band developed when the anti-tumor nucleolar antibodies reacted with the tumor nucleolar antigens. No evidence was obtained for cross-reactivity of these tumor and liver antigens. However, nucleolar antigens appeared in the Novikoff tumor that were not present in the liver and vice versa (Fig. 3).

Continued immunization revealed more antibodies. Fourteen antigens were detected by immunoelectrophoresis in nucleoli of Novikoff hepatoma ascites cells (Davis et al., 1978) (Fig. 4). The antisera to normal liver nucleoli distinguished 10 of the 14 antigens detected by the anti-Novikoff hepatoma nucleolar antiserum.

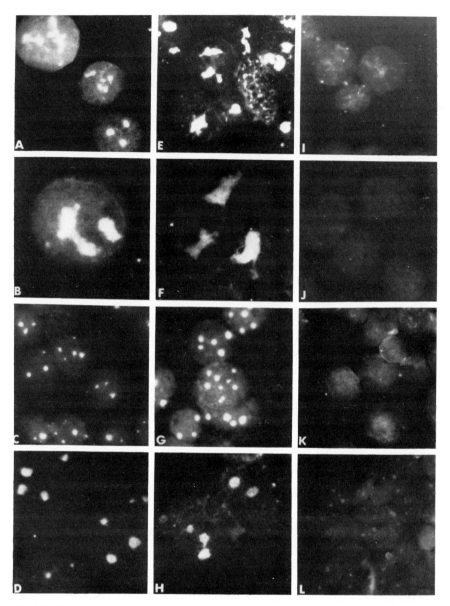

Fig. 1. Photomicrographs of nuclei analyzed with antisera or normal sera by the indirect immunofluorescent technique. Tumor anti-nucleolar antisera were incubated with Novikoff hepatoma nuclei [(A) ×1170, (B) ×4500] and liver nuclei [(C) ×1170, (D) ×4500]. Liver anti-nucleolar antisera were incubated with Novikoff hepatoma nuclei [(E) ×1170, (F) ×2925] and liver nuclei [(G) ×1170, (H) ×4500]. Normal rabbit sera were incubated with Novikoff hepatoma nuclei [(I) ×1170] and liver nuclei [(J) ×1170]. Tumor anti-nucleolar antiserum absorbed with Novikoff tumor nucleoli was incubated with Novikoff nuclei [(K) ×1170], and liver anti-nucleolar antiserum absorbed with liver nucleoli was incubated with liver nuclei [(L) ×1170]. (From Busch *et al.*, 1974, by permission of *Cancer Research.*)

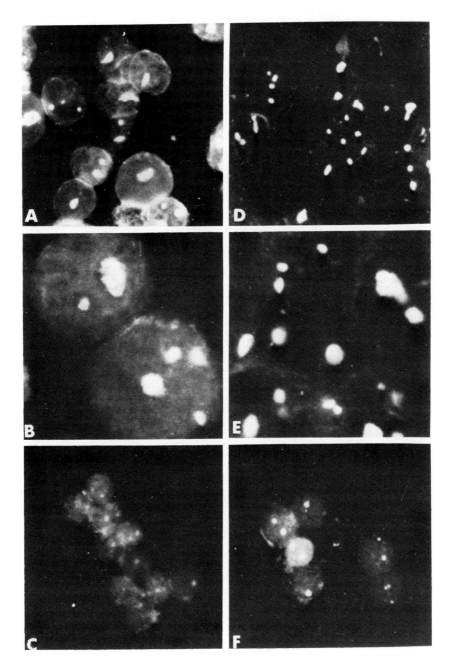

Fig. 2. Photomicrographs of nuclei analyzed with antisera by the indirect immunofluorescent technique. Novikoff hepatoma anti-nucleolar antisera were incubated with Walker nuclei [(A) ×1260, (B) ×3150] and kidney nuclei [(C) ×1260]. Liver nucleolar antisera were incubated with Walker nuclei [(D) ×1260, (E) ×3150] and kidney nuclei [(F) ×1260]. (From Busch *et al.*, 1974, by permission of *Cancer Research*.)

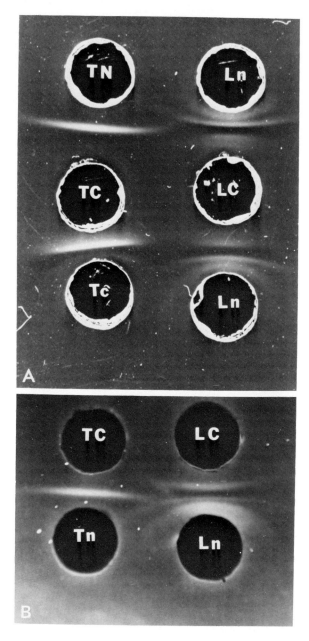

Fig. 3. (A) Immunodiffusion plate containing 0.6 M NaCl extract of nuclear chromatin (TC) in the left center well and liver chromatin (LC) in the right center well (300–400 g) as antigens. The TC antigen formed precipitin bands with tumor nuclear (TN) and tumor chromatin (Tc) antibodies. The LC antigen formed at least three precipitin bands with the liver nucleolar (Ln) antibodies. The antibody wells contained 33 µl. (B) Immunodiffusion plate containing 0.6 M NaCl extracts of tumor nuclear chromatin TC in the top left well and liver nuclear chromatin LN in the top right well (300–400 µg). The tumor chromatin antigen formed a precipitin band with the tumor nucleolar (Tn) antiserum. The LC antigen formed at least three precipitin bands with the liver nucleolar antibodies (Ln). The antibody wells contain 33 µl. (From Busch and Busch, 1977, by permission of *Tumori*.)

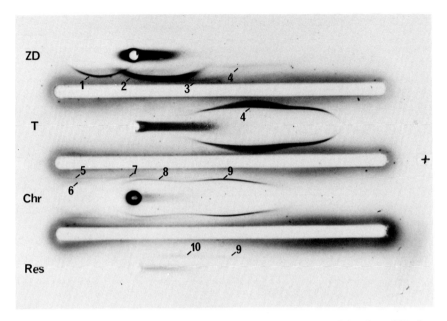

Fig. 4. Immunoelectrophoretic profile of tumor nucleolar antigens. The Zubay-Doty (ZD), low ionic strength Tris (T), and 0.6 M NaCl/5 M urea extract (Res) of Novikoff hepatoma nucleoli were analyzed by immunoelectrophoresis. The immunoelectrophoresis slide was presoaked in running buffer for 2 hours. ZD (20 µg) and 40 µg of the other antigens were placed in the antigen wells. After electrophoresis at 100 V for 30 minutes, 50 µl of anti-tumor nucleolar immunoglobulin at 80 mg protein per ml was placed in the antiserum troughs, and the precipitin arcs that formed in 24 hours were stained with Coomassie brilliant blue. (From Davis *et al.*, 1978, by permission of *Cancer Research*.)

In fetal liver, antigens common to the tumor were not present in the adult liver. This finding supported the earlier result of Yeoman *et al.* (1976) that fetal nuclear antigens in tumors are not present in normal adult tissues. Similar results in cytoplasmic enzymes, serum proteins, and a variety of "oncofetal" and "oncoembryonic" antigens have been reported in other systems (Fishman and Busch, 1979).

D. Absorption of the Antibodies

Although immunoelectrophoresis provided evidence for distinctive antigens in the Novikoff hepatoma, normal, and fetal liver, an impressive distinction was shown by absorption of the antisera with extracts of nucleoli and nuclei of both liver tissues (Fig. 5). After absorption of the antiserum with normal liver nuclear products, bright nucleolar fluorescence was observed in the Novikoff hepatoma cells, but none was seen in the normal liver nucleoli. Antisera to normal liver nucleoli still produced immunofluorescence after absorption with Novikoff hepatoma nuclear products beyond the point of nuclear fluorescence (Davis et al., 1979).

These studies with anti-nucleolar antisera, whether by immunoprecipitin band analysis, analysis of immunoelectrophoresis patterns, or evaluation of the results of absorption by immunofluorescence, indicated that antigens in the nucleoli of Novikoff hepatomas differed from those of normal liver nucleoli and vice versa.

E. Purification of Tumor Nucleolar Antigens

To conduct further studies on these antigens, they were first purified to homogeneity by isolating them from the Novikoff hepatoma and then subjecting them to a series of extractions generally employed for chromatin (Rothblum et al., 1977). First, the nucleoli were extracted with an approximately isotonic

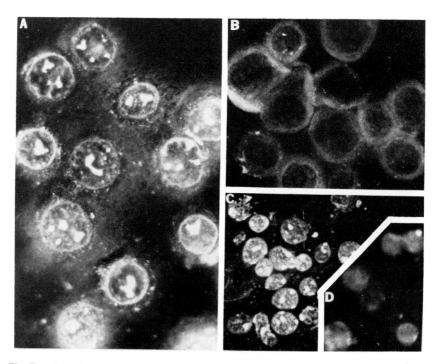

Fig. 5. Photomicrographs (×1386) of cells tested with preabsorbed anti-nucleolar antiserum incubated with Novikoff hepatoma cells (A) or with normal liver cells (D). Preabsorbed anti-liver nucleolar antiserum incubated with Novikoff hepatoma cells (B) or with normal liver cells (C). (From Davis et al., 1978, by permission of *Cancer Research*.)

extractant (0.075 M NaCl/0.025 M EDTA, pH 8) to remove divalent ions important to the maintenance of the nucleolar ultrastructure. Next, the nucleoli were exposed to a very low ionic strength solution buffer (0.01 M Tris-HCl, pH 7.8) designed to permit the chromatin to swell and to release nuclear particles entrapped in the chromatin. In addition, a number of nucleolar components, such as RNA polymerases and kinases, are apparently soluble in this low ionic strength buffer but are not soluble in isotonic buffers. In subsequent extraction procedures, a variety of salt solutions of higher ionic strengths (0.35, 0.6, and 1.0 M NaCl) were successively used as extractants. Finally, the nucleoli were extracted with 3 M NaCl/7 M urea. Although this last step does not completely solubilize all the nucleolar elements, the residue fraction was very small (Rothblum et al., 1977; Marashi et al., 1979).

In studies on the antigens of the Novikoff rat hepatoma nucleoli, less than 50% of the antigen was extracted with isotonic or low ionic strength buffers rather than with the 0.6 M NaCl extract which followed the initial extraction steps (Marashi et al., 1979). The antibodies in these antigens were used to prepare affinity columns for binding the antigen. Hydroxylapatite chromatography was used for final purification. A Coomassie blue-stained gel (Fig. 6A) showed that only a single protein was present. Its molecular weight was 60,000, and its isoelectric point was 5.0 (Marashi et al., 1979).

To prove that the antigen was indeed this protein, the area was excised from the two-dimensional gel. A distinctive "rocket" (Fig. 6B) was produced which demonstrated the presence of the antigen in this sample (Marashi et al., 1979).

F. Liver Nucleolar Antigens

The antigens in rat liver nucleoli differ from those of Novikoff hepatoma nucleoli. Abelev et al. (1979) reported that liver-specific antigens are known, but none has been characterized. R. K. Busch et al. (1979, 1981), when studying liver nuclear antigens, found three nucleolar antigens, two of which were not present in the Novikoff hepatoma (Fig. 7). One liver antigen, Ln-1, was found only in the liver.

When purified by absorption on affinity columns and eluted with 0.2 M Tris-HCl/0.05 M NaCl, pH 11, the antigens contained three RNA bands. These bands were rich in guanylic and cytidylic acids and had approximate molecular weights of 200,000 (about 600 nucleotides), thus categorizing the antigens as small nuclear ribonucleoprotein particles (snRNP's). The functions of the RNA or the snRNP's are not defined.

In an extension of previous studies on the antigens in rat liver nucleoli (R. K. Busch et al., 1979; R. K. Busch and Busch, 1977); rabbit antibodies were elicited to human liver nucleoli isolated by the sucrose-Mg^{2+} method. Fluorescent nucleoli were found in liver cryostat sections treated with rabbit anti-human

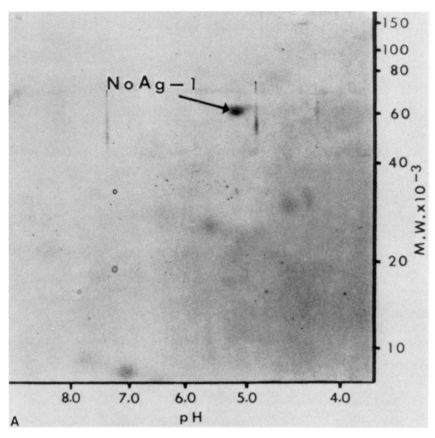

Fig. 6A. Two-dimensional gel electrophoresis of NoAg-1, the tumor antigen. NoAg-1 (10 μg) was added to lyophilized sample buffer [9 M urea/2% ampholines (3/10 Bio-lyte)/2 M dithiothreitol] and subjected to two-dimensional electrophoresis. (From Marashi et al., 1979, by permission of Cancer Research.)

liver nucleolar antibodies followed by fluorescein-conjugated goat anti-rabbit antibodies. In HeLa cells, fluroescence was distributed throughout the nucleus and in a nuclear network but was not localized to the nucleolus. In placental cryostat sections, an overall nuclear fluorescence was observed with some localization to nucleoli. Immunodiffusion analysis revealed two immunoprecipitin bands which appeared to be liver specific. Other immunoprecipitin bands were common to liver, placenta, and HeLa nuclear extracts. Rocket immunoelectrophoresis revealed two liver-specific antigens, one each migrating to the cathode and anode. These results demonstrate the presence of human liver nucleolar-specific antigens not found in HeLa and placental cells.

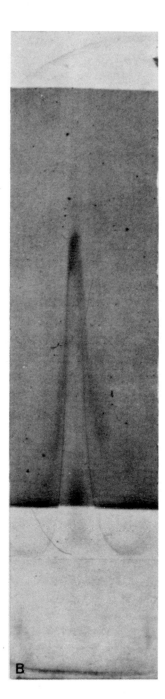

Fig. 6B. Rocket immunoelectrophoresis of NoAg-1 from the SDS-polyacrylamide slab gel. One cubic centimeter NoAg-1 spot from the SDS-polyacrylamide slab gel was placed on a blank agarose gel and electrophoresed into agarose gel containing antibody and anti-nucleolar antiserum (1.0 mg/ml). A strip of agarose gel containing 1.5% Triton X-100 was placed between the blank and antibody gel to trap SDS and release NoAg-1. (From Marashi *et al.*, 1979, by permission of *Cancer Research*.)

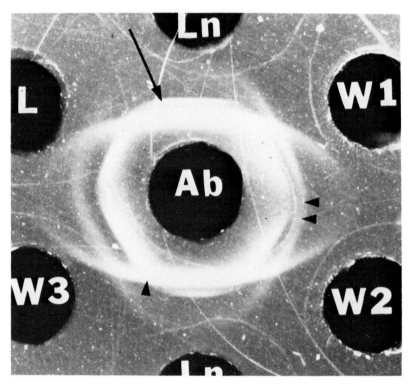

Fig. 7. Immunodiffusion plate with Ig from antiserum in liver nucleoli in the center well. The following antigens were in the outer wells: (Ln) the 0.15 M NaCl extract of liver nucleoli; the supernatants from the three consecutive 0.15 M NaCl extracts of liver pressate designated (W1), (W2), and (W3); the 0.15 M NaCl extract (1 hour) of the "washed" liver pressate (L). The following immunoprecipitin bands formed between the nucleolar antibodies (Ab) and the antigens: Ln-1 (arrow) with antigens W1, W2, W3, L, and Ln; Ln-2 (arrowhead) with antigen Ln; and Ln-3 (double arrowheads) with antigens W1, W2, W3, and L. Extraction of liver pressate with 0.15 M NaCl readily solubilized antigens Ln-1 and Ln-3. (From R. K. Busch *et al.*, 1979, by permission of the Society for Experimental Biology and Medicine.)

G. mRNA for Nucleolar Antigens

To determine whether the polysomes and the mRNA for synthesis of nucleolar antigens were unique to the tumors, the antibodies to the Novikoff hepatoma nucleolar proteins were first purified by absorption with liver nucleolar proteins (Davis *et al.*, 1978), then subjected to affinity chromatography on Sepharose-4B columns containing normal rat liver proteins which removed the antibodies that were nonspecifically absorbed to such proteins. The IgG which did not bind to the liver Sepharose column was applied to a column containing Novikoff hepatoma nucleolar proteins. The subsequent effluent was discarded

3. Human Tumor Nucleolar Antigens

and the bound IgG was eluted with $1M$ NaCl/$3M$ urea/$0.2M$ Tris-HCl, pH 7.5 and dialyzed. These IgG fractions, constituting only 1–2% of the total IgG, were then labeled with ^{125}I. The antibodies bound to the polysomes of the Novikoff hepatoma cells with a sharply defined plateau (Fig. 8), but did not bind to the polysomes of normal liver cells (Reiners *et al.*, 1980). The labeled antibodies bound to the regenerating liver polysomes at about one-third the binding rate of the Novikoff hepatoma. The tumor nucleolar and nuclear extracts effectively competed with the antibody binding to the polysomes, but the normal liver nuclear and nucleolar extracts did not (Fig. 9). These studies demonstrated a high order of specificity of the interactions of the anti-nucleolar antibody and IgG fractions with the tumor polysomes and the tumor products from nucleoli and nuclei (Reiners *et al.*, 1980).

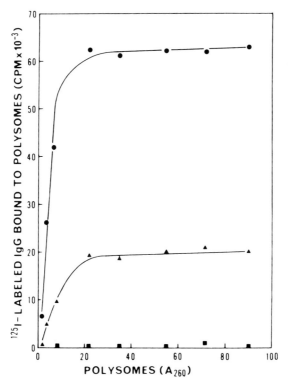

Fig. 8. Titration of ^{125}I-labeled IgG with variable amounts of polysomes. Iodinated absorbed affinity chromatography-purified anti-tumor nucleolar IgG (2.5 × 10^5 cpm) was incubated with tumor (●), regenerating liver (▲), and normal liver (■) polysomes. Over the range of polysomes titered, 5800 to 6600 cpm were precipitated in the absence of Mg^{2+} and subtracted from the values for sedimented polysome-IgG complexes. (From Reiners *et al.*, 1980, by permission of *Cancer Research*.)

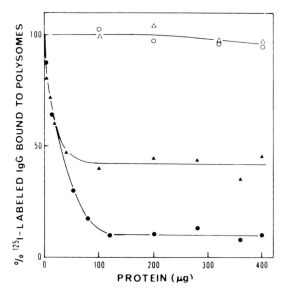

Fig. 9. Competition performed by nuclear and nucleolar extracts with tumor polysomes for iodinated absorbed affinity chromatography-purified anti-tumor nucleolar IgG (2.5×10^5 cpm). Tumor nuclear extract (●); nucleolar 0.075 M NaCl/0.025 M EDTA extract (▲); normal liver nucleolar extract (△); normal liver nuclear extract (○). One hundred percent precipitation is represented by 71,000 cpm. Data are uncorrected for the 6500 cpm (9% maximum) sedimented in the absence of Mg^{2+}. (From Reiners et al., 1980, by permission of Cancer Research.)

The binding of the antibody to the polysomes depended upon Mg^{2+} and was inhibited almost completely by puromycin but only partially by RNase, indicating that the interactions were protein–protein rather than protein-RNA. Some cross-reactivity occurred with the fetal liver nuclear extracts supporting the idea that some of the antigens were "oncofetal" or "oncoembryonic," as noted in the earlier studies of Yeoman et al. (1976) and Davis et al. (1978).

H. Nucleolar Antigens in Human Tumors

Although studies on the nucleolar antigens of rodent tumors and nontumor tissues suggested important differences in these nucleolar proteins, initial attempts to extend these studies to human neoplasms met with the problem that human tumor nucleoli did not fluoresce brightly when treated with antibodies to rat tumor nucleoli. After preliminary studies showed that these antibodies did not produce corresponding immunoreactivity with nucleolar antigens of human tumor specimens, nucleolar preparations and nuclear Tris extracts (0.01 M Tris-HCl, pH 8.0) of human HeLa cells were used as immunogens. The immunized rabbits developed anti-human tumor nucleolar antibodies, leading to the more definitive studies of Davis *et al.* (1979) and H. Busch *et al.* (1979, 1980) on human tumor nucleolar antigens.

As shown in Figs. 10 and 11, many human malignant tumors contain brightly fluorescent nucleolar antigens (H. Busch *et al.*, 1979). These tumors include carcinomas of many types, a variety of sarcomas, and many hematological neoplasms (Tables I and II). To discriminate the nucleolar fluorescence of the tumors from that found in other tissues, such as placenta, extensive absorption of nontumor antibodies from the antisera, Ig, or IgG fractions were necessary. Fetal calf serum, the initial growth medium of the HeLa cells, was an essential absorbent. Human placentas were also used, as they provided a good source of human tissue nuclei for isolation (Davis *et al.*, 1979; H. Busch *et al.*, 1979). The absorptions were performed with serum proteins or whole serum.

Although it would be highly desirable to have a precipitating antibody of the type found with the anti-Novikoff hepatoma antibodies, such precipitins have not yet been found with human tissues despite the use of several rabbits, sheep, and goats. The antigenic determinants of human nucleolar antigens differ from those of the rodent, but, as yet, no specific information is available on the determinants in either of these types of antigens.

I. Antibodies to Other Tumors

Two malignant tumors other than HeLa cells were used as sources of the antigen: the Namalwa cell line and a human prostate carcinoma grown in tissue culture. Both of these tumors contain nucleolar antigens as demonstrated in rabbits which produced anti-nucleolar antibodies with similar titers and specificities. With the prostatic carcinoma, a nucleolar preparation was used for immunization, while a Tris extract of nuclei of the Namalwa cells was used as the immunogen. The similarity of the results obtained suggested common antigens in these and in the HeLa cells.

J. Tumor Nucleolar Antigens in Nontumor Tissues

One may ask whether the antigens found reflect a neoplastic process or a normal process involved in growth, cell division, or other normal physiological effects. Our theory of carcinogenesis (Busch, 1976) noted that cancer is probably the result of misapplication of normal gene readouts as a dysplastic phenomenon involving fetal genes rather than a process involving new events, such as the integration of a viral genome or other gene aberration. Accordingly, a search has been made for the antigens in a variety of normal, growing, and fetal tissues. The nucleolar antigens were not found in a broad array of nontumor tissues which presumably exhibit slow growth or none at all.

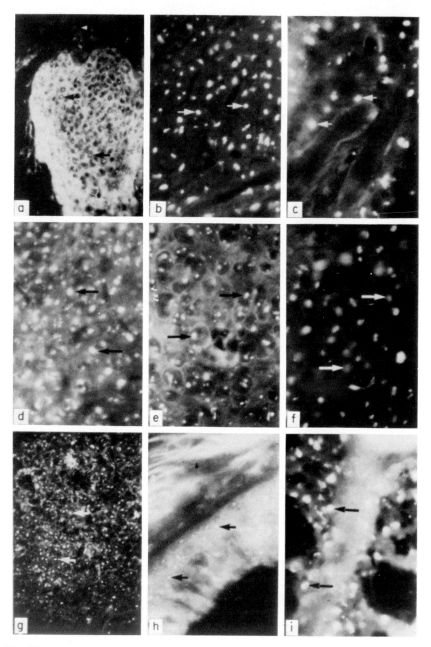

Fig. 10. Bright nucleolar fluorescence in a series of squamous cell carcinomas and adenocarcinomas. The antibody concentration was 0.3 to 0.5 mg/ml. (a) Squamous cell carcinoma, skin. ×112. (b) Squamous cell carcinoma, lung. ×280. (c) Squamous cell carcinoma, metastatic to muscle. ×441. (d) Squamous cell carcinoma, esophagus. ×280. (e) Squamous cell carcinoma, metastatic to muscle. ×280. (f) Higher power squamous cell carcinoma, lung. ×441. (g) Adenocarcinoma, prostate. ×112. (h) Adenocarcinoma, lung. ×280. (i) Adenocarcinoma, lung (high power). ×441. Arrows, positive fluorescent nucleoli; arrowheads, negative region in adjacent tissue. (From H. Busch et al., 1979, by permission of *Cancer Research*.)

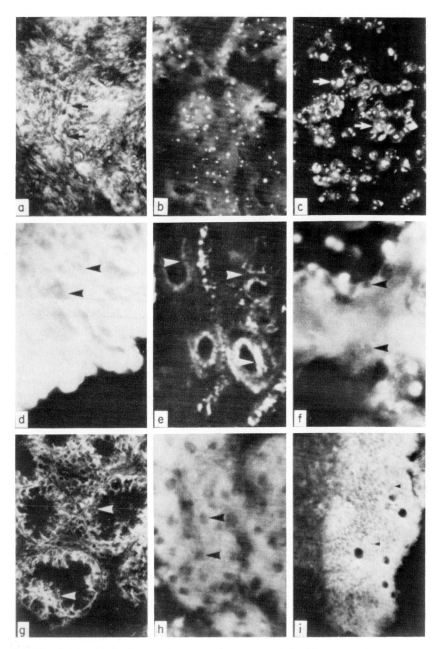

Fig. 11. Bright nucleolar immunofluorescence in tumors and lack of fluorescence in nontumorous tissues. (a) Brain, glioblastoma. ×112. (b) Adenocarcinoma of stomach, liver metastasis. ×280. (c) HeLa cells, S3, culture. ×280. (d) Bronchial epithelium, newborn. ×280. (e) Bronchial glands. ×280. (f) Thyroid adenoma. ×441. (g) Intestinal crypts. ×280. (h) Adenoma, parathyroid. ×112. (i) Gastric epithelium (hyperplastic). ×112. Arrows, positive fluorescent nucleoli; arrowheads, negative nucleus or negative regions in cells. (From H. Busch *et al.*, 1979, by permission of *Cancer Research*.)

53

TABLE I
Bright Nucleolar Fluorescence in Human Malignant Tumor Specimens[a,b]

I. Carcinomas	II. Sarcomas
1. Lung	1. Chondrosarcoma (1)
Adenocarcinoma (3)	2. Fibrosarcoma (4)
Oat cell (4)	3. Giant cell tumor (1)
Squamous cell (22)	4. Granulocytic myoblastoma (2)
2. Gastrointestinal	5. Leimyosarcoma (4)
Oral cavity (8)	6. Lymphoma (10)
Pharynx (4)	7. Meningiosarcoma (1)
Esophagus, squamous cell (5)	8. Myoblastoma (2)
Stomach, adenocarcinoma (5)	9. Osteogenic (6)
Metastasis: liver	10. Pulmonary blastoma (1)
Metastasis: liver node	11. Reticulum cell sarcoma (1)
Colon, adenocarcinoma (9)	12. Synovial sarcoma (1)
Metastasis: liver (2)	III. Hematological neoplasms
Transplantable carcinoma (GW-39)	1. Acute lymphocytic leukemia (2)
Liver, primary carcinoma (3)	2. Acute myelocytic leukemia (7)
Pancreas (4)	3. Acute monocytic leukemia (2)
3. Genitourinary	4. Chronic myelocytic leukemia (5)
Kidney (4)	5. Hodgkins disease (9)
Prostate, adenocarcinoma (22)	6. Leukemia: CLL (12), hairy cell (1)
Bladder (4)	7. Mycosis fungoides
4. CNS	8. Plasmacytomas (7)
Glioblastoma (1)	
Astrocytoma (5)	
5. Endocrine	
Breast (3)	
Cervix (4)	
Parathyroid (1)	
Thyroid (5)	
6. Skin	
Basal cell (8)	
Eccrine gland (1)	
Squamous cell (7)	
Metastasis: lymph node	

[a] From Busch et al. (1982).
[b] The numbers in parentheses are the number of slides examined for different cases for these studies.

One may also ask whether nucleolar antigens were present in normal-growing nontumor tissues. In studies on bone marrow, skin Malpighian layers, and intestinal epithelia, nucleolar antigens were not found. In studies on the bone marrow of leukemia patients, the neoplastic cells contained the antigens but their nontumor counterparts in the same maturation series did not (Smetana et al., 1979). As controls for the nontumor tissues, studies were made on cells of patients with acute infectious mononucleosis and lymphoid hyperplasia, but neither exhibited

TABLE II
Negative Nucleolar Fluorescence in Human Tissues[a]

I. Normal tissue
  1. Lung
  2. Gastrointestinal
      Stomach
      Intestine
      Small, crypts of Lieberkuhn
      Large
      Liver
      Pancreas
  3. Genital urinary
      Kidney
      Bladder
      Prostate
  4. Endocrine
      Thyroid
      Breast
      Placenta
  5. Skin
II. Hematologic
  1. Bone marrow
  2. Lymph nodes
      Lymphocytes
      Hyperplastic lymph nodes
  3. Benign growing tissues
      Thyroid, goiter
      Prostate, hyperplastic
III. Inflammatory diseases
  1. Chronic ulcerative colitis
  2. Glomerulonephritis
  3. Granuloma and fibrosis of lung
  4. Liver: cirrhosis, hepatitis
  5. Lupus profundus (mammary gland and skin)
  6. Pemphigus: bullous
  7. Ulcer, gastric

[a] From Busch et al. (1982).

bright nucleolar fluorescence under the circumstances of these studies. These results suggest that immunological analyses of whole cells, cell suspensions, or cell sections for the nucleolar antigen could be useful for immunodiagnoses of malignant disease.

In extending these studies, fetal tissues were examined partly because earlier investigations found that nucleolar antigens were expressed in the Novikoff hepatoma (Davis et al., 1978; Yeoman et al., 1976).

In more mature fetal tissues, such as 9-month fetal lung or 6-month fetal liver, no evidence was found for bright nucleolar fluorescence. However, examination of the IMR-90 and WI-38 diploid fetal fibroblast lines in tissue culture indicated brightly positive nucleolar fluorescence (H. Busch et al., 1979). Such studies indicated that nucleolar antigens in malignant tumors might reflect activation of the fetal genes important to the overall neoplastic process (H. Busch et al., 1979; Busch, 1976).

K. Breast Cancer

For a "preliminary" evaluation of the possible diagnostic use of the fluorescence assay for the human tumor antigens, it was important to find a series of a particular tumor type that would lend itself to a "blind" evaluation. At the Michigan Cancer Foundation, a prospective study on human breast cancer is in progress. As part of this study, a large series of patients with breast cancer are being analyzed for relationships between the pathology of the neoplasm and the course of the disease. Sections of normal breast tissue as well as benign and malignant tumors were analyzed for bright nucleolar fluorescence by the indirect immunofluorescence technique. These investigations were approved by the Human Research Committee of Baylor College of Medicine and the Committee for Protection of Human Subjects of the Michigan Cancer Foundation.

The cryostat sections of the human tumors were kept at $-20°C$; they were cut at 2 μm and fixed for 12 minutes in acetone at $4°C$. The tumors were evaluated by a panel of five pathologists who provided a consensus diagnosis, a designation of tumor grade, and other histopathological characteristics on the basis of hematoxylin and eosin staining. The specimens were treated with either Ig or IgG fractions prepared from antisera absorbed with human placental nuclear extracts, fetal bovine serum, and human serum.

Before the "blind" study was initiated, known samples of breast carcinomas were examined. In 19 of 20 samples (95%), bright nucleolar fluorescence was either found throughout the sections or limited to particular portions of the sections. These positive results for nucleolar fluorescence resembled those obtained by H. Busch et al. (1979) in which 81 of 84 samples (96%) were positive in a variety of malignant tumors.

Figure 12 presents micrographs of bright nucleolar fluorescence in breast carcinoma. In some cases, the whole specimen consisted of cells with large, irregular, brightly fluorescent nucleoli (Busch et al., 1981). In others, the neoplastic cells were either in focal lesions or randomly distributed, and in some, chords of tumor cells were distributed throughout the specimens. The background fluorescence ranged from negligible to moderately intense.

In some sections, the neoplastic cells were confined to small areas or were not visible. In some "false negatives," the antibodies did not penetrate the cells either as a result of sample thickness, proteolysis, or "waxy deposits." In other

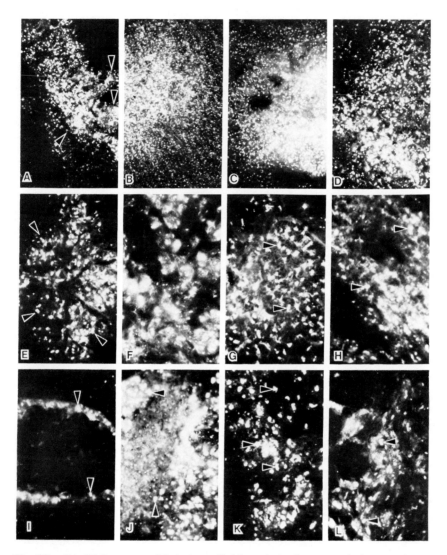

Fig. 12. (A)–(D) Low power (96×) views of bright nucleolar fluorescence in breast carcinomas illustrating clustering of cells, extensions from masses [arrowheads in (A)], and variable densities within masses. (E)–(H) High power views of bright nucleolar fluorescence of fields within breast carcinomas showing abortive pseudolobular formations (E) and (F) and nucleolar irregularity (G) and (H). ×400. (I)–(L) Ductule in a carcinoma (I), two sizes of brightly fluorescent nucleoli (J), large nucleoli (K), and irregular nucleoli (L). Arrowheads, brightly fluorescent nucleoli. ×240.

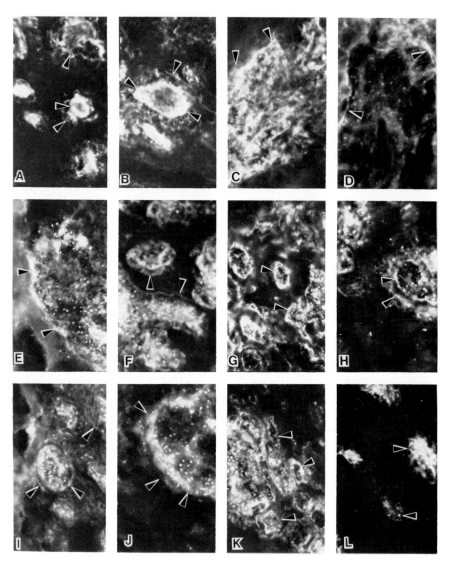

Fig. 13. Varying types of fluorescence observed in benign breast tumors and normal breast tissues. (A) and (B) Normal breast tissue. Central nonfluorescent areas are surrounded by a dense fluorescent "boundary" (arrowheads) around which are moderately fluorescent elements (arrowheads). (A) ×120. (B) ×300. (C) Adenosis. Arrowheads show dense outer "boundary." Within the structures are "microspherules" and semiparallel "fibrillar" elements. ×120. (D) Fibrocystic disease. Pointers show dense boundaries surrounding inner areas containing "Microspherules." ×120. (E) Sclerosing adenosis. Pointers show dense boundaries surrounding inner areas containing "microspherules." ×120. (F) Adenosis. Area containing "microspherules" surrounded by thin boundaries (pointers). ×120. (G) Fibrocystic disease. Dense "boundary" layer (pointers) surround-

instances, the region containing the tumor cells was lost or only at the periphery of the specimen. To avoid such problems, examination of several slides from a single specimen was necessary to define sharply the area of the neoplastic cells.

L. Normal Breast Tissue and Benign Tumors

In the normal breast tissue, with the same antibodies that were used in the studies on the known carcinomas, there were no visible brightly fluorescent nucleoli. In some sections, there were "minispots" which were either intraluminal or in extracellular spaces. The anti-nucleolar antibodies were apparently nonspecifically absorbed by basement membrane elements as fluorescent boundaries were occasionally visible around the terminal ducts as a thin or thick limiting membrane, or both (Fig. 13).

In benign tumors, the results were very similar to those observed in the normal breast tissues. The lining membranes were larger and less regular than those observed in normal breast tissues. In all cases, such "limiting membranes" were visible. In a few instances, fluorescent "spots" were seen in the benign tumors—usually, but not always, in small confined areas. Whether the antibodies bound to relatively nonspecific but "sticky" substances (such as procollagen, keratin, or vimentin), or to another nucleolus—unrelated macromolecule whose antigenic determinants are common to those of the nucleolar antigens—has not been established. Further studies are needed to improve the specificity of the antibody preparation by purifying the antibodies to monospecificity or by utilizing monoclonal preparation for monospecific antibodies.

In a few benign breast tumors, nucleolar fluorescence was apparent. These nucleoli were smaller and fewer in number than in the malignant tumors. Although such nucleolar fluorescence was found in only a few benign tumors, it led to an investigation of the sizes of the nucleoli in these benign lesions. It was possible to distinguish sharply between benign and malignant lesions on the basis of overall nucleolar diameters, but it was not possible to specify the malignancy of any one cell simply by observing the nucleolar size or fluorescence. It is not known whether the antigen in the benign tumors is the same as that found in the malignant tumors or whether another molecular species accounts for the positive nucleolar fluorescence in these benign tumors.

---

ing areas containing microspherules. ×120. (H) Normal breast tissue. Dense and less dense "boundaries" (pointers) surrounding areas containing "microspherules." ×120. (I) Sclerosing adenosis. Note boundary layers (pointers) surrounding areas containing microspherules. ×120. (J) Sclerosing adenosis. Dense and less dense "boundaries" (pointers) surrounding areas containing "microspherules." ×300. (K) Normal breast tissue. Large number of structures containing "microspherules" surrounded by "boundary" elements. ×300. (L) Normal breast tissue. Fluorescent structures (pointers) seen in normal breast tissue specimens. ×300.

M. "Blind" Study

In this study, 80 breast samples were evaluated. Of these, 55 were carcinomas and 25 were either normal breast or benign tumor specimens (Busch et al., 1982).

The code for this study was kept in Detroit and the samples were evaluated in Houston. The results were reported in four series containing 13–29 samples per group. Some were clearly positive or negative for the nucleolar antigen. In the first evaluation group (Table III), the 14 positive and 6 negative samples were correctly identified except for one negative sample—a benign tumor containing some fluorescent nucleoli. In the second group, 11 negative samples were, in fact, normal tissues or benign tumors. Of the 18 positive samples, one was a false negative. Accordingly, in the first two series, the overall correct percentage (95%) equaled the original reports from the larger, known series of different tumor types and the series of known breast carcinomas.

In the third evaluation group, 16 carcinomas and 2 benign tumors, were all identified correctly. These results were derived from increased experience and improved recognition of the criteria distinguishing normal tissues from benign lesions.

TABLE III
Results of Evaluation of Bright Nucleolar Fluorescence in Unknown Specimens (Blind Study) of Benign and Malignant Breast Specimen[a,b]

Specimens studied in order of difficulty	Positive	Negative	Percentage correct
I. Less difficult			
A. Generalized fluorescence or absence of fluorescence	14	6(1)	95
B. Less generalized fluorescence or unusual structures	18(1)	11	96
C. More localized fluorescence or more unusual structures	16	2	100
	48(1)	19(1)	97
II. More difficult			
Limited fluorescence or limited visualization of cells; questionable regions	7(2)	6(1)	77
Overall correct			94

[a] From Busch et al. (1982).

[b] Numbers in parentheses are errors, i.e., after positive are malignant tumor reported as negative, and after negative is a benign tumor reported as positive.

3. Human Tumor Nucleolar Antigens 61

The fourth evaluation group contained 7 malignant samples (including 2 which did not exhibit nucleolar fluorescence) and 6 benign tumors (including 2 which did exhibit fluorescence). Only 77% of this group was diagnosed correctly.

Overall, 52 of the 55 carcinomas and 25 of 27 benign samples were correctly identified, giving detection rates of 94.6 and 92.6%, respectively.

Among the technical problems encountered in these studies were high background fluorescence in some samples and failure to develop satisfactory fluorescence in others. In the "blind study" itself, the hematoxylin and eosin stained sections were unavailable, but they were provided in a later review of the study. Diagnostic errors were caused by antibodies failing to penetrate the samples, as shown by border fluorescence, excessive sample thickness, and specimen necrosis.

These studies indicated that considerable improvements were necessary to substantiate and expand these results and, more importantly, to improve their specificity. Inasmuch as the antibodies were not shown to be completely specific for nucleolar antigens in carcinomas, it seems essential, first, that the antigens be purified and that monospecific antibodies be provided. Second, it is necessary to evaluate the benign tumors which exhibited nucleolar fluorescence to determine whether they contain the same or different antigens and whether the nucleolar fluorescence has any prognostic significance.

N. Immune Localization of the Antigen

Studies with immunoelectron microscopy showed that the nucleolar antigens were in highest concentration in the nucleolonemas containing nucleolar RNP components. In particular, the fibrillar RNP components were most densely immunostained (Busch *et al.*, 1982). In further analysis of nuclear components containing the antigen, it was found that nucleolar "mini-particles" which are "donut" shaped also contained an antigen. The function of these particles is known, but their structures are very interesting as indicated in the recent studies by Domae *et al.* (1982).

O. Characterization of the Nucleolar Antigen

Two nucleolar antigens identified in supernatants of 0.01 M Tris-HC1 extracts of Namalwa nuclei and nucleoli are antigens 68/6.3 and 54/6.3 (MW $\times$ $10^{-3}/pI$). In addition, several antigens with molecular weights of 90, 61, 52, and 43 kd have been found in the sedimented nucleolar RNP particles. On two-dimensional "Western blots," approximately 10 antigens were detected. Although these antigens survived the "Western blotting technique," other antigens may not.

Quantitative analysis of the nucleolar antigens was done by a modified ELISA method (Kelsey *et al.*, 1981). With this assay, hybridoma antibodies (Kohler and Milstein, 1975) to the nucleolar antigens are being evaluated for their specificity for nuclear and nucleolar antigens (Son *et al.*, in manuscript). As yet, tumor-specific monoclonal antibodies have not been obtained.

III. DISCUSSION

A. Possible Functions of the Nucleolar Antigen

While there is no specific information yet on the precise role of the nucleolar p*I* 6.3 antigen, its localization has been analyzed both by light and electron microscopy. It is present in the various soluble and formed elements of the nucleolus and is concentrated in the "fibrillar" elements (approximately 70% of the antigen is in RNP particles) of the fundamental nucleolonemal structure of the nucleolus.

Several possible functions for the nucleolar antigen have been suggested. The antigen's localization is quite similar to the "silver-staining protein, C23," of rat tumor nucleoli, although evidence has not been obtained that the antigen is either silver-staining or an NOR protein. However, this needs further examination.

Another nucleolar element that is distributed in a similar fashion is the U3 RNP particle. Although the function of the U3 RNA of the nucleolus is not defined, it may serve a role similar to the "splicing" function of the U1 RNA and possibly other small nuclear RNA's. The U3 RNA is in a small RNP particle intimately associated with the large 60 S and 80 S subunit elements of the nucleolar preribosomal RNP particles. Because the nucleolar is concentrated in the particulate elements of the nucleolus, it may constitute a structural element of these small particles.

Recently, high concentrations of an intriguing ring-shaped particle about 100 Å in diameter were found in the saline extracts of both HeLa and Namalwa cells. These "miniparticles," also observed in tissue culture fluids of KB, HeLa, and CMP (a human adenocarcinoma), were reported to contain approximately 5% DNA and 10% RNA (Narayan and Rounds, 1973). Whether they are U3 RNP particles, unique molecular elements, some rhinovirus or other virus is not known.

Lowenstein *et al.* (1980) adduced that nuclei of human cancer cells contain a novel DNA polymerase, Cm, which differs from the normal DNA polymerases of human cells. Inasmuch as the nucleolar antigen might serve a special role in the replication of rDNA, it could conceivably be the same DNA polymerase or possibly an rDNA polymerase. Studies on this point are necessary.

The investigations of Higashi's group have indicated that nucleoli are essentially intact during synthesis of nucleolar DNA (Gotoh et al., 1981a,b). When synchronized L cells were pulse labeled with uridine and thymidine, both labels were incorporated into nucleolar RNA and DNA, respectively, during S phase. In these synchronized systems, DNA synthesis occured in a short phase at approximately the same time in the nucleoli and the whole nuclei. Since the rates were the same in the intact isolated nucleoli and the whole nuclei, they concluded that there was intranucleolar, rather than extranucleolar, DNA synthesis. In a prior study from the same laboratory, Hirano et al. (1981) had shown that DNA polymerases were present in nucleoli isolated from Ehrlich ascites cells. They also noted that the nucleolar number per cell was greater in S phase (1.84) than in G_1 or G_2 phase (1.46), which is further evidence against disaggregation of nucleoli during the course of DNA synthesis (Gotoh et al., 1981a,b).

Another possibility requiring study is that the antigen may reflect the presence of unusual or "fetal" subunits of RNA polymerase I, is a little-known multisubunit enzyme in human tissues, The nucleolar antigen is present in such small amounts that it may serve an enzymatic function. In the nucleolus, a major function is synthesis of rRNA, but one cannot rule out a role in any of the many synthetic, modifying, or other functions that characterize the nucleolus. The "particles" may be multiheaded enzymes.

B. The Antigen as Accelerator or Gene Control Element

The almost ubiquitous presence of nucleolar antigens in human cancers requires an explanation in terms of function. If it were transmitted epigenetically, as the data on the HeLa cells indicate (H. Busch et al., 1979), the antigen might serve to "bypass" the usual gene controls on G_1 entry. It might also serve to activate genes either directly or through feedback mechanisms that are normally nonfunctional or are normally blocked but, nevertheless, essential for growth and cell division. The "random movement" in cancer cells is clearly insensitive to extrinsic or intrinsic controls. Accordingly, an unknown system is stimulated in cancer cells to produce this antigen which is a common product of the activated genes of cancer cells. This system may be intimately related to the cancer process or it may be involved in some essential function for the growth of these cells.

C. Relation of the Nucleolar Antigens to the "src Gene" Product

When one nucleolar antigen was shown to have a molecular weight of approximately 68,000 and a pI of 6.3, it became interesting to compare the antigen to other reported proteins of similar molecular weights and isoelectric points. As noted earlier (Marashi *et al.*, 1979), the antigen in Novikoff hepatoma nucleoli had a similar molecular weight. The lower p*I* (5.1) of the Novikoff nucleoli may reflect differences in amino acid content or modifications which may also account for the differences in immunological properties by comparison to the human tumor nucleolar antigens.

Another interesting protein is the "*src* gene" product which has been studied in detail by Erikson's group (Erikson *et al.*, 1980; Collett *et al.*, 1978; Collett and Erikson, 1978). This protein has a similar molecular weight and p*I*, but differs in cellular localization, being a membrane protein (Krueger *et al.*, 1980a,b). It is primarily localized to the nuclear envelope and juxtanuclear reticular membrane structure in rat cells (Krueger *et al.*, 1980a) and on the plasma membrane of Rous sarcoma virus-transformed chicken fibroblasts. Krueger *et al.* (1980a) proposed that the "pp60src" is a membrane protein that associates with cellular membranes by hydrophobic interactions. Collett and Erikson (1978) found the *src* gene product was a protein kinase. In our studies, the human tumor nucleolar antigens were not found to be phosphorylated nor to exhibit properties of kinases.

To compare the structures of the nucleolar antigens with the "*src* gene" product, the ^{125}I-labeled antigens were subjected to partial proteolysis with Staphylococcal V8 protease and chymotrypsin. Under the conditions employed, a number of cleavage products were detected, but these did not parallel those reported for the *src* gene product (Collett *et al.*, 1978). Although the products of protein 68/6.3 and 61/6.1 were quite similar, they did not have identical cleavage patterns.

The immunological absorption techniques vary in their efficacy and in their ability to remove conflicting reactants. One would like to have a monospecific antibody for each of the antigens that would not recognize other proteins. In these studies, this goal has not yet been achieved. Accordingly, there are problems of background, problems of false positives, false negatives, etc. Quite consistently, 95% of the tumors analyzed developed bright nucleolar fluorescence, but the problems of differentiation of benign tumors and other disease states from malignant tumors have not been adequately approached.

Inasmuch as the antigen was mainly absent from nontumor cells and benign tumors, it may be related to fetal or rapid growth as discussed earlier (H. Busch *et al.*, 1979) in connection with the WI-38 and IMR-90 fetal lung fibroblast lines (Section II,J). A possible loss of the usual nucleolar matrix elements may make the antigen visible in such tissues while its detection in normal cells may be "blocked." Scheer (1980) reported the presence of a "matrix" protein (145/6.3) in *Xenopus* nucleoli which may be responsible for the nucleolar configuration. It is possible that tumors may not synthesize this protein or may substitute for the antigen in some way not yet apparent.

D. Antigens Migrating into the Nucleolus in Late G_1 Phase

Recently, two very interesting studies from Tan's group reported the presence of antigens in nucleoli of normal cells during G_1 phase of the cell cycle that are not present in interphase nucleoli (Deng et al., 1981; Takasaki et al., 1982). Deng et al. (1981) found that WiL_2 strain cells (a continuously growing human diploid fibroblast) did not have nucleolar localization of either the Sm and U1-RNP antigens at any phase of the cell cycle. These Sm and U1-RNP antigens are in small RNP particles of the nucleous which contain U-snRNA's. High concentrations of the SSB or La-antigen, which is associated with the nuclear RNP particle containing the 4.5 S RNA, were found by immunofluorescence in the nucleolus in the late G_1–S phase period. Although the function of the 4.5 S RNP is not yet known, its translocation into the nucleolus in the G_1 phase is most interesting.

Takasaki et al. (1982) found a similar result for the PCNA antigen in proliferating cells (PC) (Miyachi et al., 1978). When sera containing these autoantibodies were absorbed to remove contaminating anti-Sm antibodies, a nucleolar localization was demonstrated in both the WiL_2 and PHA-stimulated normal lymphocytes in the late G_1 phase. It is not known whether this protein is in a particle or a free state. Preliminary purification studies have suggested that its coordinates are 40/5.1. These coordinates are similar to those of actin and protein B23 (37/5.1).

An interesting feature of neoplasia is that the G_1 phase blocks seem to be bypassed. As noted earlier (H. Busch et al., 1979), some tumor cells seem to be in the G_1 phase even during telophase as indicated by the distribution and presence of silver staining proteins. The studies of Gotoh et al. (1981a,b) indicate that DNA synthesis occurs in intact nucleoli in S phase. Accordingly, the DNA polymerases and other proteins such as topoisomerases, gyrases and other preparatory proteins must also be present. No clear relationships have yet been established between the antigens that migrate into nucleoli in the G_1 phase, the human tumor nucleolar antigens, or nucleolar DNA synthesis.

ACKNOWLEDGMENTS

The original research in this report was supported by the Public Health Service, Nucleolar Antigen Grant, CA-27534, Michael E. DeBakey Fund, Pauline Sterne Wolff Memorial Foundation, William Stamps Farish Foundation, the Bristol-Myers Fund, and the Sally Laird Hitchcock Fund. We were also indebted to the National Cancer Institute, the Frederick Cancer Research Center, Drs. V. DeVita, J. Douros, and Mr. Fred Klein for providing us with the Namalwa cell line mentioned in Section II,I. Dr. F. Gyorkey and Mrs. P. Gyorkey also worked with the human prostate carcinoma culture discussed in Section II,I.

REFERENCES

Abelev, G. I., Engelhardt, N. V., and Elgort, D. A. (1979). *Methods Cancer Res.* **18,** 2–38.
Busch, H. (1976). *Cancer Res.* **36,** 4291–4294.
Busch, H., and Smetana, K. (1970). "The Nucleolus." Academic Press, New York.
Busch, H., Gyorkey, F., Busch, R. K., Davis, F. M., and Smetana, K. (1979). *Cancer Res.* **39,** 3024–3030.
Busch, H., Busch, R. K., Chan, P.-K., Daskal, Y., Gyorkey, F., Gyorkey, P., Kobayashi, M., Smetana, K., and Sudhakar, S. (1980). *Transplant. Proc.* **12,** 99–102.
Busch, H., Busch, R. K., Chan, P.-K., Kelsey, D., and Takahashi, K. (1982). *Methods Cancer Res.* **19,** 109–178.
Busch, R. K., and Busch, H. (1977). *Tumori* **63,** 347–357.
Busch, R. K., and Busch, H. (1981). *Proc. Soc. Exp. Biol. Med.* **158,** 125–130.
Busch, R. K., Daskal, I., Spohn, W. H., Kellermayer, M., and Busch, H. (1974). *Cancer Res.* **34,** 2362–2367.
Busch, R. K., Reddy, R. C., Henning, D. H., and Busch, H. (1979). *Proc. Soc. Exp. Biol. Med.* **160,** 185–191.
Chan, P.-K., Feyerbend, A., Busch, R. K., and Busch, H. (1980). *Cancer Res.* **40,** 3194–3201.
Chan, P.-K., Frakes, R. L., Busch, R. K., and Busch, H. (1982). *J. Cancer Res. Clin. Oncol.* **103,** 7–16.
Collett, M. S., and Erikson, R. L. (1978). *Proc. Natl. Acad. Sci. U.S.A.* **75,** 2021–2024.
Collett, M. S., Brugge, J. S., and Erikson, R. L. (1978). *Cell* **15,** 1363–1369.
Davis, F. M., Busch, R. K., Yeoman, L. C., and Busch, H. (1978). *Cancer Res.* **38,** 1906–1915.
Davis, F. M., Gyorkey, F., Busch, R. K., and Busch, H. (1979). *Proc. Natl. Acad. Sci. U.S.A.* **76,** 892–896.
Deng, J. S., Takasaki, Y., and Tan, E. M. (1981). *J. Cell Biol.* **91,** 654–660.
Domae, N., Harmon, F. R., Busch, R. K., Spohn, W., Subrahmanyam, C. S., and Busch, H. (1982). *Life Sci.* **30,** 469–477.
Erikson, R. L., Purchio, A. F., Erikson, E., Collett, M. S., and Brugge, J. S. (1980). *J. Cell Biol.* **87,** 319–325.
Fishman, W. H., and Busch, H., eds. (1979). "Methods in Cancer Research," Vol. 18. Academic Press, New York.
Gotoh, S., Hirano, H., Higashi, K., Shimomura, E., Nakanishi, A., and Sakamoto, Y. (1981a). *Cell Struct. Funct.* **6,** 111–119.
Gotoh, S., Shimomura, E., Nakanishi, A., Sakamoto, Y., Hirano, H., and Higashi, K. (1981b). *J. University of Occupational and Environmental Health* **3,** 91–96.
Hilgers, J., Nowinski, R. C., Geering, G., and Hardy, W. (1972). *Cancer Res.* **32,** 98–106.
Hirano, H., Nakanishi, A., Sakamoto, Y., Gotoh, S., and Higashi, K. (1981). *Biomed. Res.* **2,** 307–315.
Hnilica, L., Chiu, J., Hardy, K., Furitani, H., and Briggs, R. (1978). *In* "The Cell Nucleus" (H. Busch, ed.), Vol. 5, pp. 307–334. Academic Press, New York.
Kelsey, D. E., Busch, R. K., and Busch, H. (1981). *Cancer Lett.* **12,** 295–303.
Kohler, G., and Milstein, C. (1975). *Nature (London)* **256,** 495–497.
Krueger, J. G., Wang, E., and Goldberg, A. R. (1980a). *Virology* **101,** 25–40.
Krueger, J. G., Wang, E., Garber, E. A., and Goldberg, A. R. (1980b). *Proc. Natl. Acad. Sci. U.S.A.* **77,** 4142–4146.
Lowenstein, P. M., Lange, G. W., and Gerard, G. F. (1980). *Cancer Res.* **40,** 4398–4402.
Marashi, F., Davis, F. M., Busch, R. K., Savage, H. E., and Busch, H. (1979). *Cancer Res.* **39,** 59–66.
Miyachi, K., Fritzler, M. J., and Tan, E. M. (1978). *J. Immunol.* **121,** 2228–2234.

Narayan, K. S., and Rounds, D. E. (1973). *Nature (London), New Biol.* **243,** 146–150.
Reiners, J. J., Jr., Davis, F. M., and Busch, H. (1980). *Cancer Res.* **40,** 1367–1371.
Rothblum, L. I., Mamrack, P. M., Kunkle, H. M., Olson, M. O. J., and Busch, H. (1977). *Biochemistry* **16,** 4716–4720.
Scheer, U. (1980). *Symp. Soc. Exp. Biol.* (in press).
Smetana, K., Busch, R. K., Hermansky, F., and Busch, H. (1979). *Life Sci.* **25,** 227–234.
Son, M., Busch, R. K., Spohn, W. H., and Busch, H. (1982). In preparation.
Takasaki, Y., Deng, J. S., and Tan, E. M. (1982). In press.
Tan, E. M. (1979). *In* "The Cell Nucleus" (H. Busch, ed.), Vol. 7, pp. 457–478. Academic Press, New York.
Yeoman, L. C., Jordan, J. J., Busch, R. K., Taylor, C. W., Savage, H. E., and Busch, H. (1976). *Proc. Natl. Acad. Sci. U.S.A.* **73,** 3258–3262.

CHAPTER 4

Systematic Identification of Specific Oncoplacental Proteins

HANS BOHN
Research Laboratories of Behringwerke AG
Marburg/Lahn
West Germany

I.	Introduction	69
II.	Detection and Classification of Placental Proteins	70
III.	Purification of the Specific Oncoplacental Proteins	73
IV.	Characterization of the Specific Oncoplacental Proteins	75
V.	Localization in Normal Tissues and Cells	78
VI.	Quantitation in Placental Extracts and Normal Body Fluids	78
	A. SP_1	79
	B. PP_5	80
	C. PP_{10} and PP_{12}	81
	D. PP_{11}	81
VII.	Occurrence in Tumor Patients and Application as Markers in Oncology	81
	A. SP_1	81
	B. PP_5	82
	C. PP_{10}, PP_{11} and PP_{12}	83
	References	84

I. INTRODUCTION

The human placenta is a highly specialized organ and probably the most complex human tissue of all. It contains and produces a wide variety of biologically active compounds which are for the most part proteins; these proteins may have the function of hormones, enzymes, proenzymes, inhibitors, or receptors, to mention a few.

Some of these biologically active compounds are specific to the placenta, i.e., they normally do not occur in other human tissues. They are synthesized in the placenta and are usually also secreted into the maternal bloodstream during pregnancy. Their detection and measurement in the circulation is used to diagnose pregnancy (human chorionic gonadotropin) or as an indicator of fetoplacental well-being in the surveillance of pregnancy (human placental lactogen, oxytocinase, heat-stable alkaline phosphatase).

The developing placenta is a fast-growing organ and bears some resemblance to malignant growth. Tumor cells indeed often produce proteins which are normally found only in the placenta; this may be explained by a derepression of genes in the malignant cells. Placenta-specific proteins, therefore, are also used as markers in oncology. Detection and determination of these proteins can be helpful in screening for malignant or premalignant disease and/or for monitoring of patients, i.e., in the control of therapy as well as in the early detection of metastases or recurrence.

Most of the biologically active compounds of the placenta mentioned above have been discovered by their biological activities. But nowadays immunochemical methods are often used for their detection and determination. However, during the last 10 years we systematically investigated the human term placenta for new proteins using immunochemical methods. The aim of our work was the following.

1. To detect new placental protein antigens by immunochemical methods.
2. To isolate and characterize these antigens, i.e., to determine their physical properties and chemical composition.
3. To prepare specific antisera and develop sensitive immunochemical methods for the detection and determination of these antigens in body fluids and organs and for the immunohistochemical localization in tissues.
4. To investigate the diagnostic significance of measurements of these proteins in sera from pregnant women and from patients with tumors and other diseases.
5. To elucidate their biological function and to determine if there was a therapeutical application for these proteins or an immunotherapeutical application for antibodies to these antigens.

II. DETECTION AND CLASSIFICATION OF PLACENTAL PROTEINS

In our laboratory a large number of, for the most part, new proteins have been detected by immunochemical methods in extracts from human term placentas (Bohn, 1971, 1972a, 1979; Bohn *et al.*, 1981, 1982a). The antisera used for their detection were obtained by immunizing rabbits with different placental fractions. These antisera were first absorbed with human male serum to remove all antibodies directed against normal serum proteins. This usually resulted in

4. New Specific Oncoplacental Proteins

oligospecific antisera, i.e., antisera containing antibodies against more than one placental protein. Finally, we tried to make the antisera more or less specific against the particular proteins by absorption with selected placental fractions or if possible with purified placental proteins. The antisera obtained in this way then were used to detect the corresponding protein in placental fractions and to find the optimal conditions for the isolation of the protein by a selected combination of common fractionation procedures and/or immunoadsorption techniques.

The purified proteins were characterized by their physical properties and chemical compositions. By immunizing rabbits with the purified antigens, highly specific antisera usually could be obtained. They were used to investigate the occurrence of these proteins in body fluids and in tissue extracts by immunochemical methods, such as gel diffusion test and electroimmunodiffusion technique. According to their occurrence in pregnancy sera or in different kinds of placental extracts, the proteins were divided into the following three categories:

1. Pregnancy proteins
2. Soluble placental tissue proteins
3. Solubilized placental tissue proteins

The pregnancy proteins and the soluble placental tissue proteins could be extracted with 0.5% sodium chloride solution; the solubilized tissue proteins were obtained by using dissociating solvents (acidic buffers, chaotropic salt solutions and detergents) or by digestion with papain (Bohn, 1979). Pregnancy proteins are proteins which do not occur or occur only in trace amounts in normal sera, and which during gravidity appear in the maternal bloodstream or are strongly elevated in their concentration. Accordingly we have differentiated between pregnancy-specific and pregnancy-associated proteins. In our laboratory, four pregnancy proteins have been detected by immunochemical methods. One turned out to be identical with hPL (human placental Lactogen); the others were found to be different from all other pregnancy-specific hormones and enzymes thus far known. They have been designated as SP_1, SP_2, and SP_3 (Bohn, 1971). SP_2 was found to be identical with the steroid binding β-globulin or sex hormone binding globulin (SHBG). SP_3 turned out to be identical with the so-called pregnancy zone protein, which is now mostly designated as pregnancy-associated $α_2$-glycoprotein ($α_2$PAG). Both, SP_2 and SP_3 are pregnancy-associated proteins. They occur in trace amounts in all normal sera. Their concentrations in serum are strongly elevated not only during pregnancy but also in women taking hormonal contraceptives (Bohn, 1974a,b). SP_1, on the other hand, turned out to be specific for pregnancy. This protein was shown to have the electrophoretic mobility of a $β_1$-globulin and a high carbohydrate content and, therefore, was termed pregnancy-specific $β_1$-glycoprotein (Bohn, 1971). Comparative immunochemical studies revealed SP_1 to be identical with the "new $β_1$-globulin" of Tatarinov and Masyukevich (1970) as well as with the so-called pregnancy-associated plasma protein C (PAPP-C) of Lin et al. (1974). Later SP_1 also was designated as trophoblast-specific β_1-glycoprotein and abbreviated TBG or TSG (Tatarinov and Sokolov, 1977; Tatarinov, 1978).

The soluble placental tissue proteins are mainly found in the placental tissue with very little secreted into the maternal bloodstream. The concentration of these proteins in pregnancy sera is usually less than 1 mg/liter; therefore, they cannot be detected in sera with the common immunodiffusion tests (Bohn, 1972a). More than 30 soluble placental tissue proteins have been already detected in our laboratory in the placental extract. A great deal of them has been already isolated and characterized (Bohn et al., 1982a). The soluble placental tissue proteins have been abbreviated PP and numbered consecutively. According to their occurrence in extracts of other human tissues we have distinguished between the following.

1. Ubiquitous tissue proteins, which were found to occur in almost all human tissues in higher concentrations.
2. Solitary tissue proteins which could only be detected in certain other human tissues.
3. Placenta-specific tissue proteins which could not be detected in extracts from other normal human tissues.

"Specific" in this context is relative rather than absolute. Trace amounts of these proteins may also be present in tissues and body fluids of the nonpregnant state, especially in malignant diseases (Bohn, 1980; Inaba et al., 1980a). The group of the so-called placenta-"specific" proteins comprises the soluble placental tissue proteins PP_5, PP_{10}, PP_{11}, and PP_{12}. There are still other soluble tissue proteins present in the placenta, but their isolation and characterization has not yet been completed.

The solubilized placental tissue proteins were obtained by extracting the insoluble part of the placental tissue with solubilizing agents, after the soluble material had been removed by washing with saline. Acidic buffers, chaotropic salt solutions detergents, and digestion with papain were applied in order to dissociate the membrane-bound antigens. At least 11 different solubilized antigens could be detected (Bohn, 1979). The isolation and characterization of these proteins is now under investigation. It also remains to be investigated which of these solubilized antigens are specific to the placenta. The next sections deal with the isolation, characterization, and diagnostic significance of those of our proteins which appear to be "specific oncoplacental proteins" or, in other words, which are relatively specific to the placenta and which were found to be also present in malignant tumor tissues or in sera of patients with malignancies. This group comprises the pregnancy protein SP_1 as well as the soluble placental tissue proteins PP_5, PP_{10}, PP_{11}, and PP_{12}.

4. New Specific Oncoplacental Proteins

III. PURIFICATION OF THE SPECIFIC ONCOPLACENTAL PROTEINS

As starting material for the isolation of the pregnancy proteins and the soluble placental tissue proteins we used human term placentas. The placentas were minced in the frozen state, extracted with 0.5% NaCl solution and centrifuged. The supernatant (the extract) then was fractionated with Rivanol (Hoechst AG) (2-ethoxy-6,9-diaminoacridine lactate) and ammonium sulfate to give either 6 or 3 placental fractions.

The separation of the placental proteins into six fractions is shown by the scheme in Fig. 1. The extract was first fractionated with Rivanol. In a first step the proteins precipitating at pH 6 were removed, than the proteins precipitating at pH 8.5 were separated. The proteins which are not precipitated by Rivanol remained in the supernatant. After removal of the Rivanol with 5% NaCl solution, each of these fractions was subdivided by fractionation with ammonium sulfate in the range of 0–40 or 0–50%, and 40–70 or 50–70% saturation. Thus finally the placental fractions I to VI were obtained (Bohn, 1971).

The fractionation of the placental extract into three fractions (1 to 3) is schematically demonstrated in Fig. 2 (scheme B). In this case, the precipitate obtained with Rivanol at pH 6.0 was not divided into subfractions, and the supernatant of this precipitate was directly fractionated with ammonium sulfate (Bohn, 1976). For the isolation of our placental proteins we usually started with those placental fractions which contained the main bulk of the corresponding protein or in which the protein was already concentrated to a certain extent. Further pu-

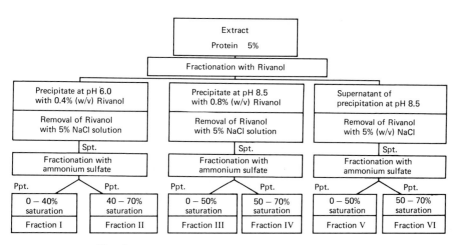

Fig. 1. Fractionation of the placental extract, scheme A.

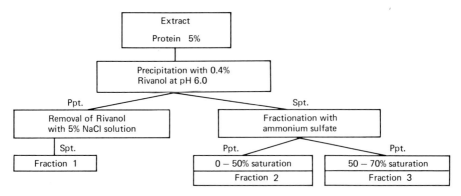

Fig. 2. Fractionation of the placental extract, scheme B.

rification was achieved by using fractionation with ethanol, euglobulin precipitation, gel filtration, ion exchange chromatography, preparative zone electrophoresis, hydroxyapatite chromatography, and immunoadsorption techniques.

The subfractionation of the placental fractions for the isolation of the placenta-specific proteins is schematically demonstrated in Fig. 3. SP_1 was first isolated from placental fraction V by common fractionation procedures, such as gel filtration, ion exchange chromatography, and preparative zone electrophoresis. The preparation thus obtained was more than 95% pure (Bohn, 1972b). It was later found that the remaining impurities can easily be removed by chromatography on hydroxyapatite; when dissolved in 0.005 M phosphate buffer pH 6.8, SP_1 passes hydroxyapatite unhindered, whereas the contaminating proteins are retained on the column (Bohn and Sedlacek, 1975). This step in addition was found to remove all the variants of SP_1 having α mobility (Bohn, 1979) and thus to result in the pure β component of SP_1. Finally a method was developed to isolate SP_1 from placental fraction 2 in good yield and in highly purified form by using an immunoadsorption technique in combination with hydroxyapatite chromatography (Bohn *et al.*, 1976). By the same method (Fig. 3) we have isolated SP_1 from retroplacental serum. Other procedures for the purification of SP_1 from pregnancy serum have been described by Tatarinov and Sokolov (1977), Engvall (1980), and more recently by Griffiths and Godard (1981). Immunoadsorption techniques have also been used in the purification procedures for the isolation of the soluble placental-specific tissue proteins PP_5, PP_{10}, PP_{11}, and PP_{12} (Fig. 3). The placental fractions which served as starting materials for the isolation of these proteins were first subfractionated by ion exchange chromatography and/or gel filtration. In case of PP_{11} and PP_{12}, part of the impurities was removed in addition by an euglobulin precipitation. The immunosorbents were prepared by coupling the immunoglobulins of the corresponding antisera to CNBr-activated Sepharose 4B. Elution from the adsorbent was performed with 3 M potassium

4. New Specific Oncoplacental Proteins

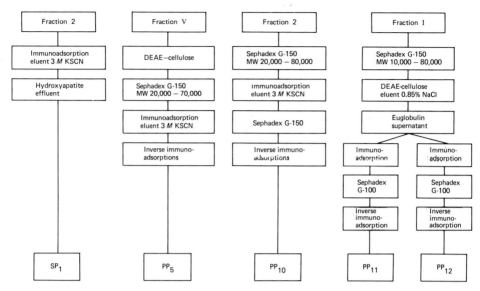

Fig. 3. Subfractionation of placental fractions for the isolation of SP_1, PP_5, PP_{10}, PP_{11}, and PP_{12}.

thiocyanate. For the final purification of the desorbed proteins, gel filtrations on Sephadex G-100 and/or inverse immunoadsorptions were applied. In the latter procedure Sepharose-bound antibodies to the contaminating proteins were used to remove residual impurities. All proteins thus were obtained in a purity >99%.

Comparative immunochemical studies revealed that PP_{12} is immunochemically identical with chorionic α_1-microglobulin (CAG-1) of Petrunin et al. (1978), which also has been designated as placenta-specific α_1-microglobulin and abbreviated PAMG-1 (Tatarinov et al., 1980a). Petrunin et al. (1980) have isolated this protein from amniotic fluid (16–25 week of pregnancy) by precipitation with lanthan chloride, ammonium and lithium sulfates, and by adsorption chromatography with calcium phosphate.

IV. CHARACTERIZATION OF THE SPECIFIC ONCOPLACENTAL PROTEINS

The analytical methods used for the determination of the physical characteristics and the chemical composition of the placental proteins have been described elsewhere (Bohn and Kraus, 1979).

The physical parameters of the specific oncoplacental proteins are summarized in Table I. The results of the amino acid and carbohydrate analyses are shown in Table II. SP_1 and PP_5 have the highest carbohydrate portions; their sugar content

TABLE I
Placenta-Specific Proteins and Their Physical Characteristics

Placental protein	Electrophoretic mobility	Isoelectric point	Sediment coefficient[c]	Molecular weight	Extinction coefficient[d]	Reference
SP_1	β_1	4.1	4.5 S	90,000[b]	11.8	Bohn et al. (1976)
PP_5	β_1	4.6	2.8 S	36,600[a]	10.6	Bohn and Winckler (1977)
PP_{10}	α_1	5.1	3.8 S	48,000[a] 65,000[b]	10.9	Bohn and Kraus (1979)
PP_{11}	α_1	5.1–5.2	3.5 S	44,300[a] 62,000[b]	13.4	Bohn and Winckler (1980)
PP_{12}	α_1	4.6–4.7	2.7 S	25,200[a] 51,000[b]	13.6	Bohn and Kraus (1980)

[a] Determined in the ultracentrifuge.
[b] Determined by SDS-polyacrylamide gel electrophoresis.
[c] $S^o_{20,w}$ in Svedberg units.
[d] $E^{1\%}_{1cm}$ (280 nm).

amounts to almost 30 and 20%, respectively, by weight; the other proteins contain less than 7% sugars. Glutamic acid, aspartic acid, leucine, serine, alanine, and glycine rank among the most abundant amino acids in these proteins. SP_1, in addition, showed a relatively high content in threonine and proline. The latter amino acid was found to be relatively high in PP_5 and PP_{12}.

A screening for certain biological functions has also been performed with the placenta-specific proteins: SP_1 appears to have immunosuppressive functions; studies have shown that at least *in vitro* SP_1 can inhibit the mixed leukocyte reaction (Bohn et al., 1976; Johannsen et al., 1976) and depress the responsiveness of lymphocytes to phytohemagglutinin (Cerni et al., 1977; Tatarinov et al., 1980b). PP_5 was found to inhibit the proteolytic activity of plasmin and trypsin (Bohn and Winckler, 1977). In addition it was shown to form complexes with heparin and to be involved with the coagulation system (Obiekwe et al., 1979; Salem et al., 1980). For PP_{10}, PP_{11}, and PP_{12} thus far no biological functions have been detected (Bohn and Winckler, 1980; Bohn and Kraus, 1980).

The immunochemical properties of the placental proteins have been investigated with use of highly specific antisera. Comparative studies revealed that immunochemically these proteins completely differ from one another, in other words, that they have no antigenic determinants in common. In studying the particular proteins immunochemical heterogenities could be detected in case of SP_1 and PP_{12}.

Molecular and immunochemical heterogenities of SP_1 have been already reported by Lin et al. (1974), Bohn (1974b), Bohn and Sedlacek (1975), and Bohn et al. (1976). Teisner et al. (1978), Horne et al. (1979), and Westergaard et al.

TABLE II
Amino Acid and Carbohydrate Compositions

	$SP_1{}^a$	$PP_5{}^b$	$PP_{10}{}^c$	$PP_{11}{}^d$	$PP_{12}{}^e$
Amino acids residues per 100 residues					
Lysine	4.37	5.10	6.51	6.26	3.96
Histidine	1.77	1.06	1.81	3.34	2.46
Arginine	5.07	6.02	3.67	3.31	4.09
Aspartic acid	9.85	12.36	9.82	10.75	7.49
Threonine	8.24	5.40	5.20	3.31	3.07
Serine	9.07	5.74	7.32	9.63	9.77
Glutamic acid	8.97	9.66	11.50	13.81	12.96
Proline	7.96	4.73	5.16	4.10	8.31
Glycine	7.07	7.13	6.97	6.23	6.91
Alanine	3.74	7.55	7.74	6.30	10.82
Cystine/2	1.57	7.12	1.94	3.37	6.66
Valine	5.84	4.40	6.32	4.53	4.08
Methionine	0.88	0.91	2.43	1.06	1.49
Isoleucine	5.79	2.60	4.09	3.60	3.13
Leucine	9.60	7.75	9.46	6.74	7.25
Tyrosine	5.75	4.89	2.96	5.90	2.72
Phenylalanine	2.12	5.61	5.44	6.06	2.60
Tryptophan	2.39	1.88	1.69	1.66	2.08
Carbohydrates (weight %)					
Hexoses	11.7	10.0	4.8	2.6	2.8
Hexosamines	10.2	4.4	1.2	1.0	0.7
Fucose	0.6	0.4	0.05	0.05	0.03
Sialic acid	6.8	5.0	0.6	0.26	0.8
Total	29.3	19.8	6.65	3.91	4.3

[a] Bohn et al. (1976).
[b] Bohn and Winckler (1977).
[c] Bohn and Kraus (1979).
[d] Bohn and Winckler (1980).
[e] Bohn and Kraus (1980).

(1979) described several variants of SP_1; one, usual the main component, having β_1 electrophoretic mobility and a molecular weight of 90,000, and the others having α_2 electrophoretic mobilities and higher weights. The composition of these α components is still unresolved, but there are indications that they are formed by a combination of $SP_1\beta$ with one or even more compounds present in normal serum or in placental extracts (Bohn, 1979; Ahmed and Klopper, 1980).

PP_{12}, on the other hand, was found to contain two completely different antigenic determinants giving rise to two separate precipitin lines (Bohn and Kraus, 1980). It is still not quite clear whether the two antigenic sites of PP_{12} reside in the same polypeptide chain or whether they belong to two different subunits or proteins which are held together by disulfide bridges or by noncovalent bonds.

V. LOCALIZATION IN NORMAL TISSUES AND CELLS

Immunohistochemical localization of the placental proteins in placentas has been studied by using an immunofluorescent technique as well as with use of a peroxidase-antiperoxidase bridge (PAP) method. Immunofluorescent studies revealed that SP_1 and PP_5 are predominantly located in the cytoplasm of the syncytiotrophoblast (Bohn and Sedlacek, 1975). With use of the PAP method the placenta-specific tissue proteins PP_{10}, PP_{11}, and PP_{12} also were found to be located in the syncytiotrophoblastic cells of placental villi (Inaba et al., 1980b). However, in case of PP_{10} and PP_{12} in addition a strong staining was observed in the cytoplasm and nuclei of histiocytes occurring in amniotic, villous and decidual tissue. The histiocytes found occasionally in the intervillous space also stained strongly for PP_{10} and PP_{12}.

These findings suggest that these proteins may also be synthesized by histiocytes. Therefore, we investigated the occurrence of PP_{10} and PP_{12} as well as of SP_1, PP_5, and PP_{11}, in the cellular constituents of normal human blood by the PAP method (Inaba et al., 1981a,b).

Erythrocytes and lymphocytes did not show positive staining for any of these proteins. Monocytes occasionally stained positive for PP_{10}, whereas granulocytes were positive for SP_1, PP_{10}, and PP_{12}. SP_1 showed positive staining only in the cytoplasm of granulocytes. In case of PP_{10} and PP_{12} staining was found to be stronger in nuclei than in cytoplasm of granulocytes. These results suggest that SP_1, PP_{10}, and PP_{12} may be synthesized by granulocytes in normal blood or accumulated there, and that these proteins are not really specific to the trophoblast.

VI. QUANTITATION IN PLACENTAL EXTRACTS AND NORMAL BODY FLUIDS

The amounts in which the placenta-specific proteins occur in extracts of human term placentas have been quantitated with an electroimmunoassay (Laurell technique). The results are summarized in Table III.

Proteins immunochemically identical or closely related to the human placental-specific proteins SP_1, PP_5, PP_{10}, PP_{11}, and PP_{12} were found to occur in extracts from placentas of monkeys and apes (Bohn and Sedlacek, 1975; Bohn et al., 1981). Estimations of the amounts of the corresponding proteins present in extracts from term placentas of cynomolgus monkeys (Macaca fascicularis) are shown in Table III for comparison. In case of the human placenta SP_1 showed the highest concentration of the placenta-specific proteins investigated here; it was followed by PP_{10}, PP_{11}, PP_{12} and finally PP_5. In the cynomolgus placenta the concentration of SP_1 was about 10-fold and of PP_{12} even more than 100-fold higher than in the human placenta. On the other hand, the values for PP_5 and PP_{10} were lower in the monkeys.

By the Laurell technique we also have investigated the concentrations of the placenta-specific proteins in normal body fluids (normal sera of males and females, maternal serum, cord blood serum, and amniotic fluid). In pregnancy sera only SP_1 and to a lesser extent also PP_{10} could be detected by the applied method (detection limit around 1 mg/liter). In normal sera and in cord blood sera none of the proteins investigated was detectable. In amniotic fluid only PP_{12} was found to occur in higher concentrations (0–146 mg/liter) especially in midpregnancy: Maximal values were found between weeks 20–30 of pregnancy; toward term there was a strong decrease in the concentration of PP_{12} in the amniotic fluid (Petrunin et al., 1978; Bohn and Kraus, 1980). To detect and determine the placenta-specific proteins in body fluids for the most part more sensitive methods were necessary. Therefore, radioimmunoassays (RIA's) and/or enzyme immunoassays (EIA's) have been developed for the measurement of these antigens. The results of these investigations are summarized in Table IV (Bohn et al., 1982b).

A. SP_1

Levels of SP_1 are high enough in sera from pregnant women to be measured with immunodiffusion tests (electroimmunoassay, radial immunodiffusion) from the tenth week of gestation onward. Circulating concentrations of SP_1 rise steadily during gravidity and tend to reach a plateau near term (Bohn, 1972b; Tatra et al., 1974; Towler et al., 1976). The shape of the SP_1 curve during pregnancy closely resembles that of hPL. Levels of SP_1, like those of hPL,

TABLE III

Quantitation of the Placenta-Specific Proteins in Extracts from Term Placentas of Man and Cynomolgus Monkey[a]

Protein	Human placenta	Cynomolgus placenta
SP_1	50	470
PP_5	2	<2
PP_{10}	33	<2
PP_{11}	18	28
PP_{12}	8	1200

[a] Amounts in extracts of term placentas (μg/g wet weight).

TABLE IV
Levels of the Placenta-Specific Proteins in Body Fluids[a]

	Concentrations (median)			
Proteins	Normal values in serum	Maternal serum (pregnancy at term)	Cord blood serum	Amniotic fluid
SP_1 (mg/liter)	<0.001	180	0.7	1.2[c]
PP_5 (mg/liter)	Not detectable	0.03	0.002	0.004[c]
PP_{10} (mU/liter)[b]	0.006	3.5	Approximately 0.01	0.6[c]
PP_{11} (mg/liter)	Not detectable	<1	<1	<1
PP_{12} (mg/liter)	♀0.019 ♂0.013	0.17	<1	40[d]

[a] From Bohn et al., 1982b.
[b] One unit of PP_{10} equals approximately 1 g.
[c] Pregnancy at term.
[d] At the twentieth week of gestation.

correlate well with the stage of gestation. Measurement of SP_1, therefore, can be used as an index of fetal well being in late pregnancy (Tatra et al., 1974; Gordon et al., 1977) and appears to be useful in the early detection of fetal growth retardation and in threatened abortion (Jandial et al., 1978; Schultz-Larsen and Hertz, 1978; Jouppila et al., 1980; Würz et al., 1981).

For the detection of SP_1 in maternal serum in early pregnancy and in other maternal compartments sensitive RIA's or EIA's have been developed (Stevens et al., 1976; Grudzinskas et al., 1977a; Tatarinov and Sokolov, 1977; Grenner, 1978). With such tests SP_1 was shown to appear in serum shortly after conception. Measurement of this protein, therefore, was found to be useful in the early detection of pregnancy, especially in hCG treated infertile women (Grudzinskas et al., 1977b; Seppälä et al., 1979a; Eiermann et al., 1981).

In sera of healthy, normal, nonpregnant subjects SP_1 usually cannot be detected; although some authors report that occasionally trace amounts of SP_1 may occur (Searle et al., 1978; Würz, 1979). This could well be the case considering the findings that SP_1 was localized in granulocytes of normal blood also (Inaba et al., 1981a,b).

B. PP_5

Quantitation of PP_5 by RIA has been performed in different laboratories (Obiekwe et al., 1979; Seppälä et al., 1979b; Nisbet et al., 1981).

In the normal sera this protein was shown to be absent. In maternal serum it appears from the sixth week of pregnancy onward, and levels increase as pregnancy processes. The highest levels are seen in weeks 37–39 of pregnancy (mean

0.03 mg per liter). Trace amounts of PP_5 could also be detected in maternal urine, in amniotic fluid, and occasionally in cord blood serum (Grudzinskas *et al.*, 1979). Recently PP_5 also was found to occur in seminal plasma (Ranta *et al.*, 1981).

C. PP_{10} and PP_{12}

Specific and sensitive radioimmunoassays have also been developed to measure the concentrations of PP_{10} and PP_{12} in body fluids. Preliminary results have already been published (Bohn *et al.*, 1981). They indicated that both proteins occur in trace amounts in normal sera (see Table IV). This is in agreement with the findings that both proteins could be localized in granulocytes of normal blood (Inaba *et al.*, 1981a). In addition the serum concentration of PP_{10} was found to be strongly (more than 100-fold) elevated during pregnancy, whereas in case of PP_{12} the differences in concentrations between normal sera and pregnancy sera were much less pronounced. In more detailed studies the concentration values of PP_{10} were recently measured by Würz *et al.* (1982a) in serum samples from pregnant women throughout pregnancy. Normal as well as pathological pregnancies have been investigated. In normal pregnancies levels of PP_{10} first increased with advancing pregnancy and then plateaued between weeks 35–40 (median 3.5 mg/liter). In pregnancies with "small-for-date" babies, a marked tendency toward reduced PP_{10} values was noted: In most cases the values were below the median, and in a large part even below the tenth percentile of the normal range. In this study PP_{10} appeared to be a better parameter in predicting those risk pregnancies than SP_1 or hPL. Measurement of PP_{12}, on the other hand, appeared to be of no value in monitoring pregnancy.

D. PP_{11}

A RIA for the determination of PP_{11} has recently been developed. This protein appears to be absent from normal sera; its concentration in sera from pregnant women and in other body fluids is still under investigation (G. Lüben, E. Spangenberg, and H. Bohn, unpublished data).

VII. OCCURRENCE IN TUMOR PATIENTS AND APPLICATION AS MARKERS IN ONCOLOGY

A. SP_1

Pregnancy-specific β_1-glycoprotein has been immunohistochemically detected in trophoblastic tumors (Tatarinov *et al.*, 1976; Horne *et al.*, 1977) as well as in nontrophoblastic neoplasms including carcinoma of the testis, breast, ovary, gastrointestinal tract, and lung (Javadpour, 1980; Horne et al., 1977; Inaba et al., 1980a).

Using an enzyme-bridge immunoperoxidase technique, Horne et al. (1976) have demonstrated the presence of SP_1 in 60% of breast cancers and in 50% of gastrointestinal cancers. In a subsequent retrospective study, they have also shown that the presence of SP_1 in the cytoplasm of breast cancers correlates with survival time:The patients whose cancers stained positively for SP_1 survive for a significantly shorter time than those whose tumors do not stain.

SP_1 also was found to occur in sera from patients with malignant tumors, especially with tumors containing trophoblastic structures. Tatarinov and Sokolov (1977) detected SP_1 levels in almost 80% of cases with trophoblastic tumors and in 15% of cases with different nontrophoblastic malignancies. The concentration of SP_1 present in such patients was in the range of 0 to 140 µg/liter. Seppälä et al. (1978), Searle et al. (1978), and Than et al. (1979) compared the conventional tumor marker hCG and SP_1 in patients with choriocarcinoma and showed that SP_1 may serve as an additional marker for such tumors. In addition measurement of circulating levels of SP_1 was found to provide an early approach to the differentiation of benign and malignant trophoblastic disease (Lee et al., 1981): Concentrations of SP_1 were found to be distinctly higher in patients with hydatidiform mole than in patients with choriocarcinoma, whereas levels of hCG did not distinguish between benign and malignant forms of these conditions.

Determination of SP_1 in serum appears also be be valuable in diagnosing and staging of testicular germ cell tumors (Lange et al., 1980; Szymendera et al., 1981). Follow-up studies of SP_1 may also be useful in patients with other nontrophoblastic malignant neoplasms which produce this protein ectopically (Würz, 1979).

B. PP_5

Seppälä et al. (1979b) studied by immunohistochemical methods the occurrence and localization of PP_5 in the normal placenta as well as in hydatidiform mole, invasive mole, and choriocarcinoma; they also investigated PP_5 level in sera from patients with trophoblastic tumors. The results demonstrated that PP_5 was expressed in the normal but not in the malignant syncytiotrophoblast. Measurement of serum PP_5, therefore, can be used for the differential diagnosis of malignant trophoblastic disease and normal pregnancy: In normal pregnancy the placental proteins PP_5, hCG, and SP_1 are all present in serum 10 weeks after the last menstrual period, whereas in malignant trophoblastic disease hCG and SP_1 are expressed in the absence of PP_5.

Lee et al. (1981) could confirm these findings of Seppälä et al. (1979b). In addition they showed that levels of PP_5, like those of SP_1, are quite distinct in hydatidiform mole and choriocarcinoma; measurement of PP_5, therefore, also can be used to differentiate between benign and malignant trophoblastic disease.

The occurrence and localization of PP_5 in nontrophoblastic tumor tissues was investigated by Inaba et al. (1980a) using an enzyme-bridge immunoperoxidase technique; positive staining was obtained in 12 out of 19 cases (63.2%) of breast cancer, in 18 out of 24 cases (75%) of testicular malignant neoplasms, and in 5 out of 12 cases (41.7%) of gastric cancers. Recently Inaba et al. (1981c) also have investigated ovarian cancer tissues; PP_5 could be detected in 28 out of 36 cases (77.8%) of cystadenocarcinomas, whereas staining for PP_5 was negative in all 5 normal ovaries tested. Measurement of PP_5 in the serum of those patients is now under investigation.

Bremner et al. (1981) studied the presence of PP_5 in homogenates of benign and malignant breast tumor tissue; PP_5 was detectable in 55% of the malignant tumors examined and in 71% of the benign tumors. The concentration of PP_5 in such tumors was generally very low. The highest values were found in one class of benign lesion, simple cystic disease. There was a relatively high frequency of PP_5 detection in homogenates of the tumors themselves, but the majority of tumor patients did not show elevated serum PP_5 levels. Serum PP_5, therefore, appeared to be not a useful marker for monitoring patients with nontrophoblastic tumors.

C. PP_{10}, PP_{11}, and PP_{12}

Localization of these proteins in tumor tissues of a variety of nontrophoblastic neoplasms has been investigated by Inaba et al. (1980a): In total 55.6% of the tumors tested were PP_{10}-positive, 38.0% PP_{11}-positive, and 31.9% PP_{12}-positive. Recently Inaba et al. (1981c) also investigated the occurrence of these proteins in the tissue of 35 different ovarian adenocarcinomas: Here the detection rate was highest in case of PP_{11} (57.1%) and much lower in case of PP_{10} (9.5%) and PP_{12} (23.8%).

Localization studies of PP_{10}, PP_{11}, and PP_{12} in trophoblastic tumor tissues have recently been performed by Wahlström et al. (1982): PP_{11} stained equally strong in neoplastic and normal syncytiotrophoblasts; whereas PP_{10} and PP_{12} were only demonstrable in the syncytiotrophoblasts of normal placentas and hydatidiform moles but not in destructive moles or choriocarcinomas.

The investigation of the occurrence of PP_{10}, PP_{11}, and PP_{12} in sera of patients with tumors has been already started. Preliminary studies indicated that circulating levels of PP_{10} and PP_{12} are often elevated above normal in tumor patients (Bohn et al., 1981). This is especially true for PP_{10}. Würz et al. (1982b) recently measured concentrations of PP_{10} and PP_{12} in sera of patients with genital carcinoma and breast cancer before treatment. All of the 26 genital carcinoma patients and around 85% of the 103 breast cancer patients tested had elevated serum PP_{10} levels. In case of PP_{12} only approximately 20% of the patients with primary carcinoma showed levels above normal. More studies have to be performed before the usefulness of one of these proteins as a "tumor marker" can be evaluated.

Measurement of PP_{11} in sera from tumor patients is now also under investigation.

REFERENCES

Ahmed, A. G., and Klopper, A. (1980). *Arch. Gynecol.* **230,** 95–108.
Bohn, H. (1971). *Arch. Gynaekol.* **210,** 440–457.
Bohn, H. (1972a). *Arch. Gynaekol.* **212,** 165–175.
Bohn, H. (1972b). *Blut* **24,** 292–302.
Bohn, H. (1974a). *Arch. Gynaekol.* **217,** 219–231.
Bohn, H. (1974b). *Blut* **29,** 17–34.
Bohn, H. (1976). *Protides Biol. Fluids* **24,** 117–124.
Bohn, H. (1979). *In* "Carcino-Embryonic Proteins" (F.-G. Lehmann, ed.), Vol. I, pp. 289–299. Elsevier/North Holland Biomedical Press, Amsterdam.
Bohn, H. (1980). *Klin. Wochenschr.* **58,** 489–492.
Bohn, H., and Kraus, W. (1979). *Arch. Gynecol.* **227,** 125–134.
Bohn, H., and Kraus, W. (1980). *Arch. Gynecol.* **229,** 279–291.
Bohn, H., and Sedlacek, H. (1975). *Arch. Gynaekol.* **220,** 105–121.
Bohn, H., and Winckler, W. (1977). *Arch. Gynaekol.* **223,** 179–186.
Bohn, H., and Winckler, W. (1980). *Arch. Gynecol.* **229,** 293–301.
Bohn, H., Schmidtberger, R., and Zilg, H. (1976). *Blut* **32,** 103–113.
Bohn, H., Inaba, N., and Lüben, G. (1981). *Oncodev. Biol. Med.* **2,** 141–153.
Bohn, H., Kraus, W., and Winckler, W. (1982a). *Placenta, Suppl.* **4,** 67–81.
Bohn, H., Dati, F., and Lüben, G. (1982b). *In* "Biology of the Throphoblast" (Y. W. Loke and A. Whyte, eds.), Vol. II, pp. 317–352. Elsevier/North-Holland Biomedical Press, Amsterdam (in press).
Bremner, R. D., Nisbet, A. D., Herriot, R., Horne, C. H. W., McArdle, C., Crawford, D., and Bohn, H. (1981). *Oncodev. Biol. Med.* **2,** 55–62.
Cerni, C., Tatra, G., and Bohn, H. (1977). *Arch. Gynaekol.* **223,** 1–7.
Eiermann, W., Albrich, W., Dati, F., Leis, E., and Eicher, W. (1981). *Geburtshilfe Frauenheilkd.* **41,** 404–406.
Engvall, E. (1980). *Oncodev. Biol. Med.* **1,** 113–122.
Gordon, Y. B., Grudzinskas, J. G., Jeffrey, D., and Chard, T. (1977). *Lancet* **1,** 331–333.
Grenner, G. (1978). *Fresenius' Z. Anal. Chem.* **290,** 99.
Griffiths, B. W., and Godard, A. (1981). *J. Reprod. Immunol.* **3,** 117–129.
Grudzinskas, J. G., Gordon, Y. B., Jeffrey, D., and Chard, T. (1977a). *Lancet* **1,** 333–334.
Grudzinskas, J. G., Lenton, E. A., Gordon, Y. B., Kelso, I. M., Jeffrey, D., Sobowale, O., and Chard, T. (1977b). *Br. J. Obstet. Gynaecol.* **84,** 740–742.
Grudzinskas, J. G., Charnock, M., Obiekwe, B. C., Gordon, Y. B., and Chard, T. (1979). *Br. J. Obstet. Gynaecol.* **86,** 642–644.
Horne, C. H. W., Reid, I. N., and Milne, G. D. (1976). *Lancet* **2,** 279–282.
Horne, C. H. W., Towler, C. M., and Milne, G. D. (1977). *J. Clin. Pathol.* **30,** 19–23.

4. New Specific Oncoplacental Proteins 85

Horne, C. H. H., Bremner, R. D., Jandial, V., Glover, R. G., and Towler, C. M. (1979). *In* "Placental Proteins" (A. Klopper and T. Chard, eds.), pp. 143–160. Springer-Verlag, Berlin and New York.
Inaba, N., Renk, T., Wurster, K., Rapp, W., and Bohn, H. (1980a). *Klin. Wochenschr.* **58**, 789–791.
Inaba, N., Renk, T., and Bohn, H. (1980b). *Arch. Gynecol.* **230**, 109–121.
Inaba, N., Renk, T., Ax, W., Schottler, S., Weinmann, E., and Bohn, H. (1981a). *Acta Haematol.* **66**, 35–38.
Inaba, N., Renk, T., Ax, W., Weinmann, E., Schottler, S., and Bohn, H. (1981b). *Blut* **43**, 315–323.
Inaba, N., Ishige, H., Ijichi, M., Satoh, N., Ohkawa, R., Sekiya, S., Takamizawa, H., Renk, T., and Bohn, H. (1981c). *Oncodev. Biol. Med.* **2**(5), P64 (Abstract).
Jandial, V., Towler, C. M., Horne, C. H. W., and Abramovich, D. R. (1978). *Br. J. Obstet. Gynaecol.* **85**, 832.
Javadpour, N. (1980). *J. Urol.* **123**, 514–515.
Johannsen, R., Haupt, H., Heide, K., Seiler, F. R., Schwick, H. G., and Bohn, H. (1976). *Z. Immunitaetsforsch.* **152**, 280–285.
Jouppila, P., Seppälä, M., and Chard, T. (1980). *Lancet* **1**, 667–668.
Lange, P. H., Bremner, R. D., Horne, C. H. W., Vessella, R. L., and Fraley, E. E. (1980). *Urology* **15**, 251–255.
Lee, J. N., Salem, H. T., Al-Ani, A. T. M., Chard, T., Huang, S. C., Ouyang, P. C., and Seppälä, M. (1981). *Am. J. Obstet. Gynecol.* **139**, 702–704.
Lin, T. M., Halbert, S. P., Kiefer, D., Spellacy, W. N., and Gall, S. (1974). *Am. J. Obstet. Gynecol.* **118**, 223–236.
Nisbet, A. D., Bremner, R. D., Herriot, R., Jandial, V., Horne, C. H. W., and Bohn, H. (1981). *Br. J. Obstet. Gynaecol.* **88**, 484–491.
Obiekwe, B. C., Pendlebury, D. J., Gordon, Y. B., Grudzinskas, J. G., Chard, T., and Bohn, H. (1979). *Clin. Chim. Acta* **95**, 509–516.
Petrunin, D. D., Gryaznova, I. M., Petrunina, Y. A., and Tatarinov, Y. S. (1978). *Byull. Eksp. Biol. Med.* **5**, 600–602.
Petrunin, D. D., Kozlyaeva, G. A., Tatarinov, A. S., and Shevchenko, O. P. (1980). *Byull. Eksp. Biol. Med.* **5**, 558–560.
Ranta, T., Siiteri, J. E., Koistinen, R., Salem, H. T., Bohn, H., Koskimies, A. I., and Seppälä, M. (1981). *J. Clin. Endocrinol. Metab.* **53**, 1087–1089.
Salem, H. T., Obiekwe, B. C., Al-Ani, A. T. M., Seppälä, M., and Chard, T. (1980). *Clin. Chim. Acta* **107**, 211–215.
Schultz-Larsen, P., and Hertz, J. B. (1978). *Eur. J. Obstet., Gynecol. Reprod. Biol.* **8**, 253–257.
Searle, F., Leake, B. A., Bagshawe, K. D., and Dent, J. (1978). *Lancet* **1**, 579–581.
Seppälä, M., Rutanen, E. M., Heikinheimo, M., Jalanko, H., and Engvall, E. (1978). *Int. J. Cancer* **21**, 265–267.
Seppälä, M., Tönnberg, L., Ylöstalo, P., and Jouppila, P. (1979a). *Fertil. Steril.* **32**, 604–607.
Seppälä, M., Wahlström, T., and Bohn, H. (1979b). *Int. J. Cancer* **24**, 6–10.
Stevens, V. C., Bohn, H., and Powell, J. E. (1976). *Am. J. Obstet. Gynecol.* **124**, 51–54.
Szymendera, J. J., Zborzil, J., Sikorowa, L., Kaminska, J. A., and Gadek, A. (1981). *Oncology* **38**, 222–229.
Tatarinov, Y. (1978). *Gynecol. Obstet. Invest.* **9**, 65–97.
Tatarinov, Y. S., and Masyukevich, V. N. (1970). *Byull. Eksp. Biol. Med.* **69**, 66–68.
Tatarinov, Y. S., and Sokolov, A. V. (1977). *Int. J. Cancer* **19**, 161–166.
Tatarinov, Y. S., Falaleeva, D. M., Kalashnikov, V. V., and Toloknov, B. O. (1976). *Nature (London)* **260**, 263.

Tatarinov, Y. S., Kozljaeva, G. A., Petrunin, D. D., and Petrunina, Y. A. (1980a). *In* "The Human Placenta" (A. Klopper, A. Genazzani, and P. G. Crosignani, eds.), pp. 35–46. Academic Press, New York.
Tatarinov, Y. S., Golovistikov, I. N., Gorlina, N. K., Petrunin, D. D., Tsagaraeva, T. M., and Shevchenko, O. P. (1980b). *Immunologia* **5,** 14–17.
Tatra, G., Breitenecker, G., and Gruber, W. (1974). *Arch. Gynaekol.* **217,** 383–390.
Than, G., Bohn, H., Csaba, I. F., Karg, N., and Mann, V. (1979). *In* "Carcino-Embryonic Proteins" (F.-G. Lehmann, ed.), Vol. II, pp. 481–486. Elsevier/North-Holland Biomedical Press, Amsterdam.
Teisner, B., Westergaard, J. G., Folkersen, J., Husby, S., and Svehag, S. E. (1978). *Am. J. Obstet. Gynecol.* **131,** 262–266.
Towler, C. M., Horne, C. H. W., Jandial, V., Campbell, D. M., and McGillivray, I. (1976). *Br. J. Obstet. Gynaecol.* **83,** 775–779.
Wahlström, T., Bohn, H., and Seppälä, M. (1982). *In* "Pregnancy Proteins: Biology, Chemistry and Clinical Application" (J. G. Grudzinskas, M. Seppälä, and B. Teisner, eds.). Academic Press, New York (in press).
Westergaard, J. G., Teisner, B., Folkersen, J., Hindersson, P., Schultz-Larsen, P., and Svehag, S. (1979). *Scand. J. Clin. Lab. Invest.* **39,** 351–359.
Würz, H. (1979). *Arch. Gynecol.* **227,** 1–6.
Würz, H., Geiger, W., Künzig, H. J., Jabs-Lehmann, A., Bohn, H., and Lüben, G. (1981). *J. Perinat. Med.* **9,** 67–78.
Würz, H., Lüben, G., Bohn, H., Künzig, H. J., and Geiger, W. (1982a). *Arch. Gynecol.* (in press).
Würz, H., Lüben, G., and Bohn, H. (1982b). *Cancer Det. Prev.* **5,** 242–243.

CHAPTER 5
Marker Expression by Cultured Cancer Cells

RAYMOND W. RUDDON
Department of Pharmacology
The University of Michigan Medical School
Ann Arbor, Michigan

I.	Tumor-Associated Antigens	88
	A. Carcinoembryonic Antigen (CEA)	88
	B. α-Fetoprotein (AFP)	90
II.	Cell Membrane-Associated Glycoproteins and Glycolipids	91
III.	Hormones	94
	A. Adrenocorticotropin (ACTH) and β-Lipotropin (β-LPH)	94
	B. Calcitonin	96
	C. Glycoprotein Hormones	96
IV.	Enzymes	98
	A. Alkaline Phosphatase	98
	B. γ-Glutamyltransferase (GGT)	99
	C. Glycosyltransferases	100
	D. Proteases	100
V.	Tumor-Derived Growth Factors	102
	References	103

The production of oncodevelopmental markers by cancer cells growing in culture is a well established phenomenon. Cell lines derived from human and experimental animal cancers frequently continue to express the same markers that are detected in cancer patients or tumor-bearing animals. This provides a powerful tool to examine the regulation of transcription and translation of oncodevelopmental genes in cancer cells and to characterize biochemically the tumor markers that are produced. Pharmacologic manipulation can also be performed more readily in cultures of cancer cells, thus facilitating study of the mechanisms regulating secretion or shedding of tumor markers by malignant cells. For these reasons, our laboratory and other laboratories have utilized cell culture systems to study tumor marker production.

In a study of 67 human tumor lines derived from a wide variety of cancers, including carcinomas of the breast, lung, kidney, testis, cervix, pancreas, colon, and thyroid as well as some lymphomas, osteosarcomas, melanomas, neuroblastomas, choriocarcinomas, and virally and chemically transformed human fibroblasts, Neuwald et al. (1980) found that 68% of these lines produced one or more oncodevelopmental markers. The markers examined included carcinoembryonic antigen (CEA), alpha and beta subunits of chorionic gonadotropin (hCG), placental alkaline phosphatase, γ-glutamyltransferase, and cystyl aminopeptidase (placental oxytocinase). There was, however, no predictable coexpression of any two of these markers in the cultured cancer cells, suggesting that the expression of these genes is not linked. Rosen et al. (1980a) reported similar data in a study of 32 cultured human neoplastic cell lines (18 carcinomas from various tissues, 3 melanomas, 3 astrocytomas, 3 glioblastomas, and 5 sarcomas) in which they examined the production of glycoprotein hormone alpha subunit, hCG-β, LH-β, FSH-β, placental lactogen, prolactin, growth hormone, α-fetoprotein (AFP) and CEA. Although certain cell types tended to produce some of the markers more often than other cell types, these authors also noted discordance in the production of these markers. Sell and Becker (1978) have noted that tumors that produce CEA do not usually produce AFP, nor are fetal liver isoenzyme patterns clearly associated with the ability of hepatomas to produce AFP. Thus, the expression of different oncodevelopmental markers does not appear to be coordinated in tumor cells. Expression of these genes may be linked in some way to the turn-on of "oncogenes" during malignant transformation. In different cell types, different sets of oncodevelopmental genes may be derepressed together with the oncogenes. The pattern of oncodevelopmental gene expression may vary from cell type to cell type, species to species, patient to patient, and with the nature of the carcinogen.

Examples of oncodevelopmental markers that are produced by cancer cells in culture and studies of the biochemical characterization of these markers will be the primary topic of this chapter.

I. TUMOR-ASSOCIATED ANTIGENS

A. Carcinoembryonic Antigen (CEA)

The production of CEA by cells in tissue culture has been described for human cell lines derived from carcinomas of the colon (Egan and Todd, 1972; Goldenberg et al., 1972; Tompkins et al., 1974; Drewinko et al., 1976; Carrel et al., 1977; Neuwald et al., 1980), liver (Rosen et al., 1980a), lung (Ellison et al., 1977; Neuwald et al., 1980; Rosen et al., 1980a), prostate (Williams et al., 1977), breast, testis, cervix, pancreas, and thyroid (Neuwald et al., 1980), as well as melanomas (Neuwald et al., 1980; Rosen et al., 1980a) and a glioblastoma (Neuwald et al., 1980).

In most of these studies CEA levels were determined by radioimmunoassay of cell lysates or culture medium. Definitive biochemical characterization of the cell culture-produced material has not been done, leaving the possibility that in some of these studies, substances immunologically cross-reactive with CEA may have been detected. Utilizing antisera specific for CEA and the two immunologically closely related substances "normal glycoprotein" (NGP; called "nonspecific cross-reacting antigen" (NCA) by others) and biliary glycoprotein 1 (BGP 1), Hammarström et al. (1977) have demonstrated a specific antibody-dependent cell-mediated lysis of tumor cells bearing different relative amounts of these antigens on their surfaces. A line of CEA-producing colon carcinoma cells was killed in the presence of anti-CEA, whereas a non-CEA-producing bladder carcinoma cell line was not. However, partial killing of the CEA-producing line was also obtained with anti-NGB and anti-BGP 1, indicating that both CEA and the normal tissue cross-reacting components were present on the surface of the colon tumor cells. Whether CEA, NGP and BGP 1 were present on the same or different cells in the tumor cell population was not clear, but it points out the difficulty in interpreting some of the data on CEA production by cultured cancer cells.

Carrel et al. (1977) have also reported antibody-dependent cell-mediated cytolysis of cultured human colon carcinoma cells induced by antisera against CEA. Target cells not containing CEA on their surface, as determined by immunofluorescence, were not lysed. Absorption of the anti-CEA antiserum with CEA blocked the killing effect, but absorption with erythrocytes of different blood groups or with normal tissues did not decrease the cytotoxic activity of anti-CEA.

Clonal variant sublines that differ in CEA production have been isolated from colon carcinomas. For example, Rosenthal et al. (1977, 1980) have isolated variant clones that produce significantly more or less CEA than the parent cell line. The level of expression of CEA appeared to be stable with continued passage in culture. Interestingly, all the lines produced tumors in nude mice, suggesting that CEA synthesis by colon tumor cells does not play a key role in their tumorigenicity.

A number of factors can modulate the expression of CEA by cultured human cancer cells. Enhanced production of CEA has been observed in clones of adenocarcinoma cells treated with bromodeoxyuridine or theophylline (Rosenthal et al., 1980). In the latter case, the effect was dose-dependent and required continual presence of the drug and continual protein synthesis. The effect of theophylline did not appear to be mediated by cyclic adenosine 3',5'-monophosphate (cAMP). In another human colon carcinoma cell line, interferon produced a two- to threefold increase in shedding of CEA from the tumor cells into the medium

(Attallah et al., 1979). In a CEA-producing human lung carcinoma line, release of CEA was shown to be directly proportional to the number of cells, to be less in rapidly dividing cells than in slowly dividing cells, and to be increased by cytotoxic changes in the pH of the growth medium (Ellison et al., 1977).

B. α-Fetoprotein (AFP)

α-Fetoprotein synthesis by hepatoma cell lines was first described by Irlin et al. (1966). Since that report, the synthesis of AFP by a variety of hepatoma-derived cell lines has been described (Bernhard et al., 1973; Tsukada et al., 1974; Rioche et al., 1974; Sell and Morris, 1974; Nishina, 1975; Becker et al., 1976; McMahon et al., 1977; Princler et al., 1978). Most of the studies have been done with mouse or rat hepatoma cultures. In cultures of Morris rat hepatoma lines, the rate of production of AFP peaks as the cells reach late exponential phase of growth and then declines as the cells reach confluency (Becker et al., 1976; de Néchaud et al., 1977). This is consistent with the observation that AFP production *in vivo* is associated with hepatocyte proliferation. For example, partial hepatectomy or chemically induced liver necrosis in rats that is followed by a burst of hepatocyte proliferation results in an increase in serum AFP (for review, see Sell and Becker, 1978). The decreased AFP synthesis observed at high cell densities is directly proportional to the level of intracellular AFP messenger RNA (Innis and Miller, 1979), suggesting that inhibition of transcription of the AFP gene occurs in nonproliferating cells. Synthesis of AFP, however, is not coupled to DNA synthesis in cultured cells and begins to occur after partial hepatectomy in hepatocytes that have not yet begun to proliferate (Guillouzo et al., 1979; Engelhardt et al., 1976).

There is marked amino acid sequence homology between AFP and albumin in the human (Ruoslahti and Terry, 1976) and in the mouse (Gorin et al., 1981; Law and Dugaiczyk, 1981). AFP is about 4% carbohydrate, and AFP from human (Rouslahti et al., 1978) and rat (Bayard and Kerckaert, 1981) amniotic fluid are electrophoretically heterogeneous due to variation in carbohydrate composition. This was determined by lectin affinity, neuraminidase digestion, and carbohydrate structural analyses. Molecular variants of AFP have also been detected in serum of rats (Smith et al., 1977) and patients (Alpert et al., 1972) with hepatomas. Cloned cell lines derived from Morris hepatoma 7777 produce two variants based on affinity for concanavalin A (Con A), and each of the Con A affinity variants has two electrophoretic variants (McMahon et al., 1977). These four molecular variants continue to be produced during serial passage in culture and are similar to the forms produced in hepatoma-bearing rats *in vivo*. The patterns of production of Con A-reactive and Con A-nonreactive forms, however, are different in various primary and transplantable rat hepatomas from that observed during normal neonatal development: The percentage of the total

AFP not reactive with Con A remains constant at 42 to 45% from birth through 4 weeks of age in neonatal rat serum, whereas the percentages of Con A-nonreactive AFP in the sera of hepatoma-bearing rats varies from 11 to 64% among different tumors (Smith et al., 1977). The percentage of the Con A-nonreactive molecular variant is characteristic for each hepatoma line. These data suggest that the oligosaccharide processing of AFP may be different in hepatomas than in normal neonatal liver.

Antibodies directed against AFP have been reported to kill AFP-producing mouse hepatoma cells in culture (Allen and Ledford, 1977). The cytotoxic effect paralleled the secretion of AFP in synchronized cultures and was specific in that the antiserum did not kill AFP-nonproducing mouse neuroblastoma cells. This is somewhat surprising because AFP is a secretory protein rather than a cell surface-associated protein. Suppression of mouse hepatoma growth *in vivo* after administration of anti-AFP has also been reported (Mizejewski and Allen, 1974).

II. CELL MEMBRANE-ASSOCIATED GLYCOPROTEINS AND GLYCOLIPIDS

A number of changes in the biochemical composition of cell membrane glycoproteins and glycolipids have been observed in malignantly transformed cells in culture. A number of these altered membrane components have the potential to be exploited as tumor markers in cancer diagnosis. Most prominent among these alterations are (i) the appearance of glycoprotein-containing sialosyl oligosaccharides unique to transformed cells, (ii) the appearance of new or altered high molecular weight glycoproteins, (iii) incomplete processing and/or synthesis of the carbohydrate portion of membrane glycoproteins and glycolipids, and (iv) the loss of certain high molecular weight glycoproteins from the cell surface. Each of these will be described in some detail.

Differences between untransformed and SV40-transformed 3T3 cells in the overall carbohydrate content of their cell surface macromolecules have been observed (Sakiyama and Burge, 1972). Some of these differences relate to the relative amounts of sialic acid or other sugars that make up the carbohydrate portion of membrane glycoproteins and glycolipids. One type of oligosaccharide chain is linked to asparagine residues in glycoproteins and another is linked to serines. Both of these chains contain terminal sialic acids that provide a certain amount of negative charge to the molecules. Thus, changes in sialic acid content could affect the overall surface charge on cell membranes. Various glycosyltransferases and glycosidases are involved in the synthesis and processing of these molecules, and alterations of any of the steps in these pathways could lead to changes in the relative carbohydrate content of glycoproteins of transformed cells.

A gradual increase of membrane-associated fucose-containing glycopeptides has been observed in clones of hamster embryo cells undergoing malignant transformation after infection with polyoma virus (Glick et al., 1974). The amount of this specific group of glycopeptides increased proportionally to the ability of transformed hamster cells to form tumors after injection into animals. Furthermore, the cells derived from progressively growing tumors formed by the injection of these cells had an abundance of this group of glycopeptides. Further analysis of glycopeptides from virally transformed hamster cells indicated that they also contained increased amounts of sialic acid, mannose, and galactose, suggesting that glycopeptides containing more carbohydrates are formed after viral transformation of cells. This could come about by an alteration in the processing pathway for the oligosaccharides of glycoproteins. Ogata et al. (1976) found that the fucose-containing large glycopeptides prepared by protease digestion of polyoma-transformed BHK cells contain more of the "complex" type of oligosaccharide chains made up of N-acetylglucosamine, galactose, and sialic acid than untransformed cells. They propose that there are more branches of these complex-type sugar chains attached to mannose residues, suggesting that the key step leading to the formation of the large glycopeptides of transformed cell membranes is the addition of entire sialic acid-Gal-GlcNAc chains to glycoproteins. Highly malignant melanoma cells also contain more neuraminidase-accessible total cell surface sialic acid and a greater concentration of sialic acid-Gal and sialic acid-GalNAc residues than their less malignant counterparts (Yogeeswaran et al., 1978). It has also been observed that mouse embryonal teratocarcinoma cells have large fucose-containing glycopeptides that disappear from the differentiated cell types that develop from the teratocarcinoma cells, suggesting that the large, fucose-containing glycopeptides are characteristic of undifferentiated cells (Muramatsu et al., 1978).

Fibronectin, a 440,000 molecular weight glycoprotein that is located on the external surface of many eukaryotic cells and forms a fibrillar matrix around cells, is decreased in content in certain cell types during malignant transformation in culture (Yamada and Olden, 1978). Some cultured tumor cells synthesize fibronectin and shed it from the cell surface at a rapid rate. For example, mouse Ehrlich ascites cells synthesize and release large amounts of fibronectin into the culture medium, and when these cells are injected into mice, plasma fibronectin levels rise in concert with proliferation of the ascites tumor cells in vivo (Zardi et al., 1979). These data are consistent with the observation that cancer patients with a variety of malignancies have elevated plasma levels of fibronectin (Parsons et al., 1979). As noted above for other tumor cell-associated glycoproteins, glycosylated fragments obtained from transformed hamster cell fibronectin are larger than those obtained from normal hamster cells in culture (Wagner et al., 1981). The fibronectin glycopeptides obtained from transformed cells have more oligosaccharide branches per core and a higher degree of sialylation than those obtained from untransformed cells.

Some unique high molecular weight species of glycoproteins have been isolated from transformed cells. Whether their appearance is due to an alteration in the protein or carbohydrate portions of the molecules or both is not yet clear, but they appear to have an unusual oligosaccharide composition. Bramwell and Harris (1978) have identified a membrane protein with a relative molecular weight of 100,000 in a wide range of tumorigenic transformed cells. The glycoprotein from transformed cells has a higher affinity for Con A than for wheat germ lectin, whereas the reverse is true for the analogous glycoprotein extracted from untransformed cells. This suggests that the tumor cell derived 100,000 MW glycoprotein has more of the "high-mannose" type oligosaccharide chains than its normal cell counterpart and, thus, may be a less fully processed molecule than that from normal cells. Koyama *et al.* (1979) have found, in highly malignant clones of a rat fibrosarcoma, a 150,000 MW glycoprotein that has less sialic acid-containing oligosaccharide than does the analogous molecule from less tumorigenic clones. These data appear to be in contrast to what was noted above from transformed cell lines that have a higher proportion of sialic acid-containing glycopeptides. The explanation for this inconsistency may be that transformed cells can have a variety of alterations in their oligosaccharide synthesis and processing pathways and, depending on the transformed cell type and its state of differentiation, different patterns may emerge. One of the earliest changes may be in the addition of extra sialic acid-containing oligosaccharides, whereas in more highly malignant, less well differentiated cells a primary alteration may be in the processing pathway such that a majority of the cells' glycoproteins contain high mannose type oligosaccharides, representing less "mature" molecules. Perhaps the rapidly dividing, highly malignant cell type does not have time to finish the processing of its oligosaccharide-containing glycoproteins and to add the terminal complex oligosaccharides to these molecules before it is time to divide again. In support of this idea, Muramatsu *et al.* (1976) have reported that proliferating human fibroblasts predominantly contain glycopeptides with the less "mature" high mannose oligosaccharides, but the cells of nonproliferating cultures have an increased percentage of complex oligosaccharides, suggesting that the processing of glycosylated proteins decreases in rapidly proliferating cells.

The most common change in glycolipids of transformed cells is the production of more simplified neutral glycolipids, gangliosides, and fucolipids (Brady and Fishman, 1974; Hakomori, 1975). This appears to come about as a result of the block of particular glycosyltransferases rather than enhanced glycosidase activity (Hakomori, 1975). A typical change in gangliosides of transformed cells is a relative accumulation of an early precursor in the gangliosides' biosynthetic pathway and a decrease in the more complex gangliosides (Brady and Fishman, 1974). Comparative studies of the neutral fucolipids and fucogangliosides of normal rat liver cells and rat hepatoma cells indicate that some rat hepatomas contain more neutral fucolipids than normal liver cells and have some species of fucolipid and fucogangliosides not seen in normal liver cells (Baumann et al., 1979). In addition, the accumulation of a novel fucolipid, α-L-fucopyranosylceramide, has been observed in human colon carcinoma cells, and its concentration per milligram of cellular protein appears to be related to degree of malignancy, since tumor tissue taken from metastatic sites had higher levels than localized colon carcinomas (Watanabe et al., 1976).

Most of the data cited above support the concept that a change in glycosylation patterns of membrane glycoproteins and glycolipids occurs during malignant transformation. In general, these alterations reflect a trend toward increased simplification of the glycosylated products, i.e., an accumulation of less fully processed glycoproteins and glycolipids. In addition, certain unique glycosylated molecules appear to be present in transformed cells. These latter products may result from an increased number of glycosylation sites in the protein or lipid or from the addition of extra oligosaccharide chains to the carbohydrate core structure.

The increased release of surface components by cells after their transformation suggests that they shed their cell surface at a faster rate than untransformed cells. This could come about because of increased cell membrane turnover, perhaps related to the increased rate of cell division and cell loss that many malignant cultures and tissues have, or because of increased degradative enzymes that many transformed malignant cells contain. Elevated glycosidases and protease activities have been found in malignant cells, and this may be one of the mechanisms by which cell surface components are released. Whatever the exact mechanism, the phenomenon is widespread enough among malignant cells to indicate that it is a common characteristic of the malignant phenotype. From a practical point of view, this effect results in the production of "biological markers" of cancer that can be taken advantage of clinically. For example, the presence of cell surface components, such as carcinoembryonic antigen, alkaline phosphatase isoenzymes, and various other glycoproteins in the blood of cancer patients, can be used to indicate the degree of tumor spread and tumor burden. These molecules are attached to the cell surface or are components of the cell membrane, and their elevation in the blood of cancer patients appears to reflect a phenomenon similar to that observed in cell culture studies.

III. HORMONES

A. Adrenocorticotropin (ACTH) and β-Lipoprotein (β-LPH)

Production of the precursor molecule containing the amino acid sequences for ACTH and β-endorphin (pro-opiomelanocortin) and various processed intermediates of this prohormone molecule have been detected in cultures of mouse

(Eipper *et al.*, 1976; Mains *et al.*, 1977; Roberts and Herbert, 1977; Giagnoni *et al.*, 1977) rat (Bourassa *et al.*, 1978), and human (Miller *et al.*, 1980) pituitary tumors cells and in cultures of human small cell lung carcinoma cells (Bertagna *et al.*, 1978a,b; Sorenson *et al.*, 1981).

In several animal species, including man, ACTH, β-LPH, endorphin, and melanotropin arise from the common precursor pro-opiomelanocortin (Mains *et al.*, 1977; Roberts and Herbert, 1977; Nakanishi *et al.*, 1977; Liotta *et al.*, 1979, 1980; Miller *et al.*, 1980). In a cultured mouse pituitary tumor line, AtT 20, that has served as a model system for many of these studies, the precursor protein of about 30,000 MW is cleaved in two places to yield ACTH intermediates and β-LPH. The ACTH intermediates are further processed to produce ACTH and a large amino-terminal fragment lacking known biological function; β-LPH is further processed to β-endorphin and γ-LPH (Mains and Eipper, 1978; Roberts *et al.*, 1978). In mouse AtT 20 cells, pro-opiomelanocortin is glycosylated, adding further complexity to the intracellular and secreted forms derived from the precursor molecule. There is evidence that two forms of ACTH are secreted from these cells: one of 4500 MW corresponding to standard ACTH and another of about 13,000 MW that is a glycosylated form of ACTH (Phillips *et al.*, 1981). In organ cultures of human pituitary adenomas, a precursor molecule similar to that of AtT 20 cells has been observed; however, glycosylation patterns are different in human adenomas in that only the amino-terminal fragment is glycosylated in human cells and not the ACTH sequence (Miller *et al.*, 1980). There appears to be a processing error in human tumors ectopically producing ACTH because an additional intermediate form of about 22,000 MW is detected in the plasma of cancer patients with the ectopic ACTH syndrome (Ratter *et al.*, 1980).

Cultured mouse pituitary tumor cells release ACTH, β-LPH, and β-endorphin, and their release appears to be coordinately regulated (Allen *et al.*, 1978). Agents that stimulate release of ACTH (e.g., hypothalamic extract and vasopressin) or inhibit release (e.g., dexamethasone) affect the release of endorphins in the same way, and the proportions of different forms of ACTH and endorphin are similar in culture medium and cell extracts. This is consistent with what is seen *in vivo* in normal animals: β-endorphin and ACTH are concomitantly secreted in increased amounts by the adenohypophysis in response to acute stress or adrenalectomy, and the secretion of both is inhibited by administration of dexamethasone (Guillemin *et al.*, 1977). A continuous line of human small cell lung carcinoma cells also secretes ACTH, LPH, and β-endorphin concimitantly into the culture medium (Bertagna *et al.*, 1978a,b). β-LPH and small amounts of β-endorphin are also present in plasma of cancer patients with ectopic ACTH syndrome, suggesting that tumor cells ectopically producing ACTH *in vivo* also release other processed products of the precursor molecule (Ratter *et al.*, 1980).

B. Calcitonin

Calcitonin (CT) is produced and secreted eutopically by medullary carcinomas of the thyroid (MTC) and ectopically by a variety of human cancers, including carcinomas of the lung, breast, and gastrointestinal tract (Silva and Becker, 1978). Cell culture lines that produce CT have been established from a variety of tumors, including rat medullary thyroid carcinoma (Boorman *et al.*, 1974; Gagel *et al.*, 1980); human breast carcinomas (Coombs *et al.*, 1976), lung carcinomas (Coombs *et al.*, 1976; Bertagna *et al.*, 1978b), osteosarcoma, fibrosarcoma, and carcinomas of the testis and urinary bladder (Neuwald *et al.*, 1980).

The predominant form of CT synthesized and secreted by cultured rat MTC cells migrates like standard calcitonin monomer on gel filtration columns, but a smaller amount of a larger immunoreactive CT form is also observed (Gagel *et al.*, 1980). Calcitonin secretion by these cells in culture is stimulated by $CaCl_2$, KCl, and glucagon. Immunoreactive CT secreted by a poorly differentiated human epidermoid lung carcinoma in culture is heterogeneous in size and has charge and immunological differences from CT monomer (Coombs *et al.*, 1976). Calcitonin also appears to be synthesized as a higher molecular weight, glycosylated precursor that is processed to the mature hormone of 3,500 MW in rat MTC cells (Amara *et al.*, 1980; Jacobs *et al.*, 1981) and in human MTC and normal thyroid tissue (Dermody *et al.*, 1981). Higher molecular weight forms of CT are detected in the media of cultured human cancer cells (Coombs *et al.*, 1976; Bertagna *et al.*, 1978b) and also circulate in the blood (Roos *et al.*, 1980; Dermody *et al.*, 1981) and are excreted in the urine (Silva and Becker, 1978) of cancer patients. In both the blood and urine, antisera directed against the C-terminal end of CT gives higher values than antisera against the N-terminal end of CT (Dermody *et al.*, 1980; Silva and Becker, 1978).

C. Glycoprotein Hormones

Production of the glycoprotein hormone α subunit, of chorionic gonadotropin (hCG) β subunit, and of complete hCG (α : β dimer) has been observed in a wide variety of cultured human cancer cells. Eutopic production occurs in cultured trophoblastic cell lines (Pattillo and Gey, 1968; Pattillo *et al.*, 1971; Bridson *et al.*, 1971; Chou, 1978a), and ectopic production has been observed in cell lines derived from human carcinomas, sarcomas, glioblastomas, melanomas, and SV40 transformed fibroblasts (Rabson *et al.*, 1973; Kameya *et al.*, 1975; Pattillo *et al.*, 1977; Braunstein *et al.*, 1978; Ruddon *et al.*, 1979a; Rosen *et al.*, 1980a). The production of the α and β subunits of hCG is discordant in many cultured human cancer cell lines, with a number of cells producing only α subunit, some producing both α and β but with excess free α subunit, and some producing balanced amounts of α and β, although the latter appears to be the exception rather than the rule (Tashjian et al., 1973; Lieblich et al., 1976; J. Y. Chou et al., 1977; Hussa, 1977; Ruddon et al., 1979a; Rosen et al., 1980a). One cell line, the glioblastoma line CBT, produces only free hCG β subunits in culture (Ruddon et al., 1980a; Rosen et al., 1980b).

A few cell lines produce the pituitary glycoprotein hormones FSH and LH (Ghosh and Cox, 1977; Ruddon et al., 1979a; Rosen et al., 1980a), but ectopic production of these hormones by human cancer cells is a much less frequent phenomenon than production of α subunit, hCG-β subunit, or complete hCG. Ectopic production of TSH β subunit or TSH by cultured human tumor cells appears to be a rare phenomenon.

The intracellular and secreted forms of hCG subunits produced by cultured human tumor cells have been characterized, and the rate-limiting step in secretion has been determined (Ruddon et al., 1979b, 1980a, 1981; Dean et al., 1980). Both eutopically and ectopically hCG-producing cells synthesize 18,000 and 15,000 MW forms of α subunit and 24,000 and 18,000 MW forms of β subunit; these forms represent intermediates in the processing pathway to mature, secreted forms of α (22,000 to 24,000 MW) and β (34,000 MW). The forms that accumulate intracellularly contain "high-mannose" oligosaccharides with eight mannose residues, and the rate-limiting step in processing and secretion appears to be at an α-mannosidase step that cleaves the high mannose oligosaccharide to a mannose$_3$-containing form (Ruddon et al., 1981). To this form are then added the terminal sugar residues, including sialic acid, prior to secretion (Ruddon et al., 1981). Two forms of α subunit are secreted by cultured human cells: a large free α subunit (~24,000 MW) and a smaller α subunit (~22,000 MW) that is immunoprecipitated as part of the hCG α:β dimer (Hussa, 1977; Benveniste et al., 1979; Dean et al., 1980; Ruddon et al., 1981). Only one form of hCG β subunit is secreted.

Secretion of hCG subunits from cultured cells can be stimulated by addition of sodium butyrate (Ghosh and Cox, 1976; Chou, 1978b; Kanabus et al., 1978; Hussa et al., 1978; Ruddon et al., 1979a), dibutyryl cAMP (Hussa et al., 1978; Barker and Isles, 1978; Ruddon et al., 1980c; Rosen et al., 1980b), phosphodiesterase inhibitors (Hussa et al., 1977), epidermal growth factor (Benveniste et al., 1978a; Huot et al., 1981), concanavalin A (Benveniste et al., 1978b), and certain inhibitors of DNA synthesis (Azizkhan et al., 1979). The eutopic production of hCG subunits by normal or malignant trophoblastic cells in culture and the ectopic production by nontrophoblastic cell lines have been reported to be differentially affected by dibutyryl cAMP and sodium butyrate. For example, dibutyryl cAMP induces eutopic secretion of hCG subunits by BeWo, JAR, and JEG-3 choriocarcinoma lines, whereas exposure of these cells to butyrate has no effect or inhibits secretion (J. Y. Chou et al., 1977; Barker and Isles, 1978; Hussa et al., 1978; Ruddon et al., 1980c). On the other hand, butyrate induces ectopic secretion of one or both hCG subunits in a variety of nontrophoblastic cell lines (Lieblich *et al.*, 1977a; J. Y. Chou *et al.*, 1977; Tralka *et al.*, 1979; Hussa *et al.*, 1978; Kanabus *et al.*, 1978; Ruddon *et al.*, 1980c)but dibutyryl cAMP induces little or no secretion by ectopic hCG-producing lines. An exception to these observations is the glioblastoma cell line CBT that produces only the hCG-β subunit in culture (Ruddon *et al.*, 1980a). In this cell line, dibutyryl cAMP stimulates β subunit secretion, whereas sodium butyrate does not (Ruddon *et al.*, 1980c; Rosen *et al.*, 1980b).

Injection of hCG subunit-producing human cell lines into nude mice results in continued production of the subunit(s) made *in vitro*, thus providing a useful *in vivo* model in which to study modulation of hCG secretion, regulation of hCG pharmacokinetics, and the relationship of hCG levels to tumor burden and metastases (Kameya *et al.*, 1976; Lieblich *et al.*, 1977b; Ruddon *et al.*, 1980b).

IV. ENZYMES

A. Alkaline Phosphatase

The isoenzymes of human alkaline phosphatase (AP) vary in their relative amounts during development and in different tissues of the adult organism. Various iosenzymes of AP are also found in malignant tumors (Fishman, 1974; Fishman *et al.*, 1976). These isoenzymes appear to be the products of one of three structural genes coding for placental, intestinal, or liver–bone isoenzymes (McKenna *et al.*, 1979). The heat stable, phenylalanine-sensitive placental form of AP, the so-called Regan isoenzyme (Fishman *et al.*, 1968), has been found at elevated levels in pregnancy sera and in the blood of patients with various types of cancer (Fishman *et al.*, 1968; Fishman, 1974). It is also expressed by several tumor-derived cell lines in culture, including cells lines derived from cancers of the cervix (Singer and Fishman, 1974; Singer *et al.*, 1980), lung (Ludueña *et al.*, 1974), nasopharynx (Ludueña and Sussman, 1976), liver (Ludueña *et al.*, 1977), gastrointestinal tract (Singer *et al.*, 1980), and urinary bladder (Herz and Koss, 1979) as well as from choriocarcinomas (Kameya *et al.*, 1975; Speeg *et al.*, 1977; Sakiyama *et al.*, 1978; Neuwald and Brooks, 1980) and osteosarcomas (Singh *et al.*, 1978).

Other developmental phase-specific alkaline phosphatases also appear in neoplastic cells. Three such isoenzymes appear at different stages of development in the human placenta (Fishman *et al.*, 1976). Phase 1 (6 to 10 weeks) has two heat-sensitive, L-homoarginine-inhibited activity bands on electrophoresis. One of the bands possesses antigenic determinants of liver–bone type AP. Phase 2 (11 to 13 weeks) has a mixture of phase 1 and phase 3 isoenzyme components. Phase 3 (14 weeks to term) exhibits 2 bands of activity characteristic of term placenta and containing the Regan isoenzyme. These three isoenzyme patterns have been found in certain human cancer cell lines (Fishman et al., 1976).

There are some biochemical properties common to the alkaline phosphatase isoenzymes that have been purified and characterized (Harkness, 1968; Ludueña and Sussman, 1976). They are glycoproteins composed of dimers of similar molecular weight. The native term placental enzyme is a dimer of 125,000 MW, consisting of two subunits of about 64,000 MW. Human liver alkaline phosphatase has also been purified and is a dimer of about 135,000 MW composed of subunits of equal size (Badger and Sussman, 1976). The liver and placental forms of AP differ in molecular weight, amino-terminal sequence, peptide maps, and amino acid composition. This suggests that they are different in primary structure and are coded by different structural genes (see also Chapter 1).

Alkaline phosphatases contained in cultured human cancer cell lines have also been biochemically characterized. For example, both HeLa cervical carcinoma cells (Tan and Aw, 1971) and KB nasopharyngeal carcinoma cells (Ludueña and Sussman, 1976) contain a form of AP that is similar to placental AP in molecular weight, electrophoretic mobility, and immunological properties. KB cells, however, contain both a 64,000 MW subunit of AP that appears to be identical in protein structure to placental AP and, in addition, a 72,000 MW subunit that has several of the properties of placental AP subunit but lacks one peptide present in the placental AP map and contains more carbohydrate than placenta AP (Ludueña and Sussman, 1976).

Alkaline phosphatase activity is inducible in a variety of human tumor lines by corticosteroids, dibutyryl cAMP, sodium butyrate, and 5-bromodeoxyuridine (BUdR) (Griffin et al., 1974; Hamilton et al., 1979). Interestingly, the induction of synthesis of placental AP, but not of liver AP, can be produced by BUdR and dbcAMP in cultured choriocarcinoma cells, indicating that the structural genes coding for placental and liver AP's are regulated in a noncoordinate manner (Hamilton et al., 1979). Moreover, both AP genes respond differently to these inducing agents than the gene for glycoprotein hormone α subunit (Hamilton et al., 1979).

B. γ-Glutamyltransferase (GGT)

γ-Glutamyltransferase appears to be an oncodevelopmental enzyme marker at least in rodents. It is present in fetal rat liver, virtually absent in adult liver, and elevated in premalignant lesions of liver and in hepatomas (Fiala et al., 1976). Moreover, GGT activity is not found in the normal epidermis of adult mice; however, it is elevated in skin carcinomas induced with 7,12-dimethylbenz[a]anthracene (DeYoung et al., 1978). In man, GGT serum levels are a sensitive indicator of hepatic metastasis (Schwartz, 1973). GGT is produced in high amounts by cultured human carcinomas of the breast, pancreas, urinary bladder, and thyroid (Neuwald et al., 1980).

In primary cultures of rat hepatocytes derived from putative premalignant and malignant lesions of chemical carcinogen-treated rats, GGT activity is detected in those hepatocytes associated with the macroscopic lesions but not in those from morphologically normal areas of liver from the same animals (Laishes et al., 1978). Huberman et al. (1979) found that GGT activity was prominent (25 to 90% positive cells) in 3 of 5 malignant rat liver cell lines, but no activity was observed in 9 fibroblast or 4 nonmalignant epithelial cell lines. GGT activity also is reported to be an accurate predictor of in vivo tumorigenicity of chemically transformed cultured rat hepatocytes and of established hepatoma lines (San et al., 1979).

C. Glycosyltransferases

As indicated in Section II above, malignant cells are characterized by alterations of cell surface glycoproteins and glycolipids. Malignancy-associated alterations of cellular glycosyltransferase activities is one of the ways that changes in the carbohydrate composition of these surface molecules could occur. Elevated serum levels of various glycosyltransferases, e.g., galactosyl-, fucosyl-, and sialytransferase, have been seen in patients with various cancers, and elevated activity of glycosyltransferases has also been observed in animal and human tumors as compared to normal tissue or cells (Bosmann and Hall, 1974; Bernacki and Kim, 1977; Plotkin et al., 1979). The increased serum levels of glycosyltransferase in cancer patients could be the result of increased production and/or release of enzyme from tumor cell surfaces. Indeed, Klohs et al. (1981) have demonstrated that the release of galactosyltransferase from cultures of transformed cell lines closely paralleled growth rate of the cultured cells. Podolsky et al. (1978) have reported that a galactosyltransferase glycopeptide acceptor purified from human malignant effusions inhibited the growth of virally transformed hamster cells and cultured human breast and pancreatic carcinoma cells but not of untransformed cells in culture. The mechanism of this inhibition is not clear, but it appears to correlate with inhibition of shedding of tumor-derived enzyme by the glycopeptide acceptor.

D. Proteases

It has long been known that certain types of malignant tumors produce and release proteolytic enzyme activities. In 1925, Fischer observed that explants of virally induced tumors in chickens caused lysis of plasma clots, whereas explants of normal connective tissues did not. A number of years later, increased fibrinolytic activity was found in human sarcomas (Cliffton and Grossi, 1955). The relationship between fibrinolytic activity and transformation has been studied in detail by Reich and his colleagues. They established that fibroblast cultures derived from chicken, hamster, mouse, and rat embryos produced an enzymatic activity capable of hydrolyzing fibrin after the cells were transformed by oncogenic DNA or RNA viruses (Unkeless et al., 1973; Ossowski et al., 1973). The fibrinolytic activity was not detected in untransformed embryo or 3T3 cell cultures, but later was found in other untransformed 3T3 lines during their log growth phase (I.-N. Chou et al., 1977). The activity was also lower in cultures of cells infected with a variety of nononcogenic viruses. It was subsequently shown that the fibrinolytic activity was produced by the interaction of the serum proteolytic proenzyme plasminogen, present in the cell culture medium, and a factor called plasminogen activator elaborated by the transformed cells (Unkeless et al., 1974; Quigley et al., 1974). Plasminogen activator (PA) is itself a protease of the serine-active site type that activates plasminogen to the active fibrinolytic enzyme plasmin by cleavage of an arginine–valine bond in the carboxy-terminal portion of the proenzyme molecule. Plasmin is a normal plasma protein that has a role in the body in dissolving fibrin clots.

Extracts from certain human tumors have higher levels of PA than extracts of adjacent control tissues (Nagy et al., 1977). Two types of PA activity have been observed in human tumor cells (Vetterlein et al., 1979; Wilson et al., 1980). One of them is about 55,000 MW and is immunologically identical to a form of PA called urokinase that is excreted in normal human urine. The other form is about 73,000 MW and differs from urokinase in its immunological specificity and in its sensitivity to inhibition by glucocorticoids. Glucocorticoid treatment of PA-producing human cells inhibits the production of urokinase-like activity but not of the 73,000 MW form (Roblin and Young, 1980). These data suggest that there may be at least two genes coding for PA-like activity in human cells.

A number of groups have attempted to correlate levels of fibrinolytic activity with the phenotypic characteristics of transformed cells. For tumors growing *in vivo*, the release of a protease activity could explain, at least in part, the invasive properties of cancer cells. Some studies have shown a correlation between release of plasminogen activator activity and other phenotypic characteristics of malignant cells. For example, it has been reported that plasminogen activator secretion by SV40 virus-transformed rat embryo cells correlates with loss of anchorage dependence and loss of intracellular actin-containing cables (Pollack and Rifkin, 1975). A link between the ability of cultured malignant human and animal cells to produce PA, grow in soft agar, and form tumors in immunosuppressed hosts has also been observed (Laug et al., 1975). Hamster embryo cells transformed by the chemical carcinogen benzpyrene also have enhanced fibrinolytic activity (Barrett et al., 1977). Christman et al. (1975) have shown that tumorigenicity and production of fibrinolytic activity are coordinately repressed in mouse melanoma cells grown in the presence of 5-bromodeoxyuridine. These data suggest a correlation between production of PA activity and cell transformation; however, such a correlation is not always present, and a number of investigators have reported discrepancies between the two (San et al., 1977; Wolf and Goldberg, 1978).

Other proteolytic activities also are produced and released by malignant cells. For example, Koono et al. (1974) have extracted a factor from rat ascites hepatoma cells that stimulates the release of a protease from rat hepatoma cells but not from normal liver cells. Intradermal injection of this protease induced extravascular migration of circulating tumor cells and formation of metastases. An additional cell surface proteolytic enzyme activity, distinct from PA or plasmin, has been found in cultured normal and transformed mouse epidermal cells, but transformed cells contained 3–4 times more surface proteolytic activity than normal cells (Hatcher et al., 1976). This surface protease activity was correlated with the rate of cell proliferation of the cultured cells, being higher in mitotic than in postmitotic cells. Finally, secretion of the lysosomal protease cathepsin B has been reported to be significantly higher in explants of human mammary carcinomas than in explants of benign fibroadenomas or normal breast tissue (Poole et al., 1978).

V. TUMOR-DERIVED GROWTH FACTORS

The discovery of tumor-derived growth factors came about as the result of experiments showing that mouse 3T3 cells transformed with murine or feline sarcoma viruses rapidly lost their ability to bind epidermal growth factor (EGF), whereas cells infected with nontransforming RNA viruses maintained normal levels of cell surface EGF receptors (Todaro et al., 1976). These initial results suggested that the sarcoma virus genome produced something that altered EGF receptors. Later, however, it was found that murine sarcoma virus-transformed mouse fibroblasts produce a polypeptide growth factor that competes for binding of EGF on cell surfaces (DeLarco and Todaro, 1978). This factor was called sarcoma growth factor (SGF). SGF is a 9000 MW, heat-stable, trypsin-sensitive polypeptide that stimulates proliferation of transformed and untransformed fibroblasts. SGF competes with EGF for binding to EGF receptors, but it has a different molecular weight and is immunologically distinct from EGF. SGF has the interesting property of being able to promote anchorage independent growth in cultures of normal fibroblasts and thus to confer on normal cells properties associated with the transformed phenotype. This phenomenon is reversible so that after the SGF is removed from the growth medium, the cells regain normal growth properties. Thus, SGF appears to be a growth factor produced specifically by transformed cells and capable of stimulating their proliferation, suggesting that neoplastic cells are capable of "autostimulation" by producing their own growth factors. In this way, they could presumably escape the negative feedback systems of the normal host that control the production and release of endogenous hormones and growth factors.

Cultured cell lines derived from human tumors also produce SGF-like growth factors that can transform fibroblast and epithelial cells in culture (Todaro *et al.*, 1980). These proteins are heat stable, produce transformation-related morphological changes when added to the growth medium of normal rat or human fibroblasts, enable cells to grow in soft agar, and bind to EGF receptors on cell surfaces. Addition of human tumor cell-derived transforming growth factors to cultures of human carcinoma cells induces phosphorylation of specific tyrosine acceptor sites in the 160,000 MW EGF receptor (Reynolds *et al.*, 1981).

Other growth factors are also produced by cultured human cancer cells. For example, cultured human fibrosarcoma cells produce multiplication-stimulating activity (MSA) (Marquardt *et al.*, 1980), and osteosarcoma cells produce a factor similar to platelet-derived growth factor (Heldin *et al.*, 1980).

It will be of interest to see whether such factors are produced in detectable amounts in cancer patients and whether they can be used clinically as tumor markers in diagnosis and therapeutic management.

REFERENCES

Allen, R. G., Herbert, E., Hinman, M., Shibuya, H., and Pert, C. B. (1978). *Proc. Natl. Acad. Sci. U.S.A.* **75**, 4972–4976.
Allen, R. P., and Ledford, B. E. (1977). *Cancer Res.* **37**, 696–701.
Alpert, E., Drysdale, J. W., Isselbacher, K. J., and Schur, P. H. (1972). *J. Biol. Chem.* **247**, 3792–3798.
Amara, S. G., Rosenfeld, M. G., Birnbaum, R. S., and Roos, B. A. (1980). *J. Biol. Chem.* **255**, 2645–2648.
Attallah, A. M., Needy, C. F., Noguchi, P. D., and Elisberg, B. L. (1979). *Int. J. Cancer* **24**, 49–52.
Azizkhan, J. D., Speeg, K. V., Jr., Stromberg, K., and Goode, D. (1979). *Cancer Res.* **39**, 1952–1959.
Badger, K. S., and Sussman, H. H. (1976). *Proc. Natl. Acad. Sci. U.S.A.* **75**, 2201–2205.
Barker, H., and Isles, T. E. (1978). *Br. J. Cancer* **38**, 158–162.
Barrett, J. C., Crawford, B. D., Grady, D. L., Hester, L. D., Jones, P. A., Benedict, W. F., and Tso, P. O. P. (1977). *Cancer Res.* **37**, 3815–3823.
Baumann, H., Nudelman, E., Watanabe, K., and Hakomori, S.-I. (1979). *Cancer Res.* **39**, 2637–2643.
Bayard, B., and Kerckaert, J.-P. (1981). *Eur. J. Biochem.* **113**, 405–414.
Becker, J. E., de Néchaud, B., and Potter, V. R. (1976). *In* "Oncodevelopmental Gene Expression" (W. H. Fishman and S. Sell, eds.), pp. 259–270. Academic Press, New York.
Benveniste, R., Speeg, K. V., Jr., Carpenter, G., Cohen, S., Lindner, J., and Rabinowitz, D. (1978a). *J. Clin. Endocrinol. Metab.* **46**, 169–172.

Benveniste, R., Speeg, K. V., Jr., Long, A., and Rabinowitz, D. (1978b). *Biochem. Biophys. Res. Commun.* **84,** 1082–1087.
Benveniste, R., Conway, M. C., Puett, D., and Rabinowitz, D. (1979). *J. Clin. Endocrinol. Metab.* **48,** 85–91.
Bernacki, R. J., and Kim, U. (1977). *Science* **195,** 577–580.
Bernhard, H. P., Darlington, G. J., and Ruddle, F. H. (1973). *Dev. Biol.* **35,** 83–96.
Bertagna, X. Y., Nicholson, W. E., Sorenson, G. D., Pettengill, O. S., Mount, C. D., and Orth, D. N. (1978a). *Proc. Natl. Acad. Sci. U.S.A.* **75,** 5160–5164.
Bertagna, X. Y., Nicholson, W. E., Pettengill, O. S., Sorenson, G. D., Mount, C. D., and Orth, D. N. (1978b). *J. Clin. Endocrinol. Metab.* **47,** 1390–1393.
Boorman, G. A., Heersche, J. N. M., and Hollander, C. F. (1974). *JNCI, J. Natl. Cancer Inst.* **53,** 1011.
Bosmann, H. B., and Hall, T. C. (1974). *Proc. Natl. Acad. Sci. U.S.A.* **71,** 1833–1837.
Bourassa, M., Scherrer, H., Pezalla, P. D., Lis, M., and Chrétien, M. (1978). *Cancer Res.* **38,** 1568–1571.
Brady, R. O., and Fishman, P. H. (1974). *Biochem. Biophys. Acta* **355,** 121–148.
Bramwell, M. E., and Harris, H. (1978). *Proc. R. Soc. London, Ser. B* **201,** 87–106.
Braunstein, G. D., Kamdar, V. V., Kanabus, J., and Rasor, J. (1978). *J. Clin. Endocrinol. Metab.* **47,** 326–332.
Bridson, W. E., Ross, G. T., and Kohler, P. O. (1971). *J. Clin. Endocrinol. Metab.* **33,** 145–149.
Carrel, S., Delisle, M.-C., and Mach, J.-P. (1977). *Cancer Res.* **37,** 2644–2650.
Chou, I.-N., O'Donnell, S. P., Black, P. H., and Roblin, R. O. (1977). *J. Cell. Physiol.* **91,** 31–37.
Chou, J. Y. (1978a). *Proc. Natl. Acad. Sci. U.S.A.* **75,** 1854–1858.
Chou, J. Y. (1978b). *In Vitro* **14,** 775–778.
Chou, J. Y., Robinson, J. C., and Wang, C. C. (1977). *Nature (London)* **268,** 543–544.
Christman, J. K., Silagi, S., Newcomb, Z. W., Silverstein, S. C., and Acs, G. (1975). *Proc. Natl. Acad. Sci. U.S.A.* **72,** 47–50.
Cliffton, E. E., and Grossi, C. E. (1955). *Cancer* **8,** 1146–1154.
Coombes, R. C., Ellison, M. L., Easty, G. C., Hillyard, C. J., James, R., Galante, L., Girgis, S., Heywood, L., MacIntyre, I., and Neville, A. M. (1976). *Clin. Endocrinol.* **5,** Suppl., 387–396.
Dean, D. J., Weintraub, B. D., and Rosen, S. W. (1980). *Endocrinology* **106,** 849–858.
DeLarco, J. E., and Todaro, G. J. (1978). *Proc. Natl. Acad. Sci. U.S.A.* **75,** 4001–4005.
de Néchaud, B., Fromont, S., and Bergès, J. (1977). *Biochem. Biophys. Res. Commun.* **79,** 789–795.
Dermody, W. C., Ananthaswamy, R., Rosen, M. A., Perini, F., and Levy, A. G. (1980). *Clin. Chem. (Winston-Salem, N.C.)* **26,** 235–242.
Dermody, W. C., Rosen, M. A., Ananthaswamy, R., McCormick, W. M., and Levy, A. G. (1981). *J. Clin. Endocrinol. Metab.* **52,** 1090–1098.
DeYoung, L. M., Richards, W. L., Bonzelet, W., Tsai, L. L., and Boutwell, R. K. (1978). *Cancer Res.* **38,** 3697–3701.
Drewinko, B., Romsdahl, M. M., Yang, L. Y., Ahearn, J. J., and Trujillo, J. M. (1976). *Cancer Res.* **36,** 467–475.
Egan, M. L., and Todd, C. W. (1972). *JNCI, J. Natl. Cancer Inst.* **49,** 887–889.
Eipper, B. A., Mains, R. E., and Guenzi, D. (1976). *J. Biol. Chem.* **251,** 4121–4126.
Ellison, M. L., Lamb, D., Rivett, J., and Neville, A. M. (1977). *JNCI, J. Natl. Cancer Inst.* **59,** 309–312.
Engelhardt, N. V., Lazareva, M. N., Abelev, G. I., Uryvaeva, I. V., Factor, V. M., and Brodsky, V. Ya. (1976). *Nature (London)* **263,** 146–148.

Fiala, S., Mohindru, A., Kettering, W. G., Fiala, A. E., and Morris, H. P. (1976). *JNCI, J. Natl. Cancer Inst.* **57,** 591–598.
Fischer, A. (1925). *Wilhelm Roux' Arch. Entwicklungsmech. Org.* **104,** 210–261.
Fishman, L., Miyayama, H., Driscoll, S. G., and Fishman, W. H. (1976). *Cancer Res.* **36,** 2268–2273.
Fishman, W. H. (1974). *Am. J. Med.* **56,** 617–650.
Fishman, W. H., Inglis, N. R., Green, S., Antiss, C. L., Ghosh, N. K., Reif, A. E., Rustigian, R., Krant, M. J., and Stolbach, L. L. (1968). *Nature (London)* **219,** 696–699.
Gagel, R. F., Zeytinoglu, F. N., Voelkel, E. F., and Tashjian, A. H., Jr. (1980). *Endocrinology* **107,** 516–523.
Ghosh, N. K., and Cox, R. P. (1976). *Nature (London)* **259,** 416–417.
Ghosh, N. K., and Cox, R. P. (1977). *Nature (London)* **267,** 435–437.
Giagnoni, G., Sabol, S. L., and Nirenberg, M. (1977). *Proc. Natl. Acad. Sci. U.S.A.* **74,** 2259–2263.
Glick, M. C., Rabinowitz, Z., and Sachs, L. (1974). *J. Virol.* **13,** 967–974.
Goldenberg, D. M., Pavia, R. A., Hansen, H. J., and Vandervoorde, J. P. (1972). *Nature (London), New Biol.* **239,** 189–190.
Gorin, M. B., Cooper, D. L., Eiferman, F., Van de Rijn, P., and Tilghman, S. M. (1981). *J. Biol. Chem.* **256,** 1954–1959.
Griffin, M. J., Price, G. H., Bazzell, K. L., Cox, R. P., and Ghosh, N. K. (1974). *Arch. Biochem. Biophys.* **164,** 619–623.
Guillemin, R., Vargo, T., Rossier, J., Minick, S., Ling, N., Rivier, C., Vale, W., and Bloom, F. (1977). *Science* **197,** 1367–1369.
Guillouzo, A., Boisnard-Rissel, M., Bélanger, L., and Bourel, M. (1979). *Biochem. Biophys. Res. Commun.* **91,** 327–331.
Hakomori, S.-I. (1975). *Biochim. Biophys. Acta* **417,** 55–89.
Hamilton, T. A., Tin, A. W., and Sussman, H. H. (1979). *Proc. Natl. Acad. Sci. U.S.A.* **76,** 323–327.
Hammarström, S., Troye, M., Wahlund, G., Svenberg, T., and Perlmann, P. (1977). *Int. J. Cancer* **19,** 756–766.
Harkness, D. R. (1968). *Arch. Biochem. Biophys.* **126,** 503–512.
Hatcher, V. B., Wertheim, M. S., Rhee, C. Y., Tsien, G., and Burk, P. G. (1976). *Biochim. Biophys. Acta* **451,** 499–510.
Heldin, C.-R., Westermark, B., and Wateson, A. (1980). *J. Cell. Physiol.* **105,** 235–246.
Herz, F., and Koss, L. G. (1979). *Arch. Biochem. Biophys.* **194,** 30–36.
Huberman, E., Montesano, R., Drevon, C., Kuroki, T., St. Vincent, L., Pugh, T. D., and Goldfarb, S. (1979). *Cancer Res.* **39,** 269–272.
Huot, R. I., Foidart, J.-M., Nardone, R. M., and Stromberg, K. (1981). *J. Clin. Endocrinol. Metab.* **53,** 1059–1063.
Hussa, R. O. (1977). *J. Clin. Endocrinol. Metab.* **44,** 1154–1162.
Hussa, R. O., Story, M. T., Pattillo, R. A., and Kemp, P. G. (1977). *In Vitro* **13,** 443–448.
Hussa, R. O., Pattillo, R. A., Ruckert, A. C. F., and Scheuermann, K. W. (1978). *J. Clin. Endocrinol. Metab.* **46,** 69–76.
Innis, M. A., and Miller, D. A. (1979). *J. Biol. Chem.* **254,** 9148–9154.
Irlin, I. S., Perova, S. D., and Abelev, G. I. (1966). *Int. J. Cancer* **1,** 337–347.
Jacobs, J. W., Lund, P. K., Potts, J. T., Jr., Bell, N. H., and Habener, J. F. (1981). *J. Biol. Chem.* **256,** 2803–2807.
Kameya, T., Kuramoto, H., Suzuki, K., Kenjo, T., Oshikiri, T., Hayashi, H., and Itakura, M. (1975). *Cancer Res.* **35,** 2025–2032.

Kameya, T., Shimosato, Y., Tumuraya, M., Ohsawa, N., and Nomura, T. (1976). *JNCI, J. Natl. Cancer Inst.* **56**, 325–332.
Kanabus, J., Braunstein, G. D., Emry, P. K., DiSaia, P. J., and Wade, M. E. (1978). *Cancer Res.* **38**, 765–770.
Klohs, W. D., Mastrangelo, R., and Weiser, M. M. (1981). *Cancer Res.* **41**, 2611–2615.
Koono, M., Katsuya, H., and Hayashi, H. (1974). *Int. J. Cancer* **13**, 334–342.
Koyama, K., Nudelman, E., Fukuda, M., and Hakomori, S.-I. (1979). *Cancer Res.* **39**, 3677–3682.
Laishes, B. A., Ogawa, K., Roberts, E., and Farber, E. (1978). *JNCI, J. Natl. Cancer Inst.* **60**, 1009–1016.
Laug, W. E., Jones, P. A., and Benedict, W. R. (1975). *JNCI, J. Natl. Cancer Inst.* **54**, 173–179.
Law, S. W., and Dugaiczyk, A. (1981). *Nature (London)* **291**, 201–205.
Lieblich, J. M., Weintraub, B. D., Rosen, S. W., Chou, J. Y., and Robinson, J. C. (1976). *Nature (London)* **260**, 530–532.
Lieblich, J. M., Weintraub, B. D., Rosen, S. W., Ghosh, N. K., and Cox, R. P. (1977a). *Nature (London)* **265**, 746.
Lieblich, J. M., Rosen, S. W., Weintraub, B. D., Sindelar, W. F., Tralka, T. S., and Rabson, A. S. (1977b). *JNCI, J. Natl. Cancer Inst.* **59**, 1285–1289.
Liotta, A. S., Gildersleeve, D., Brownstein, M. J., and Krieger, D. T. (1979). *Proc. Natl. Acad. Sci. U.S.A.* **76**, 1448–1452.
Liotta, A. S., Loudes, C., McKelvy, J. F., and Krieger, D. T. (1980). *Proc. Natl. Acad. Sci. U.S.A.* **77**, 1880–1884.
Ludueña, M. A., and Sussman, H. H. (1976). *J. Biol. Chem.* **251**, 2620–2628.
Ludueña, M. A., Sussman, H. H., and Rabson, A. S. (1974). *JNCI, J. Natl. Cancer Inst.* **52**, 1705–1709.
Ludueña, M. A., Iverson, G. M., and Sussman, H. H. (1977). *J. Cell. Physiol.* **91**, 119–130.
McKenna, J. J., Hamilton, T. A., and Sussman, H. H. (1979). *Biochem. J.* **181**, 67–73.
McMahon, J. B., Kelleher, P. C., and Smith, C. (1977). *Biochem. Biophys. Res. Commun.* **76**, 1144–1150.
Mains, R. E., and Eipper, B. A. (1978). *J. Biol. Chem.* **253**, 651–655.
Mains, R. E., Eipper, B. A., and Ling, N. (1977). *Proc. Natl. Acad. Sci. U.S.A.* **74**, 3014–3018.
Marquardt, H., Wilson, G. L., and Todaro, G. J. (1980). *J. Biol. Chem.* **255**, 9177–9181.
Miller, W. L., Johnson, L. K., Baxter, J. D., and Roberts, J. L. (1980). *Proc. Natl. Acad. Sci. U.S.A.* **77**, 5211–5215.
Mizejewski, G. J., and Allen, R. P. (1974). *Nature (London)* **250**, 50–51.
Muramatsu, T., Koide, N., Ceccarini, C., and Atkinson, P. H. (1976). *J. Biol. Chem.* **251**, 4673–4679.
Muramatsu, T., Gachelin, G., Nicolas, J. F., Condamine, H., Jakob, H., and Jacob, F. (1978). *Proc. Natl. Acad. Sci. U.S.A.* **75**, 2315–2319.
Nagy, B., Ban, J., and Brdar, B. (1977). *Int. J. Cancer* **19**, 614–620.
Nakanishi, S., Inoue, A., Taii, S., and Numa, S. (1977). *FEBS Lett.* **84**, 105–109.
Neuwald, P. D., and Brooks, M. (1981). *Cancer Res.* **41**, 1682–1689.
Neuwald, P. D., Anderson, C., Salivar, W. O., Aldenderfer, P. H., Dermody, W. C., Weintraub, B. D., Rosen, S. W., Nelson-Rees, W. A., and Ruddon, R. W. (1980). *JNCI, J. Natl. Cancer Inst.* **64**, 447–459.
Nishina, K. (1975). *Acta Med. Okayama* **29**, 17–28.
Ogata, S.-I., Muramatsu, T., and Kobata, A. (1976). *Nature (London)* **259**, 580–582.
Ossowski, L., Unkeless, J. C., Tobia, A., Quigley, J. P., Rifkin, D. B., and Reich, E. (1973). *J. Exp. Med.* **137**, 112–126.
Parsons, R. G., Todd, H. D., and Kowal, R. (1979). *Cancer Res.* **39**, 4341–4345.
Pattillo, R. A., and Gey, G. O. (1968). *Cancer Res.* **28**, 1231–1236.

Pattillo, R. A., Ruckert, A., Hussa, R., Bernstein, R., and Delfs, E. (1971). *In Vitro* **6**, 398–399.
Pattillo, R. A., Hussa, R. O., Story, M. T., Ruckert, A. C. F., Shalaby, M. R., and Mattingly, R. F. (1977). *Science* **196**, 1456–1458.
Phillips, M. A., Budarf, M. L., and Herbert, E. (1981). *Biochemistry* **20**, 1666–1675.
Plotkin, G. M., Wides, R. J., Gilbert, S. L., Wolf, G., Hagen, I. K., and Prout, G. R., Jr. (1979). *Cancer Res.* **39**, 3856–3860.
Podolsky, D. K., Weiser, M. M., and Isselbacher, K. J. (1978). *Proc. Natl. Acad. Sci. U.S.A.* **75**, 4426–4430.
Pollack, R., and Rifkin, D. (1975). *Cell* **6**, 495–506.
Poole, A. R., Tiltman, K. J., Recklies, A. D., and Stoker, T. A. M. (1978). *Nature (London)* **273**, 545–547.
Princler, G. L., McIntire, K. R., and Adamson, R. H. (1978). *JNCI, J. Natl. Cancer Inst.* **60**, 643–648.
Quigley, J. P., Ossowski, L., and Reich, E. (1974). *J. Biol. Chem.* **249**, 4306–4311.
Rabson, A. S., Rosen, S. W., Tashjian, A. H., Jr., and Weintraub, B. D. (1973). *JNCI, J. Natl. Cancer Inst.* **50**, 669–674.
Ratter, S. J., Lowry, P. J., Besser, G. M., and Rees, L. H. (1980). *J. Endocrinol.* **85**, 359–369.
Reynolds, F. H., Jr., Todaro, G. J., Fryling, C., and Stephenson, J. R. (1981). *Nature (London)* **292**, 259–262.
Rioche, M., Quelin, S., and Masseyeff, R. (1974). *Pathol. Biol.* **22**, 867–876.
Roberts, J. L., and Herbert, E. (1977). *Proc. Natl. Acad. Sci. U.S.A.* **74**, 4826–4830.
Roberts, J. L., Phillips, M., Rosa, P. A., and Herbert, E. (1978). *Biochemistry* **17**, 3609–3618.
Roblin, R., and Young, P. L. (1980). *Cancer Res.* **40**, 2706–2713.
Roos, B. A., Lindall, A. W., Baylin, S. B., O'Neil, J. A., Frelinger, A. L., Birnbaum, R. S., and Lambert, P. W. (1980). *J. Clin. Endocrinol. Metab.* **50**, 659–666.
Rosen, S. W., Weintraub, B. D., and Aaronson, S. A. (1980a). *J. Clin. Endocrinol. Metab.* **50**, 834–841.
Rosen, S. W., Calvert, I., Weintraub, B. D., Tseng, J. S., and Rabson, A. S. (1980b). *Cancer Res.* **40**, 4325–4328.
Rosenthal, K. L., Tompkins, W. A. F., Frank, G. L., McCulloch, P., and Rawls, W. E. (1977). *Cancer Res.* **37**, 4024–4030.
Rosenthal, K. L., Tompkins, W. A. F., and Rawls, W. E. (1980). *Cancer Res.* **40**, 4744–4750.
Ruddon, R. W., Anderson, C., Meade, K. S., Aldenderfer, P. H., and Neuwald, P. D. (1979a). *Cancer Res.* **39**, 3885–3892.
Ruddon, R. W., Hanson, C. A., and Addison, N. J. (1979b). *Proc. Natl. Acad. Sci. U.S.A.* **76**, 5143–5147.
Ruddon, R. W., Hanson, C. A., Bryan, A. H., Putterman, G. J., White, E. L., Perini, F., Meade, K. S., and Aldenderfer, P. H. (1980a). *J. Biol. Chem.* **255**, 1000–1007.
Ruddon, R. W., Bryan, A. H., Meade-Cobun, K. S., and Pollack, V. A. (1980b). *Cancer Res.* **40**, 4007–4012.
Ruddon, R. W., Anderson, C., and Meade-Cobun, K. S. (1980c). *Cancer Res.* **40**, 4519–4523.
Ruddon, R. W., Bryan, A. H., Hanson, C. A., Perini, F., Cecorrulli, L. M., and Peters, B. P. (1981). *J. Biol. Chem.* **256**, 5189–5196.
Ruoslahti, E., and Terry, W. D. (1976). *Nature (London)* **260**, 804–805.
Ruoslahti, E., Engvall, E., Pekkala, A., and Seppälä, M. (1978). *Int. J. Cancer* **22**, 515–520.
Sakiyama, H., and Burge, B. W. (1972). *Biochemistry* **11**, 1366–1377.
Sakiyama, T., Robinson, J. C., and Chou, J. Y. (1978). *Arch. Biochem. Biophys.* **191**, 782–791.
San, R. H. C., Rice, J. M., and Williams, G. M. (1977). *Cancer Lett.* **3**, 243–246.
San, R. H. C., Shimada, T., Maslansky, C. J., Kreiser, D. M., Laspia, M. F., Rice, J. M., and Williams G. M. (1979). *Cancer Res.* **39**, 4441–4448.

Schwartz, M. K. (1973). *Clin. Chem. (Winston-Salem, N.C.)* **19,** 10–22.
Sell, S., and Becker, F. F. (1978). *JNCI, J. Natl. Cancer Inst.* **60,** 19–25.
Sell, S., and Morris, H. P. (1974). *Cancer Res.* **34,** 1413–1417.
Silva, O. L., and Becker, K. L. (1978). In "Biological Markers of Neoplasia: Basic and Applied Aspects" (R. W. Ruddon, ed.), pp. 295–310. Am. Elsevier, New York.
Singer, R. M., and Fishman, W. H. (1974). *J. Cell Biol.* **60,** 777–780.
Singer, R. M., Leahy, E. M., and Herz, F. (1980). *Oncodev. Biol. Med.* **1,** 77–92.
Singh, I., Tsang, K. Y., and Blakemore, W. S. (1978). *Cancer Res.* **38,** 193–198.
Smith, C. J., Morris, H. P., and Kelleher, P. C. (1977). *Cancer Res.* **37,** 2651–2656.
Sorenson, G. D., Pettengill, O. S., Brinck-Johnson, T., Cate, C. C., and Maurer, L. H. (1981). *Cancer* **47,** 1289–1296.
Speeg, K. V., Jr., Azizkhan, J. C., and Stromberg, K. (1977). *Exp. Cell Res.* **105,** 199–206.
Tan, K. K., and Aw, S. E. (1971). *Biochim. Biophys. Acta* **235,** 119–127.
Tashjian, A. H., Jr., Weintraub, B. D., Barowsky, N. J., Rabson, A. S., and Rosen, S. W. (1973). *Proc. Natl. Acad. Sci. U.S.A.* **70,** 1419–1422.
Todaro, G. J., DeLarco, J. E., and Cohen, S. (1976). *Nature (London)* **264,** 26–31.
Todaro, G. J., Fryling, C., and DeLarco, J. E. (1980). *Proc. Natl. Acad. Sci. U.S.A.* **77,** 5258–5262.
Tompkins, W. A. F., Watrach, A. M., Schmale, J. D., Schultz, R. M., and Harris, J. A. (1974). *JNCI, J. Natl. Cancer Inst.* **52,** 1101–1110.
Tralka, T. S., Rosen, S. W., Weintraub, B. D., Lieblich, J. M., Engel, L. W., Wetzel, B. K., Kingsbury, E. W., and Rabson, A. S. (1979). *JNCI, J. Natl. Cancer Inst.* **62,** 45–61.
Tsukada, Y., Mikuni, M., and Hirai, H. (1974). *Int. J. Cancer* **13,** 196–202.
Unkeless, J. C., Tobia, A., Ossowski, L., Quigley, J. P., Rifkin, D. B., and Reich, E. (1973). *J. Exp. Med.* **137,** 85–111.
Unkeless, J. C., Dano, K., Kellerman, G. M., and Reich, E. (1974). *J. Biol. Chem.* **249,** 4295–4305.
Vetterlein, D., Young, P. L., Bell, T. E., and Roblin, R. (1979). *J. Biol. Chem.* **254,** 575–578.
Wagner, D. D., Ivatt, R., Destree, A., and Hynes, R. O. (1981). *J. Biol. Chem.* **256,** 11708–11715.
Watanabe, K., Matsubara, T., and Hakomori, S.-I. (1976). *J. Biol. Chem.* **251,** 2385–2387.
Williams, R. D., Bronson, D. L., Elliott, A. Y., and Fraley, E. E. (1977). *JNCI, J. Natl. Cancer Inst.* **58,** 1115–1116.
Wilson, E. L., Becker, M. L. B., Hoal, E. G., and Dawdle, E. B. (1980). *Cancer Res.* **40,** 933–938.
Wolf, B. A., and Goldberg, A. R. (1978). *Proc. Natl. Acad. Sci. U.S.A.* **75,** 4967–4971.
Yamada, K. M., and Olden, K. (1978). *Nature (London)* **275,** 179–184.
Yogeeswaran, G., Stein, B. S., and Sebastian, H. (1978). *Cancer Res.* **38,** 1336–1344.
Zardi, L., Cecconi, C., Barberi, O., Carnemolla, B., Picca, M., and Santi, L. (1979). *Cancer Res.* **39,** 3774–3779.

CHAPTER 6

The Study of Oncodevelopmental Markers in Heterotransplanted Tumors

DEREK RAGHAVAN
Department of Clinical Oncology
Royal Prince Alfred Hospital
Camperdown, N.S.W., Australia

Ludwig Institute for Cancer Research
Camperdown, N.S.W., Australia

I.	Introduction	109
II.	Oncodevelopmental Proteins	112
	A. Human Chorionic Gonadotropin (hCG)	112
	B. α-Fetoprotein (AFP)	115
	C. Carcinoembryonic Antigen (CEA)	122
III.	Hormones and Other Substances	123
IV.	Cell Surface Antigens	125
V.	Summary	126
	References	126

I. INTRODUCTION

Until recently, much of the data pertaining to the biology of human oncodevelopmental proteins had been derived from clinical studies, as summarized elsewhere in this volume. However, the limitations of human experimentation mandated the search for adequate models of marker protein production by human tumors. The culture of tumor cell lines *in vitro,* although a useful model, is limited by its "two-dimensional" nature, the lack of a physiological milieu and the sensitivity of the system to extrinsic factors, such as temperature, pH, gas concentrations, and bacterial or tumor cell contamination. The data derived from animal tumors may not always accurately represent the analogous human neoplasms, for example, the 129 strain murine teratocarcinoma compared with human germ cell tumors (O'Hare, 1978; Raghavan and Neville, 1982).

The establishment of a "xenograft"—a viable tissue transplanted from one species into another—represents an effort to create a more pathophysiological replica of human tissue (Fig. 1); thus, the transplantation of human tumors into a variety of animal species has been attempted in order to establish a bank of "human" tumor tissues for research purposes. Several "privileged" sites have been demonstrated in which heterologous tissue can survive, including the anterior chamber of the eye (Van Dooremaal, 1873; Greene, 1938), the hamster cheek pouch (Lemon et al., 1952), and the hamster brain (Greene, 1951). However, the utility of these models is restricted by the physical constraints of the sites of tumor growth, the excessive morbidity to the tumor recipients, the difficulty of access to the growing tumors, and the development of a progressive immune response with rejection of the tumors.

A more useful approach has been provided by the systemic immunosuppression of the recipient animals. The initial use of measures such as whole body irradiation (Toolan, 1953) proved to be toxic to the animals, with severe damage to the gastrointestinal tract and bone marrow, with a high mortality rate. With less extensive immunosuppressive techniques, the xenografted tumors were rapidly rejected. More recently, selective immunosuppressive procedures have

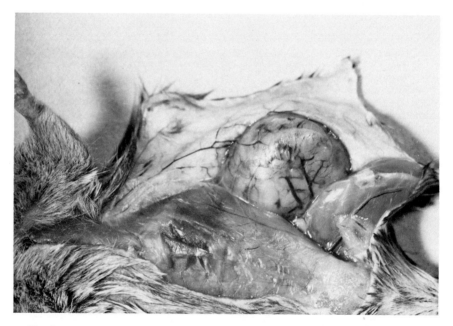

Fig. 1. Xenograft HX 57, yolk sac carcinoma. Note the blood supply derived from the host.

been used, including neonatal thymectomy, the administration of anti-lymphocyte globulin, or combinations of both with a resultant decrease in the rejection of xenografts (Osoba and Auersperg, 1966; Phillips and Gazet, 1967; Cobb, 1972). One of the most effective approaches to specific immunosuppression is achieved by combining neonatal thymectomy, whole body irradiation, and reconstitution with syngeneic bone marrow (Castro, 1972), the marrow replacement limiting the mortality rate induced by the whole body irradiation. However, an important limitation of this approach is the risk of partial immune reconstitution in association with the syngeneic marrow replacement.

An important advance that has further reduced this problem has been the use of "priming" techniques, in which cytosine arabinoside is administered 24 to 48 hours before whole body irradiation and protects the normal gut and marrow tissues of the animals against the lethal effects of whole body irradiation (Steel et al., 1978).

The concept of genetic immunosuppression has had similar applications - the hairless ("nude") mouse mutant, first described by Flanagan (1966), was later shown to be congenitally athymic (Pantelouris, 1968) and thus lacking T lymphocytes. Rygaard and Povlsen (1969) successfully xenografted a human colonic adnenocarcinoma into a nude mouse. These workers and others subsequently demonstrated the utility of the "nude" mouse for xenografting studies (Povlsen and Rygaard, 1971; Giovanella et al., 1978; Shimosato et al., 1976). The major obstacle to the use of this model was the relatively short life span of "nude" mice under conventional conditions, animals rapidly perishing from generalized infection. However, it was soon demonstrated that these animals could attain life spans in excess of 1 or 2 years if kept in sterile conditions with specific pathogen-free environments and sterilized diets and water (Giovanella and Stehlin, 1973).

Of considerable importance in assessing the use of the xenograft system as a model of human disease has been the evaluation of the similarities and differences between the animal model and the disease state. In the last 20 years, several functions have been assessed in order to resolve this issue, including morphology, growth kinetics, ultrastructure, tumor marker production, secretion of hormones, karyotype, cell surface antigenicity, metastatic potential, and sensitivity to treatment. With the possible exception of metastatic potential and, in some instances, the response to chemotherapy, there is ample documentation that xenografted tumors closely resemble their tumors of origin, as reviewed elsewhere (Fogh and Giovanella, 1978; Houchens and Ovejera, 1978).

However, it should not be forgotten that this model also has important limitations, including tumor–host size inequality, the abnormal immune function of the host, differences between human and murine metabolism, the risks of contamination of human tissues by transformed murine cells, the risks of cross-species immunological interactions, and the persistent problem of relatively low initial "take" rates.

II. ONCODEVELOPMENTAL PROTEINS

In clinical practice, the concept of the "ideal" tumor marker—diagnostic of the presence of viable tumor, with a reliable quantitative relationship to tumor mass, highly sensitive and specific—is illustrated, although imperfectly, by the production of human chorionic gonadotropin (hCG) by germ cell tumors and gestational choriocarcinomas and of α-fetoprotein (AFP) by hepatomas and germ cell tumors. Hence, one aspect of the study of neoplastic cell function, as measured by marker protein production, has been the establishment and characterization of xenografts of these tumor types, with an attempt to clarify the relationship between the production of these marker proteins and cellular morphology, ultrastructure, and therapeutic sensitivity, and thus to increase the clinical utility of these proteins.

A. Human Chorionic Gonadotropin (hCG)

In the earliest studies of marker protein production by xenografts, cortisone-treated hamsters were used as hosts (Table I). Pierce and his group (1959) demonstrated chorionic gonadotropin (CG)-like activity in 20 to 30% of hamsters bearing the xenografted embryonal cell carcinoma line "PITT 61", based on the presence of uterine hyperplasia; to exclude pituitary factors as a cause, tumor-bearing animals were hypophysectomized. The specific CG-producing cells were not identified morphologically in this study. Despite their attempts to develop a CG-secreting strain of PITT 61 by selective transplantation of hormone-secreting xenografts, the prevalence of the hormonally active tumor subline was never increased above 30%, a finding which remains unexplained.

These workers also demonstrated CG-like activity by bioassay in frogs for other lines derived from choriocarcinoma (Pierce *et al.*, 1958); the presence of CG was associated with ovarian and uterine hypertrophy in the host animals. In a subsequent study, Pierce *et al.* (1962) assessed the response of two uterine choriocarcinoma xenograft lines to methotrexate, measuring tumor regression and the extent of hypertrophy of uterus and ovary in treated and control animals. However, in this study, there was no difference between the two groups, which may have reflected the presence of methotrexate-resistant cells in the original tumor, or alternatively the selection of inappropriate doses of the drug. More recently, with the availability of double antibody radioimmunoassay (RIA) techniques, the relationship between plasma levels of hCG and tumor growth after treatment with methotrexate and actinomycin D has been studied in xenografts of human choriocarcinoma maintained in hamster cheek pouches (Lewis *et al.*, 1969; Knecht and Hertz, 1978). In these studies a progressive rise in circulating levels of hCG accompanied the increase in tumor weight with time, although this relationship was not truly stoichiometric, in that the rate of increase of tumor

TABLE I

hCG Production by Xenografts of Testicular and Other Tumors[a]

Line	Donor Histology	Donor hCG ↑	Prior treatment	Histology	Xenograft Host	hCG in blood	hCG site demonstrated	Host response to hCG	Comments	Reference
PITT61	E	+	No	E	HCP	±	—	±	hCG production occasionally	Pierce et al., 1959, 1961
PITT89	C	+?	Yes	C/E	HCP	+	—	+		
PITT94	E	—	No	E	HCP	—	—	?		
PITT100	T/E/C	+	No?	C	HCP	+	—	+		
HX36	C/E	+	No	C/E/ST	Ix	+	STGC/MC	?	Autoradiographs	Selby et al., 1979
HX39	E	+	Yes	E	Ix/N	±	MC	?	hCG production occasionally	Raghavan et al., 1981a
HX67	S/T/Y/E	+	No	S/Y/E	Ix	+	STGC/MC	?	—	Raghavan et al., 1981a
PITT146	GC	—	No	C	HCP	+	—	Yes	No growth delay	Pierce et al., 1962
PITT147	GC	+	Yes	C	HCP	+	—	Yes	No growth delay	Pierce et al., 1962
SCH	C[b]	+		C	N	+	STGC/MC	+	AP production/EM localization Growth delay—DSE	Kameya et al., 1976 Hayashi et al., 1978
OCC-MM	C[c]	+	Yes	C	N	+	?	+	—	Kim et al., 1978
CC-1	GC	+	Yes	C	N	+	?	?	No growth delay	Hayashi et al., 1978
CC-3	GC	+	?	C	N	+	?	?	HCG-RIL	Searle et al., 1981
GCC-SV	GC	+	No	C	N	+	?	+	—	Kim et al., 1978
GCC-RS	GC	+	Yes	C	N	+	?	+	—	Kim et al., 1978
BeWo	GC	?	?	C	HCP	+	?	?	Growth delay/DSE	Knecht and Hertz, 1978
Wo	GC	+	?	C	HCP	+	?	?	Growth delay	Lewis et al., 1969

[a] E, embryonal cell carcinoma; C, choriocarcinoma; TC, teratocarcinoma; S, seminoma; Y, yolk sac carcinoma; ST, syncytiotrophoblastic cells; MC, mononuclear cells; RIL, radio-immuno-localization studies; Ix, immune-suppressed; N, nude; HCP, hamster cheek pouch; STGC, syncytotrophoblastic giant cells; DSE, drug-specific effect on hCG production; AP, alkaline phosphatase; EM, electron microscope; GC, gestational choriocarcinoma.
[b] Gastric.
[c] Ovarian.

mass exceeded the rate of change of circulating hCG levels. In a preliminary study (Knecht and Hertz, 1978), it appeared that drug treatment could precipitate an increase in plasma hCG, while inhibiting tumor growth, perhaps due either to the release of hCG after tumor cell lysis or due to a drug-induced increase in hCG synthesis as suggested by studies *in vitro* (Hussa *et al.*, 1973). Surgical excision of the total tumor mass was followed by a decline in circulating hCG with a half-life of 1–2 days (Lewis *et al.*, 1969).

More extensive studies have been carried out using nude mice as hosts for xenografts of gestational, ovarian, and gastric choriocarcinoma (Kameya *et al.*, 1976; Kim *et al.*, 1978; Hayashi *et al.*, 1978). Kameya *et al.* (1975) demonstrated that a cell line (SCH), derived from a gastric choriocarcinoma, secreted hCG during cellular proliferation, and localized the production of this antigen to a small proportion of mononuclear and multinuclear cells by cytochemical and ultrastructural techniques. In addition, these workers demonstrated the production of a placenta-specific isoenzyme of alkaline phosphatase. As an extension of these studies, they established the SCH line in nude mice (Kameya *et al.*, 1976). These tumors retained the characteristic morphological and ultrastructural features of choriocarcinoma, with populations of cells resembling cytotrophoblast and syncytiotrophoblast, with a third population of "intermediate" cells. Using immunocytochemical techniques, positive staining for hCG was demonstrated in less than 5% of the tumor cells. As the tumors increased in size, they grew as fluid-filled sacs, and fluid aspirated yielded hCG concentrations as high as 6×10^4 μg/liter, more than twice the maximum value of the circulating plasma hCG at the same time. Although only a small proportion of the tumor cells stained positively for hCG, there was a linear relationship between the increasing tumor size and the levels of plasma hCG as measured by an antiserum directed against the β subunit of hCG. As with the cortisone-treated hamsters (*vide supra*), the host animals showed evidence of hCG end-organ effects, with enlarged and dilated ovarian follicles, luteal hypertrophy, hyperemia of the uteri, and enlargement of breast tissue. Similar results were documented for other xenografts of gestational and ovarian choriocarcinomas (Kim *et al.*, 1978), although these workers found a greater discrepancy between intracystic tumor concentrations and plasma levels of hCG.

Using the xenograft line SCH and another line established from a gestational choriocarcinoma (CC-1), workers at the National Cancer Center Research Institute, Tokyo, studied the effects of single agent treatment with methotrexate, actinomycin D, and vinblastine on tumor growth and marker production. Although they were unable to show specific histological evidence of tumor necrosis, a good correlation between growth delay and a decrease in hCG production was shown. However, an interesting discrepancy in these preliminary experiments was that, although vinblastine appeared to exert a greater growth inhibition, actinomycin D was associated with a more marked suppression of hCG production (Hayashi *et al.*, 1978; Shimosato *et al.*, 1978). Whether this phenomenon was due to a specific inhibition of hCG production or merely reflected necrosis of hCG-producing cells, as distinct from non-marker-producing cells, was not assessed in this report. More extensive studies of this type may explain the nature of the "release" phenomenon and the artifacts and discordant results in marker protein levels seen in clinical practice (Kohn and Raghavan, 1981).

Although the majority of the xenografted marker-producing tumors appear to retain this function through serial passages, Selby *et al.* (1979) demonstrated a reduction in hCG-producing elements beyond the fifth serial passage. In this detailed study in which the histology, immunocytochemical staining patterns, and autoradiographic appearances were compared, it appeared that the hCG-producing cells predominantly constituted a nondividing population—in the xenograft, the average labeling index was 28% (with a range of 18–42%); however, where both autoradiography and indirect immunoperoxidase staining were carried out together, little uptake of the tritiated thymidine was seen in the cells staining positively for hCG. This was demonstrated both in the xenograft and in the original tumor. The significance of these findings is not yet clear, as these results seem to be at variance with the traditional concept of trophoblastic tissues being kinetically active. There is considerable controversy in clinical practice as to whether the presence of raised circulating levels of hCG is, in itself, an adverse prognostic feature, independent of tumor bulk (Bagshawe, 1980; Kohn and Raghavan, 1981). Further studies of this type, in which tumor cell kinetics are related to marker production may yield further information, and can be performed with greater facility using the xenograft model than in clinical practice.

B. α-Fetoprotein (AFP)

The study of AFP in animal models has had a shorter history than has hCG. Its production has been characterized in animal tumors, including the mouse teratocarcinoma (Kahan and Levine, 1971; Engelhardt *et al.*, 1973) and in the rat yolk sac carcinoma (Sakashita *et al.*, 1976; Albrechtsen *et al.*, 1978). More recently, several xenografts of AFP-producing germ cell tumors and hepatomas have been established as serially transplantable lines in thymectomized and irradiated or congenitally athymic mice (Tables II and III). These xenografts have been extensively characterized, and a close correlation has been demonstrated between the original tumors and the serially transplantable lines in as many as 15–20 generations.

1. Germ Cell Tumor Xenografts

Several criteria have been evaluated in order to compare the donor and xenografted tumors. The most common histological subtype to have been suc-

TABLE II
AFP Production by Xenografts of Germ Cell Tumors[a]

	Donor				Xenograft				
Line	Histology	AFP ↑	Prior treatment	Histology	Host	AFP in blood	AFP site demonstrated	Other tumor markers	Reference
HX 39	E	±	Yes	E	Ix/N	±	—	—	Raghavan 1981
HX 53	"S"	+	Yes	S/Y	Ix/N	+	"S"/Y/U	FN, AAT	Raghavan et al., 1981a
HX 57	Y	+	Yes	Y	Ix/N	+	Y/U	FN, AAT	Raghavan et al., 1981b
HX 67	S/Y/T/E	+	No	S/Y/E	Ix	+	Y/U	?	Raghavan et al., 1980
HX 111	E	+	No	E	Ix/N	±	—	?	Raghaven et al., 1979
HX 112	Y	+	Yes	Y	Ix/N	+	Y/U	AAT	Ruoslahti et al., 1981
OE	Y	+	No	Y	N	+	Y	AAT, Tf, Alb	Yoshimura et al., 1978
TE	Y	+	No	Y	N	+	Y	AAT, Tf, Hx Pre	Hata et al., 1980
TT-1-JCK	Y	+	No	Y	N	+	Y	AAT	Kaneko et al., 1980
HYST-G1	Y	+	No	Y	N	+	—	?	Shirai et al., 1977
EST 1	Y	+	?	Y	N	+	Y	LDH$_1$, AP	Takeuchi et al., 1979
EST 2	Y	+	?	Y	N	+	Y	LDH$_1$, AP, CEA	

[a] E, embryonal cell carcinoma; S, seminoma; Y, yolk sac carcinoma; U, undifferentiated cells; T, teratocarcinoma; Ix, immune-suppressed mice; N, nude; AAT, α_1 anti-trypsin; Tf, transferrin; Hx, hemopexin; Alb, albumin; Pre, pre-albumin; LDH$_1$, lactic acid dehydrogenose isoenzyme 1; CEA, carcinoembryonic antigen; FN, fibronectin; AP, alkaline phosphatase.

TABLE III
AFP Production by Hepatic Tumor Xenografts[a]

	Donor				Xenograft			
Line	Histology	AFP ↑	Prior reaction	Histology	Host	AFP blood	AFP site demonstrated	Other tumor markers	Reference
Li-7	Hep	+	?	Hep	N	+	?	Alb, AAT, Tf	Hirohashi et al., 1979
Li-16	Hep	+	?	Hep	N	+	?	Alb, AAT, Tf, α-hCG	
Li-19	Hep	+	?	Hep	N	+	?	Alb, AAT, Tf	
Li-23	Hep	+	?	Hep	N	+	?	Alb, AAT, Tf	Shimosato et al., 1978
Li-15	H b'oma	+	?	H b'oma	N	+	?	Alb, AAT, Tf	
Li-24	H b'oma	+	?	H b'oma	N	+	?	Alb, AAT, Tf	
PLC/PRF/5	Hep	?	?	Hep	N	?	Hep	CEA	Shouval et al., 1981
Hpb-1-JCK	H b'oma	?	?	Hpb'oma	N	+	?	AAT, Alb, Tf, α_2-M	Yoshimura et al., 1978

[a] Hep, hepatocellular carcinoma (hepatoma); H b'oma; hepatoblastoma; N, nude mouse; Alb, albumin; AAT, α_1 anti-trypsin; Tf, transferrin; α-hCG: α-subunit of hCG; CEA, carcinoembryonic antigen; α_2-M: α_2-macroglobulin.

cessfully xenografted has been yolk sac carcinoma (endodermal sinus tumor), the classic pattern of microcyst formation and Duval-Schiller bodies usually being retained in serial passages (Shirai *et al.*, 1977; Hata *et al.*, 1980; Raghavan *et al.*, 1981a; Raghavan, 1981). However, the retention of the light morphological appearances of a variety of testicular tumors when xenografted has been well documented, including embryonal cell carcinoma (Fig. 2) and combined tumors with elements of embryonal cell carcinoma, teratocarcinoma, choriocarcinoma, syncytiotrophoblastic giant cells, seminomatous regions, and classic yolk sac carcinoma (Bronson *et al.*, 1980; Andrews *et al.*, 1981; Tveit *et al.*, 1980; Zeuthen *et al.*, 1980; Raghavan *et al.*, 1979, 1980, 1981a; Raghavan, 1981). No pure classical seminomas have proved to be serially transplantable, although one xenograft line of a "seminoma-like" tumor, associated with AFP production, has been described (Raghavan *et al.*, 1981b) as discussed below.

In addition to the light microscopic appearances, two detailed studies have shown that the ultrastructural patterns of germ cell tumors are retained with little change through successive transplant generations (Pierce, 1966; Monaghan *et al.*, 1982).

To date, much of the emphasis in the study of marker protein production by xenografts has been descriptive, most studies concentrating on the measurement

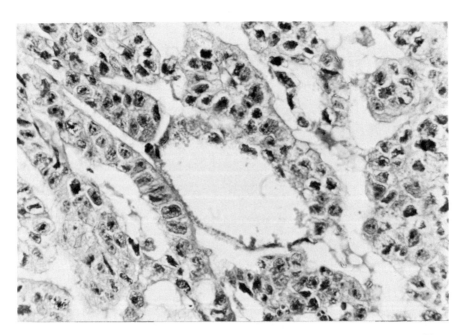

Fig. 2. Xenograft HX 111, Embryonal cell carcinoma. Note the retention of an adeno-papillary pattern and high mitotic rate.

and characterization of the antigens produced—several groups have demonstrated the similarity of normal human yolk sac, human yolk sac carcinoma, and xenografts of these tumors in the production of proteins such as AFP, albumin, pre-albumin, transferrin, α_1-anti-trypsin, hemopexin, β-lipoprotein, carcinoembryonic antigen, and lactic acid dehydrogenase (Shirai *et al.*, 1977; Yoshimura *et al.*, 1978; Takeuchi *et al.*, 1979; Kaneko *et al.*, 1980; Raghavan *et al.*, 1979, 1980, 1981a,b). Although a human variant of fibronectin with a similar structure to the human amniotic fluid isoenzyme has been shown to be produced by xenografted yolk sac carcinomas (Ruoslahti *et al.*, 1981), (Fig. 3) our investigations have suggested that laminin is not produced by these xenografts (E. Ruoslahti and D. Raghavan, unpublished), in contrast to a study in which rat yolk sac carcinomas were shown to produce this protein (Wewer *et al.*, 1981).

Since the initial emphasis on characterization of tumor lines, the xenograft model has been used to address questions of clinical relevance. The relationship between intratumor and circulating levels of AFP has been documented in a series of xenografts (Raghavan *et al.*, 1980), and, in the same study, the nature of the discordance between measurable tumor mass or volume and circulating AFP concentrations was evaluated; factors that contributed to such a discrepancy

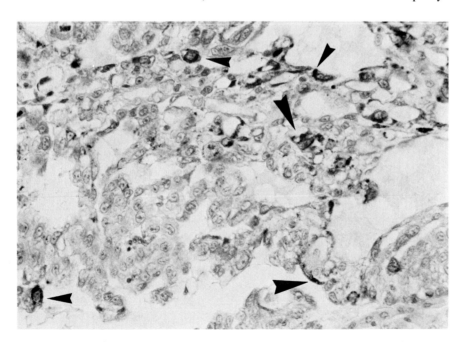

Fig. 3. Xenograft HX 57, indirect immunoperoxidase stain for human fibronectin. Arrows denote positive staining. Note the retention of a microcystic pattern. Absorbed controls were negative.

included increasing necrosis and fluid accumulation within tumors of increasing size, the heterogeneity of tumor tissue (some cells producing AFP and others being "marker-negative"), and an apparent suppression of AFP production after radiotherapy (which may merely have represented a selection process).

In clinical practice, the production of AFP has traditionally been associated with apparent or occult elements of nonseminomatous tumor; thus, in a patient with a histological diagnosis of "pure seminoma," but with elevated circulating levels of AFP, it has been assumed that an occult focus of nonseminoma is present, either in the primary tumor or in an undiagnosed metastatic deposit (Lange et al., 1980; Javadpour, 1980). We have studied this phenomenon by establishing a xenograft from such a patient (Raghavan et al., 1981b). The xenograft line HX53 was established from a lymph node deposit which developed in a patient with a histological diagnosis of seminoma but with a circulating level of AFP of 6600 µg/liter. The histology of the node metastasis was similar to that of the primary tumor, but with more prominent areas of necrosis and increased nuclear pleomorphism and mitotic activity. Extracellular eosinophilic globules were noted in association with small cystic spaces, which were more prominent than in the primary tumor. Lymphocytic infiltration was sparse and no giant cells were seen. Specimens from serial xenograft passages, in as many as 15 generations, showed retention of the basic tumor pattern with sheets or trabeculae of tumor cells with well-defined cell membranes, open clear cytoplasm, and single rounded vesicular nuclei with prominent nucleoli. Nuclear pleomorphism and mitotic activity were marked and the appearances were consistent with a diagnosis of "anaplastic seminoma." In association with this pattern, there were areas of microcyst formation, each being surrounded by tumor cells similar to those of the solid areas of the lesion. Levels of AFP greater than 100,000 µg/liter were demonstrated by RIA in the blood of tumor-bearing mice and in fluid aspirated from the tumors. Immunocytochemical staining with affinity-purified antibodies revealed AFP in the cytoplasm of tumor cells surrounding the cystic areas, in the cysts themselves, and also in some cells in the solid "seminomatous" regions. Ultrastructural characterization of this xenograft line, in addition to the light morphological and immunocytochemical data, suggested the existence of a solid variant of yolk sac carcinoma (endodermal sinus tumor), with many of the light microscopical features of pure seminoma, but with ultrastructural and functional traits suggesting varying degrees of differentiation along a continuum between classical seminoma and yolk sac carcinoma. Such a concept could explain the phenomenon of "combined tumors," with elements of seminoma and nonseminoma in the same tumor, both at gonadal and extragonadal sites. Similarly, this could explain the occurrence of nonseminomatous metastases from an apparently pure seminoma primary tumor, and the fact that up to 10% of stage I seminomas are not initially cured by orchidectomy and conventional radiotherapy. In the latter situation, it may well be that a solid yolk sac carcinoma has been mistakenly diagnosed as a pure seminoma and treated accordingly. Clinical evidence in support of this hypothesis has been presented in a report of six such tumors (Raghavan et al., 1982) and a more detailed controlled study is in progress.

At the present time, there are no detailed published data regarding the production of AFP by xenografts treated with chemotherapy or radiotherapy, although such studies are in progress. As with hCG-producing xenografts, studies of this type may well elucidate the phenomenon of tumor lysis with release of AFP and may help to clarify the true role of AFP half-life determinations as a prognostic tool (Kohn and Raghavan, 1981).

2. Hepatoma Xenografts

Serially transplantable xenografts have been established from both hepatomas (Hirohashi et al., 1979; Shimosato et al., 1976; Shouval et al., 1981) and from hepatoblastomas (Hirohashi et al., 1979; Yoshimura et al., 1978). The xenograft lines have tended to retain the light microscopic appearances of the donor tumors and many of the ultrastructural features, although the hepatomas in one study were noted to have suffered some disruption of their architecture in relation to the hepatic sinusoids (Hirohashi et al., 1979).

The transplantable hepatoblastomas (Li-15 and Li-24) consisted of neoplastic epithelial cells which morphologically resembled their tumors of origin and which ultrastructurally resembled fetal liver cells; however, the stromal component, osteoid material, and foci of extramedullary hemopoiesis were lost during serial transplantation (Hirohashi et al., 1979). Both hepatoma- and hepatoblastoma-derived xenografts were noted to produce a series of tumor antigens, including AFP, α_1-anti-trypsin, albumin, α_2-macroglobulin, and transferrin.

The human hepatoma cell line PLC/PRF/5, which synthesizes and secretes hepatitis B surface antigen, was injected into nude mice. AFP, CEA, and α_1-anti-trypsin were detected in nude mouse xenografts using the peroxidase-antiperoxidase technique. Hepatitis B virus DNA was also identified in these tumors by molecular hybridization techniques and was detected in the nude mouse serum in concentrations roughly correlated with tumor mass (Shouval et al., 1981). This study provided further support to the hypothesis linking the hepatitis B virus to hepatic oncogenesis, and the possible interrelationship to the consequent elaboration of oncofetal antigens.

In an extensive study of the production of AFP and normal serum proteins by hepatoma xenografts, Hirohashi et al. (1979) showed that the rate of production of these proteins was correlated with the level of morphological differentiation of tumor cells toward normal liver cells, that the levels correlated with tumor growth, and that the growth rate correlated with the rate of production of AFP; rapidly growing hepatoma xenografts were moderately differentiated and produced large amounts of AFP, whereas a slowly growing line (Li-19) was well differentiated and produced little AFP. A most interesting phenomenon demonstrated in this study was that the production of normal serum proteins (albumin, α_1-anti-trypsin, etc.) appeared to parallel the production of AFP, suggesting that disordered production of serum proteins by hepatoma may not be merely the result of "fetal regression," and thus contradicting the concept that the differentiation of hepatomas corresponds to the maturation of the fetal liver.

Shimosato *et al.* (1978) demonstrated a dose–response relationship for the effects of mitomycin C on the hepatoma xenograft line Li-7, serial estimations of AFP in the mouse serum correlating with tumor regression. By contrast with data derived from choriocarcinomas (*vide supra*), a good correlation was noted between therapeutic response, tumor histology, and AFP values. At a mitomycin C dose level of 1 mg/kg no major histological changes were detected in the treated xenografts, and there was no change in the level of AFP; at 3 mg/kg, marked tumor necrosis correlated with a fall in circulating levels of AFP. AFP values were unaffected by treatment with adriamycin in this study. Further studies of this type may well lead to the development of more effective regimens for the treatment of human hepatomas, and, in similar fashion to the germ cell tumors, may lead to more rational clinical use of tumor markers in patient management.

C. Carcinoembryonic Antigen (CEA)

The production of carcinoembryonic antigen (CEA) has been demonstrated in xenografted tumors in hamster cheek pouches (Goldenberg and Hansen, 1972), in thymectomized irradiated mice (Houghton and Taylor, 1978; Bailey *et al.*, 1981), and in congenitally athymic "nude" mice (Sordat *et al.*, 1974). Most of the xenograft lines associated with CEA production have been derived from colonic adenocarcinomas (Sordat *et al.*, 1974; Goldenberg and Hansen, 1972; Miwa *et al.*, 1976; Houghton and Taylor, 1978), although this antigen has also been associated with xenografts of carcinoma of the breast (Bailey *et al.*, 1981), yolk sac carcinoma (Takeuchi *et al.*, 1979), and a variety of other tumors.

The retention of CEA production, in association with stable histological patterns and growth characteristics, has been demonstrated in a series of xenografted colonic adenocarcinomas (Sordat *et al.*, 1974; Houghton and Taylor, 1978), and the sites of production have been demonstrated by immunofluorescence and immunoperoxidase techniques. Miwa *et al.* (1976) have shown that serum CEA levels correlate with tumor weight.

Similarly, in a detailed study of CEA production by xenografted mammary carcinomas, Bailey *et al.* (1981) have demonstrated persistence of the antigen in the donor and xenografted tumors, with a high level of concordance. CEA was localized to ductlike glandular structures composed of tumor cells, and was strongly positive in infiltrating duct carcinoma and mucus-secreting adenocarcinoma, with weak staining patterns found in atypical medullary carcinomas and some infiltrating duct carcinomas. In this study, concordance between donor and xenografted tumors was also demonstrated for the epithelial membrane antigen (EMA). Both antigens were able to be expressed by the same epithelial cells, both in the donor tumors and xenografts, illustrating yet another aspect of the heterogeneity of tumor cell function.

The relationship between CEA production and tumorigenicity has been evaluated in two published studies (Rosenthal et al., 1977; Brattain et al., 1981), but with opposing results—Rosenthal et al. suggested no difference in tumorigenicity between CEA-producing sublines and CEA-"negative" cells derived from a common "parent" tumor, whereas Brattain et al. found a greater "take" rate in nude mice injected with CEA-producing sublines. Further studies of this type may yield insights into the prognostic significance of CEA production by the tumors encountered in clinical practice.

Perhaps the most potentially useful clinical application of the xenograft model has been in the development of techniques or radio-immunolocalization, in which radio-labeled specific antibody to a tumor antigen is used to localize deposits of the tumor producing that antigen. Primus et al. (1973) demonstrated in vivo tumor localization by ^{125}I-labeled anti-CEA antibodies injected into male Syrian hamsters bearing the CEA-producing xenograft GW-39; tumor uptake was greater than for a ^{131}I-labeled control nonspecific IgG. Similar studies were performed in nude mice bearing xenografted colonic adenocarcinomas (Mach et al., 1974). More recently, specific uptake of affinity-purified radio-labeled anti-CEA has been demonstrated in immune suppressed mice bearing xenografts of human breast tumors (Moshakis et al., 1981a); the degree of tumor uptake is correlated with the amount of tumor CEA and is unaffected by levels of circulating CEA or of CEA/anti-CEA immune complexes. These studies have given rise to successful clinical programs for the detection of metastatic deposits of adenocarcinoma of the colon (Goldenberg et al., 1978), and the principles have been extrapolated for use with other tumor antigens (vide infra).

III. HORMONES AND OTHER SUBSTANCES

Many of the observations and principles outlined above have been demonstrated for a wide variety of tumor products elaborated by xenografted tumors. Although a detailed discussion of these data is beyond the scope of this brief chapter, the nature and variety of these tumor markers are summarized in Table IV. As can be seen, the majority of these products are either enzymes or hormones, many of which retain their function in the host animals. Thus, for example, nude mice bearing xenografts derived from bronchogenic small cell undifferentiated cancer may show adrenal hypertrophy or physiological and biochemical evidence of the inappropriate secretion of ADH.

TABLE IV
The Production of Hormones and Enzymes by Xenografts[a]

Line	Donor		Xenograft				Reference
	Histology	Tumor product	Histology	Tumor product	End-organ response	Comments	
Lu-1	SCUC lung	↑ADH, No MSH, No ACTH	SCUC	βMSH, ACTH, No ADH		NS granules, variation between passages	Shimosato et al., 1976; Kameya et al., 1977
Lu-19	SCUC lung	Calcitonin	SCUC	Calcitonin		I-C (+)r, RIA (+)	Kameya et al. 1976
—	SCUC lung	ACTH	SCUC	ACTH	Adrenal hypertrophy		Shin et al., 1977
OTUK	SCC lung	CSF, ?PTH	SCC	CSF, ?PTH	WCC ↑, Ca²⁺ ↑		Ohsawa et al. 1977
	Renal CA		Renal CA	Erythropoietin	Polycythemia		Ohsawa et al., 1977
—	SCUC lung	ADH	SCUC	Vasopressin ↑, neurophysin ↑	SIADH		Kondo et al. 1981
HEp-2	SCC larynx	?	SCC	LDH isoenzyme	—	Monitoring of response	Di Persio et al. 1979
SW480	AdenoCA colon	?	AdenoCA	LDH isoenzyme	—	Monitoring of response	Di Persio et al. 1979
Clouser	AdenoCA breast	?	AdenoCA	LDH isoenzyme	—	Monitoring of response	Di Persio et al. 1979
HeLa TRC-1	SCC cervix	?	SCC	AP isoenzymes		Variable expression with different hosts	Singer et al. 1979
—	Neuroblastoma	?	Neuroblastoma	Dopamine. β-hydroxylase			Helson et al. 1975
Me-1	Melanoma	?	Melanoma	Tyrosinase			Shimosato et al., 1976

[a] SCUC, small cell undifferentiated carcinoma; CA, carcinoma; SCC, squamous cell CA; ADH, anti-diuretic hormone; β-MSH, β-melanocyte stimulating hormone; ACTH, adrenocortical stimulating hormone; CSF, colony stimulating factor; PTH, parathyroid trophic hormone; WCC, white cell count; SIADH, syndrome of inappropriate ADH secretion; NS, neurosecretory; I-C, immunocytochemistry; RIA, radioimmunoassay; AP, alkaline phosphatase; LDH, lactic acid dehydrogenase.

In some instances, it has been shown that the circulating levels of these tumor products (for example, lactic acid dehydrogenase) correlate with changes in tumor mass in response to chemotherapy or other treatment procedures (Di Persio *et al.*, 1979).

An important concept that should be emphasized is that the expression of tumor antigens in xenograft-bearing animals may be influenced by factors within the hosts themselves that are independent of the tumors; for example, Singer *et al.* (1979) have shown the production of different isoenzymes of alkaline phosphatase by sublines of the cell line HeLa TCRC-1 when grown in the cheek pouches of hamsters which were immunosuppressed by different methods (cortisone acetate or anti-lymphocyte serum). Although the nature of this phenomenon has not yet been explained, it may be of major importance in evaluating the factors which alter tumor marker production in xenografts and in patients. Furthermore, in view of the well documented contamination of many cell lines by the HeLa strain, it is of great importance that the integrity of marker-producing cell lines be regularly monitored to avoid this major potential artefact in the system.

IV. CELL SURFACE ANTIGENS

The study of the expression of tumor-specific antigenic determinants on the surfaces of xenograft cells is in its infancy, largely because the characterization of these determinants in human tumors is still being developed. Povlsen *et al.* (1973) studied the antigenic properties of xenografts of Burkitt's lymphoma and demonstrated the retention of surface IgM and of EBV-associated membrane antigen and viral capsid antigen in up to six serial passages. A minority of mouse-passaged Burkitt tumor cells reacted with antisera against mouse immunoglobulin, indicating that some of the xenografted cells may have been coated with mouse immunoglobulins.

Furthermore, several workers have demonstrated common cell-surface antigens (the so-called "F_9" antigen and others) on preimplantation mouse embryos, 129 strain mouse embryonal carcinoma cells, human sperm, and human embryonal carcinoma cells (Artzt *et al.*, 1973; Hogan *et al.*, 1977; Holden *et al.*, 1977). More recently, using monoclonal antibodies, common cell surface antigenicity has been demonstrated between murine embryonal carcinoma cells and xenografted human germ cell tumors (Andrews *et al.*, 1981; McIlhinney, 1981; Cotte *et al.*, 1982).

These antigenic properties have been adapted for use in radio-immunolocalization techniques. Using a ^{125}I-labeled monoclonal antibody raised against a xenografted human germ cell tumor, Moshakis *et al.* (1981b) have demonstrated specific uptake of the antibody into a series of germ cell tumor xenografts in immunosuppressed mice. No uptake was demonstrated in "control" xenografts (derived from breast adenocarcinoma, bronchial adenocarcinoma, and squamous cell carcinoma). At the time of writing, a pilot study of this monoclonal antibody for use in the detection of germ cell tumor deposits in clinical practice is in progress.

V. SUMMARY

In this brief chapter, similarities and differences between xenografts and the human tumors from which they are derived have been discussed. In most instances, the morphology, ultrastructure, and tumor marker production by the xenografts closely reflect the human tumors, and, in cases where discrepancies are present, the xenografts often reflect outgrowth of one subline from a heterogeneous population of tumor cells which comprised the original tumor. Although less data are available, it appears that xenografts also reflect their "parent" tumors in cell surface antigenicity and chemosensitivity.

The xenograft model, with its function as a "tissue amplifier," which can generate large amounts of "human" tumor tissue for experimentation, thus provides a useful extension of the available animal models and tissue culture systems. As shown in the preceding sections, the retention of the functional properties of tumor cells as xenografts (production of tumor markers, such as AFP, hCG, and CEA, and of physiologically active hormones) will allow this model to be used to characterize further the nature of tumor marker production and the factors which affect it. The use of xenografts in the development of chemotherapy regimens and diagnostic imaging techniques may add a further dimension to the clinical relevance of this useful model of neoplastic disease.

REFERENCES

Albrechtesen, R., Hirai, H., Linder, D., Norgaard-Pedersen, B., and Wewer, U. (1978). *Scand. J. Immunol.* **8,** Suppl. 8, 165–169.

Andrews, P. W., Bronson, D. L., Benham, F., Strickland, S., and Knowles, B. B. (1981). *Int. J. Cancer* **26,** 269–280.

Artzt, K., Dubois, P., Bennett, D., Condamine, H., Babinet, C., and Jacob, F. (1973). *Proc. Natl. Acad. Sci. U.S.A.* **70,** 2988–2992.

Bagshawe, K. D. (1980). *Brit. J. Cancer* **41,** suppl. IV, 186–190.

Bailey, M. J., Ormerod, M. G., Imrie, S. F., Humphreys, J., Roberts, J. D. B., Gazet, J. C., and Neville, A. M. (1981). *Br. J. Cancer* **43,** 125–134.

Brattain, M. G., Fine, W. D., Khaled, F. M., Thompson, J., and Brattain, D. E. (1981). *Cancer Res.* **41,** 1751–1756.

Bronson, D. L., Andrews, P. W., Solter, D., Cervenka, J., Lange, P. H., and Fraley, E. E. (1980). *Cancer Res.* **40,** 2500–2506.

Castro, J. E. (1972). *Nature (London)* **239,** 83–84.
Cobb, L. M., (1972). *Br. J. Cancer* **26,** 183–189.
Cotte, C., Raghavan, D., McIlhinney, R. A. J. & Neville, A. M. (1983).
Cotte, C., Raghavan, D., McIlhinney, R. A. J., and Neville, A. M. *In Vitro*. (in press).
Di Persio, L., Kyriazis, A. P., Michael, J. G., and Pesce, A. J. (1979). *JNCI, J. Natl. Cancer Inst.* **62,** 375–379.
Engelhardt, N. V., Poltoranina, V. S., and Yasowa, A. K. (1973). *Int. J. Cancer* **11,** 448–459.
Flanagan, S. P. (1966). *Genet. Res.* **8,** 295–309.
Fogh, J., and Giovanella, B. C., eds. (1978). "The Nude Mouse in Experimental and Clinical Research." Academic Press, New York.
Giovanella, B. C., and Stehlin, J. S. (1973). *JNCI, J. Natl. Cancer Inst.* **51,** 615–618.
Giovanella, B. C., Stehlin, J. S., Williams, L. J., Lee, S. S., and Shephard, R. C. (1978). *Cancer* **42,** 2269–2281.
Goldenberg, D. M., and Hansen, H. J. (1972). *Science* **175,** 117–118.
Goldenberg, D. M., Deland, F., Kim, E., Bennett, S., Primus, F. J., van Nagell, J. R., Estes, N., Desimone, P., and Rayburn, P. (1978). *N. Engl. J. Med.* **298,** 1384–1388.
Greene, H. S. N. (1938). *Science* **88,** 357–358.
Greene, H. S. N. (1951). *Cancer Res.* **11,** 529–533.
Hata, J., Ueyama, Y., Tamaoki, N., Akatsuka, A., Yoshimura, S., Shimuzu, K., Morikawa, Y., and Furukawa, T. (1980). *Cancer* **46,** 2446–2455.
Hayashi, H., Kameya, T., Shimosato, Y., and Mukojima, T. (1978). *Am. J. Obstet. Gynecol.* **131,** 548–554.
Helson, L., Das, S. K., and Hajdu, S. I. (1975). *Cancer Res.* **35,** 2594–2599.
Hirohashi, H., Shimosato, Y., Kameya, T., Koide, T., Mukojima, T., Taguchi, Y., and Kageyama, K. (1979). *Cancer Res.* **39,** 1819–1828.
Hogan, B., Fellous, M., Avner, P., and Jacob, F. (1977). *Nature (London)* **270,** 515–518.
Holden, S., Bernard, O., Artzt, K., Whitmore, W. F., Jr., and Bennett, D. (1977). *Nature (London)* **270,** 518–520.
Houchens, D. P., and Ovejera, A. A., eds. (1978). "Proceedings of the Symposium on the Use of Athymic (Nude) Mice in Cancer Research." Fischer, Stuttgart.
Houghton, J. A., and Taylor, D. M. (1978). *Br. J. Cancer* **37,** 199–223.
Hussa, R. O., Patillo, R. A., Delfs, E., and Mattingly, R. F. (1973). *Obstet. Gynecol.* **42,** 651–657.
Javadpour, N. (1980). *Cancer* **45,** 2166–2168.
Kahan, B. W., and Levine, L. (1971). *Cancer Res.* **31,** 930–936.
Kameya, T., Kuramoto, H., and Suzuki, T. (1975). *Cancer Res.* **35,** 2025–2032.
Kameya, T., Shimosato, Y., Tumuraya, M., Ohsawa, N., and Nomura, T. (1976). *JNCI, J. Natl. Cancer Inst.* **56,** 325–332.
Kaneko, M., Takeuchi, T., Tsuchida, Y., Saito, S., and Endo, Y. (1980). *Gann* **71,** 14–17.
Kim, W., Takahashi, T., Nisselbaum, J. S., and Lewis, J. L. (1978). *Gynecol. Oncol.* **6,** 165–182.
Knecht, M., and Hertz, R. (1978). *Cancer Treat. Rep.* **62,** 2101–2103.
Kohn, J., and Raghavan, D. (1981). *In* "The Management of Testicular Tumors" (M. J. Peckham, ed.), pp. 50–69. Arnold, London.
Kondo, Y., Mizumoto, Y., Katayama, S., Murase, T., Yamaji, T., Ohsawa, N., and Kasaka, K. (1981). *Cancer Res.* **41,** 1545–1548.
Lange, P. H., Nochomovitz, L., Rosai, J., and 18 others (1980). *J. Urol.* **124,** 472–478.
Lemon, H. M., Lutz, B. R., and Pope, R. (1952). *Science* **115,** 461–465.
Lewis, J. L., Davis, R. C., and Ross, G. T. (1969). *Am. J. Obstet. Gynecol.* **104,** 472–478.
Mach, J. P., Carr, S., Merenda, C., Sordat, B., and Cerottini, J. C. (1974). *Nature (London)* **248,** 704–706.
McIlhinney, R. A. J. (1981). *Int. J. Androl.*, Suppl. **4,** 93–110.

Miwa, M., Sakura, H., Kawachi, T., Sugimura, T., Nagai, K., Hirohashi, S., Shimosato, Y., and Ohkura, A. H. (1976). In "Onco-Developmental Gene Expression" (W. H. Fishman and S. Sell, eds.), pp. 423–425. Academic Press, New York.

Monaghan, P., Raghavan, D., and Neville, A. M. (1982). *Cancer* **49**, 683–697.

Moshakis, V., Bailey, M. J., Ormerod, M. G., Westwood, J. H., and Neville, A. M. (1981a). *Br. J. Cancer* **43**, 575–581.

Moshakis, V., McIlhinney, R. A. J., Raghavan, D., and Neville, A. M. (1981b). *Br. J. Cancer* **44**, 91–99.

O'Hare, M. J. (1978). *Invest. Cell Pathol.* **1**, 39–63.

Ohsawa, N., Ueyama, Y., Morita, K., and Kondo, Y. (1977). *Proc. Int. Workshop Nude Mice, 2nd, 1976* pp. 365–374.

Osoba, D., and Auersperg, N. (1966). *JNCI, J. Natl. Cancer Inst.* **36**, 523–528.

Pantelouris, E. M. (1968). *Nature (London)* **217**, 370–371.

Phillips, B., and Gazet, J. C. (1967). *Nature (London)* **215**, 548–549.

Pierce, G. B., Dixon, F. J., and Verney, E. (1958). *Cancer Res.* **18**, 204–207.

Pierce, G. B., Dixon, F. J., and Verney, E. (1959). *Arch. Pathol.* **67**, 204–210.

Pierce, G. B., Midgley, A. R., and Verney, E. L. (1962). *Cancer Res.* **22**, 563–567.

Pierce, G. B., Jr. (1966) *Cancer* **19**, 1963–1983.

Povlsen, C. O., and Rygaard, J. (1971). *Acta Pathol. Microbiol. Scand.* **79**, 159–169.

Povlsen, C. O., Fialkow, P. J., Klein, E., Klein, G., Rygaard, J., and Wiener, F. (1973). *Int. J. Cancer* **11**, 30–39.

Primus, F. J., Wang, R. H., Goldenberg, D. M., and Hansen, H. J. (1973). *Cancer Res.* **33**, 2977–2982.

Raghavan, D. (1981). *Int. J. Androl., Suppl.* **4**, 79–92.

Raghavan, D., and Neville, A. M. (1982). In "Scientific Foundations of Urology" (D. I. Williams and G. D. Chisholm, eds.), 2nd ed. Heinemann, London pp. 785–796.

Raghavan, D., Gibbs, J., Neville, A. M., and Peckham, M. J. (1979). *N. Engl. J. Med.* **302**, 811.

Raghavan, D., Gibbs, J., Costa, R. N., Kohn, J., Orr, A. H., Barrett, A., and Peckham, M. J. (1980). *Br. J. Cancer.* **41**, Suppl. IV, 191–194.

Raghavan, D., Heyderman, E., Gibbs, J., Neville, A., and Peckham, M. J. (1981a). In "Thymusaplastic Nude Mice and Rats in Clinical Oncology" (G. B. A. Bastert, ed.), pp. 439–445. Fischer, Stuttgart.

Raghavan, D., Heyderman, E., Monoghan, P., Gibbs, J., Ruoslahti, E., Peckham, M. J., and Neville, A. M. (1981b). *J. Clin. Path.* **34**, 123–128.

Raghavan, D., Sullivan, A., Peckham, M. J., and Neville, A. M. (1982). *Cancer* **50**, 982–989.

Rosenthal, K. L., Tompkins, W. A. F., Frank, G. L., McCulloch, P., and Rawls, W. E. (1977). *Cancer Res.* **37**, 4024–4030.

Ruoslahti, E., Jalanko, H., Comings, D. E., Neville, A. M., and Raghavan, D. (1981). *Int. J. Cancer* **27**, 763–767.

Rygaard, J., and Povlsen, C. O. (1969). *Acta Pathol. Microbiol. Scand.* **77**, 758–760.

Sakashita, S., Hirai, H., Nishi, S., Nakamura, K., and Tsuiji, I. (1976). *Cancer Res.* **36**, 4232–4237.

Schilling, J. A., Snell, A. C., and Favata, B. V. (1949), *Cancer* **2**, 480–490.

Searle, F., Boden, J., Lewis, J. C. M., and Bagshawe, K. D. (1981). *Br. J. Cancer* **44**, 137–144.

Selby, P. J., Heyderman, E., Gibbs, J., and Peckham, M. J. (1979). *Br. J. Cancer* **39**, 578–583.

Shimosato, Y., Kameya, T., Nagai, K., Hirohashi, S., Koide, T., Hayashi, H., and Nomura, T. (1976). *JNCI, J. Natl. Cancer Inst.* **56**, 1251–1260.

Shimosato, Y., Kamey, T., Hayashi, H., Kubota, T., Iida, S., Hirohashi, S., and Mukojima, T. (1978). In "Proceedings of the Symposium on the Use of Athymic (Nude) Mice in Cancer Research" (D. Houchens and A. A. Ovejera, eds.), pp. 195–206. Fischer, Stuttgart.

Shin, S., Reid, L. C., Colburn, P. C., Kadish, A. S., and Fleischer, N. (1977). *Proc. Int. Workshop Nude Mice, 2nd, 1976* pp. 375–393.
Shirai, T., Yoshaki, T., and Itoh, T. (1977). *Gann* **68,** 847–849.
Shouval, D., Reid, L. M., Chakrabroty, P. R., Ruiz-Opazo, N., Morecki, R., Gerber, M. A., Thung, S. N., and Safritz, D. A. (1981). *Cancer Res.* **41,** 1342–1350.
Singer, R. M., Leahy, E. M., and Harrington, D. B. (1979). *In* 'Carcino-Embryonic Proteins' (F.-G. Lehmann, ed.), Vol. 2, pp. 234–240. Elsevier/North-Holland Biomedical Press, Amsterdam.
Sordat, B., Fritsche, R., Mach, J. P., Carrel, S., Ozzello, L., and Cerottini, J. C. (1974). *Proc. Int. Workshop Nude Mice, 1st, 1973* pp. 269–278.
Steel, G. G., Courtenay, V. D., and Rostom, A. Y. (1978). *Br. J. Cancer* **37,** 224–231.
Takeuchi, T., Nakayasu, M., Hirohashi, S., Kameya, T., Kaneko, M., Yokomori, K., and Tsuchida, Y. (1979). *J. Clin. Pathol.* **32,** 693–699.
Toolan, H. W. (1953). *Cancer Res.* **13,** 389.
Tveit, K. M., Fodstad, O., Brogger, A., and Olsnes, S. (1980). *Cancer Res.* **40,** 949–953.
Van Dooremaal, J. C. (1873). *Arch. Opthalmol.* **19,** 359–373 (cited in Schilling *et al.,* 1949).
Wewer, U., Albrechtsen, R., and Ruoslahti, E. (1981). *Cancer Res.* **41,** 1518–1524.
Yoshimura, S., Tamaoki, N., Ueyama, Y., and Hata, J. (1978). *Cancer Res.* **38,** 3474–3478.
Zeuthen, J., Norgaard, J., Avner, P., Fellous, M., Wartiovaara, J., Vaheri, A., Rosen, A., and Giovanella, B. B. (1980). *Int. J. Cancer* **25,** 19–32.

CHAPTER 7
Specific Antibodies in Cytopathology and Immunohistology

D. J. R. LAURENCE   A. M. NEVILLE

Ludwig Institute for Cancer Research (London Branch)
Royal Marsden Hospital
Sutton, Surrey, England

I.	Introduction	131
II.	The Improved Classification of Human Tumors	133
	A. Tumors of the Lymphoid Tissues	133
	B. Human Germ Cell Tumors	136
III.	The Detection of Micrometastases	139
IV.	Endocrine Aspects of Cancer	143
	A. Estrogen Receptor	143
	B. APUD Tumors	145
V.	Monoclonal Antibodies and Immunohistology	147
VI.	Conclusions	152
	References	152

I. INTRODUCTION

Immunocytological and immunohistochemical techniques are currently of increasing interest in the study of cancer biology and pathology. The exploration of cellular subsets within the network of cell differentiation can benefit from these methods that have an established utility in characterizing cellular variants in hematology and bacteriology. However, the subtle distinctions that are now required have prompted a considerable expansion of methodology in the last decade which now encompasses enzyme, radioisotope and fluorescence-tagged antibodies. A parallel seems to exist with the progress in studying tumor markers in the body fluids that occurred in the early 1970's following the establishment of radioimmunoassays.

Many of the problems that are now being studied at the cellular level were first investigated by means of conventional polyclonal antisera, e.g., anti-carcinoembryonic antigen (CEA) antibodies and the location and delineation of various carcinomas (Hsu *et al.,* 1972 and Fig. 1). The development of monoclonal techniques (Köhler and Millstein, 1975) has added a new refinement to all these studies, though it would be wrong to deny the considerable progress made by the older methods. Somatic hybridoma systems are free from many of the problems that deterred the more widespread use of antibodies as reagents. Monoclonal antibodies are of well defined specificity, and their characteristics can be screened at an early stage before many resources have been committed to any one antibody. Selection can, therefore, be made from a wider range of products, and it is likely that the reagents finally chosen will be appropriate to the problem to be solved. Once the antibody producing system is established, the reagent can be available in almost unlimited amounts, can be of constant quality, can be distributed internationally, and can become a standard reference. The antibodies so derived are likely to relate to a unique antigen that is sufficiently characterized by the antibody itself but is also relevant to the problem because of

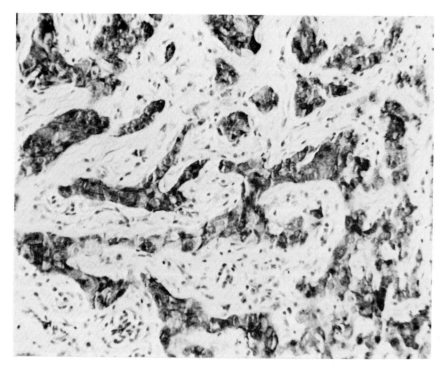

Fig. 1. Human breast carcinoma. Immunocytochemical demonstration with peroxidase-linked antibodies of carinoembryonic antigen (CEA) which may be seen in the cytoplasm and on tumor cell surfaces. ×153.

the nature of the initial screening process. The antibody can be used to isolate the antigen and, in some cases, to demonstrate its functional role.

In this chapter we have attempted to illustrate a few approaches to the value of using antibodies, both polyclonal and monoclonal, in the classification, detection, and biological understanding of some aspects of human neoplasms.

II. THE IMPROVED CLASSIFICATION OF HUMAN TUMORS

A. Tumors of the Lymphoid Tissues

In addition to classifying such disorders as being of T or B cells, leukemias have been found to be examples of blocks in the early stages of the differentiation pathway which such cells normally pursue.

1. T Cell Disorders

Previous work has established that acute lymphoblastic leukemia of the T cell type (T-ALL) resemble thymus cells rather than peripheral T cells, for example, in having the enzyme terminal deoxynucleotidyltransferase (TdT), which represents an early feature of T cell differentiation. Thymus cells and peripheral T cells also differ in their composition of surface antigens. Several groups of workers (e.g., Reinherz and Schlossman, 1980; Haynes, 1981) have developed a "panel" of monoclonal antibodies to thymus cells. These were then classified by their reactions with thymus and with peripheral T cells by cell sorting and immunocytochemical methods screening for "developmental antigens" that relate to cell mutants or functions. The "early" thymocytes have no developmental antigens in common with the peripheral cells. The "middle" thymocytes have some but not all of the antigens of the peripheral cells, while the "late" thymocytes carry all these antigens. About 10, 70–80, and 10% of thymocytes were found to fall into these groups, respectively. The antigens typical of the early cells were T9 and T10 recognized by antibodies OKT9 and OKT10. The T9 antigen has been identified as a transferrin receptor, already recognized to be a feature of actively dividing lymphoid cells (Golding and Burns, 1981). Antigen T10 is associated with the large TdT-containing blast cells in the thymus and also occurs in TdT positive cells that are found as a small percentage of the bone marrow cells which have colony-forming unit capability (Crawford *et al.*, 1981). Both T9 and T10 antigens disappear as the cells traverse the middle cell compartment.

Antigen T3 is characteristic of late cells and peripheral T cells, but is absent on early and middle cells. Antigen T3 can be used with immunoperoxidase staining as a marker for peripheral T cells and is a possible replacement for the conventional rosetting techniques (De Cock *et al.*, 1981). The antibody to T3 (OKT3) has the property of being mitogenic to T cells at a concentration of 10^{-12} M, suggesting that it can bind with high affinity to a control element in the lymphocyte surface. At a higher concentration, this antibody also inhibits cell mediated lysis (Chang *et al.*, 1981).

In addition to these antigens related to cell maturity there are other antigens related to cell function. In the thymus the middle cells acquire antigens T4 and T8 but, during passage to the late compartment, one of these two antigens is lost and the other is retained (Tidman *et al.*, 1981). Antigen T4 is a marker to T cells with "helper" function and T8 for cells with "suppressor" function.

A study of 21 patients with ALL showed that the majority (16 of 21 patients) had tumor cells characterized by the T9 and T10 antigens that correspond with the minority "early" cells of the thymus. In only one of these patients was the "late" T3 antigen expressed on the leukemic cells (Reinherz and Schlossman, 1980).

In contrast with the ALL patients, of whom the majority had cells with early antigens and with little spread of maturity in the tumor cell population, patients with T cell lymphoblastic lymphoma have a much wider spread of cell maturity in their tumor (Boumsell *et al.*, 1981). It is possible to find cells carrying the early and the late antigens in the same tumor. Similar diversity of antigen expression is seen in carcinomas (*vide infra*). The wide range of differentiation stages could account for resistance to therapy of these tumors. It is when the cellular population is more homogenous that a uniform response to chemotherapy is possibly more likely, even if the tumor cells are predominantly immature and so suggest a poor prognosis on the basis of conventional histology (McGrath, 1981).

The OKT series of antibodies spans the main course of development of the lymphocytes, but in cutaneous T cell lymphomas (Sezary syndrome), usually regarded as composed of mature T cells, a further stage is revealed by immunocytochemistry (Haynes *et al.*, 1981). The antibody 4F2 does not react with peripheral T cells unless they are activated, but it does react with the majority of Sezary cells and is an important clonal marker for these cells in the blood. Increased reaction of the lymphoma cells to 4F2 is accompanied by a loss in reactivity to antibody 3A1. This antibody reacts with the majority of the peripheral blood T cells. These results identify cutaneous lymphoma cells as a type of activated mature T cell.

It is still an open question whether the leukemic stem cells that are a potential target for chemotherapy have been identified by these techniques. Early cells with antigens T9 and T10 have a consistently high level of the enzyme TdT and carry the ALL antigen (Newman *et al.*, 1981). TdT can be detected immunochemically inside cells, and the immunological assay has about 1000 times the sensitivity of the conventional bulk assay done on cell homogenates (Okamura *et al.*, 1980). Considerable interest has been recently shown in the TdT positive cells in the bone marrow, especially since these were also found in the peripheral blood during blast crisis in chronic myelogenous leukemia. The bone marrow cells, unlike similar cells in the thymus, also carry the Ia-like antigen as do B cells. These cells can be thought of as a possible precursor of both the T and the B cell series. Changes in numbers of these cells in the bone marrow of patients with ALL during therapy have given a good indicator of disease activity (Miller *et al.*, 1981).

2. B Cell Disorders

The immediate precursor cell for the B cell series is a lymphocyte with no surface immunoglobulin and with intracellular μ heavy chains but no light chains (Schnit and Hijmans, 1980). The presence of Ia-like antigens on apparently nonfunctional "null" ALL may also suggest an early stage of B cell development. The mature B cells carry a characteristic surface immunoglobulin (SmIg). When malignancy develops the tumor cells can be shown to be a monoclonal population by the uniformity of their SmIg light chains. For example, Burkitts's lymphoma cells always have a monoclonal appearance when examined for light chains by immunocytochemistry, whereas other Epstein-Barr (EB) virus-associated lymphadenopathy lymphocytes are polyclonal (Wright, 1981). The ultimate immunological specificity is attainable by developing an antibody against the Ig idiotype or V region peptide sequence. Somatic hybridoma techniques can be used to convert B cell lymphoma cells with SmIg into a cell population secreting Ig of the same idiotype. The pulse cytofluorimeter method is useful in establishing the monoclonal nature of patient's lymphocytes by demonstrating a cell population both with a given light chain or idiotype and a given surface density of Ig. The uniformity in surface density is revealed when the light pulses on, optically exciting the anti-Ig fluorescent tag bound to the cells which are sorted in order of intensity. A monoclonal population can be detected when it is diluted with 1000 times the number of polyclonal mononuclear cells (Ligler *et al.*, 1980). The high sensitivity of this method can lead to a false interpretation if there is a concurrent secretion of Ig by the tumor. Nonmalignant mononuclear cells with Fc receptors (e.g., monocytes and L cells) will bind this monoclonal Ig and thereby imitate the appearance of a monoclonal B cell population. Chemotherapy in this case may reduce the number of these Fc receptor cells without causing any regression of the myeloma (King and Wells, 1981).

The expression of SmIg by B cells and B cell tumors provides an estimate of the sensitivity of immunofluorescence and immunoenzyme methods applied to the detection of cell surface molecules. A mature B cell has about 10^5 SmIg molecules on its surface, while a chronic lymphocytic leukemia (CLL) cell has 10^4 molecules. The SmIg methods applied to CLL cells based on immunofluorescence are at the margin of sensitivity. Some normal null cell peripheral lymphocytes react negatively with routine SmIg immunofluorescence techniques, but can be shown to have SmIg in the surface by rosetting methods. Similarly, a rosetting method using monoclonal anti-light chain antibody fixed to red blood cells can detect monoclonal SmIg in the surface of CLL cells with considerable sensitivity and specificity (Lowe *et al.*, 1981).

The advantage in sensitivity of cytochemical methods applied to the detection of tumor cells over methods that analyze body fluids is discussed in more detail in Section III.

B. Human Germ Cell Tumors

The use of immunocytochemical probes for human chorionic gonadotropin (hCG) and α-fetoprotein (AFP) has proved to be of significant value in questioning current classifications of human germ cell tumors and in suggesting modified schemes with biological, diagnostic, and therapeutic significance.

The production of hCG is classically associated with the trophoblastic elements which comprise choriocarcinomas (Bagshawe, 1969). However, elevated plasma levels of hCG have been noted in association with 15% of seminomas and 46% of nonseminomatous germ cell tumors (Mann *et al.*, 1980). This is generally regarded as being due to the presence of choriocarcinomatous elements in those tumors which, despite a meticulous search, however, are frequently not demonstrable. The application of immunocytochemical methods for hCG has shown that seminomas and nonseminomatous germ cell tumors, e.g., embryonal cell carcinomas, may contain hCG-positive cells (Fig. 2), often giant in type, forming syncitia in relation to blood vessels. Occasionally isolated single cells of varying size may also contain hCG. Thus hCG production can be associated with a germ cell tumor type which does not necessarily have the characteristic structural form of trophoblast elements.

Studies of AFP, both in the plasma and at the cellular level, also raise doubts about the accuracy of current classifications of germ cell tumors. AFP is a product of the normal yolk sac and its carcinomas (endodermal sinus tumors). Recent studies have drawn attention to the fact that embryonal cell carcinomas in the absence of histologic yolk sac elements may be associated with raised plasma AFP levels (Grigor *et al.*, 1977; Mann *et al.*, 1980). Similarly, rare examples of "pure seminomas" may be found to have raised AFP values (Raghavan *et al.*, 1981a and b, 1982).

Solutions to these problems have been derived from studies of human germ cell tumor xenografts, which maintain, through many serial passages, the structural and functional properties of the original human material. One xenograft, of especial interest, morphologically resembled the primary testicular tumor of its derivation and was a classic anaplastic seminoma but with cystic spaces. In some xenograft passages, the cysts contained occasional Schiller-Duval bodies, an

7. Specific Antibodies in Cytopathology and Immunohistology

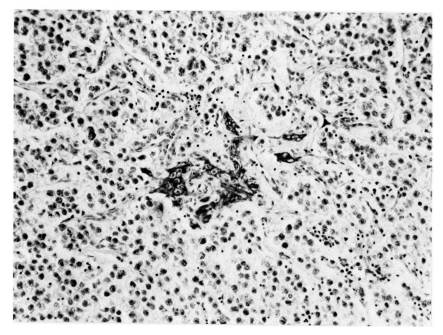

Fig. 2. Human testicular seminoma. Human chorionic gonadotropin (hCG) is demonstrable with peroxidase-linked antibodies in a large syncitial-type giant cell with several further satellite cells also being positive. The rest of the tumor cells of the seminoma are negative. ×87.6.

overt morphological feature of yolk sac-type differentiation (Raghavan et al., 1981b).

Using suitable immunocytochemical methods, AFP was demonstrated in the serum, in the cysts and their surrounding cells, and, to a lesser extent, in some of the cells in the solid seminoma areas of the tumor (Raghavan et al., 1981b). Fibronectin was also produced by this lesion (Ruoslahti et al., 1981).

Ultrastructurally the solid areas consisted of tumor cells with features of classic seminoma cells, whereas in the cystic areas the tumor cells were more like yolk sac carcinoma cells. Between those two extremes there were intermediate cells showing gradations of structure between seminoma and yolk sac (Monaghan et al., 1982). There is, therefore, an apparent ultrastructural and functional continuum between seminoma and yolk sac tumors. Moreover, the xenograft is not unique, as our further studies have confirmed that there is a group of tumors which at the light microscopic level have the appearances of seminoma but which, in fact, are more properly classified as yolk sac tumors.

Embryonal carcinoma cell xenografts also suggest that a continuum may exist between so-called embryonal cell carcinoma and yolk sac carcinoma (Monaghan

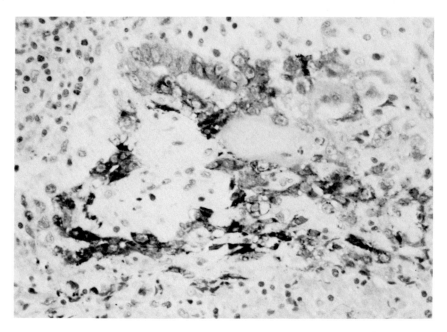

Fig. 3. Embryonal cell carcinoma of the human testis. An immunocytochemical stain for AFP using phosphatase as the indicator. AFP is present in the cytoplasm of some cells only, whose arrangement is with that of the classic yolk sac carcinoma. ×200.

et al., 1982). This conclusion has been derived from ultrastructural studies and the intracellular demonstration of AFP in foci of embryonal carcinomas (Fig. 3).

It may be, therefore, that many so-called embryonal cell carcinomas and seminomas represents tumors with blocks at different stages in their differentiation along an extraembryonic pathway (i.e., of placental and yolk sac type).

These conclusions regarding homology between different germ cell tumor types also help to explain how seminoma and teratoma may coexist; why non-seminomatous metastases may occur from tumors diagnosed as pure seminomas; why seminomatous and teratomatous elements may occur together in extragonadal sites; and the association of marker proteins such as hCG and AFP with lesions not showing the classic histological patterns associated with their production.

Germ cell tumors have, in theory, the multiple potential to form endodermal, ectodermal, mesodermal, and extraembryonic structures. In view of the limited ability of conventional histology to recognize different cell types (of AFP, hCG) and lineages, other markers are needed. The monoclonal antibody technique (Köhler and Milstein, 1975) may provide immune probes for cell surface differentiation antigens of the various cell lineages. Such requirements would then facilitate further the more accurate classification of germ cell tumors with histologic and clinical relevance.

III. THE DETECTION OF MICROMETASTASES

Of patients without known metastatic diesease at the time of initial surgery, about 1 in 4 will develop overt metastases after a follow-up of 2 years or more if the axillary lymph nodes were involved and 1 in 15 if no lymph nodes were involved. If these at-risk patients could be identified early enough, additional treatment with possible potential benefit could be given without involving the majority of patients in unnecessary therapy. The assay of tumor markers in the body fluids in breast cancer subjects fails to identify this at-risk population.

The most striking example of a tumor marker detected in the body fluids and sensitive to the presence of tumor cells is hCG as a marker for choriocarcinoma (Bagshawe, 1973). This tumor produces about 5 μg hCG per day for every milligram wet weight of tumor. Perhaps 10–20% of this will enter the urine. A recent β-hCG assay gives a background level of about 0.45 μg for a 24-hour sample of urine (Javadpour and Chen, 1981), and consequently about 1 mg tumor could be detected by this system. Conditions are rarely as favorable as this, and with about the same rate of secretion of monoclonal Ig, patients with multiple myeloma contain 50–100 g of cells before a significant elevation above the polyclonal background can be detected (Salmon and Smith, 1970).

With breast cancer, Coombes *et al.* (1981) investigated 26 potential biochemical tests of plasma or tumor that had been suggested by at least one laboratory to be potential value in monitoring breast cancer. Only one of these markers could have been regarded as breast specific, i.e., α-lactalbumin, and this was the most consistently disappointing of all the tests. Not surprisingly, few of these tests gave an abnormal level during the follow-up of patients with localized breast cancer that could be related to subsequent development of metastases. However, at least one of the markers usually showed abnormally high levels at the time that metastases became clinically apparent. The average lead time obtained from the three most sensitive markers, i.e., CEA, total alkaline phosphatase, and γ-glutamyltransferase, was 3 months, and this was only obtained in 50% of patients. As a method of anticipating recurrence at the time of initial surgery, therefore, these methods had no clinical utility.

Useful cytochemical markers for detection of residual breast cancer have been obtained by exploiting the milk fat globule membrane (MFGM) as a source of mammary epithelial cell membranes. It was established in the early 1960's that the proteinaceous coating of the fat globules in the milk is a result of the way the globules are extruded through the mammary epithelial cell surface, trapping some epithelial membrane on the globules.

Ceriani *et al.* (1977) considered that their anti-MFGM antibody was mammary cell specific, but Heyderman *et al.* (1979) described an anti-MFGM antiserum that reacted with a surface antigen of most organs, showing glandular differentiation including the salivary glands, pancreas, stomach, some bronchial mucous glands, bile ducts, endometrial and cervical glands, as well as with mammary epithelial cells. In the normal mammary tissues reaction was with the luminal surface of the epithelial cells (Fig. 4). Because of this widespread distribution, the substance that was detected by the antibody was termed the epithelial membrane antigen (EMA) (Heyderman *et al.*, 1979).

Over a hundred breast carcinomas have been examined immunocytochemically and all showed expression of EMA, including those with various degrees of differentiation. When glandular differentiation could be seen, EMA occurred in the luminal surfaces with variable cytoplasmic staining and some staining of other cell surfaces (Fig. 5). In metastases involving small aggregates or single cells, the staining was mainly cytoplasmic, but clearly recognizable as though the substance was placed in the membrane in the presence of an appropriate tissue architecture but otherwise accumulated inside the cells. No EMA-positive cells could be detected in normal lymph nodes, bone marrow, and liver par-

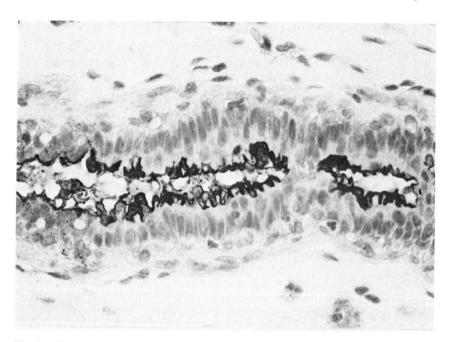

Fig. 4. Human normal breast duct. An immunoalkaline phosphatase reaction to demonstrate the epithelial membrane antigen (EMA) on the luminal face of the lining epithelial duct cells. Note the outer myoepithelial cells and the stroma are negative. ×200.

7. Specific Antibodies in Cytopathology and Immunohistology

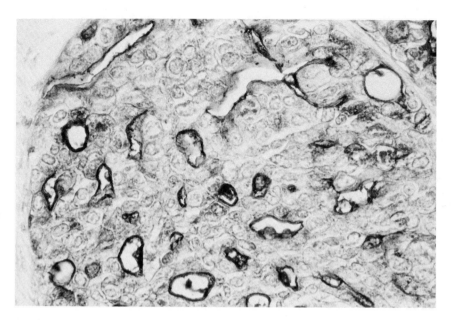

Fig. 5. Human breast carcinoma. An immunoperoxidase stain for the epithelial membrane antigen EMA showing expression of the antigen on the luminal membrane of ducts and in the cytoplasm of some tumor cells. ×224.

enchyma of patients without a history of malignant disease. The observation paved the way to examine at a cellular level bone marrows from patients with breast cancer but without overt osseous metastases to ascertain whether this immunocytochemical approach with EMA was able to detect truly micrometastic disease. A pilot study (Sloane *et al.*, 1980) was made by comparing the reports on conventional staining of marrow smears from one iliac crest done at the time of receiving the samples for EMA staining of paraffin blocks of the remaining part of the smears. A group of patients who had remained free of recurrence over 5 years after primary surgery were found to have no EMA positive cells in their bone marrow sections. EMA-positive cells in the marrow were seen in patients with lymph node involvement but with no known distant metastases. However in patients with distant metastases, EMA-positive cells were regularly found. Of major importance, however, was the finding, in seven patients, of EMA-positive cells in the marrow not detected by conventional staining. The tumor cells occurred in small groups or as isolated cells (Fig. 6).

To increase the sensitivity of the method, further studies were effected staining smears for EMA and conventional May-Grunwald Giemsa staining (Dearnaley *et al.*, 1981). In carrying out the immunocytochemical reaction for EMA, it was found necessary to separate out the nucleated cells by removal of the red blood

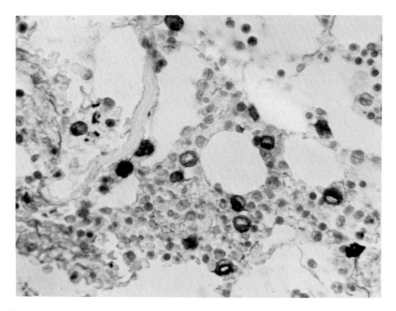

Fig. 6. Human breast carcinoma. An immunoalkaline phosphatase stain for EMA showing breast cancer cells singly and in groups in the bone marrow whose normal constituent cells are negative. ×100.

cells using Lymphoprep. The smears from 74 patients were examined in this way, and 21 of the patients' marrow samples were found to contain tumor cells as revealed by the EMA method compared with eight by conventional methods. In only one case was the conventional method positive and the EMA method negative, and this case could be attributed to a failure to obtain a sufficiently large sample of the marrow. When the reason for the success of the EMA method was analyzed it was found that either method was equally successful if the smear contained more than 100 tumor cells. With less than 100 EMA-positive cells per smear the conventional method was unable to detect any tumor cells.

The main failure of EMA immunocytochemistry in this study was in not detecting clinically apparent bone metastases, probably due to an uneven distribution of the metastases in the marrow. A single smear corresponds to a marrow sample of about 0.2 ml or perhaps 1 part in 10,000 of the total marrow space. It was considered that increasing the marrow sample with the scanning of multiple smears from both iliac crests, the sternum, and lumbar vertebra would increase the detection rate significantly. With a 5-ml marrow sample the labor of working through the smears proved excessive in terms of technician time and a method of enriching the nucleated cell population in tumor cells became necessary. This was done by a rosetting method using an antibody to bone marrow cells to remove most of the nonmalignant cells before making the smear (Buckman *et al.*, 1980).

It was confirmed that the increase in marrow volume had a significant effect in improving detection of metastatic tumor and that any reduction in volume would result in some falsely negative results. In some cases, the tumor was seen as a single cell in the 5 ml sample analyzed.

Supposing the total metastatic volume to be 1000 times the sample volume and that the tumor cells are distributed at random within this volume, it becomes possible to detect 10^3 cells in the body. This is a significant improvement over the 1 mg tumor (10^6 cells) that can be detected by the most sensitive assay for tumor marker in the body fluids. The long-term prognostic value of this advance in diagnostic sensitivity is the subject of current investigation at this Institute.

This immunocytochemical approach could also be valuable in the detection of osseous metastases from other carcinomas. Antibodies are available which may facilitate the finding of prostatic (prostatic acid phosphatase, prostate-specific antigen) Nadji *et al.*, 1979), bronchial (EMA, keratins, hormones such as ACTH and calcitonin, neurone-specific enolase) colonic (CEA, zinc glycinate marker) carcinomas, and malignant melanomas (S100 protein) (Gaynor *et al.*, 1981) or p97 (Woodbury *et al.*, 1980) at this site.

IV. ENDOCRINE ASPECTS OF CANCER

A. Estrogen Receptor

The detection of tumor cells in the body requires a uniformity of response of the cells to immunocytochemical reagents. The detection of therapeutic target cells requires a different approach, i.e., to obtain as far as possible a relevant measure of heterogeneity within the tumor. Breast carcinoma cells are unlikely to respond to hormone therapy if they lack the necessary receptors and cellular mechanisms required to generate the responses. This is confirmed in practice with a low response (<5%) to hormone therapy if the tumor appears to lack significant amounts of estrogen receptor but about 50% response if estrogen receptors are detected. The probability of response increases with the content of receptors.

Estrogen receptors are intracellular components of mammary epithelial cells. Immunocytochemical methods have been directed mainly at detecting estradiol at the binding sites of the receptor using anti-estradiol antisera. In the simplest form, the tissue section is treated with estradiol, excess is removed, and then the section stained with anti-estradiol antiserum (Ghosh *et al.*, 1978). This procedure has been the subject of some criticism, as it was felt that a specific combination of a small molecule such as a steroid with two macromolecules, i.e., the receptor and the antibody, is inherently unlikely because of steric constraints. As a modification of the method, a polymer or multifunctional complex of estradiol is made so that the same small molecule would not be involved in both complexing reactions (Taylor et al., 1981). More recently specific monoclonal antibodies directed toward the steroid receptor itself have become available (Greene et al., 1980). It is too early to judge whether these reagents are able to solve the problem of determining the location of estradiol receptors, though preliminary reports are extremely promising.

One difficulty with detection of estrogen receptors by immunocytochemistry is the relatively small number of receptor molecules in a cell. A rough calculation estimates that the level of "just positive" in the bulk assay of a breast carcinoma homogenate is 10 fmoles/mg soluble protein; there are 250 molecules of receptor per cell. It will be recalled (see Section II) that 10^4 molecules SmIg are difficult to detect by immunofluorescence. Immunoperoxidase staining can be somewhat more sensitive under the best conditions. The kinetics of horseradish peroxidase have been intensively investigated (Chance, 1950), and at 10^{-6} M enzyme concentration the time constant of oxidation of a substrate is of the order of 1 second. At 10^{-9} M corresponding to the concentration of receptor in a weakly receptor positive tumor, the reaction would be estimated to require about 15–20 minutes if a simplistic extrapolation were justified. It is likely, however, that other factors, such as endogenous peroxidases, would be a limiting factor of an unknown order of magnitude.

A good correlation has been found between the relative levels of estrogen receptor measured by histochemistry and in homogenates by the dextran-coated charcoal (DCC) procedure or between histochemical estimates and response to estrogen therapy (Pertschuk et al., 1981). Furthermore, migration of receptors from the cytoplasm to the nucleus has been observed histochemically (Barrows et al., 1980). One possible explanation for the discrepancies is that the receptors are not distributed uniformly over the cellular population, and low levels are associated with a high proportion of receptor negative cells. The concentration of estrogen receptors in a small part of the total population would increase the ease of their detection in these cells. It is the main conclusion of histochemical estrogen receptor determinations that there is a considerable heterogeneity of distribution within the tumor cell population.

Alternatively, there may be a second population of receptors with lower affinity and higher concentration than for the physiologically effective estrogen receptors (Chamness et al., 1980). The concentration of these secondary (or type II) receptors may be sufficiently correlated with the primary receptors to account for the correlations in comparisons with the DCC procedure, and these receptors may also migrate to the nucleus if suitably loaded with estrogen. The situation is complicated by the histochemical observation of progesterone receptors (Lee, 1980) for which a similar dual affinity system might need to be postulated.

The heterogeneity of the cellular populations revealed by these methods is clearly important in interpreting the response of mammary tumors to endocrine therapy. Although the response of a receptor-rich tumor is likely to occur, it is also unlikely to be of long duration. Estrogen receptor studies on cells could be helpful in explaining relapse under endocrine therapy by proliferation of the receptor-poor part of the tumor. However, metastatic breast carcinoma in general is not receptor-poor, and some metastatic deposits can contain more receptors than are found in the primary tumor.

B. APUD Tumors

The endocrine of APUD (Amine Precursor Uptake and Decarboxylation) cell tumors include a wide variety of types, i.e., medullary thyroid carcinoma, pancreatic islet cell tumors, adrenal medullary tumors, carcinoid tumors, small cell carcinoma of the lung, and possibly corticotropin-producing tumors of the pituitary. Only medullary thyroid carcinoma and the adrenal medullary tumors have a certain origin from the neural crest. The other examples are more likely to arise from endodermal cells. Many of these tumors have an origin from a minority cell population in the tissue of origin.

Medullary thyroid carcinoma is derived from the thyroid C cells, and small cell carcinoma of the lung is probably derived from the bronchial Kultschitzsky or K cells. The APUD cell tumors are characterized by secretion of a number of different peptides which are stored in characteristic granules (Fig. 7). With the pancreatic islet cell tumors there is usually one principal product, and the patient then manifests symptoms of hormone excess whether this is of insulin, gastrin, somatostatin, or vasoactive intestinal peptide (VIP). Occasionally a pancreatic tumor can be shown by immunocytochemistry to have a more diverse spread of synthesis; one example apparently making β-endorphin enkephalin, gastrin, somatostatin, and ACTH in a minority of cells of the tumor (Wilander *et al.*, 1981).

Other APUD tumors most often produce more than one product. Medullary thyroid carcinomas may contain immunocytochemically β-endorphin and somatostatin as well as calcitonin (Fig. 8), the product regarded as typical of the tissue of origin. In some cells, all of these hormones could be detected (Deftos *et al.*, 1980). Significant secretion of multiple hormones, such as calcitonin, corticotrophin ACTH, and vasopressin ADH, is commonly found in patients with small cell carcinomas of the lung.

The association of ACTH with β-endorphin and β-LPH in tumor secretions is understandable because they are synthesized together from the same messenger RNA template and separated by posttranslational proteolysis. Recently, a close examination of the 5′ end of the messenger RNA suggested that there was a message for an additional hormone named γ-MSH (gamma melanotrophin). By reading the amino acid sequence from the base sequence of the RNA, it was possible to synthesize the peptide sequence of γ-MSH, a previously unknown

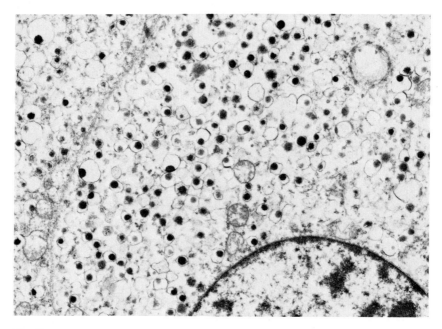

Fig. 7. Human phaeochromocytoma. The conventional ultrastructure of parts of two tumor cells are shown with the nucleus on the lower right. The cytoplasm is full of typical dense core granules with a peripheral halo typical of APUD cells. ×13,200.

substance, and develop an antibody against it. Immunocytochemical studies with this antibody were successful in locating γ-MSH in the pituitary and in a medullary thyroid carcinoma (Nakai et al., 1980). This is an example of a new trend in immunocytochemistry. As the base sequence of the DNA becomes known, it is possible to generate antibodies to each of the segments translated into peptide "code" and then examine where they are expressed in the body or in tumors. This trend is actively being pursued, for example, by workers in the virus field (Lerner et al., 1981). The use of radioactively labeled DNA's generated by a reverse transcription represents an alternative probe to supplement the results of immunocytochemistry.

The association of ACTH and calcitonin secretion is more difficult, and, at present, probably impossible to explain. The DNA coding is known for both these substances, and a short sequence up- and downstream from the genes is also known (Jacobs et al., 1981); but if these products have a relationship expressed at DNA level, this is at present not understood. That there is a physiological relationship is likely since both calcitonin and ACTH levels increase on provocative testing of patients with medullary carcinoma of the thyroid (Lamberts et al., 1980).

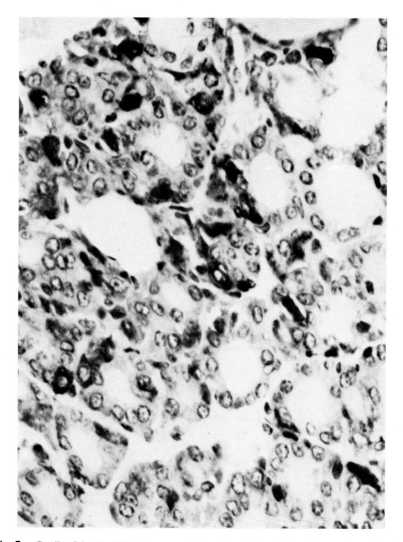

Fig. 8. C cells of the thyroid. An immunoalkaline phosphatase strain to demonstrate calcitonin in the perifollicular C cells. ×300.

V. MONOCLONAL ANTIBODIES AND IMMUNOHISTOLOGY

Several references have already been made to the potential of monoclonal antibodies as immunological probes, and their value has been illustrated in relation to leukemic disorders. There is little doubt that such reagents will be-

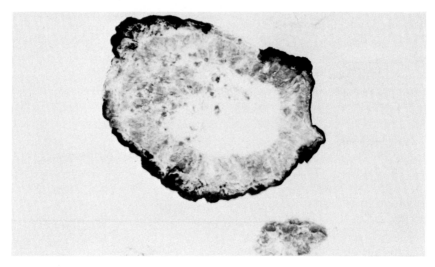

Fig. 9. Murine teratocarcinoma OC1551. An embryoid body derived from OC1551 is stained with monocloan antibody LICR-LON-22/23. Only the peripheral endoderm is positive. × 450. (From McIlhinney *et al.*, 1981, by permission of John Wiley & Sons Limited.)

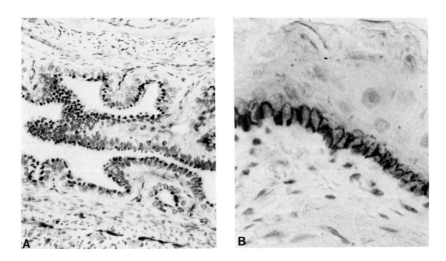

Fig. 10. Normal Mouse Oviduct (A) and Esophagus (B). The oviduct contains a supranuclear area stained by LICR-LON-22/23 which also stains a membrane antigen expressed only by the basal cells of the esophagus. Note the lack of staining of the basal cells where they rest on the basement membrane. (A) ×144; (B) ×540. (From McIlhinney *et al.*, 1981, by permission of John Wiley & Sons Limited.)

come the dominant immunological probes in cellular biology and pathology of diverse tissues and tumors in most, if not, all future studies (Yelton and Scharff, 1981). While monoclonal antibodies have been prepared to defined markers such as hormones (Cuello *et al.*, 1980), AFP and CEA (Kupchick *et al.*, 1981), blood group and red cell antigens (Anstee and Edwards, 1982), the HLA complex and related antigens (Fleming *et al.*, 1981; Howe *et al.*, 1981), receptors (Yavin *et al.*, 1981), and intermediate filament proteins (Alttmansberger *et al.*, 1981), most attention has been paid to raising antibodies to as yet uncharacterized human cell surface components with tumor, tissue, or cell specificity (Koprowski *et al.*, 1978, 1981; Steplewski, 1980; Kennett and Gilbert, 1979; Cuttitta *et al.*, 1981).

While several monoclonal antibodies have been obtained with apparent tissue or cell specificity, such conclusions have been based upon the lack of binding of a particular antibody to either cell lines other than these of the particular human type to which the antibody was raised or to extracts of other tumors or normal tissues.

This approach is inadequate to substantiate specificity claims. In all instances,

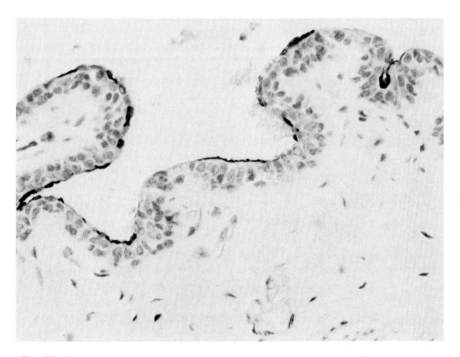

Fig. 11. Normal human breast duct. Using an immunoalkaline phosphatase stain, an anti-breast epithelial monoclonal antibody (LICR-LON-M8) is seen to detect a cell surface antigen expressed by some but not all of the luminal epithelial cells. ×240.

immunohistology is also required, screening a wide variety of normal and neoplastic tissues. This is well illustrated by our own studies of a monoclonal antibody raised to murine teratocarcinoma cells which, on the basis of lack of binding to other cell lines, appeared to be specific. An immunocytochemical screen of normal rodent tissues revealed binding to several subpopulations of normal adult cells and in the cytoplasm of certain gonadal and adnexal tissues (Figs. 9 and 10) (McIlhinney *et al.*, 1981). Absorption studies of tissue extracts would almost certainly have failed to detect this distribution. Others have also reached similar conclusions with respect to certain mammary, melanoma-related, and lymphocyte antigens (Arklie *et al.*, 1981; Murphy *et al.*, 1981; Foster *et al.*, 1982a; Woodbury *et al.*, 1980). Nevertheless, after adequate screening, some monoclonal antibodies may have been identified which have tissue and/or tumor specificity (Colcher *et al.*, 1981). Such results show the potential power of this approach to deriving meaningful biological and pathological immunological probes.

Although certain monoclonal antibodies may not be wholly specific to a par-

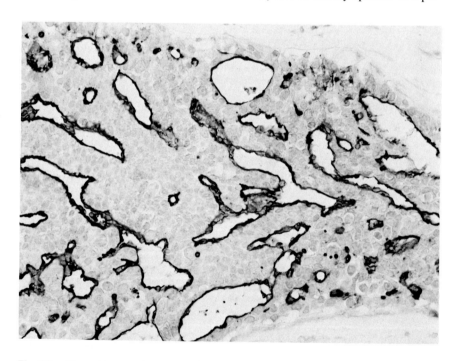

Fig. 12. Human breast carcinoma. An intraduct carcinoma is shown stained by an immunoalkaline phosphatase method using LICR-LON-M8 monoclonal antibody. Note the expression of the antigen on those tumor cells forming lumina only. ×262.5. (From Foster *et al.*, 1982b, by permission of Springer-Verlag.)

ticular normal or neoplastic tissue, such reagents may still demonstrate intratissue specificity. In this way subsets of cells within a particular cell lineage may be detected which are not recognizable by conventional morphology. While this phenomenon is already known in tissues, such as the gut, bronchus, and endocrine system, other normal tissues, such as the breast ducts, appear to be composed of a single uniform epithelial cell type. However, monoclonal antibodies can detect cell surface antigenic differences (heterogeneity) between the lining epithelial cells (Fig. 11). Moreover, there may be sequential distribution of some of the antigens which are present in some ducts and absent from the cells of others (Foster et al., 1982a). Whether this heterogeneity is due to different cell types, or to a single cell type at different stages of a functional cycle in the composition of the cell membranes, or to posttranslational cellular events that are less clearly defined than the peptide chain directed antibody control mechanisms remains to be ascertained.

However, in view of these normal breast cell data it is, therefore, not surprising to find heterogeneity of antigen expression in breast carcinomas (Figs. 12 and 13) (Foster et al., 1982b).

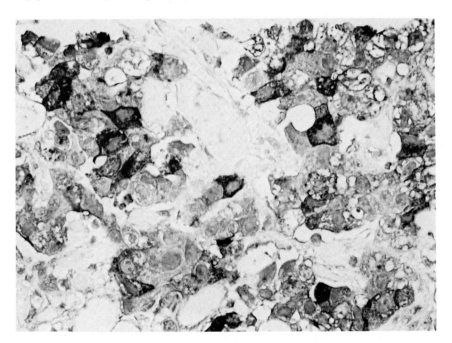

Fig. 13. Human breast carcinoma. An infiltrating duct carcinoma is stained with LICR-LON-M24 and shows the cytoplasmic location of the relevant antigen and marked heterogeneity of expression. In the normal breast this antigen is on the epithelial luminal membrane. ×300. (From Foster et al., 1982b, by permission of Springer-Verlag.)

This ability of monoclonal antibody probes to supplement conventional morphology could be of major histologic importance in studies of tissue differentiation and could in time yield reagents with prognostic and cell lineage relevance in tumors.

Regretfully, conventional processing of tissues for light morphology (e.g., formalin fixation, paraffin embedding) often results in antigen loss or modification so that antibodies which bind to unfixed cells no longer work after fixation and processing. There is, thus, a need to develop fixatives which still give good morphology but which preserve antigens to a greater extent.

VI. CONCLUSIONS

The study of cells and tissues by the use of specific antibodies to tumor markers is a rapidly expanding activity and no doubt at present we are perceiving only the first few impressions of its true potential. In defining stages of differentiation, our knowledge of the lymphoid tumors is clearly in advance of anything that can be forseen in the study of carcinomas.

However, probes for solid tissues are becoming available, and by means of cell sorting and immunocytochemistry, the path is now opening for defining cellular subset stages of differentiation and the effects of therapy and for a more precise understanding of the histogenesis of neoplasms.

REFERENCES

Altmannsberger, M., Osborn, M., Schauer, A., and Weber, K. (1981). *Lab. Invest.* **45**, 427–434.
Anstee, D. J., and Edwards, P. A. W. (1982). *Eur. J. Immunol.* **12**, 228–232.
Arklie, J., Taylor-Papdimitriou, J., Bodmer, W., Egan, M., and Millis, R. (1981). *Int. J. Cancer* **28**, 23–29.
Bagshawe, K. D. (1969). "Choriocarcinoma. The Clinical Biology of the Trophoblast and its Tumours." Arnold, London.
Bagshawe, K. D. (1973). *Adv. Cancer Res.* **18**, 231–263.
Barrows, G. H., Stroupe, S. B., and Riehm, J. D. (1980). *Am. J. Clin. Pathol.* **73**, 330–339.
Boumsell, L., Reinherz, E. L., Nadler, L. M., Ritz, J., Coppin, H., Richard, Y., Valensi, F., Dausset, J., Flandrin, G., Lemerle, J., and Scholossman, S. F. (1981). *Blood* **57**, 1105–1110.
Buckman, R., Dearnaley, D. P., Coombes, R. C., and Neville, A. M. (1980). *Cancer Chemother. Pharmacol., Suppl.* **5**, 7.
Ceriani, R. L., Thompson, K., Peterson, J. A., and Abraham, S. (1977). *Proc. Natl. Acad. Sci. U.S.A.* **74**, 582–586.
Chamness, G. C., Mercer, W. D., and McGuire, W. L. (1980). *J. Histochem. Cytochem.* **28**, 792–798.
Chance, N. (1950). *J. Am. Chem. Soc.* **72**, 1577–1581.
Chang, T. W., Kung, P. C., Gringas, S. P., and Goldstein, G. (1981). *Proc. Natl. Acad. Sci. U.S.A.* **78**, 1085–1088.

Colcher, D., Horan Hand, P., Nuti, M., and Scholm, J. (1981). *Proc. Natl. Acad. Sci. U.S.A.* **78,** 3199–3203.
Coombes, R. C., Powles, T. J., Gazet, J.-C., Ford, H. T., McKinna, A., Abbott, M., Gehrke, C. W., Keyser, J. W., Mitchell, P. E. G., Patel, S., Stimson, W. H., Worwood, M., Jones, M., and Neville, A. M. (1981). *Cancer* **48,** 310–315.
Crawford, D. H., Francis, G. E., Wing, M. A., Edwards, A. J., Janossy, G., Hoffbrand, A. V., Prentice, H. G., Sercher, D., McConnell, I., Kung, P. C., and Goldstein, G. (1981). *Br. J. Haematol.* **49,** 209–218.
Cuello, A. C., Galfre, G., and Milstein, C. (1980). *In* "Receptors for Neurotransmitters and Peptide Hormones" (G. Pepeu, M. J. Kuhar, and S. J. Enna, eds.), pp. 349–363. Raven Press, New York.
Cuttitta, F., Rosen, S., Gazdar, A. F., and Minna, J. D. (1981). *Proc. Natl. Acad. Sci. U.S.A.* **78,** 4591–4595.
Dearnaley, D. P., Sloane, J. P., Ormerod, M. G., Steele, K., Coombes, R. C., Clink, H. M. D., Powles, T. J., Ford, H. T., Gazet, J.-C., and Neville, A. M. (1981). *Br. J. Cancer* **44,** 85–90.
De Cock, W., De Cree, J., and Verhaagen, J. (1981). *J. Immunol. Methods* **43,** 131–134.
Deftos, L. J., Bone, H. G., and Parthenmore, J. G. (1980). *J. Clin. Endocrinol. Metab.* **51,** 857–862.
Fleming, K. A., McMichael, A., Morton, J. A., Woods, J., and McGee, J. O'D. (1981). *J. Clin. Pathol.* **34,** 779–784.
Foster, C. S., Edwards, P. A. W., Dinsdale, E. A., and Neville, A. M. (1982a). *Virchows Arch. A: Pathol. Anat. Histol.* **394,** 279–293.
Foster, C. S., Dinsdale, E. A., Edwards, P. A. W., and Neville, A. M. (1982b). *Virchows Arch. A: Pathol. Anat. Histol.* **394,** 295–305.
Gaynor, R., Herschmann, M. R., Irie, R., Jones, P., Morton, D., and Cochran, A. (1981). *Lancet* **1,** 869–871.
Ghosh, L., Ghosh, B. C., and Das Gupta, T. K. (1978). *J. Surg. Oncol.* **10,** 221–224.
Golding, J. W., and Burns, G. F. (1981). *J. Immunol.* **127,** 1256–1258.
Greene, G. L., Nolan, C., Engler, J. P., and Jensen, E. V. (1980). *Proc. Natl. Acad. Sci. U.S.A.* **77,** 5115–5119.
Grigor, K. M., Detre, S. I., Kohn, J., and Neville, A. M. (1977). *Br. J. Cancer* **35,** 52.
Haynes, B. F. (1981). *Immunol. Rev.* **57,** 127–161.
Haynes, B. F., Bunn, P., Mann, D., Thomas, C., Eisenharth, G. S., Minna, J., and Fauci, A. S. (1981). *J. Clin. Invest.* **67,** 523–530.
Heyderman, E., Steele, K., and Ormerod, M. G. (1979). *J. Clin. Pathol.* **32,** 35–39.
Howe, A. J., Seeger, R. C., Molinaro, G. A., and Ferrone, S. (1981). *JNCI, J. Natl. Cancer Inst.* **66,** 827–829.
Hsu, K. C., Zimmerman, E. A., Rudin, L., LoGerfo, P., Bennett, S., and Tattenbaum, M. (1972). *In* "Embryonic and Fetal Antigens in Cancer" (W. G. Anderson, J. H. Coggin, E. Cole, and J. W. Holleman, eds.), Vol. II, pp. 147–151. Oak Ridge Natl. Lab., Oak Ridge, Tennessee.
Jacobs, J. W., Goodman, R. H., Chin, W. W., Dee, P. C., and Habener, J. F. (1981). *Science* **213,** 457–459.
Javadpour, N., and Chen, H.-C. (1981). *J. Urol.* **126,** 176–178.
Kennett, R. H., and Gilbert, F. (1979). *Science* **203,** 1120–1121.
King, M. A., and Wells, J. V. (1981). *Clin. Exp. Immunol.* **45,** 552–556.
Köhler, G., and Milstein, C. (1975). *Nature (London)* **256,** 495–497.
Koprowski, H., Steplewski, Z., Herlyn, D., and Herlyn, M. (1978). *Proc. Natl. Acad. Sci. U.S.A.* **75,** 3405–3409.
Koprowski, H., Herlyn, M., and Steplewski, Z. (1981). *Science* **212,** 53–55.
Kupchik, H. Z., Zurawski, V. R., Hurrell, J. G., Zamcheck, N., and Black, P. H. (1981). *Cancer Res.* **41,** 3306–3310.

Lamberts, S. W. J., Hackeng, W. H., and Visser, T. J. (1980). *J. Clin. Endocrinol. Metab.* **50,** 565–568.
Lee, S. H. (1980). *Am. J. Clin. Pathol.* **73,** 323–329.
Lerner, R. A., Sutcliffe, J. G., and Shinnick, T. M. (1981). *Cell* **23,** 309–310.
Ligler, F. S., Smith, R. G., and Kettman, J. R. (1980). *Blood* **55,** 792–801.
Lowe, J., Hardie, D., Jefferis, R., Ling, N. R., Drysdale, P., Richardson, P., Raykundalia, C., Catty, D., Appleby, P., Drew, R., and MacLennan, C. M. (1981). *Immunology* **42,** 649–660.
McGrath, I. I. (1981). *JNCI, J. Natl. Cancer Inst.* **67,** 501–514.
McIlhinney, R. A. J., Dinsdale, E., and Neville, A. M. (1981). *Diagn. Histopathol.* **4,** 129–135.
Mann, K., Lamerz, R., Hellmann, T., Kumper, H. J., Staehler, G., and Karl, H. J. (1980). *Oncodev. Biol. Med.* **1,** 301.
Miller, W. M., Stass, S. A., Schumaker, H. R., and Bollum, F. J. (1981). *Am. J. Hematol.* **10,** 1–8.
Monaghan, P., Raghavan, D., and Neville, A. M. (1982). *Cancer* **49,** 683–697.
Murphy, G. F., Bhan, A. K., Sato, S., Harrist, T. J., and Mihm, M. C., Jr. (1981). *Lab. Invest.* **45,** 465–468.
Nadji, M., Tabei, Z., Castro, A., and Morales, A. R. (1979). *Lancet* **1,** 671–672.
Nakai, Y., Tanaka, I., Fukata, J., Nakao, K., Oki, S., Takai, S., and Imura, H. (1980). *J. Clin. Endocrinol. Metab.* **50,** 1147–1150.
Newman, R. A., Sutherland, R., and Greaves, M. F. (1981). *J. Immunol.* **126,** 2024–2027.
Okamura, S., Crane, F., Jamal, N., Messner, H. A., and Mak, T. W. (1980). *Br. J. Cancer* **41,** 159–167.
Pertschuk, L. P., Tobin, E. H., Gaetjens, E., Carter, A. C., Degenshein, G. A., Bloom, N. D., and Brigati, D. J. (1981). *Cancer* **46,** 2896–2901.
Raghavan, D., Heyderman, E., Gibbs, J., Neville, A. M., and Peckham, M. (1981a). *In* "Thymus Aplastic Nude Mice and Rats in Clinical Oncology" (G. Bastert, H. Schmidt-Mathiesson, and H. P. Fortimeyer, eds.), pp. 439–445. Fischer, Stuttgart.
Raghavan, D., Heyderman, E., Monaghan, P., Gibbs, J., Ruoslahti, E., Peckham, M. J., and Neville, A. M. (1981b). *J. Clin. Pathol.* **34,** 123–128.
Raghavan, D., Sullivan, A. L., Peckham, M. J., and Neville, A. M. (1982). *Cancer* **50,** 982–989.
Reinherz, E. L., and Schlossman, S. F. (1980). *Cell* **19,** 821–827.
Ruoslahti, E., Jalanko, H., Comings, D. E., Neville, A. M., and Raghavan, D. (1981). *Int. J. Cancer* **27,** 763–767.
Salmon, S. E., and Smith, B. A., (1970). *J. Clin. Invest.* **49,** 1114–1121.
Schnit, H. R. E., and Hijmans, W. (1980). *Clin. Exp. Immunol.* **41,** 567–572.
Sloane, J. P., Ormerod, M. G., Imrie, S. F., and Coombes, R. C. (1980). *Br. J. Cancer* **42,** 392–398.
Steplewski, Z. (1980). *Transplant. Proc.* **12,** 384–387.
Taylor, C. R., Cooper, C. L., Kurman, R. J., Goebelsmann, V., and Markland, F. S. (1981). *Cancer* **47,** 2634–2640.
Tidman, N., Janossy, G., Bodger, M., Granger, S., Kung, P. C., and Goldstein, G. (1981). *Clin. Exp. Immunol.* **45,** 457–467.
Wilander, E., Elsalhy, M., Willen, R., and Grimelius, L. (1981). *Virchows Arch. A: Pathol Anat. Histol.* **392,** 263–270.
Woodbury, R. G., Brown, J. P., Yeh, M.-Y., Hellström, I., and Hellström, K. E. (1980). *Proc. Natl. Acad. Sci. U.S.A.* **77,** 2183–2187.
Wright, D. H. (1981). *Lancet* **2,** 1194–1195.
Yavin, E., Yavin, Z., Schneider, M. D., and Kohn, L. D. (1981). *Proc. Natl. Acad. Sci. U.S.A.* **78,** 3180–3184.
Yelton, D. E., and Scharff, M. D. (1981). *Annu. Rev. Biochem.* **50,** 657–680.

CHAPTER 8
Use of Monoclonal Antibodies to Recognize Tumor Antigens

ZENON STEPLEWSKI   HILARY KOPROWSKI

The Wistar Institute of Anatomy and Biology
Philadelphia, Pennsylvania

I.	Introduction	155
II.	Hybridomas	156
III.	Monoclonal Antibody-Defined Antigens of Human Melanoma	156
IV.	Monoclonal Antibody-Defined Antigens of Human Gastrointestinal Tumors	159
V.	Oncodevelopmental Character of the Monoclonal Antibody-Defined Antigens	163
	References	164

I. INTRODUCTION

The complicated and cumbersome analysis of tumor antigens by the use of xenoantisera required extensive absorption procedures on normal human tissues and serum. This procedure quite often removed minor specific antibodies which may be of major importance. The use of monoclonal antibodies (Köhler and Milstein, 1975; Koprowski *et al.*, 1978; Steplewski, 1980) has revolutionized the approaches to the analysis of cell surface antigens. It is now possible to have pure, monoclonal reagents that react specifically with a single antigenic epitope. Through production of a large panel of monoclonal antibodies, we were able to detect and characterize antigens associated with human tumors such as melanoma (Koprowski *et al.*, 1978; Steplewski *et al.*, 1979; M. Herlyn *et al.*, 1980; Mitchell *et al.*, 1980, 1981) and gastrointestinal (GI) tract (M. Herlyn *et al.*, 1979; Koprowski *et al.*, 1979) and lung adenocarcinomas (Mazauric *et al.*, 1982).

II. HYBRIDOMAS

Phenotypic changes associated with transformation into malignant cells with unrestricted growth potential involves, among others, changes at the tumor cell surface. It is now quite firmly established that some tumors express antigens on their surfaces which differ quantitatively or qualitatively from those present on the normal cells of the same lineage. Using hybridoma technology, it now became possible to detect such antigens and to produce large amounts of monoclonal antibodies directed against single epitopes of tumor-associated antigens (Koprowski *et al.*, 1978; Steplewski, 1980; Brown *et al.*, 1981b). Application of immunochemical methods in conjunction with monoclonal antibodies permits isolation and precise analysis of these antigens. To establish monoclonal antibody-secreting hybridomas, mice are immunized with tumor cells, cell extracts, or their products. The immune splenocytes are fused with myeloma cells and hybrids selected in hypoxanthine-aminopterin-thymidine (HAT) medium (Szybalski *et al.*, 1962; Littlefield, 1964). The resulting hybridomas are cloned, and antibodies secreted by each clone are screened for their specificity (Koprowski *et al.*, 1978). Selected monoclonal antibodies are then used for the analysis of the antigens by immunoprecipitation (Mitchell *et al.*, 1980, 1981; Brown *et al.*, 1981a; Dippold *et al.*, 1980; Woodbury *et al.*, 1980) or as fractions of cell extracts such as glycolipids (Blaszczyk *et al.*, 1982; Brockhaus *et al.*, 1982; Magnani *et al.*, 1982). The large quantities of monoclonal antibodies available, representing different isotypes of immunoglobulins, are also extremely useful for the analysis of their biological functions (D. Herlyn *et al.*, 1979, 1980).

III. MONOCLONAL ANTIBODY-DEFINED ANTIGENS OF HUMAN MELANOMA

Many human antigens recognized by the mouse may be normal constituents of melanoma cells used for immunization. Some of these antigens may be expressed only by tumor cells, while others could be present on normal melanocytes or could be shared with embryonal tissues. The general strategy for the identification of these antigens by hybridoma-secreted monoclonal antibodies involves first the rigorous analysis of binding specificities of a large panel of monoclonal antibodies. As a next step, the identification of reactive cell surface components and their biochemical characterization provides necessary information for the identification of the antigens. We have shown for the first time in 1978 (Koprowski *et al.*, 1978) that it is possible to establish hybridomas secreting monoclonal antibodies with specificity for human melanoma cells. The existence of three distinct antigenic determinants on human melanoma cells has been established (Koprowski *et al.*, 1978; Steplewski *et al.*, 1979), and the molecules bearing these determinants have been immunoprecipitated (Mitchell *et al.*, 1980, 1981).

A different monoclonal antibody (Woodbury *et al.*, 1980) detects a glycoprotein p97 associated with membranes of tumor cells which was later found to be widely distributed on neoplastic tissues and in considerably smaller amounts on normal cells (Woodbury *et al.*, 1980; Brown *et al.*, 1981a,b). Further analysis of this antigen has shown that its N-terminal amino acid sequence is homologous to the N-terminal sequences of transferin and lactotransferin, and is functionally related to transferin and lactotransferin, that is, it retains iron binding potential (Brown *et al.*, 1982). Other antigens associated with melanoma cells have been reported by Yeh *et al.* (1979), Dippold *et al.* (1980), Woodbury *et al.* (1980), Carrel *et al.* (1980, 1982), Imai *et al.* (1981), Wilson *et al.* (1981), Brown *et al.* (1981a,b), Johnson *et al.* (1981), Saxton *et al.* (1982), Lloyd *et al.* (1982), Pukel *et al.* (1982), Johnson and Riethmuller (1982), Hellström *et al.* (1982), and Harper *et al.* (1982).

From these and other data (M. Herlyn *et al.*, 1980; Steplewski *et al.*, 1982), it became clear that human melanoma cells express multiple antigens with quite different distribution. In order to analyze specific distribution of melanoma antigen we have used a panel of anti-melanoma antibodies, examples of which are presented in Table I. At least six different groups of antigens are detected. Monoclonal antibodies of group I detect antigens present only on melanoma cells. These antigens are expressed only by about one-half of melanoma cell lines (Steplewski *et al.*, 1982). The antigen defined by antibody 56-1-8 has molecular weight of 40,000 and was found only on 12/22 melanoma cell lines (see Table I, Group I). Representative monoclonal antibodies directed against antigens of group II shown in Table I include antibodies Nu4B, 19-19 (Koprowski *et al.*, 1978; Steplewski *et al.*, 1979, 1982; Steplewski, 1980; Mitchell *et al.*, 1980, 1981) and antibody $J_2$60-23 (M. Herlyn *et al.*, 1982). Antibodies of this group bound to the majority of melanoma cells, but also to some astrocytomas and to some of the embryonal epitheloid cells (results not shown). Antibody Nu4B is secreted by a hybridoma which was established by immunization of mice with human melanoma–mouse fibroblast somatic cell bybrid containing only human chromosomes 14, 17, and 21. That antigen was not present on nonmalignant pigmented cells (Steplewski *et al.*, 1979) nor could it be detected on normal melanocytes in skin adjacent to primary melanomas (Thompson *et al.*, 1982). After radioiodination of melanoma cell surfaces, the target antigen was detected as a tetramer composed of three disulfide linked polypeptide chains with molecular weights of 116,000, 29,000, and 26,000 noncovalently associated with a fourth polypeptide chain of 95,000 (Mitchell *et al.*, 1981). Antibody 19-19 detects different antigen with a major component with a molecular weight of 260,000 (Mitchell *et al.*, 1982; Steplewski *et al.*, 1982). Other antibodies of this group bind to tumor cells similarly as shown in Table I (antibody $J_2$60-23). They

TABLE I
Monoclonal Antibody-Defined Antigens of Human Melanoma

	Monoclonal antibody			Ratio of cell cultures binding monoclonal antibody						
					Tumor cells				Fetal cells	
Group	Symbol	Isotype	MW of antigen defined	Melanoma	Astro-cytoma	Terato-carcinoma	Other		Melano-cytes	Fibro-blasts
I	56-1-8	IgG_1	40,000	12/22	0/6	0/4	0/13		0/3	0/6
	B_177-71	IgG_1	NP^a	10/21	0/6	0/4	0/13		0/3	0/5
II	Nu4B	IgG_{2a}	116,000/95,000	28/29	8/9	0/4	0/13		0/3	1/7
	19-19	IgG_1	260,000	15/21	5/8	0/4	0/13		0/3	1/7
	J_260-23	IgG_{2b}	NP	20/22	2/6	0/4	0/13		0/3	0/5
III	I_182-13	IgG_1	69,000	21/23	0/5	3/4	0/13		1/3	0/5
IV	9-11-24	IgG_1	84,000	10/17	1/6	0/4	1/13		7/9	0/5
	9-19-26	IgM	NP	9/16	0/4	0/4	2/13		NT	1/5
V	H_7-77-56	IgM	120,000/112,000	13/15	6/6	0/4	9/30		7/9	1/4
	691-6-2	IgG_1	84,000	16/17	7/9	NT^b	9/12		0/4	0/8
VI	13-17	IgG_1	35,000/28,000	11/15	1/6	0/4	5/12		0/9	0/3
(DR)	37-7	IgG_{2a}	35,000/28,000	11/15	1/6	0/4	5/12		0/9	0/3

[a] NP, nonprecipitable.
[b] NT, not tested.

bind to the majority of melanomas and some astrocytomas, but not to the normal cells of the same lineage. In the third group of antibodies (representative $I_1 82-13$ antibody in Table I) the binding specificity includes not only melanomas, but also teratocarcinomas (3/4) and melanocytes of fetal origin. The antigen detected by antibodies of this group is a 69,000 monomer. In group IV antibodies 9-11-24 and 9-19-26 react not only with melanomas and astrocytomas, but also with lung, colon, breast, and ovarian carcinomas. They did not, however, bind to teratocarcinomas (Table I). The antigen immunoprecipitated by antibody 9-11-24 was found to be a 84,000 monomer.

The large number of antibodies comprising group V is represented in Table I by antibodies H_7-77-56 and 691-6-2. The binding specificity of these antibodies varies. They bind not only to melanomas and astrocytomas, but also cross-react with other tumors of different origins and with some embryonal cells. Antibody $H_7$77-56 is directed against a dimer consistent of two polypeptide chains 120,000 and 112,000, while antibody 691-6-2 detects a 84,000 monomer (Table I).

Immunization of mice with melanoma cell lines permitted us also to establish a group of monoclonal antibodies directed against DR antigens. Antibodies 13-17 and 37-7 presented in Table I can precipitate DR antigens from radioiodinated melanoma cells with 35,000 and 28,000 chains.

IV. MONOCLONAL ANTIBODY-DEFINED ANTIGENS OF HUMAN GASTROINTESTINAL TUMORS

Immunization of mice with melanoma cells resulted in the establishment of hybridomas secreting antibodies against antigens, the majority of which are proteins (Steplewski *et al.*, 1982; Mitchell *et al.*, 1980, 1981). On the contrary, similar immunization with gastrointestinal cancer cells induced immune responses against antigens, most of which are glycolipids (Koprowski *et al.*, 1979; Magnani *et al.*, 1981, 1982). One of the anti-colorectal carcinoma antibodies (17-1A antibody in Table II) described in our first paper (M. Herlyn *et al.*, 1979) is of a IgG_{2a} isotype directed to the antigen which could not be immunoprecipitated as protein and was not found in glycolipid fraction. This antigen seems to be confined to the cell surface, and the cells treated with a variety of detergents retain the binding ability for antibody 17-1A, it is also protease sensitive (Ross, personal communication).

Hybridoma 3d clone 6 (3d-6) secretes IgG_1 monoclonal antibody 3d-6 (Table II) that binds only to gastrointestinal tumor cell lines (Koprowski *et al.*, 1979). This antibody immunoprecipitates, from gastrointestinal tumor cell lines and from commercial CEA preparations, a single CEA molecule with a molecular weight of 180,000 (Mitchell, 1980).

Most of the other monoclonal antibodies are directed against glycolipid antigens. IgG_1 antibody 19-9 (Table II) binds to the majority of gastrointestinal tumors, but not to other tumors or normal cells tested (Koprowski et al., 1979; Steplewski and Koprowski, 1982). The antigen defined by this antibody is a monosialoganglioside (Magnani et al., 1981, 1982) and its specific carbohydrate structure was determined to be

$$\begin{array}{c} \text{NeuNAc}\alpha 2\text{—3Gal}\beta 1\text{—3GlcNac}\beta 1\text{—3Gal}\beta 1\text{—4Glc} \\ | \\ 4 \\ | \\ \text{Fuc 1} \end{array}$$

The oligosaccharide is a sialylated lacto-N-fucopentaose II (LNF II), a hapten of the human Le^a blood group antigen (Magnani et al., 1982).

This gastrointestinal cancer antigen (GICA) is expressed by tumor cells of adenocarcinomas of stomach, pancreas, and colon and is present in meconium, the first discharge from the bowel of the newborn of material accumulated during fetal life (Magnani et al., 1982). Binding of anti-GICA monoclonal antibody to cells or cell extracts of gastrointestinal cancer (GIC) can be inhibited in solid-phase radioimmunoassay by GIC tumor cell extracts or by medium from tumor cells cultured in vitro that shed GICA (Steplewski et al., 1981). In the same type of assay, sera of patients with GIC inhibit binding of monoclonal antibody to its target (Chang et al., 1981). In the preliminary studies (Koprowski et al., 1981) sera of 64% of patients with clinically and histologically proven colorectal cancer, 92% of patients with pancreatic cancer, and 72% of patients with gastric cancer inhibited binding of the antibody to GICA-containing cancer cell extracts. By contrast, the sera of 8% of patients with malignancies other than GIC and 1.8% of healthy individuals, inhibited binding of the antibody. Among the seven patients with tumors other than GIC, whose sera inhibited binding of the antibody, three suffered from carcinoma of the liver.

It was also possible to detect the presence of GICA by reacting the monoclonal antibody to fixed, paraffin-embedded tissues in immunoperoxidase assay (Atkinson et al., 1982) since glycolipids generally resist destruction by fixatives such as formalin of Bouin's solution. The results of these studies indicate that 55% of the examined colorectal cancer tissue samples expressed GICA, as did 82% of pancreatic cancer tissue and 100% of gastric cancer tissue. Cells of one out of eight liver carcinomas and one out of five gall bladder carcinomas were positive for the presence of GICA. In normal tissue, binding of 19-9 antibody was detected only in cells of the columnar epithelium of the secretory ducts, such as those found in pancreas, salivary glands, liver, and bronchi. The antigen in the normal epithelium is a glycoprotein (Magnani, personal communication) and is not shed into circulation. GICA was not detected in normal mucosal cells of the

TABLE II
Monoclonal Antibody-Defined Antigens of Gastrointestinal Tract Cancers

Monoclonal antibody			Ratio of cultures binding monoclonal antibody							
			Tumor cells						Normal	
			Carcinomas							
Symbol	Isotype	Antigen	Gastric	Pancreatic	Colon	Rectal	Other	Fibroblasts	Meconium
17-1A	IgG$_{2a}$	Protein	1/1	1/1	8/8	3/3	0/20	0/7	NT
3d-6	IgG$_1$	180,000 (CEA)	1/1	NT	8/8	0/1	0/20	0/7	NT
19-9	IgG$_1$	Sialylated Lacto-N-/fucopentaose II	1/1	1/1	5/8	2/3	0/20	0/7	3/3
38a	IgM	Lewis b	NT[a]	NT	6/8	2/3	0/20	0/7	2/3
10-17	IgM	Lewis b	NT	NT	6/8	2/3	0/20	0/7	2/3
NS-43	IgG$_1$	Lewis b	NT	NT	6/8	2/3	0/20	0/7	2/3
WGHS-9-1	IgG$_1$	28,000/22,000	1/1	1/1	4/5	0/1	0/20	0/5	NT
WGHS-29-1	IgM	Lacto-N-/fucopentaose III	1/1	0/1	5/5	1/1	2/20	0/5[b]	NT
WCZYG-13	IgM	Lacto-N-/fucopentaose III	1/1	0/1	5/5	1/1	2/20	0/5[b]	NT

[a] NT, not tested.
[b] Binding was also found to normal lymphocytes and some cells of normal GI tract. The antigen lacto-N-fucopentacse III was also described as a stage-specific antigen of mouse embryo.

large or small intestines and in normal gastric mucosa, except for one tissue specimen in which a few stained scattered cells were observed.

In all patients with GICA detected in the serum samples, the antigen was also found in the tumor tissue. There were, however, six cases in which antigen was present in the tumor tissue but not detected in circulation (Koprowski et al., 1982). Finally, there is a small fraction of GIC patients which is negative for GICA in both, tumors and in blood.

Since GICA is a ganglioside containing sialylated lacto-N-fucopentaose II, a hapten of the human Le^a blood group antigen, we have compared the expression of Le antigen in gastrointestinal cancer patients with the presence of GICA in their serum and immunohistochemically in the tumor tissue (Atkinson et al., 1982). The preliminary results showed that of 15 colorectal cancer patients, 13 who were either Le(a$^-$b$^+$) or Le(a$^+$b$^-$) also showed the presence of GICA in their sera; two of these patients, for whom tissue blocks were available, showed the presence of GICA in the tumor tissue, and one did not. Two of the 15 subjects were Le(a$^-$b$^+$) or Le(a$^+$b$^-$) phenotype. The studies are at present performed to test a hypothesis that there is a relationship between the expression of gastric/pancreatic cancer phenotype and the Lewis blood group type (Koprowski et al., 1982).

At least three monoclonal antibodies, secreted by 3 independently established hybridomas from mice immunized with colon carcinoma cell line SW 1116, are directed against Le^b blood group antigen. All three antibodies—38a, 10-17, and Ns43—bind specifically to LNFII the Le^b blood group hapten with sugar sequence (Brockhaus et al., 1981)

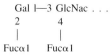

Three hybridomas, obtained from mice immunized with colon or gastric carcinoma cells freshly isolated from patients, secrete IgM antibodies that bind to glycolipids of GI tumors, granulocytes, erythrocytes, other tumors, and some normal tissues (Brockhaus et al., 1982). The oligosaccharide detected by these antibodies is a lacto-N-fucopentaose III

$$\text{Gal}\beta 1\text{—}4 \text{ GlcNac}\beta 1\text{—}3\text{Gal}\beta 1\text{—}4\text{Glc,}$$
$$\underset{\text{Fuc}\alpha 1}{\overset{3}{|}}$$

The specificity of these antibodies is the same as reported for a monoclonal antibody against a stage-specific antigen of the mouse embryo (Solter and Knowles, 1978; Hakomori et al., 1981).

Monoclonal antibody 17-1A (M. Herlyn et al., 1979) is directed against antigen(s) expressed by all GI tumors, but is not shed by tumor cells in tissue culture (Steplewski et al., 1981) and is not present in patients sera. The antigen seems to be confined to the surfaces of tumor cells, and we were not able to immunoprecipitate it. The antigen seems to be a protein, since it is sensitive to proteases, but not to neuraminidase (Ross, personal communication).

Another group of antibodies secreted by hybridomas established from mice immunized with pancreatic carcinoma cell line were shown to be specific for the human glood group B determinant (Hansson et al., 1982). One of these antibodies reacted with type 1 chain

$$Gal\alpha 1\text{—}3\ Gal\alpha 1\text{—}3GlcNAc \ldots$$
$$\underset{\alpha\ Fuc}{\overset{2}{\underset{1}{|}}}$$

and type 2 chain

$$Gal\alpha 1\text{—}3Gal\alpha 1\text{—}4GlcNAc \ldots$$
$$\underset{\alpha Fuc}{\overset{2}{\underset{1}{|}}}$$

while three others preferred a difucosyl oligosaccharide with a type 2 chain.

Of the few chemically identified antigens associated with gastrointestinal tumors, as defined by monoclonal antibodies, the majority are related to blood group antigens. This includes Le^b antigen (Brockhaus et al., 1981), sialylated hapten of Le^a antigen (Magnani et al., 1982), and B group determinants (Hansson et al., 1982). Anger et al. (1982) described another antibody which is directed against blood group H substances.

V. ONCODEVELOPMENTAL CHARACTER OF THE MONOCLONAL ANTIBODY-DEFINED ANTIGENS

Human melanoma-associated antigens, defined by murine monoclonal antibodies are very seldomly expressed on malignant melanoma cells only (M. Herlyn et al., 1982). A group of glycoprotein antigens with high molecular weight is expressed not only on melanoma cells, but also on fetal polygonal cells and astrocytomas. These antigens, defined by antibodies Nu4B (Koprowski et al., 1978; Steplewski, 1980; Mitchell et al., 1980, 1981), 19-19, 0_195-45,

$H_4$18-90, G_1, 15-43, $J_2$60-23 (M. Herlyn *et al.*, 1982; Ross *et al.*, 1982), 9.2.27 (Bumol and Reisfeld, 1982), and G-13-C6 (Carell *et al.*, 1982) have molecular weight of about 240,000, and seem to represent a group of molecules originating from neural tube.

There is also a group of antibodies (representative of which is $I_1$82-13 in Table I) which are not expressed on astrocytomas, but were found on teratocarcinomas (M. Herlyn *et al.*, 1982), thus they may belong to an oncodevelopmental group of antigens. The large group of antigens of groups IV and V (Table I) are expressed not only on melanomas and astrocytomas, but also by all types of fetal cells (M. Herlyn *et al.*, 1982).

Anti-GI tumor monoclonal antibody 19-9 is directed against a glycolipid antigen—a sialylated fucopentaose II (Magnani *et al.*, 1981, 1982). This antigen is expressed by gastrointestinal tumor cells (Atkinson *et al.*, 1982; Magnani *et al.*, 1981) of individuals carrying the Lewis gene (Koprowski *et al.*, 1982) and was also detected in meconium (Magnani *et al.*, 1981; Hansson *et al.*, 1982) thus it could also be classified as an oncofetal type of glycolipid.

ACKNOWLEDGMENTS

The data presented here were derived from experiments in which D. Herlyn, M. Herlyn, M. Blaszczyk, and A. Ross participated. Supported by Grants CA10815, CA21124, CA25298 and CA25874 from NIH.

REFERENCES

Anger, B. R., Lloyd, K. O., Oetgen, H. F., and Old, L. J. (1982). *Hybridoma* **1**, 139–147.
Atkinson, B., Ernst, C., Herlyn, M., Steplewski, Z., and Koprowski, H. (1982). *Cancer Res.* **42**, 4820–4823.
Blaszczyk, M., Karlsson, K.-A., Hansson, G., Larsson, G., Brockhaus, M., Herlyn, M., Steplewski, Z., and Koprowski, H. (1982). *Hybridoma* **1**, 202.
Brockhaus, M., Magnani, J. L., Blaszczyk, M., Steplewski, Z., Koprowski, H., Karlsson, K.-A., Larsson, G., and Ginsburg, V. (1981). *J. Biol. Chem.* **256**, 13223–13225.
Brockhaus, M., Magnani, J., Herlyn, M., Blaszczyk, M., and Ginsburg, V. (1982). *Fed. Proc., Fed. Am. Soc. Exp. Biol.* **41**, 897.
Brown, J. P., Nishiyama, K., Hellström, I., and Hellström, K.-E. (1981a). *J. Immunol.* **127**, 539–546.
Brown, J. P., Woodbury, R. G., Hart, C. E., Hellström, I., and Hellström, K.-E. (1981b). *Proc. Natl. Acad. Sci. U.S.A.* **78**, 539–543.
Brown, J. P., Hewick, R. M., Hellström, I., Hellström, K.-E., Doolittle, R. F., and Dreyer, W. J. (1982). *Nature (London)* **296**, 171–173.
Bumol, T. F., and Reisfeld, R. A. (1982). *Proc. Natl. Acad. Sci. U.S.A.* **79**, 1245–1249.
Carrel, S., Accolla, R. S., Carmagnola, A. L., and Mach, J.-P. (1980). *Cancer Res.* **40**, 2523–2528.

Carrel, S., Schreyer, M., Schmidt-Kessen, A., and Mach, J.-P. (1982). *Hybridoma* **1,** 387–397.
Chang, T. H., Steplewski, Z., Sears, H. F., and Koprowski, H. (1981). *Hybridoma* **1,** 37–45.
Dippold, W. G., Lloyd, K. O., Li, L. T. C., Ikeda, H., Oetgen, H. F., and Old, L. J. (1980). *Proc. Natl. Acad Sci. U.S.A.* **77,** 6114–6118.
Hakomori, S., Nudelman, E., Levery, S., Solter, D., and Knowles, B. (1981). *Biochem. Biophys. Res. Commun.* **100,** 1578–1586.
Hansson, G. C., Karlsson, K.-A., Larson, G., McKibbin, J. M., Blaszczyk, M., Herlyn, M., Steplewski, Z., and Koprowski, H. (1982). *J. Biol. Chem.* (in press).
Harper, J. R., Bumol, T. F., and Reisfeld, R. A. (1982). *Hybridoma* **1,** 423–432.
Hellström, I., Brown, J. P., and Hellström, K.-E. (1982). *Hybridoma* **1,** 399–402.
Herlyn, D., Herlyn, M., Steplewski, Z., and Koprowski, H. (1979). *Eur. J. Immunol.* **9,** 657–659.
Herlyn, D., Steplewski, Z., Herlyn, M., and Koprowski, H. (1980). *Cancer Res.* **40,** 717–721.
Herlyn, M., Steplewski, Z., Herlyn, D., and Koprowski, H. (1979). *Proc. Natl. Acad. Sci. U.S.A.* **76,** 1438–1442.
Herlyn, M., Clark, W. H., Mastrangelo, M. J., Guerry, D., IV, Elder, D. E., La Rossa, D., Hamilton, R., Bondi, E., Tuthill, R., Steplewski, Z., and Koprowski, H. (1980). *Cancer Res.* **40,** 3602–3609.
Herlyn, M., Steplewski, Z., Atkinson, B., Ernst, C. S., and Koprowski, H. (1982). *Hybridoma* **1,** 403–411.
Imai, K., Ng, A.-K., and Ferrone, S. (1981). *JNCI,* **66,** 489–496.
Johnson, J. P., and Riethmuller, G. (1982). *Hybridoma* **1,** 381–385.
Johnson, J. P., Demmer-Dieckmann, M., Meo, T., Hadam, M. R., and Riethmuller, G. (1981). *Eur. J. Immunol.* **11,** 825–831.
Köhler, G., and Milstein, C. (1975). *Nature (London)* **257,** 495–497.
Koprowski, H., Steplewski, Z., Herlyn, D., and Herlyn, M. (1978). *Proc. Natl. Acad. Sci. U.S.A.* **75,** 3405–3409.
Koprowski, H., Steplewski, Z., Mitchell, K. F., Herlyn, M., Herlyn, D., and Fuhrer, P. (1979). *Somatic Cell Genet.* **5,** 957–972.
Koprowski, H., Herlyn, M., Steplewski, Z., and Sears, H. F. (1981). *Science* **212,** 53–55.
Koprowski, H., Brockhaus, M., Blaszczyk, M., Magnani, J., and Steplewski, Z. (1982). *Lancet* **1,** 1331–1333.
Littlefield, J. N. (1964). *Science* **145,** 709–710.
Lloyd, K. O., Albino, A., and Houghton, A. (1982). *Hybridoma* **1,** 461–463.
Magnani, J. L., Brockhaus, M., Smith, D. F., Ginsburg, V., Blaszczyk, M., Mitchell, K. F., Steplewski, Z., and Koprowski, H. (1981). *Science* **212,** 55–56.
Magnani, J. L., Nilsson, B., Brockhaus, M., Zopf, D., Steplewski, Z., Koprowski, H., and Ginsburg, V. (1982). *Fed. Proc., Fed. Am. Soc. Exp. Biol.* **41,** 898.
Mazauric, T., Mitchell, K. F., Letchworth, G. J., Koprowski, H., and Steplewski, Z. (1982). *Cancer Res.* **42,** 150–154.
Mitchell, K. F. (1980). *Cancer Immunol. Immunother.* **10,** 1–7.
Mitchell, K. F., Fuhrer, J. P., Steplewski, Z., and Koprowski, H. (1980). *Proc. Natl. Acad. Sci. U.S.A.* **77,** 7287–7291.
Mitchell, K. F., Fuhrer, J. P., Steplewski, Z., and Koprowski, H. (1981). *Mol. Immunol.* **18,** 207–218.
Mitchell, K. F., Steplewski, Z., and Koprowski, H. (1982). *In* "Monoclonal Hybridoma Antibodies: Techniques and Applications" (G. R. Hurrell, ed.), pp. 151–168. CRC Press, Boca Raton, Florida.
Pukel, C. S., Lloyd, K. O., Trabassos, L. R., Dippold, W. G., Oetgen, H. F., and Old, L. J. (1982). *J. Exp. Med.* **15,** 1133–1147.
Ross, A. H., Mitchell, K. F., Steplewski, Z., and Koprowski, H. (1982). *Hybridoma* **1,** 413–421.

Saxton, R. E., Mann, B. D., Morton, D. L., and Burk, M. W. (1982). *Hybridoma* **1**, 433–445.
Solter, D., and Knowles, B. (1978). *Proc. Natl. Acad. Sci. U.S.A.* **75**, 5565–5565.
Steplewski, Z. (1980). *Transplant. Proc.* **12**, 384–387.
Steplewski, Z., and Koprowski, H. (1982). *In* "Methods in Cancer Research" Vol. 20 (H. Busch, ed.), pp. 285–316. Academic Press, New York.
Steplewski, Z., Herlyn, M., Herlyn, D., Clark, W. H., and Koprowski, H. (1979). *Eur. J. Immunol.* **9**, 94–96.
Steplewski, Z., Chang, T. H., Herlyn, M., and Koprowski, H. (1981). *Cancer Res.* **41**, 2723–2727.
Steplewski, Z., Mitchell, K. F., and Koprowski, H. (1982). *In* "Melanoma Antigens and Antibodies" (R. Reisfeld, ed.), pp. 365–380. Plenum, New York.
Szybalski, W., Szybalska, E. H., and Regnie, G. (1962). *Natl. Cancer Inst. Monogr.* **7**, 75–96.
Thompson, J. J., Herlyn, M. F., Elder, D. E., Clark, W. H., Steplewski, Z., and Koprowski, H. (1982). *Hybridoma* **1**, 161–168.
Wilson, B. S., Imai, K., Natali, P. G., and Ferrone, S. (1981). *Int. J. Cancer* **28**, 293–300.
Woodbury, R., Brown, J. P., Yeh, M.-Y., Hellström, I., and Hellström, K.-E. (1980). *Proc. Natl. Acad. Sci. U.S.A.* **77**, 2183–2187.
Yeh, M.-Y., Hellström, I., Brown, J. P., Warner, G. A., and Hellström, K.-E. (1979). *Proc. Natl. Acad. Sci. U.S.A.* **76**, 2927–2931.

CHAPTER 9
Radioimmunolocalization of Cancer

RICHARD H. J. BEGENT KENNETH D. BAGSHAWE

Department of Medical Oncology
Charing Cross Hospital
London, England

    I. Introduction 167
   II. Antibody Localization in Xenografts of Human Tumors in
       Experimental Animals 168
 III. Antibody Localization in Patients with Cancer 169
      A. Theoretical Considerations 169
      B. Radioimmunolocalization Procedures 171
      C. Interpretation of Results 173
      D. Clinical Results 179
 IV. Potential Improvements 184
      A. An Alternative to Subtraction 185
      B. Alternative Radionuclides 185
      C. Tomographic Imaging (Single Photon Emission
          Computerized Tomography) 185
      D. Monoclonal Antibodies 185
  V. Conclusions 186
       References 187

I. INTRODUCTION

As the focus of cancer research has moved from morphology to biochemistry so the concept has developed that diagnosis and therapy might be based on biochemical differences between the structure of normal and malignant cells. In the first decade of this century, Ehrlich popularized the idea that antibodies had the potential to recognize such differences and could be used to target therapeutic agents in the form of a "magic bullet." It was not until the 1950's that specific localization of antibodies *in vivo* in Wagner osteosarcoma and Walker carcinoma was shown by Pressman and Korngold (1953) and Bale *et al.* (1955). These experiments were made possible by the development of suitable animal tumor models, methods for partial purification of antibodies raised against the tumors, and fluorescent and radioisotopic methods for antibody labeling.

Theoretical considerations suggest that the ideal antigenic target for antibody based localization would be a tumor-specific, nonsecreted, membrane-bound protein or polypeptide. Such substances have not so far been defined for human tumors, but during the 1970's well-characterized secreted products exhibiting a degree of specificity for certain human tumors and their corresponding antisera became available, and studies in man have used these. Probably the best examples are human chorionic gonadotrophin (hCG) produced by choriocarcinoma and carcinoembryonic antigen (CEA) produced most consistently by gastrointestinal carcinomas. This chapter will be restricted to these two antigens since most clinical experience has been accumulated with them and much of it will be based on our experience of over 140 investigations in patients. Investigation of localization of these antibodies was greatly facilitated by the use of xenografts of human tumors in experimental animals.

II. ANTIBODY LOCALIZATION IN XENOGRAFTS OF HUMAN TUMORS IN EXPERIMENTAL ANIMALS

The demonstration of specific localization of partially purified antibodies to hCG in xenografts of choriocarcinoma in the cheek pouch of the golden hamster by Quinones *et al.* (1971) was the first demonstration of specific localization of antibodies in a human tumor and was followed by a similar demonstration of specific localization of antibodies to CEA in xenografts of human colon carcinoma in hamsters (Primus *et al.*, 1973; Goldenberg *et al.*, 1974). One of the problems of these experiments was that the antibodies used were only partially purified and thus some of the radiolabeled antibody had no specificity for the antigen. This was largely overcome by Mach *et al.* (1974) who used affinity-purified antibodies to CEA and also substituted nude mice for hamsters as hosts for the tumor xenografts. The superiority of affinity-purified antibodies over immunoglobulin G (IgG) fractions of antisera for localization in CEA-producing tumors was confirmed by Primus *et al.* (1977). Affinity-purified antibody to hCG has also been shown to localize satisfactorily in xenografts of human choriocarcinoma (Searle *et al.*, 1981).

III. ANTIBODY LOCALIZATION IN PATIENTS WITH CANCER

A. Theoretical Considerations

Much of the strength of the animal experiments lies in the use of a nonspecific immunoglobulin as a control (Pressman et al., 1957). This is important since the nonspecific antibody was found to accumulate in tumors in several studies (Goldenberg et al., 1974; Mach et al., 1974; Searle et al., 1981; Pressman et al., 1957). Thus when radiolabeled antibody is given to a patient it does not necessarily follow that accumulation of antibody in a tumor is because of specific reaction of antibody with its antigen. Mach et al. (1980a) gave ^{131}Iodine (I)-labeled antibody to CEA and ^{125}I-labeled normal immunoglobulin to patients with colorectal cancer prior to surgery and compared the ratios of the two antibodies in tumors and normal colon after surgical resection. Ratios in the range 2.3 : 1 to 5.8 : 1 in favor of the antibody to CEA were found 3 to 8 days after injection of antibody. These ratios are comparable to those found in xenografts of human colorectal cancer in experimental animals (Primus et al., 1973; Mach et al., 1974; Searle et al., 1980) in which there is a tendency for specificity ratios to increase with time after injection at least up to 1 week.

These results are supported by studies in which distribution of intravenously administered ^{125}I-labeled antibody to CEA is investigated by autoradiography of tissue sections of the excised human colon carcinoma xenografts. Antibody has been demonstrated to localize within the tumor specifically in tissue spaces, in necrotic areas, and in stromal connective tissue of viable tumors, but not selectively on tumor cell membranes (Lewis et al., 1982). The relatively small amounts of ^{125}I in the form of labeled antibody which it is considered safe to administer to patients has so far frustrated attempts to repeat these experiments successfully in man. Accumulation of antibody to CEA as judged by a patient with tumor was reported in 1974 by Hoffer et al. There is, however, some evidence that macromolecules can accumulate nonspecifically in tumors (Bauer et al., 1955; Goldenberg et al., 1974; Mach et al., 1974; Pressman et al., 1957; Searle et al., 1981), and it is difficult to assess the role of antibody specificity in the image obtained. Ideally, images of simultaneously administered specific and nonspecific antibodies, each labeled with a different radioisotope, would be obtained and the site of tumor determined by comparison of the two images produced. Such a comparison is likely to be particularly important when trying to image small tumors, since only 0.1% or less of an administered dose of antibody was found in tumors excised from patients (Mach et al., 1980b).

The choice of isotopic labels for the antibody imaging is clearly important. Antibodies labeled with ^{131}I can be used with currently available γ cameras,

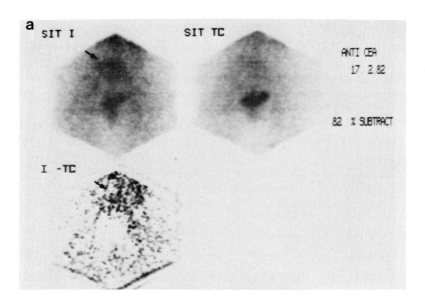

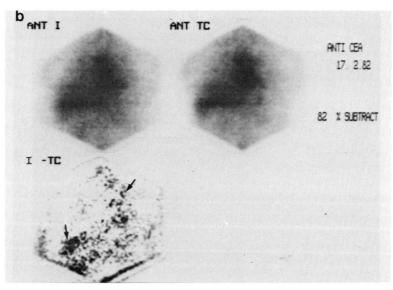

Fig. 1. RIL using ^{131}I-labeled antibody to CEA in a patient with carcinoma of the rectum. (a) Images with the patient sitting above the gamma camera. I, image of ^{131}I showing the accumulation of isotope in the bladder in the lower part of the picture and another accumulation above that which corresponds to the position of the rectum (arrow). TC, image of ^{99m}Tc subtraction medium. Accumulation in the region of the rectum is less marked than with ^{131}I. I − TC = image after

although its energy is high for optimal results. However, ^{125}I is not satisfactory for imaging purposes, and therefore the results of studies using these two isotopes in organ counting techniques cannot be reproduced in the imaging field. Iodine-123 is suitable for imaging, apart from its present cost and short half-life of 13 hours. It has been used in animal models up to 96 hours postinjection (Epenetos et al., 1982a) and in man up to 72 hours (Epenetos, 1982b).

Other possible radionuclides for dual labeling of antibodies include technetum-99m (^{99m}Tc) (Pettit et al., 1980) and indium-111 (R. Buckley, personal communication, 1982; Scheinberg et al., 1982). Neither of these has yet been shown to be satisfactory in clinical use.

B. Radioimmunolocalization Procedures

As a result of these problems with the dual antibody approach, Goldenberg et al. (1978) simulated the nonspecific distribution of antibody in the circulation and tissue spaces by giving ^{99m}TcO$_4$ 25 minutes and ^{99m}Tc-albumin 5 minutes before obtaining gamma camera images, and ^{131}I-labeled antibody to CEA was given 24 or 48 hours previously. The ^{99m}Tc image could then be subtracted from that of ^{131}I to leave an image of ^{131}I antibody which had accumulated in the tumor (Fig. 1). This method, radioimmunolocalization (RIL), was based on the observation that an image could be obtained which corresponded to nonspecific distribution of antibody rather than on similar physiological handling of ^{131}I-labeled antibody and ^{99m}TcO$_4$ with ^{99m}Tc-albumin.

Before considering clinical results, it is appropriate, therefore, that various factors influencing the results obtained should be carefully considered.

1. Antibody

Goat, rabbit, sheep, and baboon (Silva et al., 1980) polyclonal antibodies and mouse monoclonal antibodies (Mach et al., 1981) have been used successfully for radioimmunolocalization. Images have been obtained with immunoglobulin G preparations (Hatch et al., 1980; Dykes et al., 1980), but affinity purified antibodies are preferable (Mach et al., 1974; Primus et al., 1977) and have been used for the great majority of successful clinical studies. The technical

---

subtraction of the ^{99m}Tc image from that of ^{131}I, after equalization of the counts. The accumulation of ^{131}I in the region of the rectum is now seen in isolation (arrow). A primary rectal carcinoma was subsequently excised. (b) Anterior views of the thorax and upper abdomen in the same patient. The heart and liver are seen as the principal concentrations of ^{131}I (I) and ^{99m}TC (TC). After subtraction (I − TC) accumulations of ^{131}I are seen in the right lobe of the liver and lungs (arrow). The pulmonary deposits were visible on a plain radiograph but computerized tomography showed no definite abnormality in the liver. A deposit in the right lobe of the liver was eventually confirmed at laparotomy.

details of antibody preparation have been published by several authors (Mach *et al.*, 1974; Searle *et al.*, 1980; Begent *et al.*, 1980a). Hyperimmune serum is absorbed with normal tissues, then radioiodinated by modifications of the chloramine T method, submitted to gel filtration to remove immunoglobulin aggregates and fragments, and followed by Millipore filtration, pyrogenicity testing, and quantitative reconfirmation of antigen binding activity.

2. Preparation of the Patient

Iodine-131 uptake in the thyroid is blocked by oral administration of potassium iodide 60 mg three times daily starting 24 hours before administration of antibody and continued for 10 days. Lugol's iodine 5 drops twice daily may be used instead. In addition we find that oral potassium perchlorate 200 mg given every 6 hours for 4 doses, starting 30 minutes before antibody administration, reduces nonspecific uptake of ^{131}I in the stomach. Intradermal injection of 1 to 10 ug of antibody together with a saline control is given 15 minutes prior to intravenous antibody injection to test for immediate type hypersensitivity. If this is negative 0.5 to 2 mCi of ^{131}I as antibody labeled with a specific activity of between 1 and 15 mCI/mg protein is given by slow intravenous injection.

3. Adverse Reactions

In 140 studies at Charing Cross Hospital 2 patients had transient fever up to 37.5°C; 1 patient had a mild exacerbation of preexisting asthma not requiring admission to hospital; one had an acute bout of asthma and urticarial rash requiring corticosteroid and antihistamine therapy and overnight admission to hospital. This patient had a negative intradermal test with goat immunoglobulin but had a pet goat during her childhood. A further patient was not given intravenous antibody because of a positive intradermal test.

Up to 4 investigations have been performed on the same patient using the same species of antibody without corticosteroid or antihistamine prophylaxis and without any adverse effect.

4. Imaging

A large field of view gamma camera with a high energy collimater coupled to a computer suitable for storage of data, production of a digital image and subtraction techniques has been used by most of the successful groups. A rectilinear scanner can be used in place of gamma camera, but we have found this less satisfactory.

Images are obtained for ^{131}I antibody and ^{99m}Tc 24 and sometimes 48 hours after antibody administration; unlike some groups we have seldom gained any extra information from the 48 hour study. Thirty and 5 minutes before starting imaging 500 uCi ^{99m}TcO$_4$ and 500 uCi ^{99m}Tc human serum albumin are given respectively. Anterior and posterior views of the thorax and abdomen are obtained with lateral or other special views as appropriate.

5. Subtraction of ^{99m}Tc from ^{131}I Image

This is readily performed with the appropriate computer facility, and in many early studies this was performed stepwise until any salient areas of excess ^{131}I activity were clearly seen. A standardized protocol is necessary, however, for comparison of different studies and in order to try and define the limits of sensitivity. We have found equalization of the total counts in the two images to be the most satisfactory method (Jones *et al.*, 1982). The formula which gives an element of the subtracted image is $S = I - mTc$, where I is an element of the ^{131}I image, Tc is the same element of the ^{99m}Tc image, and m is the total number of counts in the ^{131}I image divided by the total number of counts in the ^{99m}Tc image. The subtraction image is displayed on a television screen with the ^{131}I and ^{99m}Tc images and inspected for discrete concentrations of radioactivity which should preferably be seen at the same anatomical site on two or more views.

Deland *et al.* (1980) have advocated equalization on counts in the cardiac region, and this is a reasonable alternative. When imaging the liver, the same group has advocated using ^{99m}Tc sulfur colloid liver scanning agent in place of ^{99m}TcO$_4$ and ^{99m}Tc albumin for subtraction (Goldenberg *et al.*, 1980a). Images of tumor deposits can certainly be obtained in this way, but may be generated by a localized defect of phagocytic capacity demonstrated by the colloid, by specific uptake of antibody, or by a mixture of the two. Thus a positive result does not carry the biochemical specificity which is the most important feature of radioimmunolocalization.

C. Interpretation of Results

1. Variations in Antibody Distribution in Normal Tissues

Immune Complexes. After intravenous administration antibody distributes rapidly in the circulation and, probably, more slowly in extravascular spaces. When the antigen for which it is specific, such as CEA or hCG, is present in the circulation immune complexes will form, and this has been demonstrated for CEA (Primus *et al.*, 1980; Searle *et al.*, 1980) and hCG (Begent *et al.*, 1980a). The amount of radiolabeled antibody retained in complexed form will vary from patient to patient, largely as a result of the amount of antigen present in the serum and extracellular fluid (Searle *et al.*, 1980). It might be predicted on the basis of studies in which immune complexes were injected into animals that these complexes would be rapidly cleared into the reticuloendothelial system,

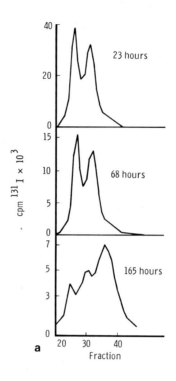
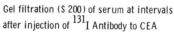
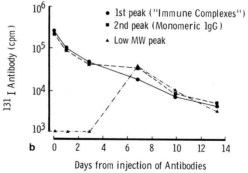

Fig. 2. Results of gel filtration of serial samples of 1 ml serum taken from a patient with metastatic carcinoma of the rectum and levels of serum CEA 224 µg/liter after intravenous administration of ^{131}I-labeled goat antibody to CEA. (a) Profile of distribution of ^{131}I-labeled antibody in fractions produced by gel filtration on Sepharyl S-200 (2.6 × 55 cm) at different times after antibody injection. The first peak represents immune complexes, the second monomeric IgG, and the third is probably a product of immunoglobulin catabolism first seen after a week. (b) Counts in the three peaks were plotted to show the clearance rates over the 2 weeks of study after correction for decay.

particularly by the Kuppfer cells of the liver (Arend and Mannik, 1971; Mannik and Arend, 1971). This was not the case up to 48 hours after injection in the studies of Primus *et al.* (1980) nor in the patient shown in Fig. 2 who was followed for 14 days.

Gel filtration was performed on serial samples taken from patients in our study so that the amount of complexed and noncomplexed antibody could be compared over a period of time. The complexed antibody was cleared at the same rate as free antibody, and the results were consistent with the findings of Primus *et al.* (1980). In general a satisfactory tumor image was obtained (Fig. 3), but unfortunately this is not invariably the case as illustrated in Fig. 4. The image shows accumulation of ^{131}I antibody diffusely in the liver, but it was subsequently found to be free from tumor. This accumulation of ^{131}I antibody may have resulted from uptake of immune complexes containing radiolabeled antibody by the Kuppfer cells. The value of radioimmunolocalization for diagnosis of diffuse involvement of the liver by tumor is limited by this finding.

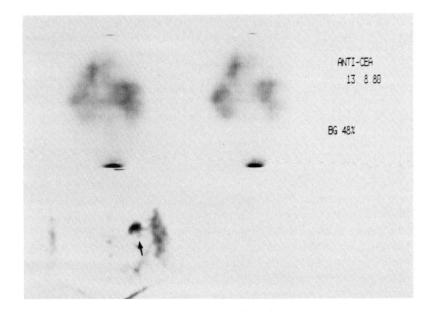

Fig. 3. RIL using ^{131}I-labeled antibody to CEA in the patient illustrated in Fig. 2. A raised serum CEA value (224 μg/liter) was found on routine follow up after resection of carcinoma of the rectum when she was clinically free from recurrence. The posterior images of the thorax and abdomen shown in top left revealed ^{131}I distributed in the heart, uppermost, stomach on the left, and liver on the right. Top right image is of ^{99m}Tc subtraction medium showing the same organs. After subtraction of ^{99m}Tc from ^{131}I (bottom left) a discrete area of accumulation of ^{131}I antibody is seen in the right lobe of the liver (arrow). Computerized tomography was performed because of this finding and confirmed the presence of a deposit at this site.

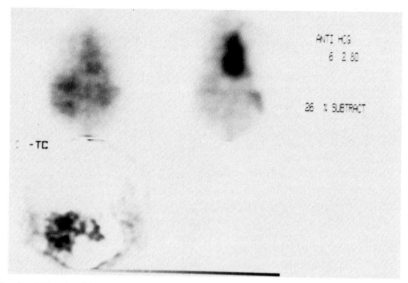

Fig. 4. RIL using ^{131}I rabbit antibody to hCG in a patient with choriocarcinoma. Anterior views of the thorax and upper abdomen showing in the top left a ^{131}I image showing heart (uppermost), liver (on right) and stomach on left. Top right shows an image of ^{99m}Tc subtraction medium showing the same organs. After subtraction (bottom left) there is accumulation of ^{131}I diffusely in the liver. Subsequent postmortem examination showed no evidence of tumor in the liver.

2. Nonspecific Accumulation of Radioisotope in Normal Organs

a. Bladder. Although free ^{131}I is removed from the administered antibody by gel filtration before administration, the catabolism of antibody (Fig. 2) results in progressive release of free ^{131}I which is excreted in the urine, appearing in the kidneys and bladder on the ^{131}I gamma camera image. ^{99m}TcO$_4$ is of course also excreted in the urine, but this dynamic process seldom results in identical ^{131}I and ^{99m}Tc images at the time of imaging so that there is often a predominance of one or other radionuclide in the urinary tract at the time of scanning (Fig. 5). This problem can be reduced by emptying of the bladder before scanning, but urine continues to accumulate while imaging. Iodine-131 is sometimes in a lower position in the bladder than ^{99m}Tc, and there is often ^{131}I remaining in the bladder after apparently complete emptying perhaps due to binding of free ^{131}I to the bladder mucosa.

b. Stomach. Uptake of ^{131}I by the gastric mucosa can usually be prevented by administration of potassium perchlorate (see Section III,B,2). Unfortunately this is not invariably the case, as shown in Fig. 6, and uptake of ^{131}I is

9. Radioimmunolocalization of Cancer

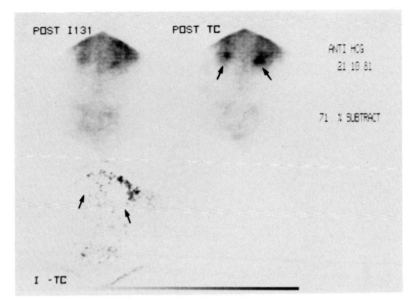

Fig. 5. RIL using ^{131}I-labeled rabbit antibody to hCG in a patient with choriocarcinoma. Posterior views of the abdomen show (I131) distribution of ^{131}I principally in the liver (at top right of image) and the kidneys and major vessels. The ^{99m}Tc image (TC) shows a relatively great concentration of ^{99m}Tc in the kidneys (arrow). After subtraction (I − TC) the renal areas have been oversubtracted (arrows) leaving reciprocal accumulation of ^{131}I in the right lobe of the liver. No evidence of liver disease was found in this patient by other means.

frequently not uniform over the stomach so that patches of excess ^{131}I uptake in the left upper quadrant of the abdomen must be assumed to be due to nonspecific uptake in the stomach until proved otherwise.

c. Thyroid. In spite of administration of iodine and potassium perchlorate administration, small amounts of ^{131}I may accumulate in the thyroid (Silva *et al.*, 1980) and produce an image of the gland.

d. Other Sites. Accumulation of ^{131}I-labeled antibody may occur at the site of a recent operation or other injury (Goldenberg *et al.*, 1978) and in our experience may persist for as long as 6 weeks. We have also observed accumulation of ^{131}I-labeled antibody to CEA in normal colon (Searle *et al.*, 1980), and this may be related to the presence of CEA in normal colon.

3. Variations in Distribution of ^{99m}Tc Albumin and ^{99m}TcO$_4$

The distribution of ^{99m}Tc may vary during the 45–90 minutes taken to obtain a full set of images, particularly as the readily diffusable ^{99m}TcO$_4$ spreads

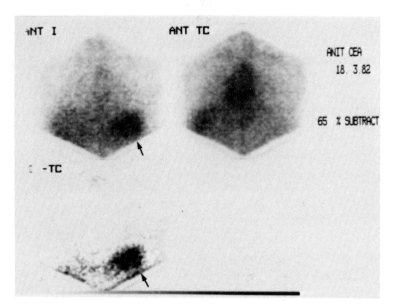

Fig. 6. RIL using ^{131}I goat antibody to CEA in a patient with raised serum CEA of unknown origin. Anterior views of the thorax and upper abdomen show accumulation of ^{131}I in the region of the stomach arrowed in the ^{131}I (I) and subtraction (I − TC) images. Gastroscopy and laparotomy showed no evidence of disease at this site. The image was probably produced by nonspecific accumulation of ^{131}I in the stomach in spite of prior administration of potassium iodide and potassium perchlorate.

into extravascular spaces and is excreted. Nonspecific uptake in the stomach and rates of urinary excretion also vary from patient to patient.

Thus differences in distribution of the two radionuclides may readily occur in the absence of tumor. An excess of ^{131}I in a particular area will, of course, produce an appearance indistinguishable from that caused by specific uptake of antibody in a tumor. Errors in reporting can be avoided if the reasons for this phenomenon listed above are considered. Excess ^{99m}Tc in a particular area can create a more difficult problem in that after subtraction "cold" areas are produced where ^{99m}Tc was present in excess and relatively "hot" regions are seen elsewhere (Fig. 7). In these circumstances there is no satisfactory basis for determining whether the hot areas are caused by small deposits of diffuse disease or are simply an artifact.

4. Imaging Characteristics of ^{99m}Tc and ^{131}I

The subtraction method depends on the ability of the gamma camera to discriminate between the different energies of the two nuclides. However the gamma camera images from the different energies are not necessarily in the same

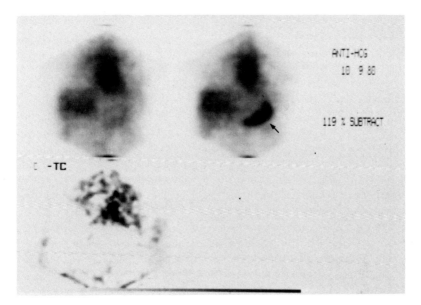

Fig. 7. RIL using ^{131}I-labeled sheep antibody to CEA in a patient with choriocarcinoma. Anterior views of the thorax and upper abdomen show more ^{99m}Tc in the stomach (arrow, top right) than in the ^{131}I image. After subtraction (I − TC) residual ^{131}I activity in the region of the heart is due to unequal distribution of the two radionuclides in normal organs. The tumor in this patient was shown to be confined to the uterus (not included in this view).

position or of the same size (the ^{131}I image being slightly larger) and do not necessarily have the same uniformity response. This can be illustrated by use of appropriate phantoms (B. F. Jones, personal communication; Jones *et al.*, 1982a) and can be minimized but not eliminated by careful and repeated adjustments of the gamma camera. Artifacts are easily produced unless these problems are recognized.

D. Clinical Results

The many problems associated with the subtraction method using ^{99m}Tc albumen and ^{99m}TcO$_4$ must be borne in mind when interpreting results of clinical studies. Nevertheless the scientific principle behind radioimmunolocalization is sound, and tumors have unequivocally been localized by several groups of workers.

1. CEA

The group at Lexington (Goldenberg *et al.*, 1978) initiated the present interest in clinical localization of CEA-producing tumors by successfully imaging tumors by the methods described above. Four patients with carcinomas originating in the colon or rectum, five of the uterine cervix, four of the ovary, and one each of the breast, bronchus, endometrium, and common duct were studied together with one case of malignant lymphoma. Tumor was found at all the positive sites by other methods, but RIL was negative at the site of a documented bony deposit, and in the non-CEA-producing lymphoma. This group have reported 142 cases of various CEA-producing tumors (Goldenberg *et al.*, 1980b; van Nagell *et al.*, 1980; Kim *et al.*, 1980) and shown that an elevation of serum CEA is not a prerequisite of a positive result. Nonspecific IgG produced negative results when used in place of antibody to CEA in a small number of patients and a positive result was produced in one patient with a benign ovarian teratoma, although 5 other benign ovarian and colonic tumors were negative. The Lexington group report true positive rates of 85–90%.

Other groups have also localized CEA-producing tumors by RIL (Mach *et al.*, 1980a,b; Dykes *et al.*, 1980; Searle *et al.*, 1980; Silva *et al.*, 1980; Begent *et al.*, 1981a; Jones *et al.*, 1982b), and there is no doubt that a wide variety of tumors can be localized by this method (Figs. 1, 3, 8, and 9). There has, however, been vigorous discussion as to whether RIL should be recommended for clinical use (Mach *et al.*, 1980a,c; Goldenberg *et al.*, 1980a; Order *et al.*, 1980; Begent *et al.*, 1980b). This argument has centered on the percentage of patients with positive results, and those from the group in Lausanne and Geneva (Mach *et al.*, 1980a,b) have had a considerably lower percentage of positive results (41% of 53 patients). This group have also reported false positive results in the liver and, in contrast to the group in Lexington, recommends that RIL should not yet enter routine clinical use but continue to be developed in a small number of specialized centers.

The important question as far as the clinician is concerned is whether RIL with antibody to CEA will provide diagnostic information which cannot be obtained by other noninvasive methods. Overall positivity rates are of little value in this context, as they will depend on the extent of disease in the sample of patients studied as well as differences in technique between centers. The group at Lexington have identified tumors as small as 2 cm in diameter (Goldenberg *et al.*, 1980b; van Nagell *et al.*, 1980), and this is in keeping with the finding at Charing Cross Hospital (Jones *et al.*, 1982b). The workers at Lausanne and Geneva estimate that tumors of 20 g (3–3.5 cm in diameter) are the smallest detectable (Mach *et al.*, 1980b). In view of the variable CEA content, vascularity, and mucus content of tumors, it would seem more likely that sensitivity will depend on factors other than size per se. Nevertheless taking 2 cm diameter as the basis for comparison with other imaging methods, one would expect RIL to offer no advantage in the lungs where computerized tomography (CT) is capable of detecting lesions a few millimeters in diameter (Husband and Golding, 1982). In the abdomen, lesions of 2 cm can be missed by CT, and van Nagell *et al.* (1980)

have demonstrated that deposits of ovarian carcinoma can be detected by RIL in patients in whom abdominal CT is negative. If this is a consistent finding, it is likely to be of benefit to patients when integrated into management protocols. This group have also reported that RIL was the sole means by which tumor was localized in a patient with carcinoma of the cervix (Goldenberg et al., 1978).

With recent appreciation that recurrent or localized metastatic disease can sometimes be eradicated by local therapy such as surgery or radiotherapy, often in combination with cytotoxic chemotherapy, a localizing method with the potential suggested above is of great interest. This is well illustrated in cancer of the colon or rectum where surgery has been used to remove metastases whose presence was suggested only by rises in serum CEA, sometimes with long-term disease-free survival (Minton and Martin, 1978; Mach et al., 1974; Ratcliffe et al., 1979). Localization of recurrent disease would assist in selection of patients for second look surgery. We have performed RIL in 23 patients in whom recurrence of colorectal cancer was predicted by a raised serum CEA, and positive results were found in 20. In 15, tumor was confirmed by surgery or conventional radiology, but only one tumor has so far been resected as a result of RIL in these circumstances (Begent et al., 1981b; Jones et al., 1982b). This would have been attempted in one other patient with an apparently solitary hepatic deposit but for his severe obstructive airways disease. He remains free from symptoms attributable to colorectal cancer more than a year later. In another patient, a localized 3 × 2 cm recurrence was identified lying adjacent to the vena cava just above its bifurcation. This proved unresectable, but a complete response on CT and repeat RIL has been achieved by a combination of cytotoxic chemotherapy and irradiation.

RIL has been useful in indicating areas worthy of further investigation, thus leading to corroboration of the findings of RIL by showing unresectable disease in 8 patients in whom laparotomy was therefore avoided. Nine patients had positive RIL which have not been corroborated by further investigations, and serum CEA remains raised in 8 of these, making it likely that they do have a recurrence. A further patient who had four investigations by RIL during follow-up after resection of a Dukes' C carcinoma of the colon had a positive area in the right lobe of the liver on each occasion but persistently normal serum CEA. At laparotomy after the fourth RIL no abnormality was found in the liver or elsewhere. An example of positive RIL in a patient with subclinical metastases is shown in Fig. 8.

By studying RIL in patients whose only evidence of recurrence was a rise in serum CEA its ability to detect tumor deposits is being tested at or beyond the limit of sensitivity of conventional imaging methods. If it is to be of benefit to patients with recurrent colorectal cancer this is the level at which it must be successful. Recurrences were detected in some patients, but the disease was usually unresectable so that the benefits achieved were small. This may be less

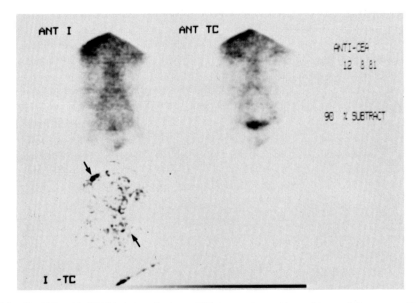

Fig. 8. RIL with ^{131}I-labeled antibody to CEA in a patient with carcinoma of the cecum. The primary tumor was resected, but 1–2 cm deposits of tumor were seen in the right lobe of the liver and paraaortic lymph nodes. Anterior views of the abdomen show these deposits (arrows) in the subtraction (I − TC) image.

the fault of RIL than of serum CEA which does not appear to be a sufficiently sensitive indicator of recurrence in our hands. Nevertheless some false positive results occurred with RIL, and these have contributed to a decision to perform a laparotomy in two patients in whom no tumor was found at operation.

By approaching the limits of sensitivity of RIL one allows the many potential artifacts inherent in the method, as discussed above, to have their greatest effect, and it is not surprising that false positive results should start to be a problem at this point. This should not inhibit active pursuit of developments of RIL in diseases such as colorectal cancer to which it brings a new dimension of investigation with much potential to assist in management.

2. RIL with Antibody to hCG

Successful localization of gestational trophoblastic tumors and hCG-producing malignant germ cell tumors has been reported (Begent *et al.*, 1980a; Goldenberg *et al.*, 1980c; Hatch *et al.*, 1980).

As with CEA, high levels of circulating hCG do not prevent localization of antibody in the tumor; positive results, however, are not reported in patients with normal serum hCG values. This is probably because of the relatively large amounts of hCG produced by these tumors. Positive results are found regularly in patients with serum hCG in excess of 500 IU/liter, indicating lesions of approximately 10^6 to 10^7 cells or more (Begent et al., 1980a). Positive or equivocal results are sometimes found with hCG as low as 10 IU/liter. In view of the relatively small number of viable tumor cells needed to produce serum hCG below 50 IU/liter, there is a possibility that some antibody might bind to hCG fixed in necrotic parts of the tumor. We have no evidence on this point with hCG, but it is interesting that antibody to CEA is concentrated in necrotic areas of xenografts of human colon carcinoma in mice (Lewis et al., 1982).

Radioimmunolocalization has been of definite value in management of patients with gestational choriocarcinoma. In more than 50 investigations at Charing Cross Hospital it was first shown that established disease could be localized, and attention has been focused on attempts to localize drug-resistant choriocarcinoma. Although the prognosis of this disease has improved greatly in the last 25 years, drug resistance still occurs in a small proportion of patients and can be fatal (Bagshawe and Begent, 1981). If the tumor is localized to resectable sites surgery can make a major contribution to tumor eradication. RIL in these circumstances in 15 patients had identified resectable pulmonary deposits in four, and resection led to complete response in each case. Abnormalties were present on CT in each case, but in one these were so minor as to be of doubtful significance and in another one deposit was shown by CT and two by RIL. Surgery confirmed the presence of two deposits, one measuring 0.5 cm in diameter (Begent et al., 1980a). An example is shown in Fig. 9. Despite the reservation indicated in Section II, D, 2 above, one of the most useful aspects of RIL in these circumstances has been to distinguish between a deposit as the source of hCG production and necrotic pulmonary deposits which may remain in the lungs as opacities on plain radiography or CT. In this study disseminated disease was demonstrated in 6 patients so that surgery could not be recommended, and false positive localization of antibody in the liver occurred in 2 patients (e.g., Fig. 4).

Smaller numbers of patients with hCG-producing germ cell tumors have been reported—11 patients in the Charing Cross series (Begent et al., 1981b) and five from the Lexington/NIH group (Javadpour et al., 1981a). Positive results are obtained in the presence of established disease, and in our experience there has been no case in which solitary deposits of drug-resistant disease were identified and excised. Javadpour et al. (1981a) reported one case in which this was achieved. They also suggested that negative RIL of the liver was of value in patient management, but this is difficult to accept in the absence of studies comparing the sensitivity of RIL with antibody to hCG and conventional imaging methods in this disease. In the same report it was also suggested that confirmation of a para-aortic mass seen by RIL permitted useful surgical resection of this mass at the start of treatment when members of the same group reported in the same year a trial showing that such surgery was of no benefit (Javadpour et al., 1981b).

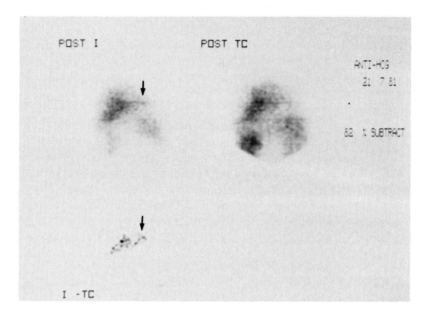

Fig. 9. RIL with ^{131}I-labeled rabbit antibody to hCG in a patient with drug-resistant choriocarcinoma. Serum hCG was 112 IU/liter, and computerized tomography (CT) showed a 1.5 cm deposit in the right lung. Posterior views of the thorax showed an area of accumulation of ^{131}I to the right of the heart in the ^{131}I image (I) and after subtraction to the right of residual cardiac activity (arrow) in I − TC image. The deposit was also seen on anterior views. This suggested that the deposit seen by CT contained viable choriocarcinoma. This was confirmed by histology after resection of the deposit. The patient remains in complete remission.

We have also shown conversion of RIL from positive to negative in patients with raised serum hCG before treatment and residual opacities in the lungs after chemotherapy has produced a return of hCG levels to normal. While interesting this is of marginal value in view of the sensitivity of serum hCG as an indicator of active trophoblastic elements. Antibodies directed against AFP have been used for localization of malignant teratoma with yolk sac elements, but this topic is outside the scope of this chapter.

IV. POTENTIAL IMPROVEMENTS

While the scientific basis for RIL is sound, the sensitivity and specificity of the method in its present form make it only of modest overall benefit in patient management. There are several ways, however, in which it might be improved.

A. An Alternative to Subtraction

Since many artifacts are attributable to the subtraction method, it is attractive to seek alternatives for removal of radiolabeled antibody which is not bound to tumor. Keep *et al.* (1981) have shown that liposomes containing a second antibody which is directed against the radiolabeled antibody will bind the radiolabeled antibody and accelerate its removal from the circulation and normal tissues as a result of uptake of liposomes into the reticuloendothelial system. This was first demonstrated in mice and rabbits, and it was shown that most of the radioactivity present in xenografts of human colon carcinomas in the mice was not removed by the liposomes. We have studied five patients with gastrointestinal tumors in the same way with similar findings (Begent *et al.*, 1982). While requiring further development this method has the potential to improve discrimination between tumor and nontumor tissue and may allow images to be obtained without subtraction. Its potential to improve the therapeutic ratio of antibody directed therapeutic agents is evident.

B. Alternative Radionuclides

The need for subtraction determines that two nuclides of different energies be used, and one inevitably has an energy which is less than ideal for modern gamma cameras. If subtraction can be avoided, ^{99m}Tc, ^{123}I, and ^{111}In would all be candidates for labeling of antibodies. Animal and human studies with ^{123}I have been reported (Epenetos *et al.*, 1982a,b), but the short half-life presents problems in regular handling for clinical purposes. Indium-111 can be linked to immunoglobulin, but it also binds to transferrin and this may lead to artifacts. Technetium-99m labeling of immunoglobulins has also been reported (Pettit *et al.*, 1980), but we are not aware of any clinical studies.

C. Tomographic Imaging (Single Photon Emission Computerized Tomography)

The three-dimensional images achieved by this method would be of considerable use in localizing tumors, but the low count rates achieved with ^{131}I-labeled antibody thwarted our early attempts (performed in conjunction with Dr. P. J. Ell). Mach *et al.* (1981) have reported tumor localization, however, using a rotating gamma camera and further developments will be observed with considerable interest.

D. Monoclonal Antibodies

The homogeneity and specificity of these reagents for a specific epitope on the immunizing antigen has led to their use in localization studies in animals and human beings (Farrands *et al.*, 1982; Epenetos *et al.*, 1982a,b) which are outside the scope of this chapter. Mach *et al.* (1981) reported successful use of monoclonal antibodies to CEA for RIL in patients and also described the use of Fab fragments of immunoglobulin.

It is by no means certain that monoclonal antibodies will eventually prove superior to affinity purified polyclonal antibodies. Several recent studies with monoclonal antibodies have shown cross-reacting sites on molecules in which conventional serology has failed to demonstrate homology. Such cross reactions may explain the unexpected binding of anti-tumor monoclonal antibodies to normal tissues which has been seen in some attempts at RIL. The polyclonal nature of conventional antibodies is such that these cross reactions are diluted out usually to the point of being undetectable. This topic has recently been reviewed by Lane and Koprowski (1982).

V. CONCLUSIONS

The method of RIL is soundly based in controlled animal and human studies, and several authors have succeeded in creating images of both hCG- and CEA-producing tumors. The small number of studies which have concentrated on situations in which RIL may make a unique contribution to management of patients have established the limits of sensitivity of the method and undoubtedly shown instances in which patients have benefited as a result of RIL. Unfortunately false positives emerge close to the limits of sensitivity and can make interpretation of results extremely difficult. Various possible reasons for this have been discussed and are sufficiently complex to make the method unsuitable for centers which are not actively involved in research in this area. Patients for whom RIL may affect management should probably be referred to such a center for the investigation to be performed and the results interpreted with considerable caution.

New and highly sensitive methods of tumor localization, such as nuclear magnetic resonance, are emerging. Each method will need to be evaluated in comparison with the alternatives in the multiplicity of distinct clinical situations which malignant disease presents. Radioimmunolocalization has much scope for improvement, and is unique in its basis in the biochemical distinctiveness of tumor tissues. Ironically, if that distinctiveness can be unequivocally defined, the therapeutic methods which would emerge might diminish the need for localization methods.

ACKNOWLEDGMENTS

We wish to thank our colleagues in the Departments of Medical Oncology and Nuclear Medicine who contributed to the work described. We are also grateful to the surgeons and physicians who referred patients. The work described in Fig. 2 was performed by Steven Per, a visitor from Thomas Jefferson University. The work was supported by the Cancer Research Campaign and the Medical Research Council.

REFERENCES

Arend, W. P., and Mannik, M. (1971). *J. Immunol.* **107,** 63–75.
Bagshawe, K. D., and Begent, R. H. J. (1981). *Gynaecol. Oncol.* **2,** 757–774.
Bale, W. F., Spar, I. L., Goodland, R. L., and Wolfe, D. E. (1955). *Proc. Soc. Exp. Biol. Med.* **89,** 564–568.
Bauer, F. K., Tubis, M., and Thomas, H. B. (1955). *Proc. Soc. Exp. Biol. Med.* **90,** 140–142.
Begent, R. H. J., Stanway, G., Jones, B. E., Bagshawe, K. D., Searle, F., Jewkes, R. F., and Vernon, P. (1980a). *J. R. Soc. Med.* **73,** 624–630.
Begent, R. H. J., Bagshawe, K. D., Stanway, G., Keep, P. A., Searle, F., Newlands, E. S., Jewkes, R. F., Jones, B. E., and Vernon, P. (1980b). *N. Engl. J. Med.* **303,** 1238.
Begent, R. H. J., Keep, P. A., Searle, F., Dent, J., Bagshawe, K. D., Jones, B. E., Jewkes, R. F., and Vernon, P. (1981a). *Oncodev. Biol. Med.* **1,** 61 (abstr.).
Begent, R. H. J., Searle, F., Stanway, G., Jewkes, R. F., Jones, B. E., Vernon, P., and Bagshawe, K. D. (1981b). *In* "Germ Cell Tumours" (C. K. Anderson, W. G. Jones, and A. Milford Ward, eds.), pp. 264–265. Taylor & Francis, London.
Begent, R. H. J., Keep, P. A., Green, A. J., Searle, F., Bagshawe, K. D., Jewkes, R. F., Jones, B. E., Barratt, G. M., and Ryman, B. E. (1982). *Lancet* **ii,** 739–742.
Deland, F. H., Kim, E. E., Simmons, G., and Goldenberg, D. M. (1980). *Cancer Res.* **40,** 3040–3049.
Dykes, P. W., Hine, K. R., Bradwell, A. R., Blackburn, J. C., Reeder, T. A., Drole, Z., and Booth, S. N. (1980). *Br. Med. J.* **280,** 220–222.
Epenetos, A. A., Nimmon, C. C., Arklie, J., Elliott, A. T., Hawkins, L. A., Knowles, R. W., Britton, K. E., and Bodmer, W. F. (1982a). *Br. J. Cancer* **46,** 1–8.
Epenetos, A. A., Britton, K. E., Mather, S., Shepherd, J., Granowska, M., Taylor-Papadimitriou, J., Nimmon, C. C., Durbin, H., Hawkins, L. R., Malpas, J. S., and Bodmer, W. F. (1982b). *Lancet* **ii,** 999–1005.
Farrands, P. A., Perkins, A. C., Pimm, M. V., Hardy, J. D., Embleton, M. J., Baldwin, R. W., and Hardcastle, J. D. (1982). *Lancet* **ii,** 397–400.
Goldenberg, D. M., Preston, D. F., Primus, F. J., and Hansen, H. J. (1974). *Cancer Res.* **34,** 1–11.
Goldenberg, D. M., Deland, F. H., Kim, E. E., Bennett, S., Primus, F. J., van Nagell, J. R., Estes, N., De Simone, P., and Rayburn, P. (1978). *N. Engl. J. Med.* **298,** 1384–1388.
Goldenberg, D. M., Kim, E. E., Deland, F. H., Bennett, S., and Primus, F. J. (1980a). *Cancer Res.* **40,** 2984–2992.
Goldenberg, D. M., Deland, F. H., and Kim, E. E. (1980b). *N. Engl. J. Med.* **303,** 1237–1238.
Goldenberg, D. M., Kim, E. E., Deland, F. H., van Nagell, J. R., and Javadpour, N. (1980c). *Science* **208,** 1284–1286.
Hatch, K. D., Mann, W. J., Boots, L. R., Tance, W., Shingleton, H. M., and Buchina, E. S. (1980). *Gynecol. Oncol.* **10,** 253–261.
Hoffer, P. B., Lathorp, K., Bekerman, G., Fang, V. S., and Refetoff, S. (1974). *J. Nucl. Med.* **15,** 323–327.
Husband, J. E., and Golding, S. J. (1982). *Br. Med. J.* **284,** 4–8.
Javadpour, N., Kim, E. E., Deland, F. H., Salyer, J. R., Shah, U., and Goldenberg, D. M. (1981a). *JAMA, J. Am. Med. Assoc.* **246,** 45–49.
Javadpour, N., Ozols, R. F., Barlock, A., Anderson, T., and Young, R. C. (1981b). *Proc. Am. Soc. Clin. Oncol.* **22,** 473.

Jones, B. E., Green, A. J., Vernon, P., Jewkes, R. F., Begent, R. H. J., Bagshawe, K. D., Searle, F., and Keep, P. A. (1982a). *Nucl. Med. Comm.* **3,** 124.

Jones, B. E., Begent, R. H. J., Jewkes, R. F., Vernon, P., Searle, F., Keep, P. A., Green, A. J., and Bagshawe, K. D. (1982b). *In* "Radioaktive Isotope in Klinik und Forschung" (R. Höfer and H. Bergmann, eds.), pp. 287–292. H. Egermann, Vienna.

Keep, P. A., Searle, F., Begent, R. H. J., Barratt, G. M., Boden, J., and Bagshawe, K. D. (1981). *Oncodev. Biol. Med.* **1,** 3 (abstr.).

Kim, E. E., Deland, F. H., Casper, S., Canagan, R. L., Primus, F. J., and Goldenberg, D. M. (1980). *Cancer* **45,** Suppl., 1243–1247.

Lane, D., and Koprowski, H. (1982). *Nature (London)* **296,** 200–202.

Lewis, J. C. M., Keep, P. A., and Bagshawe, K. D. (1982). *Oncodev. Biol. Med.* **3,** 161–168.

Mach, J.-P., Carrel, S., Merenda, C., Sordat, B., and Cerottini, J. C. (1974). *Nature (London)* **248,** 704–706.

Mach, J.-P., Forni, M., Ritschard, J., Buchegger, F., Carrel, S., Widgren, S., Donath, A., and Alberto, P. (1980a). *Oncodev. Biol. Med.* **1,** 49–69.

Mach, J.-P., Carrel, S., Forni, M., Ritschard, J., Donath, A., and Alberto, P. (1980b). *N. Engl. J. Med.* **303,** 5–10.

Mach, J.-P., Carrel, S., Forni, M., Ritschard, J., Donath, A., and Alberto, P. (1980c). *N. Engl. J. Med.* **303,** 1238–1239.

Mach, J.-P., Buchegger, F., Forni, M., Ritschard, J., Berche, C., Lumbruso, J.-D., Schreyer, M., Giradet, C., Accola, R. S., and Carrel, S. (1981). *Immunol. Today* **2,** 239–249.

Mannik, M., and Arend, W. P. (1971). *J. Exp. Med.* **134,** 195–315.

Minton, J. P., and Martin, E. W. (1978). *Cancer* **42,** Suppl. 3, 1422.

Order, S. E., Leichner, P., Ethinger, D. S., Klein, J. L., and Strand, M. (1980). *N. Engl. J. Med.* **303,** 1238.

Pettit, W. A., Deland, F. H., Bennett, S. J., and Goldenberg, D. M. (1980). *Cancer Res.* **40,** 3043–3045.

Pressman, D., and Korngold, L. (1953). *Cancer* **6,** 619–623.

Pressman, D., Day, E. D., and Blau, M. (1957). *Cancer Res.* **17,** 845–850.

Primus, F. J., Wang, R. H., and Goldenberg, D. M. (1973). *Cancer Res.* **33,** 2977–2982.

Primus, F. J., MacDonald, R., Goldenberg, D. M., and Hensen, H. J. (1977). *Cancer Res.* **37,** 1544–1547.

Primus, F. J., Bennett, S. J., Kim, E. E., Deland, F. H., Zahn, M. C., and Goldenberg, D. M. (1980). *Cancer Res.* **40,** 497–501.

Quinones, J., Mizejarski, G., and Beiewaltes, W. H. (1971). *J. Nucl. Med.* **12,** 69–75.

Ratcliffe, J. G., Wood, C. B., Burt, R. W., Malcom, A. J. H., and Blumgart, L. H. (1979). *In* "Carcino-Embryonic Proteins" (F.-G. Lehmann, ed.), Vol. II, p. 75. Elsevier/North-Holland Biomedical Press, Amsterdam.

Scheinberg, D. A., Strand, M., and Gansow, O. A. (1982). *Science* **215,** 1511–1513.

Searle, F., Bagshawe, K. D., Begent, R. H. J., Jewkes, R. F., Jones, B. E., Keep, P. A., Lewis, J. C. M., and Vernon, P. (1980). *Nucl. Med. Commun.* **1,** 131–139.

Searle, F., Boden, J., Lewis, J. C. M., and Bagshawe, K. D. (1981). *Br. J. Cancer* **44,** 137–144.

Silva, J. S., Cox, C. E., and Sullivan, D. S. (1980). *Proc. Am. Assoc. Cancer Res.* **21,** 239.

van Nagell, J. R., Kim, E. E., Casper, S., Primus, F. J., Bennett, S., Deland, F. J., and Goldenberg, D. M. (1980). *Cancer Res.* **40,** 502–506.

CHAPTER 10

Familial Cancer: An Opportunity to Study Mechanisms of Neoplastic Transformation

WILLIAM A. BLATTNER

Family Studies Section, Epidemiology Branch
National Cancer Institute
Bethesda, Maryland

I.	Introduction	189
II.	Genetic Syndromes and Cancer	190
	A. Preneoplastic States	190
	B. Familial and Hereditary Cancers	192
III.	Dysplastic Nevi and Risk for Cutaneous Malignant Melanomas	194
IV.	Defective Immune Response to EB Virus and Lymphoproliferation	196
V.	HLA-Associated Defective Immune Response Genes	197
VI.	Diverse Neoplasms in a Family and a Novel Radiation Repair Defect	198
VII.	Chromosomal Rearrangement and Cancer Risk in Families	199
VIII.	Genetic Markers Linked to Familial Cancer Risk	200
IX.	Defects in Cellular Growth Regulation and Risk for Familial Colon Cancer	201
X.	Conclusions	202
	References	202

I. INTRODUCTION

The role of genetic factors in human cancer is well established by the association of over 200 single gene disorders with high risk for cancer (Mulvihill, 1975, 1977). Familial cancer syndromes provide evidence for host genetic susceptibility, although clear single gene models are not always applicable (Fraumeni, 1982). This in part reflects the fact that cancer is relatively common in the general population and that some "familial clusters" are expected by chance alone and/or may be due to shared environmental exposures (Li, 1977;

Bishop *et al*, 1980). Claims that up to 90% of cancers are due to the environment are based on studies of international variation in cancer incidence (Haenszel and Kurihara, 1968; Higginson, 1980). Although it is likely that environmental factors do, in fact, account for the majority of cancers, the role of risk modifiers such as individual host genetic susceptibility, are clearly also important (Strong, 1982; Harris *et al.*, 1980). Why some patients, given a particular level of exposure to a known cancer causing agent, do not develop cancer while other equally exposed individuals do, points out that wide variation in individual susceptibility is important.

In this chapter, general concepts of genetic and familial factors in cancer will be summarized briefly, since they have been reviewed extensively elsewhere (Frauemeni, 1982; Heston, 1976; Lynch *et al.*, 1979a; Swift, 1982; Anderson, 1982). The major focus will be on investigative studies of cancer-prone families that have provided insights into the mechanisms of neoplastic transformation (Blattner, 1977). These clinical and interdisciplinary laboratory studies serve as a model for defining, in a highly targeted way, the oncodevelopmental mechanisms that help to explain variation in individual susceptibility.

II. GENETIC SYNDROMES AND CANCER

A. Preneoplastic States

Well-defined genetic syndromes account for a small proportion of cancer risk in the general population; however, significant attention has been given to the study of these cancer-prone disorders, despite their rarity, because they provide important models for understanding more general mechanisms of cancer risk (Strong, 1982). In Table I a partial list of genetic syndromes are summarized that are associated with heightened cancer risk. These preneoplastic states are genetic disorders characterized by the occurrence of an identifiable precursor state associated with an increased risk for cancer (Fraumeni, 1982). The hamartomatous syndromes are typified by disorders in which faulty embryonic development results in localized abnormal growth of mixed component tissues (Warkany, 1977), and this aberrant developmental pattern signals an underlying growth regulatory defect linked to a risk for cancer. The cancer-prone genodermatoses have provided important models for understanding host–environmental interaction in cancer. Xeroderma pigmentosum, in particular, has provided molecular biologists with a human analog to well-defined bacterial DNA repair deficiency models. The correlation between *in vitro* deficiency of ultraviolet-induced DNA repair and the susceptibility of such patients to the development of cutaneous neoplasms in sun-exposed areas represents the first clear example of a defined host defect resulting in susceptibility to a particular type of cancer

TABLE I
Preneoplastic States

Syndrome	Mode of inheritance[a]	Tumor types
I. Hamartomatous disorders		
Neurofibromatosis	AD	Fibrosarcoma, brain and optic glioma, meningioma, acoustic neuroma, pheochromocytoma, soft tissue sarcoma?, acute myelogenous leukemia?
Tuberous sclerosis	AD	Brain tumors (astrocytoma)
Von Hippel–Lindau syndrome	AD	Kidney cancer, pheochromocytoma, ependymoma
Multiple exostosis	AD	Chondrosarcoma
Peutz Jeghers syndrome	AD	Ovarian carcinoma
Cowden's disease	AD	Breast cancer, thyroid cancer, and meningioma
II. Genodermatoses		
Xeroderma pigmentosum	AR	Basal and squamous cell of skin, melanoma, sarcoma
Albinism	AR	Squamous cell of skin
Adult progeria (Werner's)	AR	Sarcoma
III. Chromosome breakage disorders		
Down's syndrome	—	Acute leukemia
Bloom's syndrome	—	Acute leukemia
Fanconi's syndrome	—	Acute leukemia, hepatocellular carcinoma
IV. Immune deficiency		
Ataxia telangectasia	AR	Leukemia, lymphoma, stomach, brain, lymphoreticular
Wiskott-Aldrich syndrome	XR	Lymphoreticular
"Common variable" immune deficiency	—	Lymphoproliferative, stomach
X-linked agammaglobulinemia	XR	Lymphocytic leukemia
Duncan's disease	XR	Lymphoproliferative

[a] AD, autosomal dominant; AR, autosomal recessive; XR, sex-linked.

(Cleaver, 1968). The chromosome breakage disorders are associated with a heightened risk for acute leukemia, particularly of the myelogenous type, while the immune deficiency syndromes predispose primarily to lymphatic cancers (Fraumeni, 1982). In addition, cancer of the stomach, adenocarcinoma of the lung, and cutaneous melanoma may also be linked to natural or therapeutic immunosuppression (Greene *et al.*, 1981; Spector *et al.*, 1978). The details of these syndromes and their interrelationships are summarized elsewhere (Fraumeni, 1982).

B. Familial and Hereditary Cancers

The major types of familial and hereditary cancer are listed in Table II. Hereditary versus familial cancers are distinguished by virtue of how well the underlying genetic mechanism is understood. In hereditary cancers the phenotype of the disease is so clear-cut that a single autosomal dominant gene mechanism can account for the familial occurrence. Among the familial cancer syndromes there is enough heterogeneity in the occurrence of these disorders that the mode of genetic inheritance is less clear-cut, although in the majority of cases an autosomal dominant mode is likely. This distinction reflects the complexity of sorting out host–environmental and environmental factors in defining which familial clusters result from inherited predisposition, shared environmental exposures, or combinations of these factors (Li, 1977).

The complexity of these relationships is evidenced in the fact that the common cancers of man have some tendency to cluster in families. Population-based studies have consistently shown that there is an approximate 3-fold risk for a relative of a cancer patient to develop the same neoplasm compared to the expected risk in the general population (Anderson, 1982). The following features tend to distinguish hereditary occurrences of neoplasia from those that might occur by chance alone:

1. Specificity of tumor type or site
2. A tendency for multiple occurrence of the same or different neoplasm in the same individual
3. A younger than usual age of onset
4. Distribution of tumors in a family consistent with Mendelian segregation

In hereditary cancer syndromes, the risk for neoplasia among family members may be much greater (up to 20- to 30-fold increase) than that expected for the general population (Lynch et al., 1979b; Anderson, 1982; Albert and Child, 1977).

Sometimes the familial aggregation of cancer offers an opportunity for etiologic studies that may give insights to the mechanism of cancer susceptibility and pathogenesis. We have found that an interdisciplinary approach involving the collaboration of etiologically oriented epidemiologists and laboratory investigators provides opportunities for promoting the understanding of risk mechanisms (Blattner, 1977). Since this process first begins at the bedside, a checklist of questions that helps identify high-risk patients has been developed (Mulvihill, 1975). The following questions should be considered:

TABLE II
Familial and Hereditary Cancers

Syndrome	Mode of inheritance[a]	Tumor types
I. Hereditary neoplasms		
Retinoblastoma	AD	Retinoblastoma, osteosarcoma
Nevoid basal cell carcinoma syndrome	AD	Basal cell cancer, ovarian fibromas, medulloblastoma
Multiple Endocrine Adenomatosis Type I (Werner's syndrome)	AD	Islet cell adenomas, malignant schwannoma, parathyroid, pituitary, adrenal, carcinoid
Type II (Sipple's syndrome)	AD	Medullary cancer of thyroid, parathyroid adenoma, pheochromocytoma
Type III (Mucosal neuromas and endocrine adenomatosis)	AD	Pheochromocytoma, medullary cancer of thyroid, neurofibroma, submucosal neuromas of tongue, lip, eyelid
Chemodactomas	AD	Carotid body tumor
Familial polyposis coli and Gardener's syndrome	AD	Adenocarcinoma of colon, ampulla of Vater tumors, benign fibrous tumors, pancreas, thyroid, adrenal
Tylosis with esophageal carcinoma	AD	Esophagus
Familial dysplastic nevus syndrome	AD	Malignant melanoma
II. Familial cancer syndromes		
Site-specific aggregation		
Breast	AD	Breast
Endometrium	?	Endometrium
Ovary	AD	Ovary
Prostate	?	Prostate
Lymphoreticular	?	Hodgkin's and non-Hodgkin's lymphomas, Waldenstrom's macroglobulinemia
Stomach	?	Stomach
Leukemia		CLL, AML, ALL
Colon	AD	Right-sided colon cancer
Lung	?	Lung
Bladder	?	Bladder
Kidney	?	Kidney
Nonsite specific aggregation		
Familial adenocarcinoma syndrome	AD	Colon, endometrium, breast
Turcot's syndrome	?	Brain, colon
"Li-Fraumeni syndrome" (diverse neoplasms)	AD?	Bony and soft tissue sarcomas, breast, brain, leukemia, lung, adrenocortical, other?
Torre's syndrome (multiple sebaceous gland tumors)	AR	Gastrointestinal, genitourinary cancers

[a] AD, autosomal dominant; AR, autosomal recessive.

1. What are the demographic features of the patient? Is the tumor occurring at an unusual site, or younger or older than expected?
2. Are there environmental exposures?
3. Are there identifiable host factors such as a positive family history, antecedent disease, or laboratory abnormality?

Examples of ways in which this approach has promoted our understanding of host susceptibility mechanisms are presented below.

III. DYSPLASTIC NEVI AND RISK FOR CUTANEOUS MALIGNANT MELANOMAS

In 1975, the referral of two melanoma-prone families led Dr. Mark H. Greene to the pathology laboratory of Dr. Wallace Clark. Their studies of over 14 similar families have provided important insights into the fundamental biology of cutaneous malignant melanoma (CMM) (Greene et al., 1978; Reimer et al., 1978).

Although CMM occurs more commonly in a sporadic form, hereditary cutaneous malignant melanoma (HCMM) is far from rare. It has been estimated that up to 11% of CMM is genetic in origin (Wallace et al., 1971). The studies of Dr. Greene and Dr. Clark have defined the following characteristic for HCMM:

1. Young age of onset (less than 40 years)
2. Frequent occurrence of multiple primaries
3. Equal sex ratio
4. Autosomal dominant mode of genetic transmission
5. Characteristic premalignant mole pattern, the dysplastic nevus syndrome (DNS), in approximately half of the offspring of clinically affected individuals (Clark et al., 1978)

The discovery of a clinically and pathologically distinct, premalignant mole, or dysplastic nevus, has provided important insights into the steps by which malignant transformation to CMM takes place. The accessibility of lesions, and the opportunity to follow the progression of these lesions through distinctive stages of malignant degeneration makes this syndrome a useful model for the study of the transformation process (Elder et al., 1981).

An example of a typical dysplastic nevus is shown in Fig. 1. The dysplastic nevus differs from the common acquired nevus in size, shape, pigmentation, morphology, and microscopic appearance. These lesions are characterized by an irregular outline, variable pigment, and large size in comparison to common acquired nevi. They tend to occur in an atypical distribution on the body, (e.g.,

10. Familial Cancer

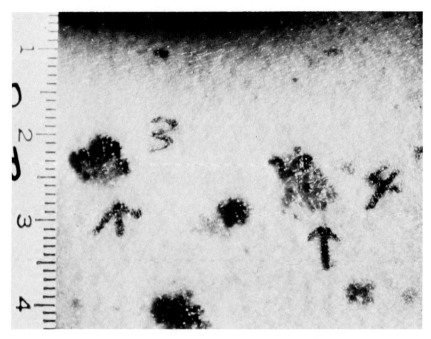

Fig. 1. Close-up view of multiple dysplastic nevi from a 25-year-old white female from NIH Family #481. Nevi are lesion to lesion characterized by variability, irregular borders, variegate pigmentation, relatively large size, multiplicity, and atypical morphology. (Photograph courtesy of Dr. M. H. Greene.)

on the scalp, buttocks, and female breast)(Clark *et al.*, 1978). Histologically, the cells show cytologic atypia and this feature points to the malignant potential of these lesions (Elder *et al.*, 1981). In fact, in the 5 years that 400 members of these 14 melanoma-prone families have been followed by Drs. Greene and Clark, 31 new primary melanomas have been diagnosed. In all 31 cases, early, surgically curable lesions have been identified through careful monitoring of these premalignant lesions.

Since the discovery of this familial syndrome, the occurrence of sporadic dysplastic nevi (i.e., those not occurring in the family setting) has been appreciated, and in one recent survey, more than one-third of all CMM's were felt to have arisen from these precursor lesions (Elder *et al.*, 1980). A study of renal transplant patients who developed malignant melanoma demonstrated that in all cases, these tumors had arisen in dysplastic nevi (Greene *et al.*, 1981). This emphasizes the importance of recognizing this lesion as premalignant, and that alterations in host immune function may play a role in the transition to malignant melanoma. Immunologic studies of 7 melanoma-prone families have identified *in vitro* immune defects in affected and high-risk family members that may contribute to the familial risk (Dean *et al.*, 1979; Vandenbark *et al.*, 1979).

Ultraviolet radiation exposure is a known risk factor for malignant melanoma, and since dysplastic nevi and CMM tend to occur in sun-exposed areas, *in vitro* studies have been done to learn how ultraviolet radiation exposure affects members of these HCMM families. These studies have revealed a defect in DNA repair that resembles that of xeroderma pigmentosum but to a less severe degree (Smith *et al.*, 1982). The molecular basis of this defect may provide important insights into the pathogenesis of malignant melanoma.

Thus, observations at the bedside have led to a series of clinical and laboratory studies that promise to provide important new insights into the pathogenesis of CMM and cancer in general. In particular, the study of this familial disorder offers the opportunity for important insights into the role of host and host–environmental interactions in the etiology of cancer.

IV. DEFECTIVE IMMUNE RESPONSE TO EB VIRUS AND LYMPHOPROLIFERATION

In 1975, David T. Purtillo and colleagues reported the clinical details of the Duncan kindred in which 3 brothers died of fatal, Epstein-Barr virus (EBV)-induced infectious mononucleosis and 2 half-brothers developed malignant lymphoma following infectious mononucleosis (Purtillo *et al.*, 1975). Maternally related cousins developed agammaglobulinemia. Since this first report, additional families with the so-called X-linked lymphoproliferative syndrome (XPL), or Duncan's disease have been registered in Dr. Purtillo's XPL registry (Purtillo, 1980). These families have provided important insights concerning the interrelationship of EBV infection and its role in the pathogenesis of malignant lymphoma in these families. In a recent update, 25 cases of EBV-associated disease (including 7 with XPL) were reviewed (Purtillo *et al.*, 1981). The clinical features of these cases include chronic or fatal infectious mononucleosis, acquired aggamaglobulinemia or aplastic anemia after infectious mononucleosis, and malignant lymphoma following EBV infection. Based on these data, Purtillo has postulated that EBV triggers a B cell proliferation which persists due to a deficient T and B cell response to the virus (Purtillo *et al.*, 1981). The presence of the virus in these immunoproliferative cases has now been documented not only by cell culture techniques but also by application of elegant molecular probes that show the presence of integrated EBV genomic material (Saemundsen *et al.*, 1981). The absence of an adequate EBV antibody response, and defects in *in vitro* cellular function, provide insights into the immunologic basis of this premalignant state (Sullivan *et al.*, 1980). Application of these approaches to immunosuppressed populations at high risk for development of lymphoma suggest that some of the same mechanisms identified through studies of XPL may be of help in explaining their lymphoma risk (Hanto et al., 1981). These clinical and interdisciplinary studies suggest that susceptibility is acting as a recessive trait in these patients, and that it should be possible to map precisely which function, coded for by an X chromosome gene, is lacking in the lymphoma-prone family members.

V. HLA-ASSOCIATED DEFECTIVE IMMUNE RESPONSE GENES

In 1973, two adolescent brothers developed acute lymphocytic leukemia within 3 weeks of each other (Blattner et al., 1978). Clinically the brothers had almost identical courses of disease. Both responded to chemotherapy and had relatively late disease relapses. One brother recently died from infectious complications of his recurrent disease. At the time that these brothers were diagnosed, it was felt, on clinical grounds, that a shared susceptibility to a common environmental exposure might be important. Although studies have attempted to identify an etiologic agent, none has been found. However, immunogenetic studies have provided important insights into the role of host susceptibility. HLA typing was performed on family members and the two leukemic sibs were found to be HLA identical. In addition, sera from mothers of leukemic children contained antibodies to B cell alloantigen markers that were linked to the major histocompatibility complex. By pedigree analysis of these reactions and the fortuitous occurrence of recombinants in one maternal haplotype, it was possible to trace these leukemia-associated reactions through the family tree and demonstrate that the sibs were homozygous for these specificities (Blattner et al., 1978). Recent follow-up confirmed that these HLA reactions are linked to a specific HLA Dr4 and a variant of Dw4 called DB31. In this case, the HLA pattern in these siblings resembles the occurrence of immunologically associated leukemia in certain strains of mice. In the murine model, homozygosity for defective *H-2* associated immune response genes, results in increased risk for acute leukemia, whereas heterozygosity allows the mouse to mount an adequate immune response (Lilly and Pincus, 1973). This recessive pattern of inheritance, in which absence of an immune response gene function is linked to leukemia susceptibility, resembles the mechanism that appears to underlie the defect in XPL (see above).

Evidence that HLA associated defective immune regulatory genes may play a role in the pathogenesis of lymphoproliferative malignancy is also seen in the recent report of a family prone to Waldenstrom's macroglobulinemia (WM) and autoimmune disease (Blattner et al., 1980). This family first came to our attention because of the occurrence of WM in 2 brothers. Laboratory testing in 22 members of the family confirmed the diagnosis of WM in a third sibling and their father, as well as clinical and subclinical autoimmunity. We found that one HLA haplotype (A2, B8, DRW3) present in all family members with WM and clinical thyroid disease and in all but one with subclinical autoantibodies. In this family, the B cell alloantigens Ia 172 and 350, previously associated with Sicca syndrome, were identified in all family members with the HLA haplotype. This is of interest in view of the exceptionally high risk of histiocytic lymphoma and WM in Sicca syndrome patients. We postulate that the susceptibility to WM and autoimmune disease in this family results from a defect in the immune regulatory system. In particular, a new class of immune suppressor genes linked to similar disorders in mice may provide a model for understanding how a common abnormality predisposes to autoimmunity and lymphoproliferative malignancy (Naor, 1979).

VI. DIVERSE NEOPLASMS IN A FAMILY AND A NOVEL RADIATION REPAIR DEFECT

In the mid-1960's, Dr. F. P. Li and Dr. J. F. Fraumeni, Jr. were intrigued by the occurrence of rhabdomyosarcoma in a child and acute leukemia in his father. Epidemiologic detective work led to the discovery of a distant family branch in which other similar cancers occurred (Li, 1977). An additional three families with a similar pattern suggested a syndrome characterized by the occurrence of bony and soft tissue sarcoma of breast and brain and leukemia (Li and Fraumeni, 1969). Other investigators have identified additional examples of this syndrome (Lynch et al., 1978).

The family shown in Fig. 2 came to our attention because of the occurrence of childhood neoplasms (osteosarcoma, bilateral malignant neurolemmoma, and acute lymphocytic leukemia) in 3 brothers (Blattner et al., 1979). Over the next few years the father of these boys developed a low grade astrocytoma and their paternal aunt developed a leiomyosarcoma. In our initial study, a geneologic search identified a total of 16 cases of cancer among the descendants of the proband's great-great-great-grandmother. This included a previously unsuspected cluster of similar neoplasms in a distant branch. The constellation of tumors in the family included bony and soft-tissue sarcomas, brain and neural tumors, leukemia, and breast carcinoma, and occurred in a pattern suggesting the action of an incompletely penetrant autosomal dominant gene with pleiotrophic effects (Blattner et al., 1979). The observation that in some cases the genetic predisposition may have interacted with environmental determinants to produce particular tumors led us to collaborate with Dr. M. C. Patterson to study the effects of γ radiation on *in vitro* DNA repair.

The first two patients to be studied were a paternal great uncle of the proband,

10. Familial Cancer

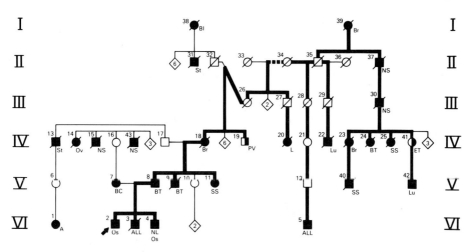

Fig. 2. Pedigree of NIH Family 0165. Squares, males; circles, females; shaded symbols, malignancy; half-shaded symbol, premalignant condition; Roman numerals, generation numbers; Arabic numerals, patient numbers; /, deceased; ↗, proband; thick lines connect descendents of case I-39 in whom malignant neoplasms developed; Os, osteosarcoma; ALL, acute lymphblastic leukemia; SS, soft tissue sarcoma; NL, malignant neurilemoma; BC, basal cell; BT, brain tumor; BR, breast cancer; PsV, polycythemia vera; L, leukemia; Lu, Lung cancer; A, idiopathic anemia; St, stomach, Ov, ovarian carcinoma; Bl, bladder carcinoma; ET, eye tumor; NS, not specified.

who had polycythemia vera and an occupation history of chemical and radiation exposure, and the brother of the proband, who developed osteosarcoma of the lumbar vertebra in the site of prior radiotherapy. When exposed to graded dosages of radiation, skin fibroblasts from these patients demonstrated an unusual survival pattern—their cells were *resistant* to the killing effects of γ radiation (Bech-Hansen *et al.*, 1981). Other relatives in the cancer-prone lineage, but not their spouses, over three generations demonstrated a similar pattern. This novel radiation phenotype could be a manifestation of a basic cellular defect, predisposing to a variety of tumors in members of this family. Thus *in vitro* radio resistance, like radiosensitivity, may be a phenotype of a mechanism that increases cancer risk in man. This hypothesis is supported by recent studies demonstrating that radioresistance may be an *in vitro* marker of aging (Licastro *et al.*, 1982).

VII. CHROMOSOMAL REARRANGEMENT AND CANCER RISK IN FAMILIES

A large body of literature exists on the relationship of cytogenetic abnormalities and cancer (Mark, 1977). Recent studies hypothesize that human cancers may not be caused by conventional mutagens alone, but are more likely to be the result of genetic transpositions (Cairns, 1981). Studies of certain childhood cancers associated with specific chromosomal deletions support this concept. In one family, Dr. L. C. Strong and colleagues were intrigued by the occurrence of retinoblastoma over four generations associated with a low penetrance and expressivity (Strong *et al.*, 1981). In the family, development of retinoblastoma was associated with a constitutional chromosome deletion [del (13)(q 13. 1q 14.5)]. The unaffected, transmitting relatives carried a balanced insertional translocation of a portion of the long arm of chromosome 13 onto the short arm of chromosome 3 (Strong *et al.*, 1981). The data on this family confirm that the predisposition to retinoblastoma is directly related to the loss of genetic material and *not* due to an alternation in gene function because of a position effect. The nature of the retinoblastoma-associated chromosome deletion in this family is consistent with data from similar deletion-associated cases (Sparkes *et al.*, 1980). Since a second mutation has been postulated as necessary for malignant transformation, the absence of a segment of chromosomal material may suggest that a mutation or recombination at the homologous locus may be involved in this process (Knudson, 1977).

Renal cell carcinoma was found in 10 adults (five men and five women) from three consecutive generations of a family (Cohen *et al.*, 1979). Seven of these cases were ascertained because of symptomatic disease, and the remaining 3 cases were diagnosed in the course of screening intravenous pyelograms undertaken because of positive family history. As part of an interdisciplinary protocol undertaken to identify risk factors in this family, 10 members were found to harbor a balanced reciprocal translocation between chromosomes 3 and 8. This translocation was present in the 5 survivors of renal cancer, a member with a renal cyst, and 4 others with no obvious renal disease. The data from this family provide support for the two-mutation model of carcinogenesis proposed by Knudson (1977). In this hereditary form of the cancer the tumor may result from a first "hit" involving a germinal, or prezygotic mutation and a second somatic mutation. In this family, a submicroscopic deletion or point mutation at the sites of chromosome rearrangement may account for the germinal mutation (Knudson, 1977).

VIII. GENETIC MARKERS LINKED TO FAMILIAL CANCER RISK

The presence of breast cancer in multiple members of a family, is a major risk factor for breast cancer (Anderson, 1976). Numerous studies have provided genetic evidence that susceptibility to breast cancer is inherited as an autosomal dominant allele (Anderson, 1982). Genetic studies by Anderson (1976) have also shown that this risk is particularly prominent in younger women. A variety of perturbations in hormonal and biochemical measures have also been identified in clinically affected and high-risk women from breast cancer-prone families (Lynch et al., 1979c).

A recent study by Dr. M. C. King in collaboration with Dr. H. T. Lynch and Dr. R. C. Elston provided the first example of a non-X-linked genetic marker of familial cancer risk (King et al., 1980). A cohort of 11 high-risk families was evaluated by segregation and linkage analysis, which showed that an autosomal dominant model best explains the distribution of cancers in these families. When a battery of genetic markers were tested for linkage to this putative autosomal dominant susceptibility gene, it was found that the gene for the enzyme glutamate-pyruate transaminase was linked to familial susceptibility (King et al., 1980). Confirmation of this finding would establish the existence of a gene which increases susceptibility to breast cancer located on the same chromosome as this enzyme. This would open the door to opportunities for potential genetic counselling in such families. Application of newer somatic cell and gene cloning approaches in the study of these families may help define more precisely the nature of the putative gene defect.

IX. DEFECTS IN CELLULAR GROWTH REGULATION AND RISK FOR FAMILIAL COLON CANCER

Familial polyposis coli and Gardner's syndrome represent one of the best characterized familial cancer syndromes (Fraumeni and Mulvihill, 1975). The characteristic features of the disease are the occurrence of numerous adenomatous polyps of the colon and rectum which usually appear during adolescence but have been documented in pre-teenagers from affected cohorts. These polyps number from hundreds to thousands in affected cases and by 45 to 50 years of age, 50% of these affected persons will be diagnosed with colon cancer (Fraumeni and Mulvihill, 1975). The syndrome is rare, occurring with a frequency of one per 8000 live births. Although the polyps occur throughout the colon, the cancers occur largely in the rectosigmoid and cecum. Gardner's syndrome combines multiple polyps with bone tumors (especially osteomas) lesions of the skin and soft tissue, (e.g., sebaceous cysts and fibromas), and increased risk of periampullary carcinoma (Pauli et al., 1980).

Investigative studies have defined a spectrum of detectable biologic abnormalities in these patients that point to underlying defects in cellular growth regulation as a concomitant of cancer susceptibility (Rasheed and Gardner, 1981; Danes et al., 1980; Lipkin et al., 1980). In the early 1960's, autoradiographic techniques were developed that showed that mice exposed to an experimental carcinogen 1,2-dimethylhydrazine experienced an expansion of the colonic crypt proliferative compartment toward the lumenal surface (Thurnherr *et al.*, 1973). *In vitro* labeling techniques were adopted for human study by Dr. E. E. Deschner who documented that a similar shift is detectable in normal mucosa of familial polyposis cases (Deschner *et al.*, 1963). This "transformed phenotype" appears to represent a fundamental defect in cellular growth regulation in these patients, since abnormalities are also detectable in skin fibroblasts of affected individuals (Kopelovich *et al.*, 1979b). The abnormal phenotype includes a low serum requirement for growth of cells, a deformation of the actin cable cytoskeleton matrix (Kopelovich *et al.*, 1980), an easy transformality by Kirsten murine sarcoma and SV40 viruses (Rasheed and Gardner, 1981), and induction of neoplastic transformation of fibroblasts by the tumor promoter 12-*O*-tetradecanoyl phorbol-13-acetate (TPA) (Kopelovich *et al.*, 1979a).

X. CONCLUSIONS

The impact of genetic host susceptibility on the overall cancer burden is not known. Studies by Swift and colleagues which demonstrate a heightened risk for cancer in heterozygous carriers of putative cancer genes from several high-risk disorders (Swift, 1982) provide evidence that "cancer genes" may be more important than previously appreciated. For example, with regard to ataxia telangectasia (AT) which occurs in 1 out of every 40,000 live births, up to 1% of the general population may be at increased cancer risk due to the fact that they are carriers of this AT cancer-associated, susceptibility gene (Swift, 1982).

Furthermore, as outlined in this chapter, interdisciplinary studies of selected high-risk families provide important opportunities to understand mechanisms of cancer risk. The future of such studies is favorable because of the advent of newer molecular biology, somatic cell, and gene cloning techniques. As these tools for investigation are improved, the interdisciplinary approach to familial cancer offers promise for defining genetically determined mechanisms that predispose to malignancy. In this respect, the approach could be considered a pilot for studying the role of host factors in human carcinogenesis with broader application to cancer etiology and environmental carcinogenesis.

REFERENCES

Albert, S., and Child, M. (1977). *Cancer* **40**, 1674–1679.
Anderson, D. E. (1976). *Cancer Detect. Prev.* **1**, 283–291.
Anderson, D. E. (1982). *In* "Cancer Epidemiology and Prevention" (D. Schottenfeld and J. F. Fraumeni, Jr., eds.), pp. 483–493. Saunders, Philadelphia, Pennsylvania.
Bech-Hansen, N. T., Blattner, W. A., Sell, B. M., McKeen, E. A., Lampkin, B. C., Fraumeni, J. F., Jr., and Patterson, M. C. (1981). *Lancet* **1**, 1335–1337.

10. Familial Cancer

Bishop, D. T., Gardner, E. J., Lalouel, J., Peto, J., Williams, R. R., and Weiss, K. M. (1980). *Banbury Rep.* **4,** 389–408.

Blattner, W. A. (1977). *In* "Genetics of Human Cancer" (J. J. Mulvihill, R. W. Miller, and J. F. Fraumeni, Jr., eds.), pp. 269–280. Raven Press, New York.

Blattner, W. A., Naiman, J. L., Mann, D. L., Wimer, R. S., Dean, J. H., and Fraumeni, J. F., Jr. (1978). *Ann. Intern. Med.* **89,** 173–176.

Blattner, W. A., McGuire, D. B., Mulvihill, J. J., Lampkin, B. C., Hananian, J., and Fraumeni, J. F., Jr. (1979). *JAMA, J. Am. Med. Assoc.* **24,** 259–261.

Blattner, W. A., Garber, J. E., Mann, D. L., McKeen, E. A., Henson, R., McGuire, D. B., Fisher, W. B., Bauman, A. W., Goldin, L. R., and Fraumeni, J. F., Jr. (1980). *Ann. Intern. Med.* **93,** 830–832.

Cairns, J. (1981). *Nature (London)* **289,** 353–357.

Clark, W. H., Reimer, R. R., Greene, M. H., Ainsworth, A. M., and Mastrangelo, M. J. (1978). *Arch. Dermatol.* **114,** 732–738.

Cleaver, J. E. (1968). *Nature (London)* **218,** 652–656.

Cohen, A. J., Li, F. P., Berg, S., Marchetto, D. J., Tsai, S., Jacobs, S. C., and Brown, R. S. (1979). *N. Engl. J. Med.* **301,** 592–595.

Danes, B. S., Bulow, S., and Svendsen, S. B. (1980). *Clin. Genet.* **18,** 128–136.

Dean, J. H., Greene, M. H., Reimer, R., LeSane, F. V., McKeen, E. A., Mulvihill, J. J., Blattner, W. A., Herberman, R. B., and Fraumeni, J. F., Jr. (1979). *JNCI, J. Natl. Cancer Inst.* **63,** 1139–1145.

Deschner, E. E., Lewis, D. M., and Lipkin, M. (1963). *J. Clin. Invest.* **42,** 1922–1928.

Elder, D. E., Goldman, L. I., Goldman, S. C., Greene, M. H., and Clark, W. H. (1980). *Cancer* **46,** 1787–1794.

Elder, D. E., Greene, M. H., Bondi, E. E., and Clark, W. H. (1981). *In* "Pathology of Malignant Melanoma" (B. V. Ackerman, ed.), pp. 185–215. Masson, New York.

Fraumeni, J. F., Jr. (1982). *In* "Cancer Medicine" (J. F. Holland and E. Frei, eds.), pp. 5–12. Lea & Febiger, Philadelphia, Pennsylvania.

Fraumeni, J. F., Jr., and Mulvihill, J. J. (1975). *In* "Cancer Epidemiology and Prevention" (D. Schottenfeld, ed.), pp. 404–415. Thomas, Springfield, Illinois.

Greene, M. H., Reimer, R. R., Clark, W. H., and Mastrangelo, M. J. (1978). *Semin. Oncol.* **5,** 85–87.

Greene, M. H., Young, T. I., and Clark, W. H. (1981). *Lancet* **1,** 1196–1199.

Haenszel, W., and Kurihara, M. (1968). *JNCI, J. Natl. Cancer Inst.* **40,** 43–68.

Hanto, D. W., Frizzera, G., Purtillo, D. T., Sakamoto, K., Sullivan, J. L., Saemundsen, A. K., Klein, G., Simmons, R. L., and Jajarian, J. S. (1981). *Cancer Res.* **41,** 4253–4261.

Harris, C. C., Mulvihill, J. J., Thorgeirsson, S. S., and Minna, J. D. (1980). *Ann. Intern. Med.* **92,** 809–825.

Heston, W. E. (1976). *Adv. Cancer Res.* **23,** 1–21.

Higginson, J. (1980). *Am. J. Med.* **69,** 811–813.

King, M. C., Go, R. C., Elston, R. C., Lynch, H. T., and Petrakis, N. L. (1980). *Science* **208,** 406–408.

Knudson, A. G. (1977). *Adv. Hum. Genet.* **8,** 1–66.

Kopelovich, L., Bias, N. E., and Helson, L. (1979a). *Nature (London)* **282,** 619–621.

Kopelovitch, L., Pfeffer, L. M., and Bias, N. (1979b). *Cancer* **43,** 218–223.

Kopelovich, L., Lipkin, M., Blattner, W. A., Fraumeni, J. F., Jr., Lynch, H. T., and Pollack, R. E. (1980). *Int. J. Cancer* **26,** 301–307.

Li, F. P. (1977). *In* "Genetics of Human Cancer" (J. J. Mulvihill, R. W. Miller, and J. F. Fraumeni, Jr., eds.), pp. 263–268. Raven Press, New York.

Li, F. P., and Fraumeni, J. F., Jr. (1969). *Ann. Intern. Med.* **71,** 747–751.

Licastro, F., Franceschi, C., Chiricolo, M., Battelli, M. G., Tabacchi, P., Cenci, M., Barboni, F., and Pallenzona, D. (1982). *Carcinogenesis (London)* **3,** 45–48.
Lilly, F., and Pincus, T. (1973). *Adv. Cancer Res.* **17,** 231–277.
Lipkin, M., Sherlock, P., and Decosse, J. J. (1980). *Curr. Probl. Cancer* **4,** 1–57.
Lynch, H. T., Mulcahy, G. M., Harris, R. E., Guirgis, H. A., and Lynch, J. F. (1978). *Cancer* **41,** 2055–2064.
Lynch, H. T., Follett, K. L., Lynch, P. M., Albano, W. A., Mailliard, J. L., and Pierson, R. L. (1979a). *JAMA, J. Am. Med. Assoc.* **242,** 1268–72.
Lynch, H. T., Guirgis, H. A., Harris, R. E., Lynch, P. M., Lynch, J. F., Elston, R. C., Go, R. C., and Kaplan, E. (1979b). *Front. Gastrointest. Res.* **4,** 142–150.
Lynch, H. T., Harris, R. E., Fishman, J., Lynch, J. F., Marrero, K., and Maloney, K. (1979c). *Cancer* **44,** 1860–1869.
Mark, J. (1977). *Adv. Cancer Res.* **24,** 165–222.
Mulvihill, J. J. (1975). *In* "Persons at High Risk of Cancer: An Approach to Cancer Etiology and Control" (J. F. Fraumeni, Jr., ed.), pp. 3–37. Academic Press, New York.
Mulvihill, J. J. (1977). *In* "Genetics of Human Cancer" (J. J. Mulvihill, R. W. Miller, and J. F. Fraumeni, Jr., eds.), pp. 137–143. Raven Press, New York.
Naor, D. (1979). *Adv. Cancer Res* **29,** 45–125.
Pauli, R. M., Pauli, M. E., and Hall, J. G. (1980). *Am. J. Med. Genet.* **6,** 205–219.
Purtillo, D. T. (1980). *Lancet* **1,** 300–303.
Purtillo, D. T., Cassel, C., Yang, J. P. S., Stephenson, S. R., Harper, R., Landing, G. H., and Vawter, G. F. (1975). *Lancet* **1,** 935–941.
Purtillo, D. T., Sakamoto, K., Saemundsen, A. K., Sullivan, J. L., Synnerholm, A., Anvret, M., Pritchard, J., Sloper, C., Sieff, C., Pincott, J., Pachman, L., Rich, K., Cruzi, F., Cornet, J. A., Collins, R., Barnes, N., Knight, J., Sandstedt, B., and Klein, G. (1981). *Cancer Res.* **41,** 4226–4236.
Rasheed, S., and Gardner, M. B. (1981). *JNCI, J. Natl. Cancer Inst.* **66,** 43–49.
Reimer, R. R., Clark, W. H., Greene, M. H., Ainsworth, A. M., and Fraumeni, J. F., Jr. (1978). *JAMA, J. Am. Med. Assoc.* **239,** 744–746.
Saemundsen, A. K., Purtillo, D. T., Sakamoto, K., Sullivan, J. L., Synnerholm, A., Hanto, D., Simmons, R., Anvret, M., Collins, R., and Klein, G. (1981). *Cancer Res.* **41,** 4237–4242.
Smith, P. J., Greene, M. H., Devlin, D. A., McKeen, E. A., and Patterson, M. C. (1982). *Int. J. Can.* **30,** 39–45.
Sparkes, R. S., Sparkes, M. C., Wilson, M. G., Towner, J. W., Benedict, W., Murphree, A. L., and Yunis, J. J. (1980). *Science* **208,** 1042–1044.
Spector, B. D., Perry, G. S., and Kersey, J. H. (1978). *Clin. Immunol. Immunopathol.* **11,** 12–29.
Strong, L. C. (1982). *In* "Cancer Epidemiology and Prevention" (D. Schottenfeld and J. F. Fraumeni, Jr., eds.), pp. 506–516. Saunders, Philadelphia, Pennsylvania.
Strong, L. C., Riccardi, V. M., Ferrell, R. E., and Sparkes, R. S. (1981). *Science* **213,** 1501–1503.
Sullivan, J. L., Byron, K., Brewster, B. K., and Purtillo, D. T. (1980). *Science* **210,** 543–545.
Swift, M. (1982). *In* "Cancer Epidemiology and Prevention" (D. Schottenfeld and J. F. Fraumeni, Jr., eds.), pp. 475–482. Saunders, Philadelphia, Pennsylvania.
Thurnherr, N., Deschner, E. E., Stonehill, E., and Lipkin, M. (1973). *Cancer Res.* **33,** 940–945.
Vandenbark, A. A., Greene, M. H., Burger, D. R., Vetto, R. M., and Reimer, R. R. (1979). *JNCI, J. Natl. Cancer Inst.* **63,** 1147–1151.
Wallace, D. C., Exton, L. E., and McLeod, G. R. C. (1971). *Cancer* **27,** 1262–1266.
Warkany, J. (1977). *In* "Genetics of Human Cancer" (J. J. Mulvihill, R. W. Miller, and J. F. Fraumeni, Jr., eds.), pp. 199–204. Raven Press, New York.

Part II
DIAGNOSTIC AND MONITORING ASPECTS

CHAPTER 11

Flow Cytoenzymology— An Update

FRANK DOLBEARE

Lawrence Livermore National Laboratory, Biomedical Sciences Division
University of California
Livermore, California

I.	Introduction	207
II.	Flow Cytoenzymology	208
	Advantages of Flow Cytoenzymology	208
III.	Approaches to Analyzing Cellular Enzymes by Flow Cytometry	209
	A. Absorbance Methods	209
	B. Fluorescent Methods	210
	C. Immunofluorescent Techniques	210
IV.	Analytical Considerations	212
	A. Tissue Dispersal	212
	B. Enzyme Heterogeneity	212
	C. Enzyme Specificity	213
	D. Substrate Diffusion	213
	E. Cell Viability	213
	F. Product Diffusibility	213
	G. Staining Artifact	214
V.	Enzymes in Oncology	214
	A. Enzymes as Markers in Oncology	214
	B. Marker Enzymes Assayable by Flow Cytometry	214
VI.	Future State of the Art	216
	A. Instrumentation	216
	B. Histochemical Techniques	216
	C. Tissue Dispersal and Heterogeneity	217
	D. Metastatic Potential	217
	References	217

I. INTRODUCTION

A flow cytometer is an electro-optical instrument that measures some physical property of cells as they flow one by one through a light beam. The signal from each cell (e.g., fluorescence, scattered light, or reduced light intensity due to absorbance) is detected by a photomultiplier. The signal from the photomultiplier is then digitized and stored in a multichannel analyzer, as a signal intensity distribution (histogram). One of the earliest instruments, described by Kamentsky et al. (1965), was based on light absorbance as cells flowed through a laser beam in a stream surrounded by a liquid sheath. Subsequently, two fluorescence instruments were developed. One used an argon ion laser to excite cellular fluorescence as cells flowed through the beam (Van Dilla et al., 1969). The second instrument analyzed cells as they flowed toward a light beam produced by a 100-W mercury lamp (Göhde et al., 1968; Dittrich and Göhde, 1969). More recently, multilaser instruments have been developed (Stöhr, 1976; Shapiro et al., 1977; Steinkamp et al., 1979; Dean and Pinkel, 1978), some with the capability to monitor fluorescence and light scatter at two wavelengths, making possible eight combinations of parameters of information on a single cell.

Uses of flow cytometers include differential cell analyses, analysis of cellular DNA content, detection of tumors by ploidy differences, and cell analysis by immunologic markers.

II. FLOW CYTOENZYMOLOGY

Flow cytoenzymology is the technique of using a flow cytometer to quantify enzyme activities in individual cells at a rapid rate (10,000 to 20,000 cells per minute). The method permits discrimination of cells with varying levels of histochemical stain (absorbance or fluorescence). Cells can be discriminated with respect to enzyme content if their content differs by more than 3- or 4-fold.

Flow techniques offer several advantages over bulk analysis of either serum or biopsy samples for a specific enzyme.

Advantages of Flow Cytoenzymology

1. Use of Single-Cell Suspensions

Microchemical assay of homogenized biopsy material yields average values for the broad range of enzyme activities in a tissue. Several advantages can be gained by assaying cells one by one rather than analyzing a tissue homogenate. The important advantage of single-cell assay is that significant increase in activity in a small number of cells will not be detected as a significant change in the bulk sample. By examining individual cells for levels of the same enzyme, we may show that a small population of cells has a large increase in specific activity.

2. Simultaneous Analysis of More Than One Parameter

Additionally, flow cytometric analysis permits the investigator to look at several enzymes or variables simultaneously. For example, two enzymes have been used simultaneously (Ornstein and Ansley, 1974; Smith and Dean, 1979; Watson, 1980; Malin-Berdel and Valet, 1980). Other second variables include light scatter (which approximates cell size) (Ansley and Ornstein, 1971; Kaplow, 1975, 1977; Dolbeare and Smith, 1977) and time (Martin and Swartzendruber, 1980). A three-parameter analysis with two enzymes and light scatter was done by Malin-Berdel and Valet (1980).

3. Cell Sorting

Cells may be sorted following staining by any of a number of fluorescent techniques and then examined for morphologic characteristics or analyzed biochemically, e.g., cells sorted on the basis of one enzyme may be analyzed for a second enzyme (Dolbeare, 1981).

III. APPROACHES TO ANALYZING CELLULAR ENZYMES BY FLOW CYTOMETRY

With present instrumentation and histochemical techniques we can analyze enzyme level or activity three ways—by absorbance staining, with fluorogenic substrates and inhibitors, and by immunofluorescent labeling.

A. Absorbance Methods

Absorbance measurements, while not permitting quantitation of the amount of substrate hydrolyzed in individual cells, do permit one to detect differences between individual cells. Absorbance measurements on enzyme histochemical stains were used as the basis of a leukocyte differential analyzer, the Hemalog D (Ansley and Ornstein, 1971; Mansberg et al., 1974) which can differentiate leukocytes by differences in α-naphthyl butyrate esterase (monocytes), myeloperoxidase (eosinphils and neutrophils), Alcian Blue staining (basophils), and light scatter (lymphocytes, which were not stained by the above methods). Subsequently, Kaplow and co-workers (Kaplow, 1975; Kaplow and Dauber, 1975; Kaplow et al., 1976) developed staining techniques for leukocyte peroxidase and esterases which could be analyzed with the Cytograph. They were able to discriminate normal leukocytes from one another and also showed that leukemic monocytes may have high levels of esterase activity.

B. Fluorescent Methods

Fluorogenic assay offers greater sensitivity with a higher signal-to-noise ratio than obtained by absorbance techniques. Fluorogenic staining reactions can generally be limited to brief, i.e., several minute staining on either fixed or unfixed cells; however, some reagents, especially coupling reagents, diffuse slowly into unfixed cells (Ansley and Ornstein, 1971). A number of fluorogenic substrates can be used for flow cytometric analysis. Fluorescent enzyme measurements on single cells in flow cytometers were initiated by the work of Hulett *et al.* (1969) using the accumulation of fluorescein in cells incubated for brief periods with fluorescein diacetate. This compound, which may serve as substrate for a several neutral esterases, is hydrolyzed to highly fluorescent fluorescein which then accumulates in the cell. This substrate was shown to discriminate viable from damaged or dead cells (Rotman and Papermaster, 1966). Sernetz (1973), Sengbusch *et al.* (1976), and Watson *et al.* (1977) measured the rate of hydrolysis of this substrate by flow cytometry. Watson also introduced the monophosphate ester of methyl fluorescein to assay membrane surface enzymes and intracellular activity. Methyl umbelliferone phosphate (Watson *et al.*, 1977) and naphthol AS (3-hydroxy-2-naphthanilide) derivatives (phosphate and β-glucuronide) (Dolbeare and Phares, 1979; Dolbeare *et al.*, 1980) were introduced for fluorescent assay of esterases in single cells. Coupling techniques were required for the flow cytometric assay of proteinases (Dolbeare and Smith, 1977) and for the assay of alkaline phosphatase with naphthol phosphates (Dolbeare, 1978; Dolbeare *et al.*, 1980). The fluorescent product formed in these reactions diffused rapidly from the cell and required a trapping agent to precipitate them within the cell. Diazo coupling was used to give insoluble fluorescent products for alkaline phosphatase measurements. The proteinase technique used peptide derivatives of 2-naphthylamine or of 4-methoxy-2-naphthylamine, which were liberated during the proteolytic action. 5-Nitrosalicylaldehyde was added to the reaction mixture to trap the product in the form of a Schiff base.

C. Immunofluorescent Techniques

Several situations prevent the use of either absorption or fluorescent catalytic methods for the assay of enzymes in a single-cell analysis by flow cytometry. These occur when the enzyme is on the outer surface of the plasma membrane, e.g., 5'-nucleotidase, or when fluorogenic and chromogenic substrates are not available to assay specific enzymes, e.g., terminal deoxynucleotidyltransferase and pyruvate kinase. Immunofluorescent assay, while not a measure of enzyme activity, can be used to quantify the number of enzyme molecules (active or inactive). A combination of immunofluorescent technique, which measures the level of the enzyme, and fluorogenic technique, which measures the activity of the enzyme, permits us to determine activation and inactivation of various enzymes. Immunofluorescent techniques may prevail clinically over catalytic methods, especially with the development of monoclonal antibodies to specific cellular antigens. Furthermore, the identification of meta-

TABLE I
Enzymes Which Have Been Measured by Flow Cytometry

Enzyme	Method	
Acid phosphatase	Methyl umbelliferone phosphate	Watson, 1980; Dolbeare, 1979
	3-O-Methyl fluorescein phosphate	Watson et al., 1979
	Naphthol AS-BI phosphate	Dolbeare and Phares, 1979
	Hydroxyflavone phosphate	Smith and Dean, 1979
Alkaline phosphatase	α-Naphthol phosphate	Kaplow, 1977
	Naphthol AS-BI[a] phosphate + fast red TR[b]	Dolbeare et al., 1980
	Naphthol AS-BI phosphate	Swartzendruber et al., 1980
Cathepsin B	CBZ-Ala-Arg-Arg—MNA[c] + NSA[d]	Dolbeare and Smith, 1977
Dihydrofolate reductase	Fluoresceinated methotrexate	Kaufmann et al., 1978
Dipeptidyl aminopeptidase I	Pro-Arg-MNA + NSA	Dolbeare and Smith, 1977
Dipeptidyl aminopeptidase II	Lys-Ala-MNA + NSA	Smith and Dean, 1979
Nonspecific esterase	Fluorescein diacetate	Hulett et al., 1969; Sernetz, 1973; Sengbusch et al., 1976; Watson et al., 1977
	Naphthol AS-D acetate	Ansley and Ornstein, 1971
	α-Naphthyl acetate + Fast blue BB[e]	Kaplow et al., 1976
β-Glucuronidase	Naphthol AS-BI -β-D-glucuronide	Dolbeare and Phares, 1979
γ-Glutamyl transpeptidase	γ-Glutamyl-4-methoxy-2-naphthylamide + NSA	Vanderlaan et al., 1979
Lactic dehydrogenase	Sodium lactate/NAD	Dolbeare, 1979
Myeloperoxidase	H_2O_2 + 4-chloro-1-naphthol	Ansley and Ornstein, 1971
Peroxidase	H_2O_2 + benzidine dihydrochloride	Kaplow, 1979
Terminal deoxynucleotidyltransferase	Fluoresceinated antibody	McCaffrey et al., 1981

[a] Naphthol AS-BI 7-Bromo-3-hydroxy-2-naphtho-o-anisidide.
[b] Fast Red TR 4-Chlorotoluene-1,5′-diazonium naphthalene disulphonate.
[c] MNA 4-Methoxy-2-naphthylamide.
[d] NSA 5-Nitrosalicylaldehyde.
[e] Fast blue BB 4-Benzoylamino-2,5-diethoxy-benzene diazonium chloride.

static cells may require antibodies that have tissue specificity as well as antigen specificity, e.g., antibodies to prostatic acid phosphatase can be used to identify prostate metastases to bone (Foti *et al.*, 1977). Table I shows a summary of enzymes that have been assayed by flow cytoenzymology, along with the methods used.

IV. ANALYTICAL CONSIDERATIONS

A number of complications must be addressed before quantitative measurements of histochemically stained cells by flow cytometry are attempted.

A. Tissue Dispersal

Flow cytometric analysis can only be done with adequately dispersed cells in suspension. Except for peripheral blood and bone marrow, all tissue from biopsy requires some physical or enzymatic manipulation to disperse the cells. Physical techniques such as syringing or pipetting produce membrane damage with enzyme loss, clumping, and cellular debris. Cells may then be separated from clumps, nuclei, and cellular debris by centrifugal elutriation. Trypsin (Rinaldini, 1958), collagenase (Lasfargues, 1973), and pronase (Gwatkin, 1964) have been used to disperse normal and tumor tissue. Biopsy suspensions containing several hundred thousand cells can then be conveniently assayed by flow cytometry for one or several enzymes.

B. Enzyme Heterogeneity

Cellular diversity within a tumor may account for a wide range of enzyme activity, depending on the location of the cell within the tumor. Metabolic activity may range from very high in rapidly proliferating cells at the edge of the tumor to near zero in necrotic cells at the center of the tumor mass. Enzyme levels may also vary several-fold within a single tissue, depending on the growth rate and differentiation of that tissue, so that within a small segment of tissue we observe large heterogeneities of activities (Knox, 1972). The significance of quantitative differences in enzyme levels from one cell to another, or from tumor to normal or to other tumor is difficult to assess and may indeed by highly variable (Weber, 1973; Knox, 1977).

Often the investigator resorts to searching for qualitative differences, e.g., an isozyme that can be discriminated from normal cellular enzymes by the use of specific assay conditions or inhibitors. A knowledge of the specific enzymes in tumor and surrounding tissue is necessary to choose the appropriate marker enzyme to discriminate tumor from normal host cell. Fortunately, qualitative differences in the form of isozymes can permit ready identification of the tumor cell. Selective conditions may permit preferential assay of the fetal isozyme over the normal tissue enzyme. Tissue heterogeneity may also be monitored by looking at several parameters: DNA, light scatter, or other enzymes may be used as additional markers of the cell types present in normal tissue.

C. Enzyme Specificity

Many enzymes nonspecifically hydrolyze common substrates used in staining. Choosing specific incubation conditions, using substrates with specific functional groups, e.g., galactosides or peptide sequences, or using inhibitors can provide specificity.

D. Substrate Diffusion

Many substrates and coupling agents require some sort of cellular fixation to permit entry into the cell. Attempts to measure certain enzyme levels in heterogeneous cell populations may give artifactual staining based on differential uptake of the stains. This is true of many ionized substrates and most tetrazolium salts and diazo compounds. Clearly, fixation with formaldehyde or glutaraldehyde or pretreatment of the cells with cold aqueous acetone or ethanol may be the only solution.

E. Cell Viability

Cells obtained by enzymatic or physical dispersal of tissue may have large numbers of damaged or nonviable cells that stain differently from the viable cell population. A selection of a second stain to serve as a marker for viability (Stohr and Vogt-Schaden, 1980) permits elimination of damaged cells from the analysis. When the cell preparation containing a large number of damaged cells is fixed prior to analysis, a greater dispersion is found in the distribution of the histogram.

F. Product Diffusibility

Most fluorogenic substrates produce highly diffusible products where trapping agents are not available. The use of complex fluorophores with low water solubility—some complex napthols, e.g., napthol AS-TR (4'-chloro-3-hydroxy-2-naphtho-o-toluidide) and substituted coumarins—may retard diffusion. The use of coupling agents, e.g., diazo salts, can block diffusion of reaction products.

G. Staining Artifact

Many absorbance stains produce high background levels due to nonspecific staining of other cellular components. Impure diazo salts are notorious for producing a number of secondary staining reactions within the cell. Fluorescence measurements done with impure or degraded diazo salts also lead to artifactual results.

V. ENZYMES IN ONCOLOGY

A. Enzymes as Markers in Oncology

Serum levels of enzymes have served as potential diagnostic indicators of neoplasia; however, blood enzyme levels may reflect a number of pathologic states besides neoplasia. A marker enzyme for histochemical detection may be selected for several reasons. (1) The enzyme is qualitatively specific for tumor, e.g., a fetal isozyme is used for the detection of tumor cells. (2) The enzyme is greatly elevated and permits discrimination of tumor tissue from host surrounding tissue. (3) The enzyme is used as a marker to determine level of tumor differentiation, e.g., acute lymphocytic leukemia versus acute myelogenous leukemia. (4) The enzyme is tissue specific and therefore can be used to determine the tissue of origin of circulating enzyme or metastatic cell.

Major advances in the cytochemical characterization of tumor cells have been in the area of hematopoietic pathology and leukemia because of the ease of obtaining and identifying specific cell types in blood and bone marrow.

B. Marker Enzymes Assayable by Flow Cytometry

A list of all the marker enzymes used for neoplastic detection is beyond the scope of this report. However, the following is a selected list of marker enzymes that can be assayed by flow cytometry by one of the three histochemical methods already discussed.

1. Alkaline Phosphatase

With the discovery of fetal phenotypes of alkaline phospatase in human tumors (Fishman *et al.*, 1968), various isozymes of alkaline phosphatase were reported in neoplasia and other pathologic disorders (Fishman, 1974; Posen and Doherty, 1981). Alkaline phosphatase have been used to aid in distinguishing several subclasses of acute lymphoblastic leukemia (Andreewa *et al.*, 1978; Catovsky *et al.*, 1981; Huhn *et al.*, 1981). Selective inhibition of specific isozymes or the use of fluorescently labeled antibodies may be used to discriminate isozymes of alkaline phosphatase for flow cytometric analysis.

2. Acid Phosphatase

High serum levels of this enzyme have been used to diagnose prostate cancer. In recent years, however, immunological techniques have improved the efficiency of early diagnosis of the disease (Foti et al., 1977). Fluoride sensitivity has been used to identify certain bone marrow and blood monocytes by acid phosphatase staining (Beckmann et al., 1974).

3. Cathepsin B and Thiol Proteinases

Cathepsin B has been shown to be elevated in sera from patients with squamous carcinoma of the uterus (Pietras et al., 1978). This group analyzed sera from women and demonstrated elevation of a thiol proteinase (cathepsin B) in those patients with invasive carcinoma of the uterus. Graf et al. (1981) have demonstrated the localization of cathepsin B at the invasion front of a carcinoma in experimental animals. Increases in cathepsin B has also been found at the edges of human breast carcinomas and fibroadenomas (Poole et al., 1978). Cathepsin B or a similar proteinase may indeed activate other proteinases such as plasminogen to enhance migration of the tumor border.

4. γ-Glutamyl Transpeptidase

This enzyme, normally present at much higher levels in fetal liver, has been used as a marker enzyme for preneoplastic lesions in the liver of mice and rats (Fiala and Fiala, 1973). Increased levels of the enzyme have been reported for colon cancer (Fiala et al., 1977) and mammary tumors (Dawson et al., 1979).

5. Naphthyl Acetate and Naphthyl AS Acetate Esterases

Several enzymes in leukocytes hydrolyze this general class of naphthol esters. Investigators have attempted to characterize the differences between the nonspecific esters by electrophoresis (Li et al., 1973; Kass and Peters, 1978; Sweetman and Ornstein, 1974; Beckmann et al., 1974). The esterase differences between various leukocytes has been the basis of the differential blood cell analyzers (Ansley and Ornstein, 1971; Kaplow and Eisenberg, 1975).

6. Terminal Deoxynucleotidyltransferase (TdT)

TdT is recognized as a specific intracellular marker for immature lymphocytes in the hematopoietic system (Bollum, 1979). TdT is localized in a major fraction of cortical thymocytes and in a small population of bone marrow lymphocytes in mammals (Bollum, 1979). The enzyme was found to associated with acute lymphoblastic anemia in children (McCaffrey *et al.*, 1973), but has been found elevated in an adult case of chronic myelogenous leukemia (CML) (Coleman *et al.*, 1974). A fluorescent antibody technique (Kung *et al.*, 1978) has been used for the flow cytometric analysis of TdT-containing cells in patients with various forms of leukemia (McCaffrey *et al.*, 1981).

VI. FUTURE STATE OF THE ART

The present state of the art has been aimed at: (1) using standard histochemical staining methods to differentiate leukocytes and other specific cell types, (2) experimentation with enzyme kinetics, and (3) the use of fluorogenic substrates on single cultured cells. However, the need is to analyze dispersed human tissue and tumors for marker enzymes that can be suitably measured by existing histochemistry. Increased purity of reagents and standardization of methods will permit automation. To make enzymologic techniques (particularly flow cytoenzymologic techniques) acceptable on a clinical scale, several areas need to be assessed.

A. Instrumentation

While at the time of this writing, at least a half-dozen commercial instruments are available for automated cytology using the techniques discussed above (Van Dilla and Mendelsohn, 1979), these are priced above $50,000. Inexpensive microprocessors and alternate light sources (arc lamps) are being used to make flow cytometers that will range between $30,000 and $40,000.

B. Histochemical Techniques

Currently, over 200 histochemical techniques are available for single-cell quantification of enzyme levels (or activity). Many standard histochemical techniques developed before 1970 are suitable for light extinction measurements of enzymes. Fluorescent methods are still in the infancy of development; new coupling techniques are needed to convert reaction products into stable fluorescent end products. Standardization is needed now for both types of assays.

Metabolic enzymes, DNA synthetic enzymes, and plasma membrane enzymes can be measured by immunofluorescent techniques. With the rapidly growing library of monoclonal antibodies to surface and intracellular antigens, we shall soon have immunofluorescent probes to many enzymes not assayable by present histochemical techniques.

C. Tissue Dispersal and Heterogeneity

Present methods are not sufficient to yield a representative sampling of cells from biopsies of solid tumor masses. Improved enzymologic techniques will be needed to yield cells in a state similar to that which exists *in situ*.

D. Metastatic Potential

A need exists in clinical oncology for a knowledge of marker enzymes and the possession of chemical probes to detect circulating tumor cells that have the potential to metastasize to a new tissue and become the focus of a secondary tumor.

ACKNOWLEDGMENTS

Work performed under the auspices of the United States Department of Energy by the Lawrence Livermore National Laboratory under Contract number W-7405-ENG-48.

REFERENCES

Andreewa, P., Huhn, D., Thiel, E., and Rodt, H. (1978). *Blut* **36,** 299–305.
Ansley, H. R., and Ornstein, L. (1971). *Adv. Autom. Anal.* **1,** 437–446.
Beckman, H., Neth, R., Gaedicke, G., Landbeck, G., Schöch, G., Wiegars, U., and Winkler, K. (1974). *In* "Modern Trends in Human Leukemia" (R. Neth, R. C. Gallo, S. Spiegelman, and F. Stohlman, eds.), pp. 26–35. Lehmanns Verlag, Munich.
Bollum, F. J. (1979). *Blood* **54,** 1203–1215.
Catovsky, D., Cardullo, L. de S., O'Brien, M., Morilla, R., Costello, C., Geneshaguru, K., and Hoffbrand, V. (1981). *Cancer Res.* **41,** 4824–4832.
Coleman, M. S., Hutton, J. J., De Simone, P., and Bollum, F. J. (1974). *Proc. Natl. Acad. Sci. U.S.A.* **71,** 4404–4408.
Dawson, J., Smith, G. D., Boak, J., and Peters, T. J. (1979). *Clin. Chim. Acta* **96,** 37–42.
Dean, P. N., and Pinkel, D. (1978). *J. Histochem. Cytochem.* **26,** 622–627.
Dittrich, W., and Göhde, W. (1969). *Z. Naturforsch., B: Anorg. Chem., Org. Chem., Brocbem., Biophys., Biol.* **24B,** 360–361.
Dolbeare, F. A. (1978). *In* "Biological Markers of Neoplasia: Basic and Applied Aspects" (R. Ruddon, ed.), pp. 581–586. Elsevier/North-Holland, New York.
Dolbeare, F. A. (1979). *J. Histochem. Cytochem.* **27,** 1644–1646.
Dolbeare, F. A. (1981). *In* "Modern Fluorescence Spectroscopy" (E. L. Wehry, ed.), Vol. 3, pp. 251–293. Plenum, New York.
Dolbeare, F. A., and Phares, W. (1979). *J. Histochem. Cytochem.* **27,** 120–124.
Dolbeare, F. A., and Smith, R. E. (1977). *Clin. Chem. (Winston-Salem, N.C.)* **23,** 1485–1491.
Dolbeare, F. A., Vanderlaan, M., and Phares, W. (1980). *J. Histochem. Cytochem.* **28,** 419–426.
Fiala, S., and Fiala, E. S. (1973). *J. Natl. Cancer Inst.* **51,** 151–158.

Fiala, S., Fiala, A. E., Keller, R. W., and Fiala, E. S. (1977). *Arch. Geschwulstforsch.* **47,** 117–122.
Fishman, W. H. (1974). *Am. J. Med.* **56,** 617–650.
Fishman, W. H., Inglis, N. R., Stolbach, L. L., and Krant, H. J. A. (1968). *Cancer Res.* **28,** 150–154.
Foti, A. G., Cooper, J. F., Herschman, H., and Malvaez, R. R. (1977). *N. Engl. Med. J.* **297,** 1357–1361.
Göhde, W., and Dittrich, W. (1969). *Acta Histochem. Suppl.* **X,** 42–51.
Graf, M., Baici, A., and Sträuli, P. (1981). *Lab. Invest.* **45,** 587–596.
Gwatkin, R. B. (1964). *J. Reprod. Fertil.* **7,** 99–105.
Huhn, D., Thiel, E., Rodt, H., and Andreewa, P. (1981). *Scand. J. Haematol.* **26,** 311–320.
Hulett, H. R., Bonner, W. A., Barret, J., and Herzenberg, L. A. (1969). *Science* **166,** 747–749.
Kamentsky, L. A., Melamed, M. R., and Derman, H. (1965). *Science* **150,** 630–631.
Kaplow, L. S. (1975). *Am. J. Clin. Pathol.* **63,** 451.
Kaplow, L. S. (1977). *J. Histochem. Cytochem.* **25,** 990–1000.
Kaplow, L. S. (1979). In "Flow Cytometry and Sorting" (M. R. Melamed, P. F. Mullaney, and M. L. Mendelsohn, eds.), pp. 531–545. Wiley, New York.
Kaplow, L. S., and Dauber, H. (1975). *J. Histochem. Cytochem.* **23,** 318–319.
Kaplow, L. S., and Eisenberg, M. (1975). *Int. Symp. Pulse-Cytophotom. [Proc.], 1st, 1974,* pp. 262–274.
Kaplow, L. S., Dauber, H., and Lerner, E. (1976). *J. Histochem. Cytochem.* **24,** 363–372.
Kass, L., and Peters, C. L. (1978). *Am. J. Clin. Pathol.* **69,** 57–61.
Kaufmann, R. J., Bertino, J. R., and Schimke, R. T. (1978). *J. Biol. Chem.* **253,** 5852–5860.
Knox, W. E. (1972). "Enzyme Patterns in Fetal, Adult and Neoplastic Tissues." Karger, Basel.
Knox, W. E. (1977). In "Biological Markers of Neoplasia: Basic and Applied Aspects" (R. W. Ruddon, ed.), pp. 547–558. Elsevier, New York.
Kung, P. C., Long, J. C., McCaffrey, R. P., Ratliff, R., Harrison, T. A.,and Baltimore D. (1978). *Am. J. Med.* **64,** 788–794.
Lasfargues, E. Y. (1973). In "Tissue Culture: Methods and Applications" (P. F. Kruse, Jr. and M. K. Patterson, eds.), pp. 45–50. Academic Press, New York.
Li, C. Y., Lam, K. W., and Yam, L. T. (1973). *J. Histochem. Cytochem.* **21,** 1–12.
McCaffrey, R., Smoler, D. F., and Baltimore, D. (1973). *Proc. Natl. Acad. Sci. U.S.A.* **70,** 521–525.
McCaffrey, R., Lillquist, A., Sallan, S., Cohen, E., and Osband, M. (1981). *Cancer Res.* **41,** 4814–4820.
Malin-Berdel, J., and Valet, G. (1980). *Cytometry* **1,** 222–228.
Mansberg, H. P., Saunders, A. M., and Groner, W. (1974). *J. Histochem. Cytochem.* **22,** 711–724.
Martin, J. C., and Swartzendruber, D. E. (1980). *Science* **207,** 199–201.
Ornstein, L., and Ansley, H. R. (1974). *J. Histochem. Cytochem.* **22,** 453–469.
Pietras, R. J., Szego, C. M., Mangan, C. E., Seeler, B. J., Burtnett, M. M., and Orevi, M. (1978). *Obstet. Gynecol.* **52,** 321–327.
Poole, A. R., Tiltman, K. J., Recklies, A. D., and Stoker, T. A. M. (1978). *Nature (London)* **273,** 545–547.
Posen, S., and Doherty, E. (1981). *Adv. Clin. Chem.* **22,** 163–245.
Rinaldini, L. M. J. (1958). *Int. Rev. Cytol.* **7,** 386–389.
Rotman, B., and Papermaster, B. W. (1966). *Proc. Natl. Acad. Sci. U.S.A.* **55,** 134–141.
Saunders, A. M. (1972). *Clin. Chem. (Winston-Salem, N.C.)* **18,** 783–788.
Schmalzl, F., and Braunsteiner, H. (1971). *Acta Haematol.* **45,** 209–217.
Sengbusch, G. V., Couwenbergs, C., Kühner, J., and Müller, U. (1976). *Histochem. J.* **8,** 341–350.

Sernetz, M. (1973). *In* "Fluorescence Techniques in Cell Biology" (A. A. Thaer and M. Sernetz, eds.), pp. 243–254. Springer-Verlag, Berlin and New York.
Shapiro, H. M., Schildkraut, E. R., Curbelo, E. R., Turner, R., Webb, R. B., Brown, D. C., and Block, M. J. (1977). *J. Histochem. Cytochem.* **25**, 836–844.
Smith, R. E., and Dean, P. N. (1979). *J. Histochem. Cytochem.* **27**, 1499–1504.
Steinkamp, J. A., Orlicky, D. A., and Crissman, H. A. (1979). *J. Histochem. Cytochem.* **27**, 273–376.
Stöhr, M. (1976). *Pulse-Cytophotom., Int. Symp., 2nd, 1976* pp. 39–45.
Stöhr, M., and Vogt-Schaden, M. (1980). *In* "Flow Cytometry IV" (O. D. Laerum, T. Lindmo, and E. Thorud, eds.). Universitetsforlaget, Bergen.
Swartzendruber, D. E., Cox, K. Z., and Wilder, M. E. (1980). *Differentiation* **16**, 23–30.
Sweetman, F., and Ornstein, L. (1974). *J. Histochem. Cytochem.* **22**, 327–339.
Vanderlaan, M., Cutter, C., and Dolbeare, F. (1979). *J. Histochem. Cytochem.* **27**, 114–119.
Van Dilla, M. A., Trujillo, T. T., Mullaney, P. F., and Coulter, J. R. (1969). *Science* **163**, 213–1214.
Van Dilla, M. A., and Mendelsohn, M. L. (1979). *In* "Flow Cytometry and Sorting" (M. R. Melamed, P. F. Mullaney, and M. L. Mendelsohn, eds.), pp. 11–37. Wiley, New York.
Watson, J. V. (1980). *Cytometry* **1**, 143–151.
Watson, J. V., Chambers, S. H., Workman, P., and Horsnell, T. S. (1977). *FEBS Lett.* **81**, 179–182.
Watson, J. V., Workman, P., and Chambers, S. H. (1979). *Biochem. Pharmacol.* **28**, 821–827.
Weber, G. (1973). *In* "The Molecular Biology of Cancer" (H. Busch, ed.), pp. 487–521. Academic Press, New York.

CHAPTER 12
Tumor Markers of Medullary Thyroid Carcinoma

ROBERT F. GAGEL

Departments of Medicine and Cell Biology
Baylor College of Medicine
Houston, Texas
Veterans Administration Medical Center
Houston, Texas

    I. Calcitonin Secretion by the Normal C Cell...... 222
   II. Calcitonin as a Tumor Marker...... 222
  III. The Histologic Evolution of Medullary Thyroid Carcinoma...... 227
  IV. Other Tumor Markers...... 229
      A. Dopa Decarboxylase...... 229
      B. Histaminase...... 230
      C. Carcinoembryonic Antigen...... 231
      D. Hormonal Tumor Markers...... 231
   V. The Rat Model of Medullary Thyroid Carcinoma...... 233
  VI. The Clonal Nature of Hereditary Medullary Thyroid Carcinoma...... 234
 VII. Control of Gene Expression in MTC...... 235
VIII. A Model for the Future Study of Hereditary Medullary Thyroid Carcinoma...... 236
      References...... 236

Williams (1966) first postulated that medullary thyroid carcinoma (MTC) was composed of C cells and predicted that the elevation of plasma calcitonin (CT) concentration might be a marker for this tumor. Since that time MTC has received attention out of proportion to its frequency in the general population because of several unique features. First, the disease is hereditary (transmitted as an autosomal dominant in its familial form). This limits the number of individuals at risk and allows for more intensive study of the tumor and afflicted individuals. Second, the measurement of the tumor marker CT has facilitated early diagnosis of the tumor and in many cases has permitted the removal of an abnormal focus of C cells in a premalignant state. Third, the C cell goes through a predictable series of histologic changes from the normal to the malignant state. Fourth, several other tumor markers (histaminase, dopa decarboxylase and carcinoembryonic antigen) are produced by this tumor, and the evolution of expression of several of these tumor markers has been associated with specific histologic stages in the development of the tumor. Together these features constitute an interesting and unique model of human carcinogenesis. In this chapter I will outline what I believe to be important studies performed to define the clinical syndrome of hereditary MTC, the natural history of the disease, and the association of tumor markers with this malignancy. I will also attempt to portray MTC as a useful model for studying hereditary malignancy and point out model systems and newer techniques which may provide direction for future research.

I. CALCITONIN SECRETION BY THE NORMAL C CELL

The parafollicular cell of the thyroid had been identified as a distinct cell in the late nineteenth century (Baber, 1876), but it was not until 1964, one year after the discovery of CT, that Foster *et al.* (1964) suggested that the C cell was the site of CT production. Immunohistochemical studies by Wolfe *et al.* (1974) and McMillan *et al.* (1974) have provided the basis for understanding the anatomic location in which MTC develops. These studies showed that the greatest number of C cells and CT content in the normal thyroid gland was located at the juncture of the upper and mid-portions of the lateral lobes along a hypothetical central axis.

The development of the CT radioimmunoassay and its application to measurement of plasma CT concentration has provided a tool to detect MTC in its earliest stages. CT normally circulates in plasma in concentrations between 30 and 100 pg/ml (Parthemore *et al.*, 1974; Wright *et al.*, 1977). The early discoveries that calcium (Melvin *et al.*, 1972) and pentagastrin (Hennessy *et al.*, 1974) stimulated CT release have proved useful for the diagnosis of early C cell disease. Normal subjects have a peak plasma CT concentration of less than 500 pg/ml after either calcium or pentagastrin (Graze *et al.*, 1978) (normal responses vary with individual assays), whereas patients with an increase in C cell mass have a greater response. These testing procedures have been used to separate normal subjects from patients with early C cell disease.

II. CALCITONIN AS A TUMOR MARKER

In 1970 Melvin *et al.* (1972) first described the clinical usefulness of CT measurements for the prospective diagnosis of MTC. Twelve patients (11 of the

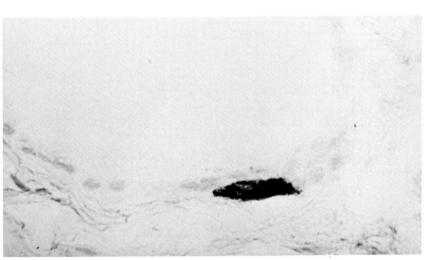
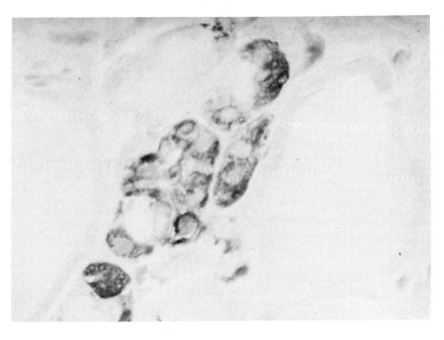

Fig. 1. C cells in the normal human thyroid gland (left) and in the thyroid gland from a patient with C cell hyperplasia (right). Stained by an immunoperoxidase reaction with anticalcitonin (methyl green counterstain). × 300. (From Gagel *et al.*, 1975, by permission of *Transactions of the American Association of Physicians.*)

12 had no clinical evidence of thyroid disease) who had abnormal plasma CT concentrations after a 4-hour calcium infusion were found to have macroscopic foci of MTC. The tumor was bilateral in each case and was located in each lateral lobe at the juncture of the upper and middle third along a hypothetical central axis, the point where the greatest number of C cells would be expected in the normal thyroid. After that initial study each member of the family (the J kindred) and several other kindreds were screened on a yearly or biyearly basis, and 3 years later Wolfe et al. (1973) defined the syndrome of C cell hyperplasia in three patients who converted from a negative to a positive test result. A careful histologic analysis of the thyroid glands removed from each of these individuals showed foci of hyperplastic C cells (similar to that shown in Fig. 1) in the characteristic location in each lobe. In none of these individuals was there evidence of local or distant metastatic disease. After the operation the plasma CT became and has remained nondetectable in each of these individuals. Over the next 11 years all members of the J-kindred were prospectively studied with a yearly calcium infusion or pentagastrin injection. Twenty-two members were found who converted from a negative to a positive test result (Fig. 2) (Gagel et al., 1975; Graze et al., 1978), and the thyroid was removed in each. The histologic lesion found in these thyroids was either C cell hyperplasia (12 patients) or C cell hyperplasia and macroscopic MTC (9 patients). In one patient we were unable to detect any abnormality. Her status is at present unclear, but for study purposes we have classified her test result as a false positive result. There was no evidence of either intra- or extrathyroidal metastasis in any of these patients, and postoperatively the plasma CT fell into the nondetectable range, although not immediately in all (Fig. 3) (Graze et al., 1978). At this time there is optimism that prospective screening and thyroidectomy (Leape et al., 1977) has resulted in a cure of the disease in this group of patients. Histologic and hormonal findings support this conclusion, but one point of some concern is the failure of the stimulated CT concentration to fall into the nondetectable range in the immediate postoperative period in all patients (Fig. 3). If all thyroidal C cells were removed at the time of surgery, the plasma CT would be expected to fall into the nondetectable range within hours. There are several possible explanations. First CT production by cells outside the thyroid has been demonstrated in recent years. These cells include the Kulchitsky cells of the lung, the pituitary gland, adrenal chromaffin cells, and the thymus (Baylin and Mendelsohn, 1980). It is possble immunoreactive CT came from one of those sources. A second possibility which cannot be ignored is that metastasis may have already occurred. Neither explanation satisfactorily explains the eventual return of the serum CT to the nondetectable range, unless one hypothesizes that removal of the thyroid eliminated a tropic substance for CT production or has in some unknown way altered the body's immune response so that malignant cells could be eliminated. It is reasonable, for several reasons to consider these patients cured of their

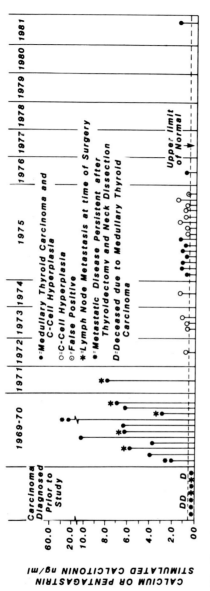

Fig. 2. Highest abnormal values detected during provocative screening tests of the J kindred carried out at yearly intervals and pathological findings in the thyroid gland. Note that there has been a striking fall in serum CT level at initial diagnosis and no demonstrated local metastasis after the initial screening period in which 12 patients with MTC were detected (1970 and 1971). The dotted line shows the upper limit of normal. (Updated and reprinted by permission of *The New England Journal of Medicine* [**299**, 980–985, 1978], from Graze *et al.*, 1978.)

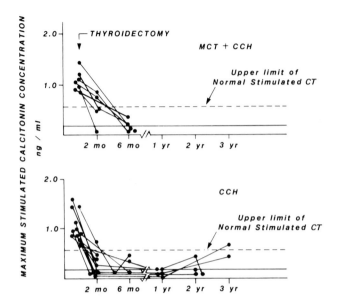

Fig. 3. Time course of decline in serum CT concentrations after thyroidectomy in patients with either C cell hyperplasia (CCH) or the combination of C cell hyperplasia and microscopic MTC. The shaded area denotes undetectable values. (Reprinted by permission from *The New England Journal of Medicine* [**299,** 980–985, 1978], from Graze *et al.*, 1978.)

disease. First, after 1–8 years of followup we have not found metastatic disease in these 22 patients. Second, in earlier studies survival curves for patients with MTC, with no local metastasis (Wang, 1978), did not differ significantly from that of the normal population. Continued followup of the J kindred will be important to determine whether they have been cured.

Prospective screening has had a dramatic effect on the age of diagnosis of C cell abnormalities. In a review of 11 kindreds at risk for MTC (MEN IIa) who have been prospectively screened we found that the mean age of conversion (38 converters) was 15 years of age (Gagel *et al.*, 1982). In contrast, the mean age of diagnosis of disease by other clinical means (prior to the availability of CT measurements) or by a positive test result on the first testing was 34 years of age (61 patients). In one decade of intense work by many groups (Wells *et al.*, 1978a; Sizemore *et al.*, 1977; Block *et al.*, 1980) MTC has been converted from a disease of middle age with frequent metastatic disease, to a curable disease of the pediatric age group.

During the course of these studies there has been a change in the method of testing for MTC. The standard 4-hour calcium infusion test has largely been replaced by the more rapid and convenient pentagastrin injection (Hennessey et al., 1974). Although pentagastrin is not always superior to the calcium infusion (Graze et al., 1978), the ease of administration and patient acceptance make it the test of choice at this time. Wells et al. (1978a) have suggested that the combination of a 1-minute calcium infusion followed by a pentagastrin injection is a more effective CT secretagogue than pentagastrin alone. Confirmative studies are in progress. We currently recommend that all patients at risk be tested with a pentagastrin injection at least once a year starting at age 5 and continuing until age 35. Thereafter, testing should continue but at less frequent intervals. We have not seen a conversion after the age of 33.

III. THE HISTOLOGIC EVOLUTION OF MEDULLARY THYROID CARCINOMA

Extensive studies in both the human and the rat indicate that the C cell progresses through a predictable sequence of histologic changes from the normal to the malignant state. Normally one would expect to see 4–10 C cells per high power field with clusters of 0–3 cells associated with the thyroid follicle (Figs. 1 and 4). The overall density of C cells is determined by the location within the thyroid gland (see Section I). Ljungberg (1972), in early pathologic studies of MTC, noted there were areas of C cell hyperplasia surrounding foci of MTC in the hereditary syndrome, but Wolfe et al. (1973) first defined C cell hyperplasia as a precursor state for hereditary MTC. In this premalignant histologic lesion there are up to 100 cells per high power field with clusters of 10–20 cells (Figs. 1 and 4). In both the human and rat forms of C cell hyperplasia, DeLellis et al. (1979) have shown that the follicular basement membrane is intact and the C cells are contained within the thyroid follicle, findings confirmed by electron microscopy (DeLellis et al., 1977). During the histologic stage of C cell hyperplasia there is no clinical or histologic evidence of either local or distant metastatic disease. As the disease progresses, C cells fill up the thyroid follicle (nodular hyperplasia) and eventually broach the follicular basal lamina (microscopic MTC) and thereby develop the potential to metastasize (Fig. 4).

This model of progression of disease (Fig. 4) has been pieced together by careful studies of thyroids (DeLellis et al., 1977) from many family members with this disease. Although each thyroid examined represents one point in time, this sequence of events has been supported by studies in a rat model of this disease (DeLellis et al., 1979). In subsequent sections I will use this histologic categorization of disease progression as a framework for describing the expression of tumor markers. It appears that there is a correlation between histologic stages and the appearance of specific tumor markers.

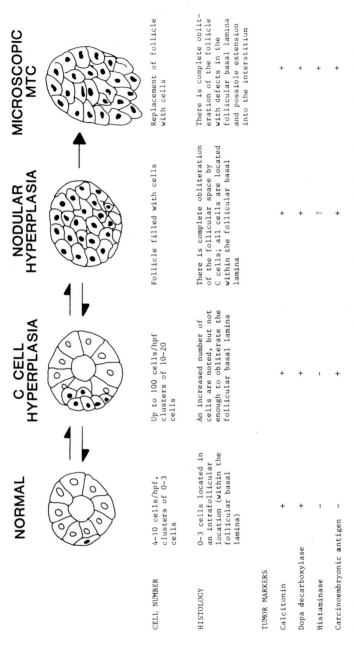

Fig. 4. The histologic evolution of medullary thyroid carcinoma. This schematic figure shows the histologic progression of C cell abnormalities in hereditary MTC (DeLellis *et al.*, 1977) and the association of the tumor markers CT, dopa decarboxylase (Baylin *et al.*, 1979b), histaminase (Baylin *et al.*, 1979a), and carcinoembryonic antigen (CEA) (DeLellis *et al.*, 1978a; Cox *et al.*, 1979) with each histologic stage. Note that even though CEA is depicted as positive for C cell hyperplasia and nodular hyperplasia, there is disagreement between DeLellis *et al.* (1978a) and Cox *et al.* (1979) (see Section IV of text). (hpf = high power field.)

IV. OTHER TUMOR MARKERS

Calcitonin was the first tumor marker associated with MTC. It is an example of a tumor marker that is normally produced by the parent cell (eutopic production) and becomes a tumor marker only because of an increase in mass of the parent cell type with a resultant increase in total protein production. Dopa decarboxylase production by the C cell is another such example, whereas histaminase and carcinoembryonic antigen (CEA) are not normally produced by the C cell (undetected by currently available techniques) and appear only after malignant transformation of the C cell (ectopic production). Hormonal markers that could be categorized as ectopic include ACTH, β-endorphin, somatostatin, vasoactive intestinal peptide, and neurotensin, although all these hormones are produced by members of the APUD system of cells (see below). In subsequent sections I will outline the current status of each of these markers and describe the studies performed to give these tumor markers relevance within the framework of histologic evolution of the tumor outlined above.

A. Dopa Decarboxylase

The C cell is a member of the class of cells which Pearse (Pearse and Takor, 1979) defined as APUD cells (amine precursor uptake and decarboxylation). These cells possess the ability to take up amine precursors (L-dopa and L-5-hydroxytryptophan) and decarboxylate these compounds (by an aromatic acid decarboxylase) to their respective amines (dopamine or 5-hydroxytryptamine). Storage of dopamine or 5-hydroxytryptamine in the C cells from a variety of species correlates with the amount of enzyme activity present (Hakanson et al., 1971), and exposure of C cells, which normally contain little dopamine, to L-dopa results in a striking accumulation of dopamine by the cell. Atkins et al. (1973) first showed that dopa decarboxylase was present in elevated concentrations in MTC. It is important, however, to point out that this elevation was relative to normal thyroid tissue containing relatively few C cells and not to a pure population of C cells. For the primary tumor (C cell hyperplasia, microscopic MTC, or macroscopic MTC in the thyroid) it seems likely that the concentration of dopa decarboxylase is appropriate for the number of C cells. Striking elevations of dopa decarboxylase have been reported in some metastatic tumors. Trump et al. (1979) have described a patient with MTC who had a large amount of metastatic tumor with an increased dopa decarboxylase content and a very low tumor and plasma CT concentration. Why the cell expressed one tumor marker in preference to another is not clear, although Trump et al. (1979) suggested dopamine might be inhibiting CT production.

A functional role for dopa decarboxylase in the C cell has been more difficult to define. Hakanson *et al.* (1971) had suggested a role for L-dopa in the control of CT secretion. In this regard, studies have yielded somewhat conflicting results. Baylin *et al.* (1979a) reported that L-dopa lowered plasma CT concentration in patients with MTC and described *in vitro* studies showing inhibition of CT secretion by L-dopa in MTC tissue slices. Spiler *et al.* (1980) showed L-dopa lowered the plasma CT concentration in only 3 of 10 patients with metastatic MTC, whereas 9 of these 10 patients had an appropriate rise in serum growth hormone concentration, indicating an effective concentration of the drug had been absorbed and converted to dopamine. In a continuous line of rat MTC cells (Gagel *et al.*, 1980) (see Section V), L-dopa and L-5-hydroxytryptophan had no effect on basal or calcium-stimulated CT secretion (Fig. 5). At this time it is unclear whether dopa decarboxylase has a functional role for controlling peptide hormone secretion in the C cell. I am not aware of studies performed to examine the effect of this enzyme and its products on other differentiated functions of the C cell such as growth, synthesis of CT, and protein or cyclic nucleotide metabolism.

B. Histaminase

Histaminase was the first tumor marker discovered not normally produced by the C cell. Baylin *et al.* (1970) first described elevated histaminase activity in tumor extracts and plasma from patients with MTC. More recently Baylin *et al.* (1979b) have performed correlative studies comparing CT and histaminase staining of C cells. The normal C cell stains positive for CT, but there is no detectable histaminase by staining or direct measurement. Histaminase is not detected in the histologic lesion of C cell hyperplasia, but becomes detectable in microscopic MTC and frank MTC (Fig. 4). The expression of histaminase in MTC could occur by several mechanisms. First and most likely is the possibility the C cell contains the structural gene for histaminase, but that it is not normally expressed. During the transformation process nuclear events occur which allow for the

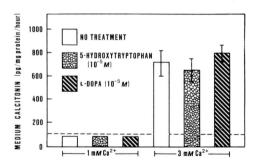

Fig. 5. Lack of effect of L-dopa or 5-hydroxytryptophan on basal (calcium 0.5 m*M*) or stimulated (calcium 3.0 m*M*) CT secretion in a 1-hour incubation. The experiment was performed as described in Gagel *et al.* (1980). Each bar represents the mean ± SE of 4 replicate plates.

transcription and subsequent translation of the structural gene for histaminase. Another explanation, although less likely, is that mRNA for histaminase is normally transcribed, but that there are significant posttranscriptional or posttranslational modifications which alter or destroy mRNA or the translated protein. At present there is no information available to answer this question, but the availability of antisera to placental histaminase, which cross-reacts with MTC histaminase, should allow for cloning of the structural gene for histaminase which in turn could be used to answer these questions in a suitable model of MTC. (See Section V).

Studies by Mendelsohn *et al.* (1978) indicate histaminase is a cytoplasmic protein not confined to secretory granules. Studies we have performed show that pentagastrin, a CT secretagogue, does not stimulate release of histaminase into the plasma of patients with frank MTC (Gagel, unpublished results). In general, measurement of histaminase in plasma has not proven to be useful as a clinical marker for C cell hyperplasia or early MTC because plasma concentrations are not significantly elevated in the early forms of the disease.

C. Carcinoembryonic Antigen

Carcinoembryonic antigen (CEA) is an oncofetal protein not usually found in the C cell. Its presence in MTC was first documented by Isaacson and Judd (1976) and Ishikawa and Hamada (1976). DeLellis *et al.* (1978a) and Cox *et al.* (1979) have extended these observations and have shown that CEA, detected by immunohistochemical techniques, first appears in the C cell (Fig. 4) during the histologic stage of hyperplasia (Cox *et al.*, 1979) or microscopic MTC (DeLellis *et al.*, 1978b). (The reasons for the different results are not clear at this time.) In frank MTC most cells stain positive for CEA. Staining for CEA is located at the cell membrane and is diffusely present throughout the cytoplasm (DeLellis *et al.*, 1978a; Hamada and Hamada, 1977). There is no evidence to suggest that CEA appears in secretory granules, and studies by DeLellis *et al.* (1978a) indicate that the staining pattern of CEA is different from CT. CEA is released into the plasma, but its release is not stimulated by CT secretagogues (Calmettes *et al.*, 1977; Wells *et al.*, 1978b), additional evidence that CT and CEA do not share a common secretory granule. Plasma CEA concentrations are generally not elevated in early MTC; it has not proven to be a useful diagnostic tool for detecting early MTC.

D. Hormonal Tumor Markers

The other tumor markers associated with MTC are predominantly peptide hormones not normally thought to be produced by the C cell. These include

ACTH, β-endorphin, somatostatin, vasoactive intestinal peptide, and neurotensin, all peptides normally produced by cells with APUD-like characteristics.

1. ACTH and β-Endorphin

ACTH and β-endorphin are posttranslational cleavage products of the proopimelanocortin precursor. It has been shown that the proopiomelanocortin precursor may be processed to one of several products including ACTH, CLIP (corticotropin-like intermediate peptide), β-endorphin, β-lipotropin, and melanocyte stimulating hormone (α, β, and γ) (Eipper and Mains, 1980). The pathways for the processing of the proopiomelanocortin precursor have been clarified in the rat pituitary gland. In the anterior pituitary, ACTH, β-endorphin and β-lipotropin are the primary products of processing, whereas in the intermediate lobe the same precursor molecule is processed to β-endorphin, α-MSH, and CLIP. The evidence that the transformed C cell produces proopiomelanocortin is compelling. Cushings syndrome secondary to production of ACTH by MTC has been periodically reported (Rosenberg et al., 1978). In addition, Sanchez-Franco et al. (1975) and others (Birgenhager et al., 1976; Abe et al., 1977) have reported immunoreactive ACTH in the plasma and tumor from patients with MTC without any evidence of glucocorticoid excess. We have found immunoreactive ACTH in extracts (R. F. Gagel and L. Kapcala, unpublished observations) from a rat MTC cell line (see Section V.) Goltzman et al. (1979) have shown (by immunohistochemical techniques) that ACTH and CT appear to be in the same cell. More recently immunoreactive β-endorphin in tumor extracts of MTC has been reported (Shibasaki et al., 1979). Taken together these results suggest proopiomelanocortin is a common product of the transformed C cell, but it is likely that processing of the precursor molecule follows a pattern similar to that of the intermediate lobe of the pituitary gland. A change in the processing to favor ACTH production would explain the occasional patient with biologically active ACTH who develops Cushings syndrome. Although no direct evidence is currently available to prove this hypothesis, it seems likely future studies will demonstrate these pathways.

2. Somatostatin

Several studies have demonstrated elevated somatostatin concentrations in plasma and tumor from patients with MTC (Berelowitz et al., 1980). Studies by Berelowitz et al. (1980) demonstrated that immunoreactive somatostatin-like material from a rat medullary thyroid carcinoma has size characteristics and biological activity similar to authentic somatostatin. It has been shown that somatostatin containing cells (? C cells) are normally found in the thyroid in areas adjacent to the normal C cells, and the suggestion has been made that local somatostatin production is important for controlling CT secretion (O'Briain et al., 1979).

3. Vasoactive Intestinal Peptide

Elevated plasma and tumor concentrations of vasoactive intestinal peptide have been reported in MTC (Said and Faloona, 1975). The presence of this peptide in plasma from patients with MTC is variable, and few studies have been performed to characterize the cell of origin.

4. Neurotensin

Neurotensin (NT) is a small peptide originally isolated from the hypothalamus by Carraway and Leeman (1973). In a somewhat fortuitous discovery we found large quantities of immunoreactive NT in a continuous cell line of C cells derived from a rat medullary thyroid carcinoma (Zeytinoglu et al., 1980b) (see Section V). Further characterization of the immunoreactive material showed it to have identical size and chromatographic and biologic characteristics as synthetic NT. Blackburn et al. (1981) have assayed plasma from patients with MTC and have not found plasma concentrations of NT to be elevated. We found no detectable NT in two tumor extracts of human MTC (Gagel and Hammer, unpublished observations). At this time it is unclear whether neurotensin is produced by human medullary thyroid carcinoma.

V. THE RAT MODEL OF MEDULLARY THYROID CARCINOMA

Rats from several strains including the Long Evans (DeLellis et al., 1979) and the WAG/Rij strains (Boorman et al., 1974) develop MTC. Twenty-five to 45% of rats from these strains over the age of 2 years will develop MTC. The C cells, in these rats, progress from the normal state through hyperplasia, nodular hyperplasia, microscopic MTC, and then frank MTC (DeLellis et al., 1979). The C cell abnormalities, like those in the human, are bilateral and multicentric and provide evidence to support the deduced sequence of events in human C cell disease. Development of C cell hyperplasia or microscopic MTC is associated with elevated plasma CT concentrations in these rats (DeLellis et al., 1979).

A transplantable tumor line, derived from geriatric WAG/Rij rats, has been developed. Rats carrying this tumor, which can be transplanted into either the renal capsule or into subcutaneous tissue, develop striking elevations in plasma CT concentration (Gagel et al., 1980; Roos et al., 1979) which can be further increased by acutely increasing the plasma calcium concentration of the rat. Using this tumor model as a starting point we have successfully established C cells in continuous tissue culture (Zeytinoglu et al., 1980a; Gagel et al., 1980). The tumor model and the cell culture system derived from the tumor have already proved useful for studies of the control of CT synthesis and secretion (Gagel et *al.*, 1980) and as starting material for the cloning of the structural gene for CT (Jacobs *et al.*, 1981; Amara *et al.*, 1980). No studies have been performed to look at nonhormonal tumor markers in this model system. Because of the rarity of human MTC, it is likely this model will be used more extensively in future investigations of CT and other tumor markers.

More recently Leong *et al.* (1982) have described a human MTC cell line which has been in culture for 2 years and continues to produce CT and carcinoembryonic antigen. It seems likely this model system will be useful for studying expression of both hormonal and nonhormonal tumor markers.

VI. THE CLONAL NATURE OF HEREDITARY MEDULLARY THYROID CARCINOMA

Knudson (1971) proposed that the development of retinoblastoma was a two-mutational event. He suggested that for hereditary retinoblastoma the first mutation was transmitted genetically and the second mutation, which occurred in genetically susceptible somatic cells, was expressed by transformation of the sensitized cell. In the nonhereditary form he suggested that both mutations occurred in somatic cells. Corollaries of this theory predicted that the age of onset for the hereditary form of the disease would be lower than the nonhereditary form because cells have been made susceptible by the first mutation, and bilaterality and multicentricity would be found in the hereditary form because *many* cells have been sensitized by the first mutation. Jackson *et al.* (1979) have provided evidence that hereditary MTC occurs as a result of a two mutational event. Using techniques similar to those of Knudson they showed that the age of onset of hereditary MTC was earlier than that seen in a comparable group with nonhereditary MTC. They and others have shown that the incidence of bilaterality and multicentricity of the tumor approaches 100% in patients with the hereditary form (Block *et al.*, 1980), whereas these features are unusual in the nonhereditary form. It is unclear at which histologic stage the final mutation occurs.

Baylin *et al.* (1976, 1978) have provided convincing evidence to show that for each focus of MTC within the thyroid gland the final mutation occurs in a single C cell. They showed that individual tumors from patients who were heterozygotes for glucose-6-phosphate dehydrogenase (i.e., whole tissue contains both the A and B forms of the enzyme but an individual cell produces either the A or B form) contained only a single form of the enzyme, indicating the tumor had arisen from a single cell rather than from multiple cells. In subsequent studies they showed that small tumors arising in separate areas of the thyroid contained either the A or B form of the enzyme but not both, thereby demonstrating each tumor had arisen from a mutation in a single cell. They concluded each of these lesions developed as a final mutation in one of multiple susceptible clones.

VII. CONTROL OF GENE EXPRESSION IN MTC

Medullary thyroid carcinoma is derived from a single clonal mutation, and, therefore, one would expect that all cells derived from that single mutation would be phenotypically identical. This view has been supported by immunohistochemical studies in which it has been shown that CT staining in individual lesions (C cell hyperplasia or microscopic MTC) is homogeneous. However, Trump et al. (1979) have described a patient with a very large tumor mass of MTC with very low plasma CT values. More careful examination of the primary and metastatic tumor from this patient showed a population of cells which stained heterogeneously for CT (some positive and some negative). They postulated the plasma CT was low because the CT negative cells were not producing CT and suggested the cells had evolved from a single clone, but had altered their phenotypic expression.

A similar type of phenomenon has been noted in the WAG/Rij rat medullary thyroid carcinoma tumor model and has provided the clues necessary for the discovery of a novel method of processing CT mRNA such that either CT or a new peptide named calcitonin-related gene product (CGRP) may be produced by transcription of the calcitonin gene. The discovery was prompted by the observation that successive passages of rat medullary thyroid carcinoma tumors (Roos et al., 1979; Gagel et al., 1980) resulted in a diminution in the tumor content of CT. Rosenfeld et al. (1981), using cloned CT cDNA as a probe, studied hybridizable mRNA from high- and low-CT-producing tumors and found low-producing tumors contained a striking increase in several large molecular weight species of mRNA and one new mRNA species 150–250 nucleotides longer than the normal CT mRNA. Subsequent studies showed that the translation product of one of these new mRNA species, which hybridized to CT cDNA, did not contain CT, but rather a new peptide which they named CRGP (Amara et al., 1982). Structural studies of CGRP mRNA showed that it was identical to the CT mRNA at the 5' end, but that the 3' end contained only a sequence coding for CGRP. These investigators have postulated that the coding regions for CT and CGRP are located within separate exons of a single gene. After transcription the resultant mRNA, which presumably contains both exons, is processed to excise one or the other exon. In the C cell, processing of the mRNA normally results in inclusion of only a sequence coding for the CT precursor peptide, whereas in the hypothalamus, the same mRNA is processed to include only the CGRP precursor peptide. The control mechanisms for this processing are unknown. This new method of mRNA processing provides a mechanism by which a single gene could code for different proteins in different cells. These observations may be of fundamental importance for understanding how ectopic proteins are produced in malignant cells.

Several groups have studied chromosomal patterns in patients with hereditary

MTC and have detected no abnormal chromosomal patterns (Li *et al.*, 1974). However, Van Dyke *et al.* (1981) recently described a deletion on the short arm of chromosome 20 in patients with the hereditary form of MTC. These studies are at present unconfirmed, but if true will provide an excellent marker for the diagnosis of the disease and focus attention on chromosome 20 for future studies to define the genetic defect in this disease.

VIII. A MODEL FOR THE FUTURE STUDY OF HEREDITARY MEDULLARY THYROID CARCINOMA

A synthesis of the preceding material provides a model for the development of hereditary MTC and expression of its tumor markers. In individuals at risk (50% of the population at risk), there is an inherited trait or mutation in many or all C cells which render them susceptible to malignant transformation. In a number of susceptible cells there is a second mutation. It is unclear at what histologic stage of the disease this mutation occurs. At various points within the histologic evolution of this disease specific tumor markers are expressed, each appearing in a characteristic histologic stage. Little is known about the nuclear or cytoplasmic events leading up to the abnormal expression of these proteins. In coming years the availability of cloned DNA complementary to a variety of protein tumor markers of MTC will facilitate studies of mRNA synthesis and turnover and allow us to determine when each of these abnormal proteins is expressed. *In situ* hybridization with cDNA using light or electron microscopic techniques will permit us to characterize the cell of origin of these tumor markers and may allow direct studies of the genome, which might localize the mutation(s) to a specific site. Knowledge of this sort would have important implications for the diagnosis and treatment of this disease.

ACKNOWLEDGMENTS

The author acknowledges the help and support of Drs. Seymour Reichlin, Armen H. Tashjian, Jr., Ronald A. DeLellis, Hubert J. Wolfe, Fusun N. Zeytinoglu, and Harry H. Miller and Mrs. Zoila T. Feldman during the period of this study.

REFERENCES

Abe, K., Adachi, I., Miyakawa, S., Tanaka, M., Yamaguichi, K., Tanaka, N., Kameya, T., and Shimasato, T. (1977). *Cancer Res.* **37,** 4190.

Amara, S. G., David, D. N., Rosenfeld, M. G., Roos, B. A., and Evans, R. M. (1980). *Proc. Natl. Acad. Sci. U.S.A.* **77,** 4444–4448.

Amara, S. G., Jonas, V., Rosenfeld, M. G., Ong, E. S., and Evans, R. M. (1982). *Nature* **298,** 240–244.
Atkins, F. L., Beaven, M. A., and Keiser, H. R. (1973). *N. Engl. J. Med.* **289,** 545–548.
Baber, E. C. (1876). *Philos. Trans. R. Soc. London* **166,** 557.
Baylin, S. B., and Mendelsohn, G. (1980). *Endocr. Rev.* **1,** 45–77.
Baylin, S. B., Beaven, M. A., Engelman, K., and Sjoerdsma, A. (1970). *N. Engl. J. Med.* **283,** 1239–1244.
Baylin, S. B., Gann, D. S., and Hsu, S. H. (1976). *Science* **193,** 321–323.
Baylin, S. B., Hsu, S. H., Gann, D. S., Smallridge, R. C., and Wells, S. A. (1978). *Science* **199,** 429–431.
Baylin, S. B., Hsu, T. H., Stevens, S. A., Kallman, C. H., Trump, D. L., and Beaven, M. A. (1979a). *J. Clin. Endocrinol. Metab.* **48,** 408–414.
Baylin, S. B., Mendelsohn, G., Weisburger, W. R., Gann, D. S., and Eggleston, J. C. (1979b). *Cancer* **44,** 1315–1321.
Berelowitz, M., Cibelius, M., Szabo, M., Frohman, L., Epstein, S., and Bell, N. H. (1980). *Endocrinology* **107,** 1418–1424.
Birgenhager, J. C., Upton, G. V., Seldenrath, H. J., Krieger, D. T., and Tashjian, A. H., Jr. (1976). *Acta Endocrinol. (Copenhagen)* **83,** 280–292.
Blackburn, A. M., Bryant, M. G., Adrian, T. E., and Bloom, S. (1981). *J. Clin. Endocrinol. Metab.* **52,** 820–822.
Block, M. A., Jackson, C. E., Greenawald, K. A., Yott, J. B., and Tashjian, A. H., Jr. (1980). *Arch. Surg. (Chicago)* **115,** 142–148.
Boorman, G. A., Heersche, J. N. M., and Hollander, C. F. (1974). *JNCI, J. Natl. Cancer Inst.* **53,** 1011–1015.
Calmettes, C., Moukhtar, M. S., and Milhaud, G. (1977). *Biomedicine* **27,** 52–54.
Carraway, R., and Leeman, S. E. (1973). *J. Biol. Chem.* **248,** 6854–6861.
Cox, C. E., Van Vickle, J., Froome, L. C., Mendelsohn, G., Baylin, S. B., and Wells, S. A. (1979). *Surg. Forum* **30,** 120–121.
DeLellis, R. A., Nunnemacher, G., and Wolfe, H. J. (1977). *Lab. Invest.* **36,** 237–248.
DeLellis, R. A., Rule, A. H., Spiler, I., Nathanson, L., Tashjian, A. H., Jr., and Wolfe, H. J. (1978a). *Am. J. Clin. Pathol.* **70,** 587–594.
DeLellis, R. A., Wolfe, H. J., Rule, A. H., Reichlin, S., and Tashjian, A. J., Jr. (1978b). *N. Engl. J. Med.* **299,** 1082.
DeLellis, R. A., Nunnemacher, B. A., Bitman, W. R., Gagel, R. F., Tashjian, A. H., Jr., Blount, M., and Wolfe, H. J. (1979). *Lab. Invest.* **40,** 140–154.
Eipper, B. A., and Mains, R. E. (1980). *Endocr. Rev.* **1,** 1–27.
Foster, G. V., MacIntyre, I., and Pearse, A. G. E. (1964). *Nature (London)* **203,** 1029–1030.
Gagel, R. F., Melvin, K. E. W., Tashjian, A. H., Jr., Miller, H. H., Feldman, Z. T., Wolfe, H. J., DeLellis, R. A., Cervi-Skinner, S., and Reichlin, S. (1975). *Trans. Am. Assoc. Physicians* **88,** 177–191.
Gagel, R. F., Zeytinoglu, F. N., Voelkel, E. F., and Tashjian, A. H., Jr. (1980). *Endocrinology* **107,** 516–523.
Gagel, R. F., Jackson, C. E., Block, M. A., Feldman, Z. T., Reichlin, S. R., Hamilton, B. P., and Tashjian, A. H. (1982). *J. Pediatrics* **101,** 941–946.
Goltzman, D., Huang, S., Browne, C., and Solomon, S. (1979). *J. Clin. Endocrinol. Metabl.* **49,** 364–369.
Graze, K., Spiler, I. J., Tashjian, A. H., Jr., Melvin, K. E. W., Cervi-Skinner, S., Gagel, R. F., Miller, H. H., Wolfe, H. J., DeLellis, R. A., Leape, L., Feldman, Z. T., and Reichlin, S. (1978). *N. Engl. J. Med.* **299,** 980–985.
Hakanson, R., Owman, C., and Sundler, F. (1971). *Biochem. Pharmacol.* **20,** 2187–2190.
Hamada, S., and Hamada, S. (1977). *Br. J. Cancer* **36,** 572–576.

Hennessey, J. F., Wells, S. A., Ontjes, D. A., and Cooper, C. W. (1974). *J. Clin. Endocrinol. Metab.* **39,** 487–495.
Isaacson, P., and Judd, M. A. (1976). *Lancet* **2,** 1016–1017.
Ishikawa, N., and Hamada, S. (1976). *Br. J. Cancer* **34,** 111–115.
Jackson, C. E., Block, M. A., Greenawald, K. A., and Tashjian, A. H., Jr. (1979). *Am. J. Hum. Genet.* **31,** 704–710.
Jacobs, J. W., Goodman, R. H., Chin, W. H., Dee, P. C., and Habener, J. F. (1981). *Science* **213,** 457–459.
Knudson, A. G. (1971). *Proc. Natl. Acad. Sci. U.S.A.* **68,** 820–823.
Leape, L. L., Miller, H. H., Graze, K., Torres-Feldman, Z., Gagel, R. F., Wolfe, H. J., DeLellis, R. A., Tashjian, A. H., Jr., and Reichlin, S. (1977). *J. Pediatr. Surg.* **5,** 831–837.
Leong, S. S., Horoszewicz, J. S., Shimaoka, K., Friedman, M., Kawinski, E., Song, M. J., Zeigel, R., Chu, T. M., Baylin, S., and Mirand, E. A. (1982). *Proc. Int. Colloq. Thyroid Neoplasia, 1981* (in press).
Li, F. P., Melvin, K. E. W., Tashjian, A. H., Jr., Levine, P. H., and Fraumeni, J. F. (1974). *JNCI, J. Natl. Cancer Inst.* **52,** 285–287.
Ljungberg, O. (1972). *Acta Pathol. Microbiol. Scand.* **80,** 589–599.
McMillan, R. J., Hooker, W. M., and Deftos, L. J. (1974). *Am. J. Anat.* **140,** 73–79.
Melvin, K. E. W., Tashjian, A. H., Jr., and Miller, H. H. (1972). *Recent Prog. Horm. Res.* **28,** 399–470.
Mendelsohn, G., Eggleston, J. C., Weisburger, W. R., Gann, D. S., and Baylin, S. B. (1978). *Am. J. Pathol.* **92,** 35–43.
O'Briain, D. S., DeLellis, R. A., Wolfe, H. J., Reichlin, S., Bollinger, J., and Tashjian, A. H., Jr. (1979). *Lab. Invest.* **40,** 275.
Parthemore, J. G., Bronzert, D., Roberts, G., and Deftos, L. J. (1974). *J. Clin. Endocrinol. Metab.* **39,** 198–111.
Pearse, A. G. E., and Takor, T. T. (1979). *Fed. Proc., Fed. Am. Soc. Exp. Biol.* **38,** 2288–2294.
Roos, B. A., Yoon, M. J., Frelinger, A. L., Pensky, A. E., Birnbaum, R. S., and Lambert, P. W. (1979). *Endocrinology* **105,** 27–32.
Rosenberg, E. M., Hahn, T. J., Orth, D. N., Deftos, L. J., and Tanaka, K. (1978). *J. Clin. Endocrinol. Metab.* **47,** 255–262.
Rosenfeld, M. G., Amara, S. G., Roos, B. A., Ong, E. S., and Evans, R. M. (1981). *Nature (London)* **290,** 63–65.
Said, S. I., and Faloona, G. R. (1975). *N. Engl. J. Med.* **293,** 155–160.
Sanchez-Franco, F., Papapetrou, P. D., Gagel, R. F., Wolfe, H. J., and Feldman, Z. (1975). *Program Annu. Meet. Endocr. Soc.* p. 379.
Shibasaki, T., Deftos, L., and Guillemin, R. (1979). *Biochem. Biophys. Res. Commun.* **90,** 1266–1273.
Sizemore, G. W., Carney, J. A., and Heath, H. H., III (1977). *Surg. Clin. North Am.* **57,** 633–645.
Spiler, I. J., Kapcala, L. P., Graze, K., Gagel, R. F., Feldman, Z. T., Biller, B., Tashjian, A. H., Jr., and Reichlin, S. (1980). *J. Clin. Endocrinol. Metab.* **51,** 806–809.
Trump, D. L., Mendelsohn, G., and Baylin, S. B. (1979). *N. Engl. J. Med.* **301,** 253–255.
Van Dyke, D. L., Jackson, C. E., Babu, V. R. (1981). *Clin. Res.* **29,** 37A.
Wang, C. (1978). *In* "The Thyroid" (S. C. Werner and S. H. Ingbar, eds.), pp. 567–568. Harper and Row, Hagerstown, Maryland.
Wells, S. A., Baylin, S. B., Linehan, W. M., Farrell, R. E., Cox, E. B., and Cooper, C. W. (1978a). *Ann. Surg.* **188,** 139–141.
Wells, S. A., Haagensen, D. E., Linehan, W. M., Farrell, R. E., and Dilley, W. G. (1978b). *Cancer* **42,** 1498–1503.
Williams, E. D. (1966). *J. Clin. Pathol.* **19,** 114–118.

Wolfe, H. J., Melvin, K. E. W., Cervi-Skinner, S., Al Saadi, A. A., Juliar, J. P., Jackson, C. E., and Tashjian, A. H., Jr. (1973). *N. Engl. J. Med.* **289,** 437–441.

Wolfe, H. J., Voelkel, E. F., and Tashjian, A. H., Jr. (1974). *J. Clin. Endocrinol. Metab.* **38,** 688–694.

Wright, D. R., Voelkel, E. F., and Tashjian, A. H., Jr. (1977). *In* "Handbook of Radioimmunoassay" (G. E. Abraham, ed.), pp. 391–423. Dekker, New York.

Zeytinoglu, F. N., DeLellis, R. A., Gagel, R. F., Wolfe, H. J., and Tashjian, A. H., Jr. (1980a). *Endocrinology* **107,** 509–515.

Zeytinoglu, F. N., Gagel, R. F., Tashjian, A. H., Jr., Hammer, R. A., and Leeman, S. E. (1980b). *Proc. Natl. Acad. Sci. U.S.A.* **77,** 3741–3745.

CHAPTER 13
Markers for Germ Cell Tumors of the Testis

PAUL H. LANGE
Department of Urologic Surgery
University of Minnesota College of Health Sciences
Minneapolis, Minnesota
Veterans Administration Medical Center
Minneapolis, Minnesota

I.	Introduction	241
II.	Clinical Aspects of Testicular Cancer	242
III.	Background of α-Fetoprotein and Human Chorionic Gonadotropin	243
	A. α-Fetoprotein	243
	B. Human Chorionic Gonadotropin	244
IV.	Sensitivity and Specificity of α-Fetoprotein and Human Chorionic Gonadotropin in Testicular Cancer	245
V.	α-Fetoprotein and Human Chorionic Gonadotropin in the Diagnosis of Testicular Tumors	247
VI.	α-Fetoprotein and Human Chorionic Gonadotropin in Clinical Staging of Nonseminomatous Germ Cell Tumors	247
VII.	α-Fetoprotein and Human Chorionic Gonadotropin in the Clinical Monitoring of Nonseminomatous Germ Cell Tumors	248
VIII.	Tumor Markers in Seminoma	249
IX.	Newer Applications of α-Fetoprotein and Human Chorionic Gonadotropin in Monitoring Nonseminomatous Germ Cell Tumors	251
X.	Other Tumor Markers for Nonseminomatous Germ Cell Tumors	252
XI.	Radioimmunolocalization of Testicular Tumors	253
XII.	Immunohistochemistry of Testicular Tumor Markers	254
XIII.	The Future	254
	References	255

I. INTRODUCTION

Much has been learned from the experience with several oncodevelopmental antigens in human testicular cancer about the clinical applications and misapplications of tumor markers. This is especially true of α-fetoprotein (AFP) and human chorionic gonadotropin (hCG), which are among the best serum markers available in clinical oncology, although they are far from ideal. Now, with the promise that new techniques such as somatic cell fusion will touch off an explosive growth in the number of potential tumor markers, it is critical that we learn the lessons of the history, applications, and potential of oncodevelopmental substances in testicular cancer.

The use of AFP and hCG in testicular cancer has been described extensively (Waldmann and McIntire, 1974; Scardino and Skinner, 1979; Javadpour, 1980; Lange et al., 1980a; Nørgaard-Pedersen and Raghavan, 1980). This chapter reviews the salient clinical features of these markers, emphasizing those most relevant to the clinical application of tumor markers in general. It also mentions other oncodevelopmental markers that have recently demonstrated their value in testicular cancer, and it speculates about the future of tumor markers in this disease.

II. CLINICAL ASPECTS OF TESTICULAR CANCER

Our discussion requires a brief clinical review of testicular cancer (for a more extensive review, see Fraley et al., 1979). More than 95% of these tumors belong to a heterogenous group called germ cell tumors because it is widely believed that they arise in primordial germ cells. Such tumors are classified either as seminomas or as nonseminomatous tumors, which are classified further as embryonal carcinoma, teratoma (mature or immature), or choriocarcinoma. These tumor types may occur alone or in various combinations; mixtures of embryonal carcinoma and teratoma often are called teratocarcinoma.

In men, germ cell tumors nearly always appear first as an intra-scrotal mass; the definitive diagnosis is made after the mass has been extirpated by radical orchiectomy. Thereafter, the disease is "staged": tests are performed to determine whether the tumor has spread. Because testicular tumors usually metastasize first to the lymph nodes along the great vessels in the retroperitoneum and then to the lungs, other lymph nodes, and viscera, staging tests commonly include full lung tomography, computer-assisted tomography or ultrasonography of the abdomen, and, in some hospitals, bipedal lymphangiography. A tumor confined to the testis and spermatic cord is said to be in stage I, one that has metastasized to the retroperitoneum alone in stage II, and one that has spread outside the retroperitoneum in stage III.

When the primary tumor is pure seminoma, the accepted postorchiectomy treatment is irradiation of the retroperitoneal lymph nodes and perhaps of the mediastinum and supraclavicular nodes, depending on the stage of the disease. Cure rates exceed 90% except in patients who present with advanced cancer.

When the primary cancer is partly or entirely nonseminomatous germ cell tumor (NSGCT), the best course of treatment is a matter of considerable debate. In many medical centers, especially American ones, patients without evidence of extraretroperitoneal metastases are treated by retroperitoneal lymphadenectomy, both because it is believed that methods for evaluating the retroperitoneum are not sufficiently accurate and because operation is thought to offer the best chance for cure, particularly if adjuvant chemotherapy is used in patients who appear to have a high risk of residual disease or recurrence. However, in other (particularly European) medical centers, patients without stage III disease traditionally have been given radiation in doses higher than those used in seminoma. This debate has often been heated, but it now appears safe to say that with the advent of extraordinarily effective chemotherapy and of more accurate staging methods (including tumor markers), both surgeons and radiotherapists are reconsidering their approaches to stages I and II NSGCT.

When the patient presents with stage III disease, he usually is treated primarily by chemotherapy, with debulking operations or radiation added if the tumor does not respond completely to the drugs.

As recently as 10 years ago, NSGCT was often fatal. Now, however, even those patients who present with disseminated cancer usually are cured. Much of this success is attributable to the advances in chemotherapy, but the clinical application of AFP and hCG has been critical.

III. BACKGROUND OF α-FETOPROTEIN AND HUMAN CHORIONIC GONADOTROPIN

A. α-Fetoprotein

α-Fetoprotein was one of the first oncodevelopmental substances to be identified; in fact, its existence contributed to the formulation of that concept (Nørgaard-Pedersen, 1976; Ruoslahti and Hirai, 1978; Wepsic and Kirkpatrick, 1979). It is a single-chain glycoprotein (molecular weight approximately 70,000) produced in many species of mammals by the fetal yolk sac, liver, and gastrointestinal tract. In human beings, it is an important fetal serum protein and reaches a peak concentration at about the twelfth week of gestation. The level then declines, and after the age of 1 year, AFP usually is not detectable in the serum by the older, less-sensitive assays such as immunodiffusion. (When more sensitive assays became available in the early 1970's, "normal" concentrations of AFP as high as 16 ng/ml were observed routinely.) The function of AFP in the fetus is unknown, although it may act as an albumin-like serum protein (Wepsic and Kirkpatrick, 1979).

In 1963, Abelev found AFP in the serum of mice with chemically induced hepatomas. Soon thereafter, it was detected in the sera of many patients with hepatomas and a few with testicular teratocarcinomas. When sensitive radioimmunoassays for AFP were introduced 8 years later, it became apparent that elevated serum concentrations are as common in patients with testicular NSGCT as in those with hepatomas (approximately 70%). (The disparity between these and the earlier findings resulted from the insensitivity of the older assays and the tendency of hepatomas to produce much larger amounts of AFP than do NSGCT's.) Nevertheless, it was almost 5 more years before clinical use was made of AFP in the management of patients with testicular cancer.

The level of sensitivity necessary for an AFP assay to be useful in testicular tumors is still being debated, although 5–10 ng/ml (5.15–10.3 IU/ml) usually is satisfactory (Nørgaard-Pedersen and Raghavan, 1980). However, with the increasingly sensitive AFP assays has come decreased specificity; abnormal levels now can be detected in patients with ataxia telangiectasia or hereditary tyrosinemia; in some with pancreatic, biliary, or other gastrointestinal cancers; and in a few with nonmalignant conditions in which the liver is regenerating or who have nongastrointestinal cancers that have metastasized to the liver (Nørgaard-Pedersen, 1976; Wepsic and Kirkpatrick, 1979; Lange *et al.*, 1980a).

In normal persons, the metabolic half-life of AFP is approximately 5 days, although values as short as 3 days have been reported (Mann *et al.*, 1981). The clearance rate has not been examined rigorously in patients with NSGCT, but clinical observations are consistent with a value of 5 days (P. Lange, unpublished data).

Like many glycoproteins, AFP has molecular variants. In man, two variants have been demonstrated by virtue of differences in their binding to lectins such as concanavalin A (Con-A). Thus, approximately 95% of the AFP produced by the fetal liver binds to Con-A, whereas approximately 50% of the AFP from the yolk sac does (Ransohoff and Feinstein, 1978; Smith and Kelleher, 1980).

B. Human Chorionic Gonadotropin

Human chorionic gonadotropin is a glycoprotein composed of two dissimilar polypeptide chains, α and β, and having a molecular weight of approximately 38,000. The α subunit appears to be common to several of the pituitary hormones, including luteinizing hormone (LH) and follicle-stimulating hormone (FSH), whereas the β chain, especially its carboxy-terminal end, is unique to hCG. As a result, some antibodies raised against purified β chains cross-react very little with LH and so can be used in specific radioimmunoassays for hCG ("β-hCG assays''). Nevertheless, almost all hCG antisera cross-react with LH to some extent, so one must be cautious in interpreting borderline-abnormal serum hCG levels (Vaitukaitis *et al.*, 1976; Vaitukaitis, 1979a,b). To be useful for measuring hCG concentrations in patients with testicular tumors, an assay should be sensitive to 1–2 ng/ml (5–10 mIU/ml).

It has been known since 1930 that some patients with germ cell tumors have abnormal hCG levels. However, as in the case of AFP, 5 years passed between the introduction of the sensitive β-hCG radioimmunoassays and the recognition of the high prevalence (40–60%) of elevated hCG levels in patients with NSGCT. Abnormal hCG levels also appear in some patients with seminoma or with cancers of the liver, breast, stomach, pancreas, or lung (Vaitukaitis, 1979b).

The metabolic half-life of the intact hCG molecule is 24–36 hours, but the polypeptide chains, which may circulate independently, have much shorter half-lives: 20 minutes for the α chain and 45 minutes for the β chain (Vaitukaitis *et al.*, 1976; Vaitukaitis, 1979a). Neoplasms differ in their patterns of secretion of hCG and its subunits: some secrete only one of the chains, whereas others produce the intact molecule or abnormal forms of the molecule and its subunits (Papapetrou *et al.*, 1980). This fact complicates interpretations of the hCG metabolic decay rate after treatment.

As in the case of AFP, molecular variants of hCG have been identified by lectin-binding assays (Yoshimoto *et al.*, 1977).

IV. SENSITIVITY AND SPECIFICITY OF α-FETOPROTEIN AND HUMAN CHORIONIC GONADOTROPIN IN TESTICULAR CANCER

In the field of tumor markers, the terms "sensitivity," "specificity," "false negative," and "false positive" have precise definitions that can be cumbersome (see Ransohoff and Feinstein, 1978) and so will simply be adumbrated here. I already have used the word "sensitive" to describe the limit of detectability by an assay; the word also is applied to the ability of AFP and hCG to indicate the presence of testicular tumor, and here the measurement of sensitivity becomes more complicated. The first proviso is that both AFP and hCG must be measured, because approximately 40% of patients with NSGCT have elevated levels of only one of them and because the changes in marker levels do not always parallel each other in the course of the disease. When both markers are measured, approximately 80% of patients with active testicular cancer will have an elevated level of at least one (Lange *et al.*, 1980a). This impressive figure is misleading, however, because it applies principally to patients with NSGCT rather than seminoma and because the percentage varies with the stage of the cancer and the point in the clinical course. Thus, many factors affect the sensitivity of AFP and hCG in testicular cancer; I will discuss them when we consider the value of markers in diagnosis, staging, and monitoring.

The definition of "specificity" in relation to AFP and hCG in testicular cancer also is complicated. As I mentioned, several other conditions can be responsible for elevated levels of AFP or hCG, so in this respect they are not specific markers for NSGCT or testicular cancer. In practice, however, these markers are very specific; in a patient believed or known to have a testicular tumor, it is rare for an elevated AFP or hCG level to be caused by anything else. As a result, it is often said that there are no false-positive marker levels* (Scardino and Skinner, 1979). There are occasional exceptions—for example, a patient of ours had an unexplained elevated AFP level that proved to be the result of hepatitis—but always routine clinical evaluations have been able to rule out these other causes.

There are several other qualifiers to the statement that "there are no false-positive findings in patients with known or suspected testicular cancer." First, a marker level (particularly an AFP level) may be elevated in a tumor-free patient if the serum specimen is obtained too soon after tumor removal for the substance to have been cleared metabolically. [Methods for determining this, including those for calculating the metabolic half-life in a particular patient, have been reported (Lange et al., 1980a) and require serial serum samples; attempts to interpret marker levels from a single sample are risky.] Second, borderline elevations in the serum hCG concentration (1–3 ng/ml) may be recorded in some patients who actually have elevated LH levels and normal hCG levels. This usually happens when one of the less-sensitive commercial hCG assays is being used and is most likely to involve a patient who has hypogonadism induced by chemotherapy or orchiectomy and who therefore has a high LH level. Often, study of serial samples or reference to the clinical context will show whether it is the LH or the hCG level that is elevated. Simultaneous measurement of the serum testosterone and LH concentrations also may resolve the issue, although most LH assays cross-react extensively with hCG, so it still may not be possible to determine which hormone is being measured. In a really troublesome case, the patient can be given a short course of testosterone injections to reduce the LH level physiologically and then retested for his hCG level (Catalona et al., 1979). Third, marijuana smokers have been said to have increased circulating hCG levels (Garnick, 1980), an observation yet to be verified but which could become important if the marijuana derivative tetrahydrocannabinol (THC) is widely accepted as an antiemetic for patients receiving cancer chemotherapy.

---

*Note that "false positive" and "false negative" refer to the tumor, not the assay. That is, designation of a result as "false negative" means that a tumor apparently does not produce the marker, not that marker is present that the assay failed to detect.

V. α-FETOPROTEIN AND HUMAN CHORIONIC GONADOTROPIN IN THE DIAGNOSIS OF TESTICULAR TUMORS

Marker assays are usually of little value in the differential diagnosis of testicular masses because, although the levels are normal when a mass is benign or a non-germ-cell tumor such as lymphoma, they also are normal in most men with seminomas and some with NSGCT. For example, approximately 40% of patients with stage I NSGCT have normal AFP and hCG levels before orchiectomy. (Approximately 12% of patients with untreated stage III NSGCT also have normal marker levels, but in these cases the diagnosis is rarely in doubt.) In other words, AFP and hCG are useful diagnostically only if the levels are elevated, in which case it is very likely that the patient has a testicular tumor. These considerations also explain why AFP and hCG are not practical for screening large populations for testicular cancer.

VI. α-FETOPROTEIN AND HUMAN CHORIONIC GONADOTROPIN IN CLINICAL STAGING OF NONSEMINOMATOUS GERM CELL TUMORS

The utility of markers in staging NSGCT also needs qualification and continued reappraisal; certainly in the past it has been overemphasized. If the marker levels are truly elevated after orchiectomy (that is, serum marker does not represent substances produced by the primary tumor that have not yet been metabolically cleared), then metastatic disease is present. However, false-negative marker results are common—the patient has normal prelymphadenectomy marker levels and pathologically proved retroperitoneal metastases. This is true of at least 30% of patients with stage II disease; the precise frequency is still debated.

Because of recent refinements and developments in noninvasive staging techniques, including bipedal lymphangiography, abdominal ultrasonography, and computer-assisted tomography, the pertinent question actually is, "What contribution can marker determinations make to correction of the errors in other staging methods?" At this point, the answer is not clear, although marker assays are part of nearly all suggested staging algorithms. Currently, even the best combination of methods is accurate in only 80–85% of the cases, and we therefore believe that retroperitoneal lymphadenectomy remains the most reliable staging method for patients without demonstrable extraretroperitoneal disease. Of course, if staging accuracy can be improved by refinements in noninvasive techniques, including marker assays, this policy will require reevaluation (Lange and Raghavan, 1983).

VII. α-FETOPROTEIN AND HUMAN CHORIONIC GONADOTROPIN IN THE CLINICAL MONITORING OF NONSEMINOMATOUS GERM CELL TUMORS

The greatest value of serum AFP and hCG determinations in patients with testicular cancer is in the monitoring of chemotherapy and in follow-up after the initial treatment(s). These markers have become indispensible by virtue of their ability to reflect or predict the progression or remission of disease, often before other methods can. This is especially advantageous in testicular cancer, where the many types of cellular differentiation, each with its own marker production capabilities, and the expansion and diversification of the clinical spectrum of the disease under the impact of chemotherapy have greatly complicated clinical decision-making about the presence of recurrent tumor or the time and completeness of a remission.

Despite this accolade, there are many nuances in the application of AFP and hCG that must be recognized that probably will be relevant to the use of markers in other cancers. Accordingly, I will discuss them in some detail.

First, confidence in the specificity of AFP and hCG is so strong that an unequivocal elevation (i.e., not caused by unmetabolized marker from an already-resected tumor or by cross-reaction with LH) is sufficient evidence of the presence of cancer for chemotherapy to be started or resumed without pathologic confirmation if that would necessitate significant surgery. Occasionally, rising marker levels are the *only* immediate sign of recurrence, yet it is possible to begin treatment secure in the knowledge that the patient does indeed have growing cancer somewhere.

Second, consistently normal marker levels in patients with untreated testicular cancer or in those who have had treatment but who have persistent or growing tumor-like masses present a dilemma. Approximately 12% of patients with untreated stage III NSGCT have falsely negative markers. In patients who have completed chemotherapy, the absence of markers may signal that the mass is treatment-induced fibrosis or necrosis, a benign lesion such as lymphocele, or mature teratoma or even seminoma. Approximately 50% of patients who have only a partial response to chemotherapy have normal marker levels, but only approximately 16% will still have viable NSGCT. In other words, 66% will have something other than NSGCT, and of those who do have NSGCT, only half will have falsely negative markers. In clinical terms, these figures mean that a biopsy or debulking operation is needed before chemotherapy is started or resumed in order to identify the mass, although an exception might be made if the patient has had falsely negative markers from the beginning.

Third, markers are not something that NSGCT cells must produce to survive. Thus it often happens that serum marker levels return to normal during chemotherapy weeks or months before the tumor disappears clinically. This does *not* mean that treatment can be discontinued (Lange and Raghavan, 1983).

Fourth, analysis of the rate of decline of AFP and hCG serum levels in response to treatment (particularly chemotherapy) has been proposed as a prognostic indicator. For example, patients who respond well will, as a group, have a faster decline in their marker levels than will those who respond poorly; the question is, "Can this be used in an individual patient?" Could one predict, for example, that a patient will respond poorly to chemotherapy, even though his marker levels return to normal, by the rate at which they decline? Some retrospective data from our institution (Lange et al., 1982a) and elsewhere (Kohn, 1979; Thompson and Haddow, 1979) suggest that this is possible. However, we believe that the platinum-based chemotherapy introduced since most of those data were gathered will curtail the predictive value of the marker decay rate (N. J. Vogelzang and P. H. Lange, unpublished data). An interesting discovery made during these studies is the "release phenomenon," a temporary, often dramatic, rise in marker levels immediately after chemotherapy is started (Vogelzang et al., 1980), which probably results from sudden lysis of many marker-producing tumor cells.

Fifth, and a most controversial issue, are elevated AFP and hCG levels a prognostic variable by themselves? That is, is a marker level simply a function of tumor bulk, or does the production of markers by a tumor indicate that it is more aggressive or less sensitive to treatment than is a tumor that is marker negative? Early reports often seemed to suggest that marker production was a prognostic indicator, but when more sophisticated statistical analyses that include corrections for tumor bulk and histologic type are applied to the data, the conclusions become less clear (Lange and Raghavan, 1983). One recent study that included such analytical methods suggested that AFP is an adverse prognostic sign, whereas hCG is not (Bosl et al., 1981). Also, an elevation of either marker before a debulking operation usually is associated with poor results, so it may be prudent to perform such operations during a time when marker levels are normal whenever this is possible (Lange et al., 1982c).

VIII. TUMOR MARKERS IN SEMINOMA

Despite their common germ cell origin, seminomas and NSGCT's have long been thought to follow different developmental paths. Thus it is often said that although seminoma and NSGCT are sometimes found in the same tumor, seminoma does not differentiate into any of the histologic types of NSGCT and vice versa. Another difference traditionally attributed to seminoma and NSGCT is one of relative curability: for at least 20 years, seminoma has been called a very curable cancer. To a considerable extent this is true, because this tumor often is slow to metastasize, so that most patients present with low-stage disease, and because seminomas usually are very sensitive to radiation. However, since the introduction of the newer chemotherapies, high-stage NSGCT is cured more often than is high-stage seminoma (Lange et al., 1980a). Also, the application of AFP and hCG assays in patients with seminoma has altered thinking about the pathology of this type of tumor.

The accumulated clinical experience with tumor markers in seminoma is less than that in NSGCT, but several facts have emerged nonetheless. First, if the patient has an elevated AFP level, NSGCT is present either in the primary tumor or in manifest or occult metastases, and the patient should be treated for NSGCT even if the pathology report says that the tumor is pure seminoma. Elevated levels of hCG in these patients also can mean that NSGCT is present but undiscovered, but they also can mean that the patient has pure seminoma that produces hCG. Approximately 10–20% of seminomas produce this marker, and the significance of this finding is still in question. Our experience suggests that if the patient has low-stage seminoma, an elevated hCG level has no prognostic implications, whereas if he has high-stage seminoma, the elevated marker level suggests a poor prognosis, especially if radiation is used as the principal treatment (Lange et al., 1980b). The fact that many high-stage seminomas now are being treated by chemotherapy rather than radiation may obscure the prognostic meaning of hCG, but the marker's presence in tumors that meet the pathologic criteria for classic seminoma suggests that this tumor and NSGCT are less different than had been thought (Lange et al., 1980b).

The intense discussion of hCG in seminoma in some of the recent literature must not be allowed to obscure the fact that most patients with this tumor do not have elevated marker levels. It is for this large majority of patients that new tumor markers are needed, and one with great promise is placental alkaline phosphatase (PlAP). This fetal isoenzyme, which can be distinguished biochemically from the adult PlAP isoenzymes (bone, liver, etc.), has been found in sera from patients with various malignancies and was suggested as a marker for seminoma. The results of early studies with enzymatic and immunological assays were not impressive, however. Recently, more sensitive assays have made it appear that this oncodevelopmental enzyme will prove an important marker for seminoma. For example, in a retrospective study with a sensitive and specific enzyme-linked immunosorbent assay (ELISA), we found elevated PlAP levels in approximately 45% of men with active testicular tumors. Some of them had elevations of both PlAP and hCG, whereas others had elevated levels of only one; but 68% had elevated levels of one or both. More important, the PlAP levels in serial samples sometimes provided clinical information not available from either AFP/hCG determinations or other clinical measures (Lange *et al.*, 1981, 1982d). Confirmatory experience is needed, but the data permit one to hope.

IX. NEWER APPLICATIONS OF α-FETOPROTEIN AND HUMAN CHORIONIC GONADOTROPIN IN MONITORING NONSEMINOMATOUS GERM CELL TUMORS

More sensitive methods of measuring hCG take advantage of the fact that the kidney readily clears the intact molecule from the serum; special urine-concentrating techniques thus yield a specimen that is assayed for hCG. Because such urinary concentrates normally contain a significant amount of LH, a very specific, although less sensitive, radioimmunoassay is required that utilizes an antibody to the carboxy-terminal peptide of the hCG β chain. The method increases the sensitivity to hCG production threefold, and there is some evidence that it will be valuable clinically in the follow-up of patients with NSGCT (Javadpour and Chen, 1981).

I noted earlier that lectin-binding assays detect variants of human AFP and hCG. In adults, the percentage of AFP from the liver that binds to Con-A is consistently more than 70%, whereas less than 70% of AFP of yolk sac origin binds to the lectin (Smith and Kelleher, 1980). This fact makes it possible to determine whether unexplained AFP in a patient with a history of NSGCT originates in the liver or in recurrent tumor. This is seldom necessary, but we have treated three patients in whom Con-A binding assays for AFP were helpful clinically (Vessella and Lange, unpublished data). However, our studies showed that although AFP from patients with NSGCT varies considerably in its Con-A-binding patterns within the range typical of yolk sac AFP, this does not produce clinically useful information (Vessella *et al.*, unpublished data). Perhaps other lectins or assays will reveal relevant patterns, however.

Lectin-binding-defined variants of hCG have been identified in other hCG-producing tumors (Vaitukaitis, 1979b) but have not been sought in NSGCT. However, like the normal placenta, some tumors secrete not only intact hCG but also free α and β chains and chain fragments (Vaitukaitis *et al.*, 1976). In patients with gestational trophoblastic tumors, there is evidence that the proportion of the various forms has prognostic significance (Vaitukaitis and Ebersole, 1976), and studies are in progress to determine whether this is also true in patients with NSGCT.

X. OTHER TUMOR MARKERS FOR NONSEMINOMATOUS GERM CELL TUMORS

The success of AFP and hCG has spurred a search for new markers that would help in those cases in which the established markers are falsely negative. Several substances proposed as markers have failed to fulfill their initial promise: for example, carcinoembryonic antigen (CEA) is produced by teratomas with gut-like differentiation, but the levels do not correlate with the behavior of the tumor. Likewise, human placental lactogen, α_1-antitrypsin, and ferritin, although occasionally found in patients with testicular tumors, are either nonspecific or not sufficiently prevalent to be important clinically (Lange and Raghavan, 1983).

The enzyme lactic dehydrogenase (LDH) is not generally considered to be an oncodevelopmental substance, and it is not specific for testicular cancer, but it is of some use in monitoring patients with NSGCT. Occasionally, rising serum concentrations of LDH are the only clinical evidence of persistent or recurrent tumor. We routinely measure this enzyme and consider it helpful if nonspecific. We do not attach the importance to it that we do to AFP and hCG, however, and do not trust it as strongly (Lange and Raghavan, 1983). We also have studied the developmental isoenzyme of LDH known as LDH-x, which appears in the germinal epithelium of mammals at the onset of spermatogenesis and disappears when that process ceases. With an immunological assay, we were unable to detect LDH-x in sera or tissues from patients with testicular tumors (Vogelzang et al., 1982).

The protein SP_1 (from *Schwangerschaftsproteine* No. 1) is produced by the syncytiotrophoblastic cells of the placenta and is detectable in the serum of pregnant women as early as 7 days after ovulation and of most patients with trophoblastic tumors. Abnormal levels of SP_1 also have been detected in the serum of patients with testicular tumors, often in association with elevated hCG levels (Rosen et al., 1979; Lange et al., 1980c; Szymender et al., 1981). We find that serial determinations of SP-1 levels provide unique clinical information in some cases (Lange et al., 1980c). Anyone studying the literature will quickly note widely differing figures for the prevalence of elevated SP_1 levels in patients with testicular tumors. These probably result from differences in assay technique and the designated limits of normal and emphasize the need for refinements in the procedure and further studies to define the value of SP_1.

I have discussed the value of PlAP as a marker for pure seminoma. Curiously, this enzyme is seldom found in patients with NSGCT, even when the tumor contains seminoma, and when it *is* found, changes in its level during the clinical course do not reveal useful information (Lange et al., 1981).

Murine teratocarcinomas have unique oncodevelopmental antigens such as F9 (named for the cell line on which it was discovered), which is present on the germ cells throughout life but which disappears from all somatic cells after the second week of gestation. This antigen is also found on human germ cell tumor cell lines and on normal human spermatozoa (Artzt et al., 1973; Holden et al., 1977). Other F9-like antigens have been defined by monoclonal antibodies (which are produced by hybridomas) against murine (e.g., SSEA-1, SSEA-3) or human (LICR-LON-HT13) germ cell tumors (Solter and Knowles, 1978; Knowles et al., 1978; Edwards et al., 1980). None of these antigens has been found in the blood of patients with testicular tumors, but they are present on cell lines and fresh tissues from human germ cell tumors (Andrews et al., 1980; Bronson et al., 1980; McIlhinney, 1981; Damjanov et al., 1982). The importance of this discovery for the pathology of testicular cancer is unclear.

XI. RADIOIMMUNOLOCALIZATION OF TESTICULAR TUMORS

The existence of oncodevelopmental antigens eventually may make it possible to locate tumor foci. For this process, known as radioimmunolocalization (RIL) or radioimmunodetection (RAID), the tumor host is injected with an antibody (e.g., anti-AFP) tagged with a radiolabel (usually ^{125}I) and then examined by scintigraphy. Much of the experimental work on RIL in animals and man has been done by Goldenberg and his co-workers with anti-CEA antibody and colorectal tumors (Goldenberg, 1980; Goldenberg et al., 1980a). The technical difficulties are immense and the clinical value of the method in dispute (Mach et al., 1980), but unsuspected tumor foci have been detected by RIL, even in patients with normal serum CEA levels.

Radiolabeled anti-hCG and anti-AFP antibodies have been used in attempts at RIL of trophoblastic disease and germ cell tumors, and the early results are promising. In some cases, tumor could be detected in patients with normal marker levels (Goldenberg et al., 1980b; Begent et al., 1980; Javadpour et al., 1981; Halsall et al., 1981; Hatch et al., 1980). The F9-like antigens also were studied as targets for RIL. Labeled anti-SSEA-1 accumulated in small teratocarcinomas in mice (Ballou et al., 1979), and monoclonal antibody to a human testicular tumor accumulated in human germ cell tumor xenografts in immunosuppressed mice (Moshakis et al., 1981). These studies are now being extended to patients.

In my opinion, RIL must still be considered an investigational procedure in human testicular cancer. Only a few patients have been studied, and in many of the reports it is not clear whether the information gained was useful clinically. In our experience with seven patients, we encountered false-positive and false-negative scans as well as clinically important data (Lange et al., 1982b). Furthermore, as I said earlier, there are formidable technical problems: the required purity of the antibody, the subjectiveness of scan interpretation, and the need to eliminate background "noise" resulting from trapped antigen–antibody complexes and the small amounts of oncodevelopmental antigens present on normal tissue. Also, the use of xenogeneic antibodies poses a theoretical threat of allergic reactions, although so far this has not happened. Nevertheless, RIL is an exciting and promising new application of oncodevelopmental antigens in testicular cancer, and I suspect it will become important clinically.

XII. IMMUNOHISTOCHEMISTRY OF TESTICULAR TUMOR MARKERS

Histochemical methods, particularly immunofluorescence and immunoperoxidase staining, have identified the cells of origin of many markers associated with germ cell tumors. For example, hCG has been demonstrated in the syncytial giant cells of seminoma, NSGCT, and trophoblastic tissue, while AFP was located in cells with the morphology of classic yolk sac carcinoma, in undifferentiated cells, and in cells with features common to embryonal carcinoma and yolk sac carcinoma, and SP_1 is located in cytotrophoblastic cells. Several other markers, such as CEA, human placental lactogen, fibronectin, PlAP, and ferritin, have been found in a variety of cell types (Javadpour *et al.,* 1980; Lange and Raghavan, 1983).

It has been suggested that histochemical analysis will produce a new pathological classification system for testicular tumors and define more accurately the prevalence of each marker, its potential usefulness in a particular patient, and its prognostic significance. Although there is no doubt that histochemical studies have increased our understanding of which cell types produce markers and may aid in the identification of certain cells, so far they have not changed diagnosis or classification significantly except in a few cases (Richardson *et al.,* 1981). One reason is the technical problems: antigens can be denatured during purification, causing false-negative results, and artifacts of staining may cause either false-positive or false-negative results. False-negative results also can be caused by sampling error or inadequate antisera, and thus failure to demonstrate markers in tumor tissue does not always preclude their detection in a patient's serum. Techniques for increasing the sensitivity and specificity no doubt will reveal other antigens, possibly confined to tissue, that must be tested histochemically and evaluated clinically.

XIII. THE FUTURE

Tumor markers, particularly the oncodevelopmental antigens, will continue to be prominent in research on and management of testicular cancer. Their use in serology, pathology, and radioimaging of this relatively rare tumor already has provided ideas that will be valuable in the application of other markers to other tumors. Hybridoma techniques undoubtedly will reveal many other antigens, most of them oncodevelopmental, that will make testicular tumor markers even more fascinating, not only to the scientist and clinician absorbed with germ cell tumors but also to those concerned with developmental biology. For example, several human testicular tumor cell lines, with various morphological characteristics and marker-producing capabilities, have been established (Bronson et al., 1980, 1982), and monoclonal antibodies are being developed to their antigens (D. Solter and W. C. DeWolf, personal communications). Monoclonal antibodies to sperm or embryonic antigens also may be useful. Finally, antibodies tagged with cellular toxins such as ricin or antitumor drugs may be useful to deliver effective treatment with fewer side effects (Gilliand et al., 1980).

The cure rate for patients with stage III NSGCT is now 70–80%. If the other patients are to be cured, judicious and creative application of these new discoveries will be required.

REFERENCES

Andrews, P. W., Bronson, D. L., Benham, F., Strickland, S., and Knowles, B. B. (1980). *Int. J. Cancer* **26**, 269–280.
Artzt, K., Dubois, P., Bennett, D., Condamine, H., Babinet, C., and Jacob, F. (1973). *Proc. Natl. Acad. Sci. U.S.A.* **70**, 2988–2992.
Ballou, B., Levine, G., Hakala, T. R., and Solter, D. (1979). *Science* **206**, 844–846.
Begent, R. H. J., Stanway, G., Jones, B. E., Bagshawe, K. D., Searle, F., Jewkes, R. F., and Vernon, P. (1980). *J. R. Soc. Med.* **73**, 624–630.
Bosl, G. J., Cirrincione, C., Geller, N., Vugrin, D., Whitmore, W., and Golbey, R. (1981). *Proc. Am. Soc. Clin. Oncol.* **22**, 393.
Bronson, D. L., Andrews, P. W., Solter, D., Cervenka, J., Lange, P. H., and Fraley, E. E. (1980). *Cancer Res.* **40**, 2500–2506.
Bronson, D. L., Clayman, R. V., and Fraley, E. E. (1982). *In* "The Biology of Human Teratomas" (I. Damjanov, D. Solter, and B. B. Knowles, eds.). Humana Press, Clifton, New Jersey (in press).
Catalona, W. J., Vaitukaitis, J. L., and Fair, W. R. (1979). *J. Urol.* **122**, 126–128.
Damjanov, I., Fox, N., Knowles, B. B., Solter, D., Lange, P. H., and Fraley, E. E. (1982). *Am. J. Pathol.* **108**, 225–230.
Edwards, P., Foster, C., and McIlhinney, R. A. J. (1980). *Transplant. Proc.* **12**, 398–402.
Fraley, E. E., Lange, P. H., and Kennedy, B. J. (1979). *N. Engl. J. Med.* **301**, 1370–1377, 1420–1426.
Garnick, M. B. (1980). *N. Engl. J. Med.* **303**, 1177.
Gilliand, D. G., Steplewski, Z., Collier, R. J., Mitchell, K. F., Chang, T. H., and Koprowski, H. (1980). *Proc. Natl. Acad. Sci. U.S.A.* **77**, 5439–5443.
Goldenberg, D. M. (1980). *Cancer Res.* **40**, 2957–2959.
Goldenberg, D. M., Kim, E. E., Deland, F. H., Bennett, S., and Primus, F. J. (1980a). *Cancer Res.* **40**, 2984–2992.
Goldenberg, D. M., Kim, E. E., Deland, F. H., van Nagell, J. R., Jr., and Javadpour, N. (1980b). *Science* **207**, 1284–1286.

Halsall, A. K., Fairweather, D. S., Bradwell, A. R., Blackburn, J. C., Dykes, P. W., Howell, A., Reeder, A., and Hine, K. R. (1981). *Br. Med. J.* **283,** 942–944.

Hatch, K. D., Mann, H. J., Jr., Boots, L. R., Tauxe, W. N., Shingleton, H. M., and Buchina, E. S., Jr. (1980). *Gynecol. Oncol.* **10,** 253–261.

Holden, S., Bernard, O., Artzt, K., Whitmore, W. F., Jr., and Bennett, D. (1977). *Nature (London)* **270,** 518–520.

Javadpour, N. (1980). *Cancer* **45,** 1755–1761.

Javadpour, N., and Chen, H.-C. (1981). *J. Urol.* **126,** 176–178.

Javadpour, N., Utz, M., and Soares, T. (1980). *J. Urol.* **125,** 615–616.

Javadpour, N., Kim, E. E., Deland, F. H., Salyer, J. R., Shah, U., and Goldenberg, D. M. (1981). *JAMA, J. Am. Med. Assoc.* **246,** 45–49.

Knowles, B. B., Aden, D. P., and Solter, D. (1978). *Curr. Top. Microbiol. Immunol.* **81,** 51–53.

Kohn, J. (1979). *In* "Carcino-Embryonic Proteins" (F.-G. Lehmann, ed.), Vol. II, pp. 383–386. Elsevier/North Holland Biomedical Press, Amsterdam.

Lange, P. H., and Raghavan, D. (1983). *In* "Testis Tumors" (J. P. Donohue, ed.). Williams & Wilkins, Baltimore, Maryland (in press).

Lange, P. H., McIntire, K. R., and Waldmann, T. A. (1980a). *In* "Testicular Tumors: Management and Treatment" (L. H. Einhorn, ed.), pp. 69–81. Masson, New York.

Lange, P. H., Nochomovitz, L. E., Rosai, J., Fraley, E. E., Kennedy, B. J., Bosl, G., Brisbane, J., Catalona, W. J., Cochran, J. S., Comisarow, R. H., Cummings, K. B., deKernion, J. B., Einhorn, L. H., Hakala, T. R., Jewett, M., Moore, M. R., Scardino, P. T., and Streitz, J. M. (1980b). *J. Urol.* **124,** 472–478.

Lange, P. H., Bremner, R. D., Horne, C. H. W., Vessella, R. L., and Fraley, E. E. (1980c). *Urology* **15,** 251–255.

Lange, P. H., Millan, J., Stigbrand, T., Ruoslahti, E., and Fishman, W. H. (1981). *Proc. Soc. Oncodevel. Biol. Med.* **1,** P49.

Lange, P. H., Vogelzang, N. J., Goldman, A., Kennedy, B. J., and Fraley, E. E. (1982a). *J. Urol.* **128,** 708–711.

Lange, P. H., Vogelzang, N. J., Fraley, E. E., Kennedy, B. J., Deland, F. H., and Goldenberg, D. M. (1982b). *77th Annu. Meet. Am. Urol. Assoc., 1982.*

Lange, P. H., Vogelzang, N. J., Anderson, R. W., Fraley, E. E., and Allen, D. W. (1982c). *Proc. Am. Soc. Clin. Oncol.* Abstract C-267.

Lange, P. H., Millan, J.-L., Stigbrand, T., Vessella, R. L., Ruoslahti, E., and Fishman, W. H. (1982d). *Cancer Res.* **42,** 3244–3247.

Mach, J.-P., Carrel, S., Forni, M., Ritschard, J., Doneth, A., and Alberto, P. (1980). *N. Engl. J. Med.* **303,** 5–10.

McIlhinney, R. A. J. (1981). *Int. J. Androl., Suppl.* **4,** 93–110.

Mann, K., Lamerz, R., Hellman, T., Jumper, H. J., Staehler, G., and Karl, H. J. (1981). *Oncodev. Biol. Med.* **1,** 301–305.

Moshakis, V., McIlhinney, R. A. J., Raghavan, D., and Neville, A. M. (1981). *J. Clin. Pathol.* **34,** 314–319.

Nørgaard-Pedersen, B. (1976). *Scand. J. Immunol., Suppl.* **4,** 7–45.

Nørgaard-Pedersen, B., and Raghavan, D. (1980). *Oncodev. Biol. Med.* **1,** 327–358.

Papapetrou, P. D., Sakarelou, N. P., Braouzi, H., and Fessas, P. H. (1980). *Cancer* **45,** 2583–2592.

Ransohoff, D. F., and Feinstein, A. R. (1978). *N. Engl. J. Med.* **299,** 926–930.

Richardson, R. L., Schoumacher, R. A., Fer, M. F., Hande, K. R., Forbes, J. T., Oldham, R. K., and Greco, F. A. (1981). *Ann. Intern. Med.* **94,** 181–186.

Rosen, S. W., Javadpour, N., Calvert, I., and Kaminska, J. (1979). *JNCI, J. Natl. Cancer Inst.* **62,** 1439–1441.

Ruoslahti, E., and Hirai, H. (1978). *Scand. J. Immunol., Suppl.* **8,** 3–26.

Scardino, P. T., and Skinner, D. G. (1979). *Surgery (St. Louis)* **86**, 86–94.
Smith, C. J. P., and Kelleher, P. C. (1980). *Biochim. Biophys. Acta* **605**, 1–32.
Solter, D., and Knowles, B. B. (1978). *Proc. Natl. Acad. Sci. U.S.A.* **75**, 5565–5569.
Szymender, J. J., Zborzil, J., Sikorowa, L., Kaminska, J. A., and Gadek, A. (1981). *Oncology* **38**, 222–229.
Thompson, D. K., and Haddow, J. E. (1979). *Cancer* **43**, 1820–1829.
Vaitukaitis, J. L. (1979a). *N. Engl. J. Med.* **301**, 324–326.
Vaitukaitis, J. L. (1979b). *In* "Carcino-Embryonic Proteins" (F.-G. Lehmann, ed.), Vol. II, pp. 447–456. Elsevier/North-Holland Biomedical Press, Amsterdam.
Vaitukaitis, J. L., and Ebersole, E. R. (1976). *J. Clin. Endocrinol. Metab.* **42**, 1048–1055.
Vaitukaitis, J. L., Ross, G. T., Braunstein, G. D., and Rayford, P. L. (1976). *Recent Prog. Horm. Res.* **32**, 289–331.
Vogelzang, N. J., Lange, P. H., Bosl, G. J., Fraley, E. E., Johnson, K., and Kennedy, B. J. (1980). *Proc. Am. Assoc. Cancer Res.* **21**, 431.
Vogelzang, N. J., Lange, P. H., and Goldberg, E. (1982). *Oncodev. Biol. Med.* **3**, 269–272.
Waldmann, T. A., and McIntire, K. R. (1974). *Cancer* **34**, 1510–1515.
Wepsic, H. T., and Kirkpatrick, A. (1979). *Gastroenterology* **77**, 777–796.
Yoshimoto, Y., Wolfsen, A. R., and Odell, W. D. (1977). *Science* **197**, 575–577.

CHAPTER 14

Biochemical Markers of Human Small (Oat) Cell Lung Carcinoma— Biological and Clinical Implications

STEPHEN B. BAYLIN
Oncology Center, Department of Medicine
The Johns Hopkins University School of Medicine
Baltimore, Maryland

I.	Introduction	259
II.	Biology of Human SCC	260
	A. Biochemical Markers Associated with SCC	260
	B. Questions Concerning the Differentiation Lineage of Human SCC	261
	C. Cell Cultures of Human Lung Cancer and Implications for the Histogenesis of SCC	266
III.	The Biochemistry of SCC and the Implications for Clinical Behavior	270
	A. Implications of SCC Biochemistry for the Classification of Human Lung Cancers	271
	B. Possible Implications for the Treatment of SCC and Non-SCC Lung Cancer	273
IV.	Summary	274
	References	275

I. INTRODUCTION

The last several years have seen an increasing interest, both among biologists and clinicians, in the biochemistry of the human lung cancer, small (oat) cell carcinoma (SCC). For biologists, the markers produced by this neoplasm raise pivotal questions concerning the position of human endocrine cells in the differentiation of epithelial cells (Baylin and Mendelsohn, 1980; Gazdar *et al.*, 1981a; Pearse and Takor-Takor, 1979). Among clinicians, the exciting advances made in the treatment of this tumor by combination chemotherapy (Greco and Oldham, 1979, 1981) have prompted the search for biochemical markers which would allow for precise monitoring of tumor mass in patients with SCC. These multidisciplinary areas of investigation are closely intertwined, and the quest to elucidate the precise histogenesis of SCC goes hand in hand with attempts to improve the clinical management of this disease.

The purpose of the present chapter is to review current thinking regarding differentiation relationships between SCC (and its related endocrine biochemical activity) and bronchial epithelial cells (normal and neoplastic) which lack endocrine features. Implications of these differentiation relationships for the clinical behavior of SCC will be explored. The status of biochemical markers for clinical use in patients with SCC will also be discussed. Finally, some directions for future research into the biology and clinical aspects of SCC will be examined.

II. BIOLOGY OF HUMAN SCC

A. Biochemical Markers Associated with SCC

Much of the recent focus on SCC biochemistry has involved the long appreciated propensity for this tumor to engage in endocrine activity. SCC was among the first tumors for which it was recognized that secretion of hormones from the neoplastic tissue was responsible for the appearance of endocrine syndromes. Thus, overactivity of the cortex of the adrenal gland secondary to tumor production of ACTH (Brown, 1928; Meador *et al.*, 1962) and altered renal function due to tumor secretion of antidiuretic hormone (Amatruda *et al.*, 1963; Schwartz *et al.*, 1960) were among the first so-called "ectopic tumor syndromes" described. The fact that SCC, among all human cancers, has the highest incidence for producing such peptide hormone mediated syndromes led to the realization, at a histologic and biochemical level, that the cells in this neoplasm often have neuroendocrine characteristics (Bensch *et al.*, 1968; Hattori *et al.*, 1972). A long list of peptide hormones which have been identified either directly in SCC tissue or in the circulation of patients with SCC has now been compiled (see Table I).

Of course, the biochemical activity of SCC is not limited to endocrine expression. Other markers (Table I), most of which are found in other types of human cancers, are also associated, to a variable extent, with SCC. For example, high levels of carcinoembryonic antigen (CEA) have been described in the circulation of patients with SCC (Waalkes *et al.*, 1980). Likewise, specific isoenzyme forms of creative kinane (CK) have been found to characterize SCC lung carcinomas as opposed to non-SCC lung tumors (Gazdar *et al.*, 1981b). It remains, however, the endocrine activity of the tumor which continues to receive

TABLE I

Some Peptides Identified in Human Small (Oat) Cell Lung Cancer[a]

A. Peptide hormones
   1. ACTH (Azzopardi and Williams, 1968; Meador et al., 1962; Yalow, 1979)
   2. β-Lipotropin (Odell et al., 1979).
   3. β-Endorphin (Berger et al., 1981; Bertagna et al., 1978)
   4. ADH (Amatruda et al., 1963; Bower et al., 1964)
   5. Bombesin (Erisman et al., 1982; Moody et al., 1981; Wood et al., 1981)
   6. Calcitonin (Abe et al., 1977; Baylin et al., 1978; Hansen and Hummer, 1979; Roos et al., 1980; Silva et al., 1974)
   7. Somatostatin (Wood et al., 1981; Roos et al., 1981)
   8. Neurotensin (Wood et al., 1981)
B. Some other protein markers
   1. L-Dopa decarboxylase activity (Baylin et al., 1978, 1980; Gazdar et al., 1980)
   2. Creatine phosphokinase activity (BB isoenzyme form) (Gazdar et al., 1981b)
   3. Histaminase (diamine oxidase) activity (Baylin et al., 1975, 1978)
   4. Carcinoembryonic antigen (CEA) (Waalkes et al., 1980)
   5. Neuron-specific enolase (Marangos et al., 1982; Tapia et al., 1981)

[a] The listing in this table describes some of the hormones and other proteins most commonly found in SCC tissues. The references given are not meant to be all inclusive or to attempt in each case to define the first workers to describe each hormone or other protein in SCC. Rather, selected papers which discuss each marker are provided.

the most attention. This biochemical feature has suggested a specific cell of origin for SCC and a means to provide a battery of biomarkers for monitoring the course of this disease.

B. Questions Concerning the Differentiation Lineage of Human SCC

Attempts to explain the presence of neuroendocrine differentiation features in human SCC involve critical questions in the complex area of epithelial cell differentiation. Specifically, the resolution of the precise cellular origin of SCC could shed light on the pathways involved in the development of endocrine cells in general. Studies of human SCC may, in fact, provide one of the most important models for investigations of this aspect of cellular differentiation.

Three major types of differentiated lung cancer (SCC, squamous cell carcinoma, and adenocarcinomas) arise from the complex series of cells which populate the bronchial mucosa (Carter and Eggleston, 1980). A fourth form, large cell carcinoma, is considered an undifferentiated type of lung cancer. Each of the differentiated forms, in their purest histologic state, carry differentiation features of the different types of normal cells which constitute this epithelial cell system (Fig. 1). Thus, squamous cell carcinomas express features of normal

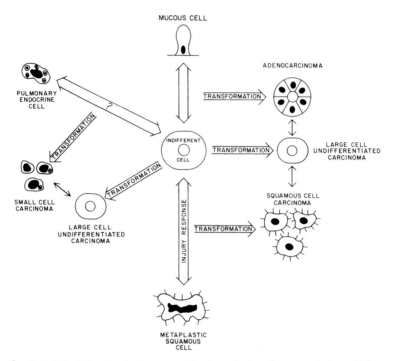

Fig. 1. Potential relationships between normal and neoplastic cell types in the bronchial mucosa. Interactions between cells with and without neuroendocrine features (normal and neoplastic) are postulated. A stem cell element (the indifferent cell) is depicted as capable of differentiation toward metaplastic squamous cells (during injury response) and to mucous cells (during injury and/or normal differentiation). The bidirectional arrows indicate that indifferent cells may possibly be derived from the mucous and/or metaplastic cells. These relationships are based on observations of McDowell *et al.* (1979) concerning repopulation of bronchial epithelium after injury. Thus, formation of adeno- and squamous carcinomas is depicted as occurring in a closely related population of cells along a spectrum of differentiation stages. This postulation fits data from several investigators (McDowell *et al.*, 1979; Yesner, 1978) and is consistent with the frequent recognition of mixed adeno- and squamous tumors (Carter and Eggleston, 1980). A form of large cell undifferentiated cancer may then derive directly from transformation of "indifferent cells" and be closely related through differentiation steps (including those ongoing in the neoplasm) to adeno- and squamous carcinomas. This postulation fits the long held hypothesis that a form of large cell undifferentiated lung cancer is closely related to adeno- and/or squamous carcinoma. Data discussed in the text underlie the depicted relationships among pulmonary endocrine cells, SCC, and the nonendocrine bronchial mucosal cells including non-SCC tumors. The question marks raise the issue as to whether the same "indifferent" stem cell can either be generated from cells with endocrine features or accounts for formation of the endocrine and nonendocrine cell elements, and whether the form of large cell undifferentiated cancer (left side of figure) which derives from SCC *in vitro* (and probably *in vivo*) is of the same lineage as the large cell undifferentiated carcinoma related to adeno and squamous carcinomas (right side of figure).

bronchial epithelial cells which have been injured or damaged and have responded with the formation of so-called "squamous metaplasia" (McDowell et al., 1978). Lung adenocarcinomas resemble the mucin-producing mucous cells of the bronchial epithelium (McDowell et al., 1978). SCC cells, with their neuroendocrine features, bear resemblance to the sparse numbers of endocrine cells which are found in the human bronchial epithelium and in the bronchial mucosa of other mammalian species as well (Cutz et al., 1975; Feyrter, 1953; Capella et al., 1978; Terzakis et al., 1972). This variable resemblance of SCC to normal endocrine cells has raised the issue that this tumor represents a completely separate biologic entity from the other types of human lung cancer. Several pieces of evidence support this hypothesis.

First, the presence of a variable number of neurosecretory granules in the cytoplasm of SCC (Bensch et al., 1968; Hattori et al., 1972) and the presence of L-dopa decarboxylase activity in this tumor (Baylin et al., 1978; Baylin and Mendelsohn, 1980) are consistent with the inclusion of SCC among the so-called "APUD" group of endocrine cells as defined by Pearse and his colleagues (Pearse, 1969). The non-SCC types of human lung cancers classically are not thought to contain these neuroendocrine differentiation features.

Second, a specific cytogenetic abnormality has recently been described for SCC, as opposed to non-SCC, lung cancer cells. Deletions in the short arm of chromosome 3 appear to be a hallmark of SCC cells, both *in vivo* and in culture (Whang-Peng et al., 1982).

Third, the clinical behavior of SCC is distinctly different from that for the classic forms of non-SCC lung cancers. SCC tends to metastasize early and widely (Bunn et al., 1977; Cohen and Matthews, 1978; Greco and Oldham, 1979). Also, this tumor demonstrates an initial sensitivity to radiation and/or chemotherapy which is much greater than that generally found for the non-SCC tumors (Cohen and Matthews, 1978; Greco and Oldham, 1979).

All of these above features have rightly led to the concept that SCC is a separate biological and clinical entity from the other types of human lung cancers. It is attractive then to speculate that a separate stem cell of origin might underlie the formation of this tumor. Indeed, in one important hypothesis for the histogenesis of SCC, the endocrine cell of the human bronchus is envisioned as the origin for this neoplasm. In turn, investigators have speculated that the neuroendocrine characteristics of both the normal endocrine elements of the human bronchial mucosa and of SCC might indicate that these cells have a separate embryologic origin in the neural crest. This neural origin has also been proposed for endocrine cells located in other epithelial systems such as the intestinal tract and the pancreatic islets (Pearse, 1969; Pearse and Polak, 1971).

While a separate neural origin for SCC has not been ruled out by experimental data, there are a number of factors which suggest against this and which, in reality, suggest a much more complex interaction between normal and neoplastic endocrine cells and nonendocrine epithelial cells in the bronchial mucosa. First, the number of endocrine cells in the normal human bronchus is sparse (Capella *et al.*, 1978; Cutz *et al.*, 1975; Terzakis *et al.*, 1972). Direct origin of SCC from such cells would not explain the very high incidence (20%) of SCC among the major human lung cancers (Greco and Oldham, 1981; Weiss, 1981). It is certainly unlikely that SCC arises directly from a mature, terminally differentiated endocrine cell. Origin from a stem cell, which retains proliferative capacity and is committed to endocrine differentiation, is possible. Whether such a stem cell has a separate neural origin or, as discussed below and in Fig. 1, could be endodermally derived is a critical question.

Second, our laboratory group and others have recognized that neuroendocrine-related biomarkers separate SCC from non-SCC lung cancers only quantitatively, and not qualitatively. Perhaps, most disturbing for the separate neural crest theory of origin for SCC is the fact that non-SCC lung cancer tissues can contain high amounts of L-dopa decarboxylase activity (Baylin and Mendelsohn, 1980; Berger *et al.*, 1981). This enzyme is the "D" in the APUD acronym and designates that a portion of the biogenic amine synthesis pathway is present. This property of amine precursor decarboxylation is thought to characterize peptide hormone-secreting cells which could be of neural origin as suggested by Pearse and colleagues (Pearse, 1969). While levels of L-dopa decarboxylase (DDC) are certainly higher in SCC tissues as a population versus non-SCC lung cancer tissues, much overlap is apparent, and some non-SCC lung cancers, particularly adenocarcinomas, can have levels as high as the top values in SCC tissues (Berger *et al.*, 1981). Similarly, some of these adenocarcinoma lesions have cells which stain positively for the presence of neurosecretory-type granules (Berger *et al.*, 1981), another property thought to characterize endocrine cells. These morphological characteristics of endocrine differentiation have been well recognized in non-SCC lung tumors by other investigators (McDowell *et al.*, 1981). Thus, the biogenic amine synthesis feature of cells with neuroendocrine differentiation is not restricted to SCC among the major types of human lung cancer.

Third, a number of investigative groups have recognized that, by immunoassay determinations, concentrations of small polypeptide hormones are not restricted to SCC among the human lung cancers. Thus, immunoreactivities for ACTH (Gewirtz and Yalow, 1974), calcitonin (Becker *et al.*, 1978; Berger *et al.*, 1981; Roos *et al.*, 1980; Schwartz *et al.*, 1979), β-endorphin (Berger *et al.*, 1981), and β-lipotropin (Odell *et al.*, 1979) have been found to be equally distributed throughout the major forms of human lung cancer. The peptide hormone component of the APUD concept is apparently not restricted to SCC among the major types of human lung neoplasms.

Fourth, pathologists are increasingly recognizing that single lesions of human lung cancers may simultaneously contain both SCC and non-SCC histologic features (Abeloff et al., 1979; Abeloff and Eggleston, 1981; Carter and Eggleston, 1980; Yesner, 1978). Furthermore, it is now well recognized that a percentage of patients with an initial diagnosis of SCC may, after treatment failure, have completely non-SCC tumor histology at autopsy (Abeloff et al., 1979; Abeloff and Eggleston, 1981; Brereton et al., 1978; Matthews, 1979). In these non-SCC tumors, the endocrine-related biochemistry is generally much lower than that found in classic SCC tissues (Abeloff et al., 1979; Abeloff and Eggleston, 1981). It is then apparent that SCC and non-SCC differentiation can be ongoing in the same tumor lesion. Whether such mixed tumors consist of cells which originate in separate primary lesions or whether the same group of cells simultaneously expresses two pathways of differentiation is not yet known. However, the characteristics of these mixed tumors suggest to pathologists that, in many instances, separate tumor origins would not be the best explanation for the presence of the mixed populations of cells. The progressive appearance of different cell types, even in tumors of single clone origin, is now well appreciated (for review, see Nowell, 1976).

Finally, studies of endocrine cells in nonpulmonary epithelial systems now suggest that a separate neural crest origin may not best explain the presence of "APUD" features of such cells. Studies by Andrew (1975, 1976) and by Fontaine and LeDouarin (1977) have failed to document a neural crest origin for pancreatic islet cells and intestinal mucosal endocrine cells. Similarly, Pictet et al. (1976) have shown that extirpation of neural crest during early embryonic development of the chick fails to prevent the appearance of pancreatic islet cells which produce insulin. It is the consensus of this group of investigators that the endocrine cells of the pancreas and the gastrointestinal tract do not arise in the neural crest. Similarly, Cheng and LeBlond (1974) suggest that a common precursor stem cell, of endodermal origin, in the intestine actually gives rise to both the differentiated endocrine and nonendocrine cells in the intestinal mucosa. Sidhu (1979), from analyses of morphologic studies, has hypothesized that pulmonary endocrine cells, likewise, have such an endodermal origin.

In summary, then, there appears to be a mounting body of evidence which suggests that human SCC may be more closely related to the non-SCC lung cancers than data of the past 10 to 15 years might have suggested. The possibility that the same endodermal cell lineage gives rise to both normal and neoplastic bronchial cells with neuroendocrine features and to the nonendocrine differentiated bronchial epithelial cells is strongly suggested. This is an important biologic consideration which deals with fundamental aspects of epithelial cell differentiation. Documentation of the precise relationships between the different types of bronchial epithelial cells could shed much light on pathways involved in the formation of endocrine cells. Studies of human lung cancers appear to provide a promising model for such investigations.

C. Cell Cultures of Human Lung Cancer and Implications for the Histogenesis of SCC

Recently, *in vitro* work with well-established cultures of human lung carcinomas has been shedding further light on possible relationships between SCC and non-SCC lung cancers. Many of the results obtained to date appear to have strong parallelisms to the *in vivo* events discussed above. It is thus worth detailing some of the more recent data and their implications for the biology of SCC.

1. Biochemistry of Human Lung Cancer in Culture

SCC and non-SCC human lung cancers seem to be more biochemically distinct *in vitro* than *in vivo*. Thus, in a series of studies with well-established culture lines, a number of neuroendocrine-related biomarkers distinguish SCC from the non-SCC neoplasms (Baylin *et al.*, 1980; Baylin and Gazdar, 1981; Gazdar *et al.*, 1980; Pettengill and Sorenson, 1981). Table II summarizes a list of endocrine-related biochemical parameters which have been found in high levels in cultures of SCC and in much lower levels in the non-SCC lung cancer lines. Three markers have been particularly impressive in this regard. First, levels of L-dopa decarboxylase (DDC) are generally very high in SCC culture lines especially during the early passages (Baylin *et al.*, 1980; Gazdar *et al.*, 1980). Presence of high activity of this key component of the "APUD" concept seems to be a hallmark feature of SCC in culture, and correlates with the presence of other neuroendocrine differentiation features such as cytoplasmic neurosecretory granules (Gazdar *et al.*, 1980; Pettengill and Sorenson, 1981).

Second, high levels of the peptide hormone, bombesin, have recently been found to characterize multiple lines of SCC in culture (Moody *et al.*, 1981). Although, as for DDC, concentrations of this hormone seem to vary over a wide range among multiple cultures, significant levels have been found in each SCC

TABLE II

Endocrine-Related Biochemistry and Ultrastructure in Established Cultures of Human Lung Cancer[a]

Tumor type	Markers			
	L-Dopa decarboxylase	Neurosecretory granules	Formaldehyde-induced fluorescence (biogenic amines)	Small polypeptide hormones
SCC	High	Present	Present	Frequent
Non-SCC	Low to absent	Absent	Absent	More variable

[a] Data derived from Baylin *et al.*, 1980; Gazdar *et al.*, 1980; Pettengill and Sorenson, 1981.

line investigated to date (Moody *et al.*, 1981). Continued studies must be done to see if bombesin may be the most constant peptide hormone marker for SCC and thus provide further clues to the cell of origin for this neoplasm. High levels of bombesin and positive immunostaining have also been found in tissue samples from patients with SCC (Moody *et al.*, 1981; Wood *et al.*, 1981) and in nude mouse heterotransplants of this tumor (Erisman *et al.*, 1982).

Third, in tissues from patients with SCC, and especially in culture lines, levels of neuron-specific enolase activity are high (Marangos *et al.*, 1982; Tapia *et al.*, 1981). This enzyme is present at specific stages of neural development and in APUD endocrine cells (Marangos *et al.*, 1982; Tapia *et al.*, 1981).

In summary, the biochemistry of SCC in culture documents the neuroendocrine features of this human neoplasm. The conditions of cell culture appear initially to favor retention of neuroendocrine expression by cells cultured from typical SCC lesions. Hence, there appears to be a wider separation between SCC and non-SCC lung cancer in culture in terms of levels of neuroendocrine-related biochemical expression. The culture lines of SCC established by several laboratories appear to provide a unique resource for studying neuroendocrine biochemistry in human tumor cells. Events in neuroendocrine differentiation, such as the packaging and processing of peptide hormones, as well as the genetic events resulting in the expression of small polypeptide hormones and biogenic amine synthesis may be clarified through investigations of these important neoplastic cells.

2. Evidence for Transition between SCC and Large Cell Undifferentiated Lung Cancer in Culture

While the culture lines discussed above suggest a rather clear separation between SCC and non-SCC lung cancer cells *in vitro*, they have also provided recent data which further links SCC and non-SCC lung cancers. Gazdar and colleagues (1981a) have observed that established lines of SCC, with time (average equals 22 months), demonstrate a tendency to lose neuroendocrine features. Thus, concentrations of L-dopa decarboxylase activity and of several polypeptide hormones decrease with time, and the loss of these properties correlates with a decrease of neurosecretory granules in the cytoplasm and loss of positive stains for formaldehyde-induced fluorescence (Gazdar *et al.*, 1981a). This diminished endocrine expression is accompanied by a distinct change in histology of heterotransplants grown from the cultured cells in nude mice (Gazdar *et al.*, 1981a). The early culture passages, which maintain endocrine features, form classic SCC tumors, while the late passage cells, which lack neuroendocrine features, form typical large cell undifferentiated lung cancers (the fourth major classification of lung cancer in humans).

The above changes in culture thus involve a dramatic change in tumor histology. Recent data indicate, however, that the loss of neuroendocrine features may precede a frank change from SCC to non-SCC histology *in vitro* and that such losses of endocrine activity may herald important biologic changes in SCC cells in culture. In one established line of SCC, OH1, our laboratory has noted a twelvefold drop in L-dopa decarboxylase activity which occurs rather suddenly after approximately 16 months of culture (Goodwin and Baylin, 1982). Neurosecretory granules are lost from the cytoplasm concomitantly with this drop in DDC. These changes in neuroendocrine differentiation are associated with only a subtle morphologic alteration in the cultured cells. There is a moderate decrease in cell–cell adhesion within the suspended aggregates of cells which constitute the typical growth pattern for line OH1 (Luk *et al.*, 1981) and for multiple other established lines of SCC (Gazdar *et al.*, 1980, 1981a). Cytologic analyses of the early and late passage cells detect no diagnostic differences, however. Also, heterotransplants in the nude mouse show both cell types to form SCC histology. Most importantly, however, *in vitro* studies of radiation sensitivity (Goodwin and Baylin, 1982) show a marked relative resistance in the late passage cells as compared to early passage cells (Fig. 2).

Thus, changes in the differentiation status of SCC cells, as monitored by biochemical and ultrastructural features of neuroendocrine differentiation, may have profound implications for the behavior of this tumor and even its response to therapy. These clinical implications are discussed in Section III below.

Do the *in vitro* transitions seen between SCC and a form of large cell undifferentiated lung cancer have precise counterparts in the *in vivo* growth patterns of these tumors? While this remains to be proven, there are important parallelisms between the culture data and *in vivo* data which cannot be ignored. First, a number of investigators have recently commented upon the findings of mixed SCC and large cell tumors in patients (Abeloff *et al.*, 1979; Matthews and Hirsch, 1981). Second, recent cell culture data from our own laboratory (Goodwin *et al.*, 1983) firmly suggest that a form of large cell undifferentiated lung cancer is closely related to SCC *in vivo*. We established a culture line of cells (OH2) from a patient with histologically proven SCC whose autopsy tissues, despite the SCC histology, lacked all of the usual endocrine markers which we have come to associate with SCC (Berger *et al.*, 1981). Cells placed into culture from pleural fluid immediately displayed non-SCC features *in vitro*. Instead of growing as suspended cell aggregates, the cells formed monolayers. Biochemical markers such as DDC were lacking, and heterotransplants of the tumors in nude mice showed large cell undifferentiated cancer.

Although the details of studies with the OH2 cells are to be published elsewhere, there is strong molecular evidence that establishes the SCC lineage of these cells. Both metabolic labeling studies with [^{35}S]methionine and cell surface labeling studies with radioiodine indicate that key SCC proteins (Baylin *et al.*, 1982) are clonally retained by these cells (Goodwin *et al.*, 1983). The

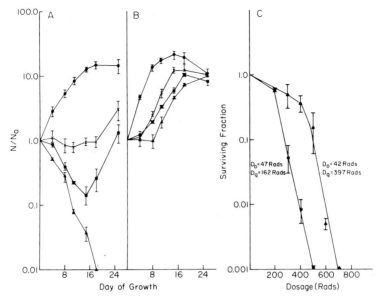

Fig. 2. Response of the cultured human small cell lung carcinoma cells (OH1) to γ irradiation. Growth curves of early (A) and late (B) passage cells following irradiation are represented as follows: control, ●; 300 rads, x; 400 rads, ■; and 500 rads, ▲. Brackets represent standard errors of the mean ($n = 2$ to 4 for each point). Dose–response curves (C) were constructed (early passage cells, ●; late passage cells, ▲) by calculating surviving fraction values from back extrapolations to day 0 from the exponential regrowth portions of multiple growth curves including those shown in (A) and (B) and identical types of curves for 200 and 600 rads (not shown). For 500 rads, in early passage cells, which produced total cell kill, the surviving fraction was assigned a value of 0.001. Brackets for each point depict the SEM ($n = 2$ to 3 except $n = 1$ for 700 rads). The D_q value was estimated as the radiation dose corresponding to the point of intersection between a line extrapolated from the linear portion of the survival curves and a survival fraction of 1.0. The D_o value reflects the slope of the exponential portions of the curves in (C), and represents the dose increment necessary to kill 63% of the existing cell population. (From Goodwin and Baylin, 1982, by permission of *Cancer Research*.)

retention of these proteins, which are not found on the surfaces of non-SCC lung cancer cells in culture, firmly suggest that line OH2 represents a large cell undifferentiated cancer of SCC origin.

In summary, there are now important *in vitro* data to suggest that a form of large cell undifferentiated lung cancer may be closely related to SCC. The precise nature of the transition between these two tumor types remains to be clarified. However, these *in vitro* data, coupled with the *in vivo* data discussed in above sections, strengthen the hypothesis that SCC and non-SCC lung cancer cells may be related to one another within a spectrum of cellular differentiation in the bronchial mucosa. Such a working hypothesis to outline the known events and the questions raised is depicted in Fig. 1. In this scheme, large cell undifferentiated lung cancer is depicted as a potential stem cell for the three differentiated types of human lung cancer. It remains to be determined whether there might be a separate large cell undifferentiated cancer related to SCC and another to adeno- and squamous lung carcinomas.

The proposal of a unitary cell of origin theory for the major forms of lung cancer is not a new one. Several investigators have suggested that differentiation links exist between each of the tumors. For example, Yesner (1978) has proposed that a form of large cell undifferentiated carcinoma may link SCC to the other differentiated lung neoplasms. He has speculated that SCC is, in reality, an undifferentiated form of tumor which may give rise to large cell undifferentiated cancers having the potential to differentiate into either adeno- or squamous carcinomas (Yesner, 1978). While this is an intriguing possibility, especially in light of the *in vitro* data reviewed above, the theory remains unproven as yet.

The link between large cell undifferentiated lung carcinoma and non-SCC differentiated lung cancers (adeno- and squamous carcinomas) is a strong one. The presence of mucin in a significant proportion of large cell undifferentiated lung cancers has led these to be classified as undifferentiated lung adenocarcinomas (Carter and Eggleston, 1980).

Furthermore, data from other investigators have suggested a close relationship between adeno- and squamous lung carcinomas; these tumors may represent two different lines of differentiation which may emerge in a closely related set of parent cells (of which large cell undifferentiated cencer cells are the neoplastic counterpart) (Gazdar *et al.*, 1981a; McDowell *et al.*, 1978). This postulation may well explain why mixed adeno- and squamous lung carcinomas are the most common type of mixed histology among the human lung neoplasms.

The culture lines under discussion may present a situation in which the transition from SCC to large cell undifferentiated cancer and subsequently to the other types of mature differentiated lung neoplasms might be observed. Such studies are ongoing in our laboratory and those of other investigators.

III. THE BIOCHEMISTRY OF SCC AND IMPLICATIONS FOR CLINICAL BEHAVIOR

The above sections have reviewed some of the current thinking about the implications of endocrine biochemistry for the biology of SCC and other lung cancers. While it will take more time to determine the practical implications of the findings for treating patients with SCC, there are a number of important possibilities which merit discussion. Our increased knowledge about the biochemical features of SCC and the relationships of this tumor to the other types of human lung cancer hold promise for developing a more sophisticated use of certain biochemical markers in better classifying the different types of human lung cancer and, perhaps, for monitoring the course of tumor in patients. This section will discuss briefly some of these possibilities and the directions that clinical research may take to further explore them.

A. Implications of SCC Biochemistry for the Classification of Human Lung Cancers

We have discussed the fact that endocrine-related biochemistry seems distributed throughout the spectrum of human lung cancer and that, quantitatively, concentrations of multiple markers are generally higher in SCC tissues (Berger *et al.*, 1981). However, we have pointed out that occasional non-SCC lung cancers can have very high values of multiple endocrine markers (Berger *et al.*, 1981). The implications for the presence of endocrine biochemistry in an individual lung tumor have not yet been defined. However, as we have pointed out in the section on cell culture, the loss of endocrine markers in at least one established culture line of SCC has correlated with the development of *in vitro* radiation resistance.

The development, *in vitro*, of resistance to a defined treatment modality for SCC has important parallelisms to the behavior of the tumor in the host. Although SCC is often initially sensitive tumor to radiation and/or combination chemotherapy (Cohen and Matthews, 1978; Greco and Oldham, 1979), the course of the tumor inevitably involves a return of the cancer and the emergence of treatment resistance. Hence, despite the exciting work with response of SCC to combination chemotherapy, this tumor remains essentially a lethal one with a median life expectancy on treatment of approximately 13 months (Abeloff *et al.*, 1979; Greco and Oldham, 1979; Weiss, 1978). The mechanism(s) for the development of treatment resistance is not known. However, the fact that a certain percentage of SCC tumors appear to develop histologic changes during the course of the illness (Abeloff *et al.*, 1979) which often involve the emergence of non-SCC histology, could be most important. The non-SCC tumors are known to be much more resistant to irradiation and/or combination chemotherapy. A histologic change in this direction might, therefore, impart an element of resistance.

What may be far more important, is that biochemical events in these tumors could define a change from SCC toward non-SCC tumors at a much more subtle level. Hence, the type of change we described in culture, wherein endocrine elements are lost although the cytology of the cells and the histology of nude mouse heterotransplants changes little, could have great bearing on the resistance problem in patients with SCC. The possibility is raised that monitoring of tissue markers and/or circulating markers during the course of a patient's illness might define therapeutically important stages of the disease. In turn, these markers might help predict the sensitivity of tumor tissue at a given point in time.

Furthermore, the question must be raised as to the clinical meaning of high endocrine activity in a non-SCC lung cancer. For example, does a lung adenocarcinoma which has high levels of DDC and of multiple polypeptide hormones have more likelihood to respond to therapy as an SCC tumor or as a typical adenocarcinoma? Again, there are no definitive data to answer this question at present. However, studies are ongoing in our laboratory and those of other investigators to match the findings for biochemistry in tissue and blood with the treatment response of patients with each of the major types of human lung cancer.

Certain caveats must be considered in postulating that endocrine biochemistry might play an important role in classifying lung cancers as to cell population content and potential responsiveness to therapy. The problem of tumor cell heterogeneity is of utmost importance. It is now known that SCC, like many other human and nonhuman tumors, consists of heterogeneous populations of cells (Baylin et al., 1978; Baylin and Gazdar, 1981). Thus, for any given biochemical marker considered, there may be populations of cells which are rich in a given parameter and those which contain little or none of the particular biomarker (Baylin et al., 1978; Baylin and Gazdar, 1981). Furthermore, as tumor progression occurs in the host, there is a dynamic shifting of cell populations such that some markers may appear and others disappear during the course of tumor growth (Baylin et al., 1975). Obviously, the shifting cell populations will have great bearing on the concentration of a given marker at a given time in a patient with SCC. As we, and others, have pointed out, tumor mass may increase in a given patient while a particular tumor marker under study actually declines (Baylin et al., 1975). We must, thus, continue to study the constituent cell populations of SCC and non-SCC lung cancers to further understand their implications for tumor growth and sensitivity to treatment modalities. Combinations of markers may prove far superior to the use of any marker alone; this is evident by the fact that no one marker, for example, quantiatively separates SCC from non-SCC lung cancer, but a battery of markers often gives an index of higher values for the SCC tumors (Berger et al., 1981).

Despite the problems with cell heterogeneity, there is an important possibility that measurements of tissue and circulating endocrine-related biomarkers in patients with lung cancer may provide the following: (a) a useful adjunct to routine histology in precisely classifying human lung cancers—thus, the presence of SCC elements in a non-SCC lung cancer and the absence of neuroendocrine differentiation features in a lung cancer with otherwise typical SCC histology could come to have predictive value for tumor behavior; (b) a way to clarify the "waste basket" category of large cell undifferentiated lung cancers—endocrine markers and recently defined cell surface markers (Baylin et al., 1982; Goodwin et al., 1983) may predict SCC lineage and define a subset of large cell undifferentiated lung cancers which are actually closely related to SCC. Such classification of large cell undifferentiated lung cancers may allow individualized treatment approaches to this tumor based on the differentiation or commitment status of a given neoplasm; (c) a battery of endocrine-related biochemistry to be used as circulating biomarkers which may allow for a more refined monitoring of SCC tumor burden in patients undergoing various treatment protocols.

Continued clinical investigative efforts which take into account the growing interest in the biology of SCC and non-SCC lung cancer should help define the use of biomarkers in SCC. It is my prejudice that continued emphasis on measuring such biomarkers directly in tumor tissue may prove particularly valuable, just as quantitation of the estrogen receptor has helped refine the treatment approach to human breast cancer. The coming years should prove exciting in terms of translating our increased knowledge of the biology of SCC into practical consequences for better treatment of patients with both SCC and non-SCC lung tumors.

B. Possible Implications for the Treatment of SCC and Non-SCC Lung Cancer

In Section II of this chapter, we have placed considerable emphasis on the questions concerning the precise differentiation lineage of SCC to the non-SCC lung cancers. The possibility that SCC tumors can change directly into the other types of tumors has been speculated upon. This type of differentiation link could have implications for the future treatment of lung cancer. There has been a growing interest in the chemical promotion of differentiation among a number of types of tumor cells, and particularly for hematopoietic cell neoplasms (Breitman et al., 1980; Huberman and Callaham, 1979; Lotem and Sachs, 1979; Rovera et al., 1979) and in neuroblastoma cells (Pahlman et al., 1981; Prasad, 1975). The fact that tumor cells can indeed differentiate and respond to various chemical manipulations *in vitro* has always been intriguing for the possibility of exploiting this type of manipulation for *in vivo* therapy. To date, this has, to our knowledge, not been achieved, but it is worth at least considering some possibilities for lung cancer.

SCC tumors, as we have previously discussed, metastasize early and widely. The non-SCC tumors, while less responsive to current chemical treatment modalities, metastasize less widely and tend to grow larger local lesions. Thus, surgery is seldom employed for the treatment of SCC because of the probability of metastases at the time tumor is discovered, while surgery can prove curative, if administered early, for the non-SCC cancers. If, under appropriate situations, the differentiation of SCC could be changed toward that of non-SCC tumors, particularly after initial chemotherapy may have eliminated a large proportion of the SCC present, the possibility could be considered that local excision of remaining lesions might prove beneficial to patients. While such possibilities remain speculative and for future exploration, they should be kept in mind as an increasing number of investigators work with the differentiation of human lung cancer *in vitro*. Again, the monitoring of any treatment procedures designed to influence the differentiation of SCC would have to depend not on histology alone, but on biochemical markers which should help designate the precise status of a given tumor lesion. It is my prejudice, again, that these possibilities are exciting for considering new approaches to the treatment of lung cancer. Hopefully, time and further investigation might justify some of the potential for this type of approach.

IV. SUMMARY

The present chapter has reviewed some aspects of the biochemistry which characterizes human SCC. The implications of present data for questions concerning the position of cells with neuroendocrine features in the differentiation pathways of epithelial cells has been stressed. The possibility that SCC and normal cells with neuroendocrine features in the bronchial mucosa arise within the same cell lineage that produces the nonendocrine types of bronchial epithelial cells has been considered. The use of endocrine-related biochemistry, including new parameters such as cell surface markers, to define the population of cells which constitute SCC and non-SCC lung tumors has been stressed. The possibility that SCC may be linked to the major types of non-SCC lung cancer, through a large cell undifferentiated stem cell, has also been explored. The observation that *in vitro* most SCC cancers, with time, move toward the large cell undifferentiated phenotype with a concomitant loss of endocrine features has been discussed, and the role that such a change would play in the emergence of treatment resistance in the host has also been reviewed.

The clinical implications of the biology of SCC have also been considered. The possibility that measurements of tumor tissue concentrations of endocrine-related biomarkers could form a useful adjunct to histology in deriving therapeutically meaningful classifications of human lung cancer has been suggested. Also, the impact of changes in SCC which involve a loss of endocrine features with time has been stressed for considerations of monitoring patients with measurements of circulating endocrine-related biomarkers. Finally, the potential for using the understanding of cell differentiation within the spectrum of human lung cancers to consider future therapies which might be based on manipulation of the cell populations involved has been briefly speculated upon.

Future investigations of SCC and other human lung cancers should prove to be of critical interest from both a biological and clinical standpoint. Events in cell differentiation, especially as they involve the origins of endocrine cells in nonendocrine epithelial systems, are an integral part of the behavior of these human cancers. Established culture lines of human lung cancer, which appear to manifest some of the differentiation changes which are ongoing *in vivo*, may prove to be critical models for studies of these important questions. Molecular events underlying the genetic control of epithelial cell differentiation may be probed in these cultured cells. The results of these biology investigations should have great impact on the understanding of the clinical aspects of human lung cancer and for the future management of patients with these aggressive neoplasms.

ACKNOWLEDGMENTS

Portions of studies cited were supported by NIH Grant CA-18404, ACS Grant PDT-108, and a gift from the W. W. Smith Foundation.

REFERENCES

Abe, K., Adachi, I., and Miyakawa, S. (1977). *Cancer Res.* **37,** 4190–4194.
Abeloff, M. D., and Eggleston, J. C. (1981). In "Small Cell Lung Cancer" (F. A. Greco, R. K. Oldham, and P. A. Bunn, Jr., eds.), pp. 235–259. Grune & Stratton, New York.
Abeloff, M. D., Eggleston, J. C., Mendelsohn, G., Ettinger, D. S., and Baylin, S. B. (1979). *Am. J. Med.* **66,** 757–764.
Amatruda, T. T., Mulrow, P. J., Gallagher, J. C., and Sawyer, W. H. (1963). *N. Engl. J. Med.* **269,** 544–549.
Andrew, A. (1975). *Gen. Comp. Endocrinol.* **26,** 485–495.
Andrew, A. (1976). *J. Embryol. Exp. Morphol.* **35,** 577–593.
Azzopardi, J. G., and Williams, E. D. (1968). *Cancer* **22,** 274–286.
Baylin, S. B., and Gazdar, A. F. (1981). In "Small Cell Lung Cancer" (F. A. Greco, R. K. Oldham, and P. A. Bunn, Jr., eds.), pp. 123–143. Grune & Stratton, New York.
Baylin, S. B., and Mendelsohn, G. (1980). *Endocrinol. Rev.* **1,** 45–77.
Baylin, S. B., Abeloff, M. D., Wieman, K. C., Tomford, J. W., and Ettinger, D. S. (1975). *N. Engl. J. Med.* **293,** 1286–1290.
Baylin, S. B., Weisburger, W. R., Eggleston, J. C., Mendelsohn, G., Beaven, M. A., Abeloff, M. D., and Ettinger, D. S. (1978). *N. Engl. J. Med.* **299,** 105–110.
Baylin, S. B., Abeloff, M. D., Goodwin, G., Carney, D. N., and Gazdar, A. F. (1980). *Cancer Res.* **40,** 1990–1994.
Baylin, S. B., Gazdar, A. F., Minna, J. D., Bernal, S. D., and Shaper, J. H. (1982). *Proc. Natl. Acad. Sci. U.S.A.* **79,** 4650–4654.
Becker, K. L., Snider, R. H., Silva, D. L., and Moore, C. F. (1978). *Acta Endocrinol. (Copenhagen)* **89,** 89–99.
Bensch, K. G., Corrin, B., and Pariente, R. (1968). *Cancer* **22,** 1163–1177.
Berger, C. L., Goodwin, G., Mendelsohn, G., Eggleston, J. C., Abeloff, M. D., Aisner, S., and Baylin, S. B. (1981). *J. Clin. Endocrinol. Metab.* **53,** 422–429.
Bertagna, X. Y., Nicholson, W. E., Sorenson, G. D., Pettengill, O. D., Mount, G. D., and Orth, D. N. (1978). *Proc. Natl. Acad. Sci. U.S.A.* **75,** 5160–5164.
Bower, B. F., Mason, D. M., and Forsham, P. H. (1964). *N. Engl. J. Med.* **271,** 934–938.
Breitman, T. R., Selonick, S. E., and Collins, S. J. (1980). *Proc. Natl. Acad. Sci. U.S.A.* **77,** 2936–2940.

Brereton, H. D., Matthews, M. M., Costa, J., Kent, H., and Johnson, R. E. (1978). *Ann. Intern. Med.* **88,** 805–806.
Brown, W. H. (1928). *Lancet* **2,** 1022–1023.
Bunn, P. A., Cohen, M. H., and Ihde, D. C. (1977). *Cancer Treat. Rep.* **61,** 333–342.
Capella, C., Hage, E., and Solcia, E. (1978). *Cell Tissue Res.* **186,** 25–37.
Carter, D., and Eggleston, J. C. (1980). "Tumors of the Lower Respiratory Tract," 2nd Ser., Fasc. 17. Armed Forces Inst. Pathol., Washington, D.C.
Cheng, H., and LeBlond, C. P. (1974). *Am. J. Anat.* **141,** 537–562.
Cohen, M. H., and Matthews, M. J. (1978). *Semin. Oncol.* **5,** 234–243.
Cutz, E., Chan, W., and Wong, V. (1975). *Cell Tissue Res.* **158,** 425–437.
Erisman, M. D., Linnoila, R. I., Hernandez, O., DiAugustine, R. P., and Lazarus, L. H. (1982). *Proc. Natl. Acad. Sci. U.S.A.* **79,** 2379–2383.
Feyrter, F. (1953). "Uber die peripheren endokrinen (parakrinen) Drusen des Menschen." Wilhelm Mandrich.
Fontaine, J., and LeDouarin, N. M. (1977). *J. Embryol. Exp. Morphol.* **41,** 209–222.
Gazdar, A. F., Carney, D. N., Russell, E. K., Sims, H. L., Baylin, S. B., Bunn, P. A., Jr., Guccion, J. G., and Minna, J. D. (1980). *Cancer Res.* **40,** 3502–3507.
Gazdar, A. F., Carney, D. N., Guccion, J. G., and Baylin, S. B. (1981a). *In* "Small Cell Lung Cancer" (F. A. Greco, R. K. Oldham, and P. A. Bunn, Jr., eds.), pp. 145–175. Grune & Stratton, New York.
Gazdar, A. F., Zweig, M. H., Carney, D. N., van Steirteghen, A. C., Baylin, S. B., and Minna, J. D. (1981b). *Cancer Res.* **41,** 2773–2777.
Gewirtz, G., and Yalow, R. S. (1974). *J. Clin. Invest.* **53,** 1022–1032.
Goodwin, G., and Baylin, S. B. (1982). *Cancer Res.* **42,** 1361–1367.
Goodwin, G., Shaper, J. H., Abeloff, M. D., Mendelsohn, G., and Baylin, S. B. *Proc. Natl. Acad. Sci. (U.S.A.),* in press, 1983.
Greco, F. A., and Oldham, R. K. (1979). *N. Eng. J. Med.* **301,** 355–358.
Greco, F. A., and Oldham, R. K. (1981). *In* "Small Cell Lung Cancer" (F. A. Greco, R. K. Oldham, and P. A. Bunn, eds.), pp. 353–379. Grune & Stratton, New York.
Hansen, M., and Hummer, L. (1979). *In* "Lung Cancer: Progress in Therapeutic Research" (F. Muggia and M. Rozencweig, eds.), pp. 199–207. Raven Press, New York.
Hattori, S., Matsuda, M., and Tateishi, R. (1972). *Cancer* **30,** 1014–1024.
Huberman, E., and Callaham, M. F. (1979). *Proc. Natl. Acad. Sci. U.S.A.* **76,** 1293–1297.
Lotem, J., and Sachs, L. (1979). *Proc. Natl. Acad. Sci. U.S.A.* **76,** 5158–5162.
Luk, G., Goodwin, G., Marton, L. J., and Baylin, S. B. (1981). *Proc. Natl. Acad. Sci. U.S.A.* **78,** 2355–2358.
McDowell, E. M., Becci, P. J., Barrett, L. A., and Trump, B. F. (1978). *In* "Pathogenesis and Therapy of Lung Cancer" (C. C. Harris, ed.), pp. 445–519. Dekker, New York.
McDowell, E. M., Becci, P. J., Schürch, W., and Trump, B. F. (1979). *JNCI, J. Natl. Cancer Inst.* **62,** 995–1008.
McDowell, E. M., Wilson, T. S., and Trump, B. F. (1981). *Arch. Pathol. Lab. Med.* **105,** 20–28.
Marangos, P. J., Gazdar, A. F., and Carney, D. N. (1982). *Cancer Lett.* **15,** 67–71.
Matthews, M. J. (1979). *In* "Lung Cancer: Progress in Therapeutic Research" (F. Muggia and M. Rozencweig, eds.), pp. 115–165. Raven Press, New York.
Matthews, M. J., and Hirsch, F. R. (1981). *In* "Small Cell Lung Cancer" (F. A. Greco, R. K. Oldham, and P. A. Bunn, Jr., eds.), pp. 35–50. Grune & Stratton, New York.
Meador, C. K., Liddle, G. W., Island, D. P., Nicholson, W. E.,Lucas, C. P., Nuckton, J. G., and Leutscher, J. A. (1962). *J. Clin. Endocrinol. Metab.* **22,** 693–703.
Moody, T. W., Pert, C. B., Gazdar, A. F., Carney, D. N., and Minna, J. D. (1981). *Science* **214,** 1246–1248.

Nowell, P. C. (1976). *Science* **194**, 23–28.
Odell, W. D., Wolfsen, A. R., Bachelot, I., and Hirose, F. M. (1979). *Am. J. Med.* **66**, 631–637.
Pahlman, S., Odelstad, L., Larsson, E., Grotte, G., and Nilsson, K. (1981). *Int. J. Cancer* **28**, 583–589, 1981.
Pearse, A. G. E. (1969). *J. Histochem. Cytochem.* **17**, 303–313.
Pearse, A. G. E., and Polak, J. M. (1971). *Gut* **12**, 783–788.
Pearse, A. G. E., and Takor-Takor, T. (1979). *Fed. Proc., Fed. Am. Soc. Exp. Biol.* **38**, 2288–2294.
Pettingill, O. S., and Sorenson, G. D. (1981). *In* "Small Cell Lung Cancer" (F. A. Greco, R. K. Oldham, and P. A. Bunn, Jr., eds.), pp. 51–77. Grune & Stratton, New York.
Pictet, R. L., Rall, L. B., Phelps, P., and Rutte, W. J. (1976). *Science* **191**, 191–192.
Prasad, K. N. (1975). *Biol. Rev. Cambridge Philos. Soc.* **50**, 129–265.
Roos, B. A., Lindall, A. W., Baylin, S. B., O'Neil, J., Frelinger, A. L., Birnbaum, R. S., and Lambert, P. N. (1980). *J. Clin. Endocrinol. Metab.* **50**, 659–666.
Roos, B. A., Lindall, A. W., Ellis, J., Elde, R., Lambert, P. W., and Birnbaum, R. S. (1981). *J. Clin. Endocrinol. Metab.* **52**, 187–194.
Rovera, G., O'Brien, T. G., and Diamond, L. (1979). *Science* **204**, 868–870.
Schwartz, K. E., Wolfsen, A. R., Forster, B., and Odell, W. D. (1979). *J. Clin. Endocrinol. Metab.* **49**, 438–444.
Schwartz, W. B., Tassel, D., and Bartter, F. C. (1960). *N. Engl. J. Med.* **262**, 743–748.
Sidhu, G. S. (1979). *Am. J. Pathol.* **96**, 5–20.
Silva, O. L., Becker, K. L., and Primack, A. (1974). *N. Engl. J. Med.* **290**, 1122–1124.
Tapia, F. J., Polak, J. M., Barbosa, A. J., Bloom, S. R., Marangos, P. J., Dermody, C., and Pearse, A. G. E. (1981). *Lancet* **1**, 808–811.
Terzakis, J., Sommers, S., and Andersson, B. (1972). *Lab. Invest.* **26**, 127–132.
Waalkes, T. P., Abeloff, M. D., Woo, K. B., Ettinger, D. S., Ruddon, R. W., and Aldenderfer, P. H. (1980). *Cancer Res.* **40**, 4420–4427.
Weiss, R. B. (1978). *Ann. Intern. Med.* **88**, 522–531.
Weiss, W. (1981). *In* "Small Cell Lung Cancer" (F. A. Greco, R. K. Oldham, and P. A. Bunn, Jr., eds.), pp. 1–34. Grune & Stratton, New York.
Whang-Peng, J., Kao-Shan, C. S., Lee, E. C., Bunn, P. A., Carney, D. N., Gazdar, A. F., and Minna, J. D. (1982). *Science* **215**, 181–182.
Wood, S. M., Wood, J. R., Ghatei, M. A., Lee, Y. C., O'Shaugnessy, D., and Bloom, S. R. (1981). *J. Clin. Endocrinol. Metab.* **53**, 1310–1312.
Yalow, R. S. (1979). *In* "Lung Cancer Progress in Therapeutic Research" (F. M. Muggia and M. Rozencweig, eds.), pp. 209–216. Raven Press, New York.
Yesner, R. (1978). *Pathol. Annu.* **13**, 217–240.

CHAPTER 15
Tumor Markers in Prostatic Cancer

MANI MENON   WILLIAM J. CATALONA
Division of Urology
Washington University School of Medicine
St. Louis, Missouri

I.	Introduction	279
II.	Serum Markers	283
	A. Acid Phosphatase	283
	B. Carcinoembryonic Antigen (CEA)	289
	C. α-Fetoprotein	290
	D. Lactic Dehydrogenase Isoenzymes (LDH)	290
	E. Creatine Kinase (CK-BB)	290
	F. Ribonuclease	291
	G. Seromucoid	291
	H. Steroid Hormones and Human Chorionic Gonadotropin	291
	I. Prostate-Specific Antigen (PA)	291
III.	Urinary Markers	292
	A. Total and Nonesterified Cholesterol	292
	B. Carcinoplacental Isoenzyme (Regan N)	293
	C. Urinary Carcinoembryonic Antigen	293
	D. Hydroxyproline	293
	E. Isoleucine	294
	F. Spermidine	294
	G. Fibronectin	294
IV.	Prostate Fluid Proteins	295
V.	Leukocyte Adherence Inhibition (LAI) and Leukocyte Migration Inhibition (LMI)	295
VI.	Summary and Conclusions	296
	References	296

I. INTRODUCTION

Prostatic cancer constitutes the second most commonly diagnosed cancer in American males. It was estimated that in 1982, 73,000 new cases of prostatic cancer will be diagnosed, and 23,300 men will die of this disease (Silverberg, 1982). Since the prospects for cure are favorable for patients with localized prostatic cancer, but are poor for those with extraprostatic tumor spread, it is important that prostatic cancer be diagnosed as early as possible in its clinical evolution. The 5-year survival rate for patients with localized prostatic cancer is approximately 70%, but is only about 35% for those with disseminated cancer (American Cancer Society, 1979). The death rate from prostatic cancer has remained remarkably constant over the past 30 years, being 13.5 per 100,000 during the period 1949–1951, and 13.9 per 100,000 during 1974–1976. Because prostatic cancer is a disease of the aged, and because the average age of the population in the United States is increasing, it is likely that the detection and monitoring of prostatic cancer will continue to be an important medical problem. Over the years, several tests have been evaluated as potential means for the early detection and monitoring of patients with prostatic cancer. In this chapter, these tests are reviewed and their clinical utility is evaluated.

The most commonly used staging system for prostate cancer is that described by Whitmore (1956, 1973). The local lesion is staged on the basis of digital rectal examination. Prostatic cancer usually manifests itself as an area of induration in an otherwise soft, compressible prostate gland. In the Whitmore system, stage A refers to incidental or clinically unsuspected carcinoma diagnosed on histologic examination of tissue removed at operation from a gland that was benign to palpation. Stage B refers to a palpable tumor deemed to be confined to the prostatic capsule. Stage C refers to a tumor that has extended to the periprostatic tissues, and stage D refers to a tumor that has metastasized. For the purposes of this discussion, stages A and B will be considered together as *localized* cancer and stages C and D will be considered together as *extensive* cancer.

The relative proportions of the various stages of prostatic cancer and their survival rates based on a recent survey conducted by the American College of Surgeons Commission on Cancer involving over 17,500 patients are shown in Table I (Murphy, 1981). It is of interest that in the 1960's, the Veterans Administration Urological Research Group (VACURG) (1967) reported that less than 25% of prostatic cancers were detected in the localized stages, as compared to 55% in the recent survey, while disease was extensive at the time of detection in over 75% of patients, as compared to 45% in the recent survey. It appears, therefore, that prostatic cancer is currently being detected in its early stages more frequently.

Paradoxically, the detection of all cases of early prostatic cancer can pose a therapeutic dilemma. Clinical and autopsy studies have demonstrated that microscopic foci of cancer can be detected in about 10% of the prostate glands of men aged 55 years and older (Moore, 1935; Rich, 1935). In 1978, there were approximately 20 million men over the age of 55 residing in the United States (U.S.

TABLE I
Relative Incidence and Survival for Prostatic Carcinoma[a]

Stage	Relative proportion (%)	5 Year survival (%)	10 Year survival (%)
A	25	75	55
B	30	65	35
C	15	50	30
D	30	20	5

[a] Calculated from Murphy (1981).

Bureau of the Census, 1978). If 10% of these men had prostatic cancer, the total prevalence of prostatic cancer would be about 2 million cases. Yet, during this time only 64,000 cases of prostatic cancer were detected clinically and only 21,000 deaths could be attributed to prostate cancer. Thus, based on these figures, only 3.2% of all prostate cancers became clinically manifest, and only 1% were fatal. If means were available to detect all of these incidental carcinomas, it would be difficult to determine how best to treat a cancer that would not become clinically manifest during the lifetime of 97% of the patients who have it and would prove fatal in only 1% of those afflicted.* Therefore, the desirability of early detection of prostatic cancer must be coupled with an accurate method of predicting the biological behavior of the tumor, i.e., identifying which tumors will remain dormant and which will progress and metastasize.

With this introduction, we shall examine the role that the measurement of biochemical markers plays in the detection and management of prostatic cancer. Conceptually, markers can be helpful in several ways. They may be instrumental in screening patients, in establishing the diagnosis of cancer, in staging the tumor, in monitoring the response to therapy, and in determining prognosis. Of these, they are least effective in screening for tumor.

The concept of mass screening for a disease by using biochemical tests has lost appeal after the disappointing experience with screening for phenylketonuria. Galen and Gambino (1975) have critically assessed the demanding requirements for screening tests, and Watson and Tang (1980) have specifically addressed the issue of screening for prostatic cancer. Three factors determine the efficiency of a screening test: (1) the sensitivity of the test (true positive ratio, i.e., the proportion of patients with disease who will have a positive test), (2) its specificity (true negative ratio, i.e., the proportion of people without disease who will have a negative test), and (3) the prevalence of disease in the population tested.

*These calculations are based on the incidence rather than the prevalence of prostatic cancer. Figures of prevalence are difficult to obtain, but with modern methods of surveillance, incidence approaches prevalence (Goldberg et al., 1956).

Since tests that are low in sensitivity and specificity will in all probability not even be evaluated clinically, for practical purposes disease prevalence is the most critical factor that determines the effectiveness of a screening test. A hypothetical example will serve to illustrate this concept. Assume that prostate cancer has a prevalence of 10%. A population of 100,000 is screened with a test having a sensitivity of 95% and a specificity of 95%. The following matrix can be constructed.

	T+	T−	Total
D+	9,500	500	10,000
D−	4,500	85,500	90,000
Total	14,000	86,000	100,000

where $D+$ and $D-$ indicate the presence and absence of disease, and $T+$ and $T-$ indicate the positivity and negativity of the test. The predictive value of a positive test (the chance that someone in whom the test is positive actually has prostate cancer) is given by the evaluation

$$P+ = \text{(true positive/total positive)} \times 100$$
$$= [9500/(9500 + 4500)] \times 100 = 68\%.$$

The predictive value for a negative test (the probability that a patient with a negative test does not have cancer) is given by the evaluation

$$P- = \text{(true negative/total negative)} \times 100 = (85,500/86,000) \times 100 = 99\%.$$

Stated differently, a subject with a positive test has a 68% chance of having cancer, and a subject with a negative test has a 99% chance of not having cancer. Let us now use the same test to screen a population where the prevalence of prostate cancer is 35 per 100,000, approaching the actual prevalence of clinically manifest cancer, as stated earlier. Then

	T+	T−	Total
D+	33	2	35
D−	4,998	94,967	99,965
Total	5,031	94,969	100,000

$P+$ becomes a 0.656% and $P-$ becomes 99.998%. Thus, only one of 152 patients with a positive test would actually have cancer. Viewed in this light, the limitations of screening for prostate cancer are readily apparent.

If markers have little value in screening for prostatic cancer, then what exactly can their value be? They can be used to help stage the tumor. It is generally accepted that the treatment for localized prostatic cancer is either radical surgery or radiation therapy, whereas the treatment for extensive disease is palliative endocrine therapy. Yet, up to 30% of patients with apparently localized disease and a negative metastatic workup will have lymph node metastases found at the time of surgical exploration. Patients with regional lymph node metastases have survival rates approaching those of patients with distant metastases. A biochemical marker that would detect microscopic lymph node metastases would obviate surgery in about one-third of the patients currently undergoing exploration. Markers also will be helpful in monitoring the results of therapy. About 70% of patients with stage D prostatic carcinoma will receive substantial palliation with endocrine therapy, whereas 30% will not (Menon and Walsh, 1979). If the clinical outcome of endocrine therapy could be predicted by alterations in levels of tumor markers, alternate forms of therapy such as cytotoxic chemotherapy could be implemented earlier in patients who are expected to fail hormone therapy.

II. SERUM MARKERS

A. Acid Phosphatase

1. Isolation and Localization

Of all the markers studied, the greatest effort has been in the measurement of serum acid phosphatases. Although a group of enzymes hydrolyzing phosphates was first detected in the blood in 1924 (Martland et al., 1924), the clinical implications of the measurement of the activity of these enzymes were neglected until 1938, when Gutman and Gutman demonstrated a marked increase in the enzymatic hydrolysis of organic phosphates at acid pH's (acid phosphatases) in the serum of patients with metastatic carcinoma of the prostate.

It is now known that acid phosphatases are a heterogeneous group of isoenzymes with different substrate specificities and immunochemical properties. Biochemical and immunohistochemical studies demonstrate acid phosphatase in the lysosomal fraction of cells. Acid phosphatases are present in various blood elements as well as parenchymatous organs, such as the spleen, kidney, and liver. The molecular basis for their heterogeneity is the number of sialic acid residues present on the molecule. Human prostatic acid phosphatase is an immologically specific (isoenzyme II) glycoprotein with a molecular weight of about 100,000 (Pontes, 1981). The human prostate also produces a small amount of acid phosphatase isoenzyme IV, which is also produced by leukocytes. Acid phosphatases are produced by prostatic epithelial cells and are secreted into the prostatic fluid in concentrations of 500 to 1000 units/ml. Normally, little acid phosphatase is absorbed into the serum (< 1 unit/ml). Tissue levels of acid phosphatase are high in hypertrophied and in normal prostate, but may be as low as 25% of normal in carcinomatous prostate (Woodard, 1952; Loor *et al.*, 1981). In contrast, serum acid phosphatase levels are elevated in about 75% of patients with metastatic prostatic cancer. Elevations of serum acid phosphatase that occur in prostatic cancer patients are probably due to the fact that cancer cells that have lost their connection with the prostatic ductal system secrete acid phosphatase directly into the circulation. Moreover, prostatic ductal obstruction from an expanding tumor may cause back diffusion of acid phosphatase into the circulation. Serum acid phosphatase levels generally fall with androgen-deprivation therapy, and this fall correlates with clinical evidence of tumor regression. Characteristically, alkaline phosphatase activity rises transiently (flare reaction) and then declines toward normal levels.

2. Techniques for Measuring Acid Phosphatase

Because the initial acid phosphatase isolated by Gutman and Gutman was not specific for the prostate, several methods have been used to improve the tissue specificity. One technique was the use of substrates that are preferentially hydrolyzed by the prostate acid phosphatase and not by other acid phosphatases (Table II). Of the various substrates used, α-naphthyl phosphate, β-glycerophosphate, and thymolphthalein monophosphate appear to be considerably more specific than the original substrate (phenyl phosphate). Nevertheless, all these substrates exhibit a certain degree of nonspecificity. A second method of increasing specificity was the use of organ-specific phosphatase inhibitors. For instance, L-tartrate inhibits the prostatic enzyme while having no effect on the erythrocyte enzyme; whereas, formaldehyde inhibits the erythrocyte enzyme, but not the prostatic enzyme (Abul-Fadl and King, 1949). However, the specificity of these inhibitors is diminished by the fact that acid phosphatases from other sources are also partially inhibited.

During the past decade, immunochemical methods have been developed for measuring prostatic acid phosphatase (Section II,A,5). These include radioimmunoassays and counterimmunoelectrophoresis. These assays are somewhat more sensitive than enzymatic assays and also are more specific for measuring acid phosphatase of prostatic origin, but they do exhibit some degree of cross-reactivity with other acid phosphatases.

3. Collection of Samples

Since prostatic acid phosphatase can be degraded by changes in pH and temperature, conditions for collection and storage of specimens are critical. Maximum stability is obtained at pH 6.2 and refrigeration or freezing. Samples should be collected on ice, serum should be removed, and the specimens should be refrigerated promptly. The heat lability of prostatic acid phosphatase is less of

15. Tumor Markers in Prostatic Cancer

TABLE II
Enzymatic Assay of Acid Phosphatase Activity[a]

Technique	Substrate	Specificity	Unit of activity[b]
Gutman	Phenyl phosphate	None	1.0 mg of phenol per hour at pH of 4.9 (King-Armstrong units)
Bodansky	β-Glycerophosphate	Prostate and other nonerythrocyte enzymes	1.0 mg of phosphorus per hour at pH of 4.9
Bessey–Lowry–Brock	p-Nitrophenyl phosphate	None	1.0 mM p-nitrophenol per hour at pH of 4.9
Huggins–Talalay	Phenophthalein phosphate	None	1.0 mg of phenophthalein per hour at pH of 4.9
Roy	Thymolphthalein monophosphate	Prostate	1.0 μM of thymolphthalein per minute at pH of 6.0 (international units)
Seligman	β-Naphthyl phosphate	None	10 mg of β-naphthol per hour at pH of 4.9
Babson–Read	α-Naphthyl phosphate	Prostate, platelet	1.0 mg of α-naphthol per hour at pH of 5.2

[a] Reproduced with permission from Henneberry et al. (1979).
[b] Quantity of product released from substrate by enzyme activity per unit time.

a problem with radioimmunoassays because they measure the antigenic rather than the enzymatic activity of the molecule, but the counterimmunoelectrophoresis assay relies upon both the enzymatic and antigenic properties of the acid phosphatase molecule. Lipemic serum can produce false positive radioimmunoassays.

4. Clinical Relevance of Acid Phosphatase Measurements

The interpretation of acid phosphatase measurements in an individual patient is complicated by several factors: (1) Not all patients with elevated acid phosphatase levels have prostatic cancer; (2) not all patients with prostatic cancer have elevated acid phosphatase levels; (3) the correlation between elevated acid phosphatase levels and tumor stage is imprecise; (4) there are considerable fluctuations in acid phosphatase levels during a 24-hour period both in prostatic cancer patients and normal men; and (5) diagnostic and therapeutic manipulations such as rectal examination or urethral instrumentation may transiently elevate acid phosphatase levels.

Elevations of acid phosphatase levels can occur in a variety of clinical disorders other than prostatic cancer, including obstructive benign prostatic hyperplasia, prostatic infarction, lymphoreticular diseases, carcinomas of nonprostatic origin, Gaucher's disease, and thromboembolic disease. In fact, in most large clinical studies of acid phosphatase, approximately 10% of various control groups are found to have elevated acid phosphatase levels (Byar, 1977). In a review of results from the Veterans Administration Cooperative Urological Research Group Studies on prostatic cancer, Byar reported that acid phosphatase levels were elevated in 1433 of 4103 (35%) patients with prostatic cancer. Of those with elevated acid phosphatase, 510 patients (36%) had bone metastases and 55 (4%) had soft tissue metastases. Thus, of all patients with elevated acid phosphatase (defined as more than 1.0 King–Armstrong units using the phenyl phosphate substrate) 60% had no evidence of metastases when seen initially. Conversely, acid phosphatase was elevated in 85% of 609 patients with bony metastases and in 65% of 84 patients with soft tissue metastases (82% of patients with either bony or soft tissue metastases). In a subsequent communication, the upper limit of normal acid phosphatase was redefined as 2.0 King–Armstrong (KA) units, because a substantial proportion of patients with values between 1.0 and 2.0 KA units did not develop metastases on follow-up (Byar et al., 1981). With this new definition, about one-half (49%) of patients with elevated acid phosphatase had evidence of metastatic disease, and 72% of patients with metastases had elevated acid phosphatase. The probability of metastases increased as a function of the initial acid phosphatase levels (Fig. 1). Thus, in the VACURG study in patients with documented prostatic cancer, the probability that an elevated acid phosphatase reflected the presence of detectable metastases was 49%, and the probability that a normal acid phosphatase indicated the absence of metastases was 94%.

An elevation of serum acid phosphatase is neither necessary nor sufficient to establish the presence of metastases in patients with known prostatic cancer. As indicated above, a significant proportion (30 to 40%) of patients with obvious metastases have normal acid phosphatases. However, most patients with clinically localized prostatic cancer with elevated acid phosphatases probably do have occult metastases in lymph nodes and/or bones, as is suggested by studies demonstrating a higher incidence of lymph node metastases and poorer prognosis in patients with elevated acid phosphatases as compared to their counterparts with normal acid phosphatases (Ganem, 1956; Freiha et al., 1979; Paulson et al., 1980). In fact, in VACURG study 2, the acid phosphatase levels within the normal range in patients with stages A and B disease correlated with subsequent tumor progression (Byar et al., 1981).

5. Immunoassays for Prostatic Acid Phosphatase

Because conventional acid phosphatases are not completely tissue specific and are elevated primarily in the presence of metastatic disease, immunological techniques have been used recently in an effort to improve specificity and in-

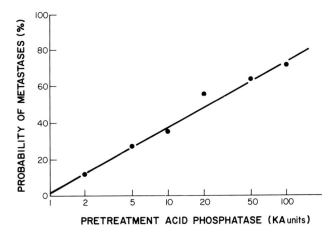

Fig. 1. Relationship of pretreatment acid phosphatase and metasases. (Plotted from data reported by Byar, 1977.)

crease sensitivity. These techniques fall into two categories: competitive binding radioimmunoassay, and enzyme immunoassay (Pontes, 1980, 1981). In the radioimmunoassay (RIA), the sample to be analyzed is mixed with a standard containing radioiodinated prostatic acid phosphatase and incubated with an antibody to human prostatic acid phosphatase. The amount of radioactive acid phosphatase displaced by the sample is measured and is an indication of the concentration of prostatic acid phosphatase present in the specimen. Enzyme immunoassay techniques are based on the formation of a complex between the antigen and antibody. The complex formed by the reaction is measured colorimetrically or by spectrophotometry.

The use of immunodetection of prostatic acid phosphatase attracted considerable attention following the reports of Foti and colleagues (1975, 1977). They reported that approximately 50% of patients with early prostatic cancer had elevations of prostatic acid phosphatase measured by radioimmunoassay and suggested that this test could be used to screen for prostatic cancer. Cooper (1980) has recently summarized their results using this technique used as a screening test. Of 6320 patients attending a general medical–surgical clinic for any cause who had radioimmunoassays for prostatic acid phosphatase performed, 443 (7%) were found to have hyperphosphatasemia. Of these, 62 had biopsy-proven prostatic cancer. Thus, 14% of all subjects with elevated prostatic acid phosphatase by radioimmunoassay had prostatic cancer and 86% did not.

Several authors have published their experience with radioimmunoassays of prostatic acid phosphatase in patients with known prostatic cancer (Griffiths, 1980; Quinones et al., 1981; Klein and Shapiro, 1981; Fair et al., 1982) (Table III). Typical of the studies reported are those of Bruce and colleagues (1980,

TABLE III
Elevated Prostatic Acid Phosphatase by Radioimmunoassay

Reference	Controls (%)	Stage A (%)	Stage B (%)	Stage C (%)	Stage D (%)
Foti et al. (1977)	4–11	33	79	71	92
Mahan and Doctor (1980)	2.5[a]	13	26	30	94
Griffiths (1980)	9	12	32	47	80
Bruce et al. (1981)	2.5[a]	14	29	24	89
Fair et al. (1982)	7.5	6	9	35	65
Klein and Shapiro (1981)	12	0	9	71	100
Quinones et al. (1981)	9	25	14	44	78

[a] 97.5 percentile.

1981). In a prospective trial, they measured prostatic acid phosphatase using three different radioimmunoassays (their own, and commercial assays from Mallinckrodt and New England Nuclear) in 90 male and 25 female volunteers, 75 patients with benign prostatic hyperplasia, and 125 patients with untreated prostatic carcinoma. Most patients with prostatic carcinoma also had chemical determinations of acid phosphatase. All three radioimmunoassays had similar sensitivities (about 55%) in detecting the presence of prostatic cancer. Levels of prostatic acid phosphatase were elevated in 22% of patients with localized disease, in 24% of patients with stage C disease, and in 89% of patients with stage D disease. Using the enzymatic assays, acid phosphatases were elevated in about 75% of patients with bony metastases. In general, these results are similar to those reports by others.

The most comprehensive study evaluating the enzyme immunoassay (counter immunoelectrophoresis) was reported by Wajsman and colleagues (1979). In a national field trial, 38% of 64 patients with stage A, 35% of 178 patients with stage B, 49% of 235 patients with stage C, and 69% of 485 patients with stage D prostatic carcinoma had elevated acid phosphatases using this technique. Interpretation of clinical studies such as these is made difficult by differences among assays, uncertainty about the accuracy of staging of prostate cancer patients, and uncertainty about whether controls were in fact free of prostate cancer.

The role of serial determinations of radioimmunoassay of prostatic acid phosphatase in the follow-up of prostatic cancer has been reported by Vihko and associates (1981). They measured prostatic acid phosphatase by chemical and immunologic techniques in 8 patients with metastatic and 18 patients with nonmetastatic prostatic carcinoma. All patients were treated with endocrine therapy. All patients with metastatic disease had elevated acid phosphatases by RIA; whereas, only 5 (65%) had chemical hyperphosphatasemia. In 6 of 8 patients,

RIA values were normalized following therapy, and in 5, clinical response or stabilization of disease occurred. The decline in acid phosphatase levels preceded the clinical response, and could be detected within 1 month of initiation of therapy. Such declines were not seen using enzymatic assays for acid phosphatase. The RIA acid phosphatase was elevated in 12 of 18 patients with nonmetastatic cancer, in one of whom increasing values predicted the occurrence of bony metastases. Based on these results, it appears that response to hormone therapy in patients with prostatic cancer can be more effectively monitored by the radioimmunoassays than the chemical assays for prostatic acid phosphatase.

Enzymatic methods have been known to be unreliable for measuring prostatic acid phosphatase in bone marrow aspirates which are rich in acid phosphatases of nonprostatic origin. Although radioimmunoassays are more specific for measuring prostatic acid phosphatase in marrow aspirates, there is still some residual cross-reactivity. To correct for this, in most studies the upper limit of normal for acid phosphatase in marrow aspirates has been set higher than that of serum. Belville and associates (1981) reported that 36% of patients having elevated marrow prostatic acid phosphatases by radioimmunoassay developed distant metastases within 2 years as compared with only 3% with normal marrow acid phosphatase levels.

In summary, most of the available information suggests that the more expensive and time-consuming immunochemical methods for measuring prostatic acid phosphatase are more sensitive and more specific than enzymatic methods, but tend to yield more false positives when used clinically. No assay for acid phosphatase has proved to be valuable as a screening test for prostatic cancer, but the radioimmunoassay may provide a better means of monitoring patients.

B. Carcinoembryonic Antigen (CEA)

Carcinoembryonic antigen (CEA) is a cancer-specific antigen originally discovered in tissue extracts from adenocarcinoma of the colon and in fetal colonic mucosa (Gold and Freedman, 1965). Subsequently, the antigen has been detected in about 70% of patients with entodermally derived tumors. Elevated CEA levels also occur in about 5% of smokers, and in some patients with inflammatory bowel disease. In patients with prostatic cancer, serum CEA levels have been reported to be elevated in 23 to 59% of patients (mean equals 42%) (Reynoso *et al.*, 1972; Neufeld *et al.*, 1974; Kane *et al.*, 1976; Fleisher *et al.*, 1977). It has been suggested that patients with advanced prostatic cancer are more likely to have elevated CEA titers than patients with localized disease. In the series of Guinan and associates (1973), 80% of patients with active prostatic cancer had elevated serum levels of CEA as compared to 47% of patients in remission.

Kane *et al.* (1976) correlated CEA and acid phosphatase levels with disease status in 28 patients with metastatic prostatic carcinoma who were under treatment with cytotoxic chemotherapy. In 19 of 29 patient visits (66%) in which the patient's clinical status was improving, CEA levels were decreasing, whereas, in 59% of these visits, acid phosphatase levels were decreasing. In 59% of patient visits in which clinical deterioration was seen, CEA levels were increasing and in 57% acid phosphatase levels were increasing. Thus, determination of serum CEA appeared to be only marginally superior to measurement of acid phosphatase in monitoring the disease status.

C. α-Fetoprotein

In the only study reported to date on prostatic cancer patients (Guinan *et al.*, 1975), α-fetoprotein levels were normal in all 54 patients tested with prostatic cancer. Most of these patients had metastatic disease.

D. Lactic Dehydrogenase Isoenzymes (LDH)

Denis and Prout (1963) determined total serum lactic dehydrogenase (LDH) and its individual isoenzymes in 21 patients with metastatic prostatic cancer and 11 controls. In 16 of the 21 prostate cancer patients, LDH isoenzymes 4 and 5 were elevated. The activity of the LDH enzymes was higher in patients with progressive disease and was lower in patients whose disease was in remission. No correlation was observed between LDH isoenzyme levels and prostatic acid phosphatase levels.

Grayhack *et al.* (1979, 1981) measured the LDH isoenzymes in prostatic fluid and suggested that prostatic fluid levels may be a better indicator of the activity of prostatic cancer than serum levels. Approximately 80% of patients with prostatic cancer and less than 15% with BPH had a prostatic LDH5:LDH1 ratio of over 2.

E. Creatine Kinase (CK-BB)

Creatine kinase (CK) occurs primarily as three isoenzymes: CK-BB, which is present in all tissues, but predominantly in the brain, gastrointestinal tract, and the genitourinary tract, CK-MM, which is found in skeletal and cardiac muscles, and CK-MB, which is present in cardiac muscle. The creatine kinase found in the sera of healthy individuals is said to be composed entirely of the MM isoenzymes.

Elevated CK-BB enzyme concentrations, as determined by radioimmunoassay, have been demonstrated in the sera of patients with breast and prostatic cancer. Fair *et al.* (1980) reported that CK-BB was elevated in only 8 of 42

(19%) of patients with nonmetastatic prostatic cancers. Zweig and Van Steirteghem (1981) measured CK-BB levels in 23 patients with metastatic prostatic cancer and reported that CK-BB levels were elevated in 6 (26%). Prostatic acid phosphatase determined by radioimmunoassay was elevated in 48 to 55% of these patients. No patient with a normal PAP had an elevated CK-BB.

F. Ribonuclease

Serum ribonuclease (RNAse) activity is elevated in about 70% of patients with prostatic cancer (Catalona et al., 1973; Chu et al., 1977). Five patients with stage A carcinoma of the prostate had normal levels. Serum RNAse activity was elevated in 2 patients with metastatic cancer and normal prostatic acid phosphatase. However, RNAse levels are also elevated in patients with other malignancies and in association with renal insufficiency.

G. Seromucoid

Cameron and Campbell (1964) reported that seromucid levels were elevated in 30 of 37 patients (81%) with carcinoma of the prostate, while, acid phosphatase was elevated in only 21 (57%). In 11 patients, the acid phosphatase levels were normal, but seromucoid was elevated, while the converse was true in only 2 cases. However, seromucoid was elevated in 9 of 53 (17%) patients with benign prostatic hyperplasia, and is also elevated in a variety of inflammatory and neoplastic disorders.

H. Steroid Hormones and Human Chorionic Gonadotropin

Broder and associates (1977) measured plasma estradiol, testosterone, α human chorionic gonadotropin (hCG), and human placental lactogen levels in 16 patients with metastatic carcinoma of the prostate. One patient had an elevation of serum hCG; however, the possibility of an occult hCG-producing tumor of nonprostatic origin was not ruled out. As discussed by Dr. Lange in Chap. 13, hCG levels are elevated in patients with testicular tumors, and also in some patients with gastrointestinal and urothelial cancers.

I. Prostate-Specific Antigen (PA)

This antigen is a glycoprotein with a molecular weight of approximately 34,000 which is distinct from acid phosphatase and is histiotypic of human prostate. PA has been localized to the cytoplasm of the ductal cells of the prostate gland (Wang et al., 1979). Immunoperoxidase staining has shown the presence of the PA in 73 primary prostatic carcinomas and 49 metastatic tumors; whereas, none of 78 nonprostatic malignancies stained positively for PA (Nadji et al., 1981). Early reports using a rocket immunoelectrophoresis, which is capable of detecting prostatic antigen at a concentration of 0.5 µg/ml, demonstrated PA in the sera of 17 of 219 (8%) patients with advanced prostatic cancer. This antigen was not found in the sera of any of 175 patients with nonprostatic cancer or in 40 normal controls (Papsidero et al., 1980). Utilizing more sensitive enzyme-linked immunoabsorbent assay (ELISA) techniques, PA at concentrations of as low as 0.1 ng/ml can be measured (Kuriyama et al., 1980). With the ELISA technique, 296 of 344 (86%) patients with stage D prostatic cancer were found to have elevated serum PA levels. However, PA levels were elevated in 10% of normal men, and in two-thirds of men with benign prostatic hyperplasia. Between 60 and 80% of patients with stages A and B prostatic cancer have elevated levels of PA. No significant difference between serum levels of PA were seen between patients with benign prostatic hyperplasia and those with localized prostatic cancer. Thus, BPH and early prostatic cancer cannot be distinguished by assaying circulating PA levels (Wang et al., 1981).

In 96 patients with metastatic prostatic cancer entered into the various chemotherapy protocols of the National Prostatic Cancer Project, pretreatment levels of serum PA correlated inversely with survival time. In 5 patients with localized prostatic cancer who underwent curative therapy and subsequently developed metastatic disease, elevations of serum PA values preceded the clinical detection of disease recurrence in one patient by 68 weeks, and were present at the time of clinical detection of recurrence in the other 4 patients (Kuriyama et al., 1981).

III. URINARY MARKERS

A. Total and Nonesterified Cholesterol

Chu et al. (1975) and Juegnst and their associates (1979) reported that urinary cholesterol levels were elevated from 52 to 63% of patients with prostatic carcinoma. In Chu's study, patients with pulmonary metastases from prostatic carcinoma were evaluated. Elevations of urinary cholesterol were detected in 62%, while acid phosphatase was elevated in 67% of these patients. When a combined determination of these two assays was done, values were elevated for either or both substances in 86% of patients. Juengst et al. reported that 52% of patients with localized prostatic cancer had elevated total cholesterol levels; however, elevations also were observed in patients with benign prostatic hyperplasia and residual urine, other benign and malignant disorders of the urinary tract, and in patients with testicular cancer.

Acevedo and associates (1973) reported that urinary nonesterified cholesterol levels were elevated in 29 of 32% of patients studied. Most of the patients had metastatic disease. Juengst *et al.* (1979) and Belis and Canedella (1979) reported elevations in 45 and 54%, respectively, of patients with clinically localized prostatic cancer. Again, false positive elevations of urinary nonesterified cholesterol can be detected in up to 25% of patients over 45 years old, in some patients with benign prostatic hyperplasia, and in other disorders of the urinary system.

B. Carcinoplacental Isoenzyme (Regan N)

In 1968, Fishman *et al.* reported on the presence of an isoenzyme of alkaline phosphatase resembling that of placental origin in tumor cells of a patient with bronchogenic carcinoma. Subsequently, this enzyme has been detected in the sera of patients with ovarian and testicular tumors and in some with tumors of the gastrointestinal tract (Fishman *et al.*, 1975). In a study performed by the National Prostatic Cancer Project, 14 of 98 (14%) patients with advanced prostatic cancer revealed elevated Regan isoenzyme activity (Slack *et al.*, 1981). Elevated levels of the isoenzyme appeared to occur in patients with metastatic disease that were clinically in progression as well as in those who apparently were responding to chemotherapy.

C. Urinary Carcinoembryonic Antigen

Fleisher and associates (1977) reported elevations of urinary CEA in 3 of 14 (21%) patients with metastatic prostatic carcinoma, but in none of 10 with localized disease. Urinary CEA levels were not elevated in the presence of urinary tract infection. There was no correlation between serum and urinary CEA levels in patients with prostatic carcinoma. The authors concluded that urinary CEA levels were of no practical value in detecting prostatic carcinoma.

D. Hydroxyproline

Urinary excretion of hydroxyproline, a reliable marker of bone matrix turnover, is increased in a variety of conditions in which collagen breakdown is excessive. Thus, patients with prostatic carcinoma with osseous metastases have increased urinary excretion of hydroxyproline. Bishop and Fellows (1977), Mundy (1979), and Unni Mooppan *et al.* (1980), reported elevations of urinary hydroxyproline in 12 to 35% of patients studied. In general, urinary hydroxyproline levels are normal in the absence of bony metastases. No overlap was observed between groups. However, although hydroxyproline is a good indicator of bony metastases, in no instance did it precede the conversion of a negative bone scan to positive. Because urinary hydroxyproline levels are normal in patients with arthritis, hydroxyproline determinations may be of some value in distinguishing between arthritis and cancer in patients with equivocal bone scans.

E. Isoleucine

McGregor and Johnson (1976), measured the excretion of urinary amino acids obtained from acid hydrolysates of urine in 16 normal subjects and 48 patients with carcinoma of the prostate. While there was no difference between the groups in the secretion of glycine, leucine, aspartic acid, methionine, and glutamic acid, patients with prostatic cancer had detectable amounts of isoleucine in the urine. Isoleucinuria was not detected in the normal controls. Of the patients studied, 16 had stage B and C tumors, while 32 had stage D disease.

F. Spermidine

Fair and associates (1973) determined the excretion of urinary polyamines in 18 patients with prostatic carcinoma and 13 age-matched controls. There was no difference between the groups in the excretion of the polyamines, putrescine, and spermine; however, urinary excretion of spermidine was elevated in 13 of 18 patients (72%) with prostatic carcinoma as compared with 3 of 13 (23%) control patients. Only 1 of 10 specimens from individuals with grade 1 lesions had an elevated urinary spermidine level. In contrast, all 12 urine specimens from patients with grade 3 or 4 lesions had increased urinary spermidine levels. The mean level of urinary spermidine was highest in patients with grade 3 prostatic cancer. The one patient with a stage A prostatic cancer had normal urinary spermidine excretion. However, increased urinary spermidine levels were seen with about similar frequency in patients with stage B, C, and D disease.

G. Fibronectin

Fibronectin is a high molecular weight (220,000) glycoprotein that is normally located on the cell surface. Increased prostatic cellular growth and invasiveness may cause shedding of fibronectin from the cell surface and its resultant appearance in the urine. Webb and Lin (1980) measured the excretion of urinary fibronectin in 32 patients with prostatic carcinoma, in 14 normal volunteers, and in 16 patients with benign urologic disease. Mean levels of urinary fibronectin were significantly increased in patients with prostatic carcinoma in comparison with both control groups. In 8 patients with metastatic prostatic cancer, in whom sequential determination of urinary fibronectin were performed, levels were increased on at least one occasion in all. In the one patient with stage A prostatic cancer in whom this measurement was performed, levels were similarly elevated.

IV. PROSTATE FLUID PROTEINS

Grayhack and associates (1979, 1981) determined concentrations of the proteins IgG, IgA, IgM, complement components C3 and C4, and transferrin in the prostatic fluid of various subjects. There were 195 specimens from 99 patients, including 14 with carcinoma of the prostate, 41 with benign prostatic hyperplasia, 32 with prostatitis, and 12 with other urological disorders not related to the prostate. Prostatic fluid specimens were obtained by rectal massage and analyzed by immunoelectrophoresis and radial immunodiffusion. No differences were detected between concentrations of IgA, IgG, and IgM among the various groups studied; whereas, concentrations of C3, C4, and transferrin were significantly elevated in patients with prostatic carcinoma. Levels of both C3 and LDH5 : LDH1 ratios were elevated in 9 of 10 patients with prostatic cancer, in 4 of 23 (70%) of patients with prostatitis and in none of the 16 patients with benign prostatic hyperplasia. Of the 14 patients with carcinoma of the prostate studied, 9 had localized disease. In one patient with clinical diagnosis of benign prostatic hyperplasia, elevated C3 and transferrin levels led to repeated biopsy of the prostate and the detection of malignancy. Elevations of C3 and transferrin levels have been noted in the sera of patients with other malignancies, suggesting that these changes are the result of nonspecific metabolic changes associated with cancer (Verhaegen *et al.*, 1976).

V. LEUKOCYTE ADHERENCE INHIBITION (LAI) AND LEUKOCYTE MIGRATION INHIBITION (LMI)

In the LAI test, leukocytes lose their ability to adhere to glass surfaces when they are cultured in the presence of an antigen to which they have been sensitized. Evans and Bowen (1977) and Bhatti and associates (1979) reported nonadherence of leukocytes in 75 and 90% of prostate cancer patients and in 18 and 0% of control subjects, respectively. In contrast, other investigators have reported positive LAI tests in only 14% of patients with advanced prostatic cancer. Pretreatment of leukocytes with prostaglandin E2 increased the sensitivity of the test to 61% (Kaneti *et al.*, 1982).

In the leukocyte migration inhibition assay, leukocytes from the patients are incubated with or without tumor and control antigens. The suspensions are drawn into capillary tubes and incubated in migration chambers. The migration patterns of the leukocytes are then measured by microscopy and the area of migration determined. The migration index is a ratio of the migration in the presence of antigen to the migration in the absence of antigen. Inhibition of leukocyte migration was detected in 61% of patients with prostatic carcinoma, in 37% of patients with benign prostatic hyperplasia, in 26% of patients with nonprostatic cancers, and in 10% of normal subjects (Wright *et al.*, 1981).

VI. SUMMARY AND CONCLUSIONS

We have discussed the various biochemical tests that have been used to detect prostatic cancer and to monitor response to treatment. Because of the low prevalence of clinically manifest cancer and the unpredictable biological behavior of incidental prostatic carcinoma, screening biochemical tests are presently of little practical value (see Section I). The time-honored measurements of prostatic acid phosphatase, perhaps with newer immunological modifications, appear to be of the greatest potential benefit in the management of patients with prostatic carcinoma (see Section II, A).

REFERENCES

Abu-Fadl, M. A. M., and King, E. J. (1949). *Biochem. J.* **45,** 51.
Acevedo, H. F., Campbell, E. A., Saier, E. L., Frich, J. C., Merkow, L. P., Hayeslip, D. W., Barstok, S. P., Grauer, R. C., and Hamilton, J. L. (1973). *Cancer* **32,** 196.
American Cancer Society (1979). "American Cancer Society Facts and Figures," p. 7. Am. Cancer Soc., New York.
Babson, A. L., and Read, P. A. (1959). *Am. J. Clin. Pathol.* **32,** 88.
Belis, J. A., and Cenedella, R. J. (1979). *Cancer* **43,** 1840.
Belville, W. D., Mahan, D. E., Sepulveda, R. A., Bruce, A. W., and Miller, C. F. (1981). *J. Urol.* **125,** 809.
Bessey, D. A., Lowry, O. H., and Brock, M. J. (1946). *J. Biol. Chem.* **164,** 321.
Bhatti, R. A., Ablin, R. J., and Guinan, P. D. (1979). *J. Reticuloendothel. Soc.* **25,** 389.
Bishop, M. C., and Fellows, G. J. (1977). *Br. J. Urol.* **49,** 711.
Bodansky, A. (1933). *J. Biol. Chem.* **101,** 93.
Broder, L. E., Weintraub, B. D., Rosen, S. W., Cohen, M. H., and Tejada, F. (1977). *Cancer* **40,** 211.
Bruce, A. W., Mahan, D. E., and Belville, W. D. (1980). *Urol. Clin. North Am.* **7,** 645.
Bruce, A. W., Mahan, D. E., Sullivan, L. D., and Goldenberg, L. (1981). *J. Urol.* **125,** 357.
Byar, D. P. (1977). *In* "Urological Pathology" (M. Tannenbaum, ed.), pp. 241–267. Lea & Febiger, Philadelphia, Pennsylvania.
Byar, D. P., Corle, D. K., and Veterans Administration Cooperative Urological Research Group (1981). *Urology* **17,** Suppl. 4, 7.
Cameron, E., and Campbell, A. (1964). *Br. J. Urol.* **36,** 257.
Catalona, W. J., Chrétien, P. B., Matthews, W. B., and Tarpley, J. L. (1973). *Urology* **11,** 577.
Chu, T. M., Shukla, S. K., Mittelman, A., and Murphy, G. P. (1975). *Urology* **6,** 291.
Chu, T. M., Wang, M. C., Kuciel, R., Valenzuela, L., and Murphy, G. P. (1977). *Cancer Treat. Rep.* **61,** 193.
Cooper, J. F. (1980). *Urol. Clin. North Am.* **7,** 653.
Denis, L. J., and Prout, G. R., Jr. (1963). *Invest. Urol.* **1,** 101.
Evans, C. M., and Bowen, J. G. (1977). *Proc. R. Soc. Med.* **70,** 417.
Fair, W. R., Wehner, N., and Brorsson, U. (1973). *J. Urol.* **114,** 88.
Fair, W. R., Catalona, W. J., Ratliff, T. R., and Heston, W. D. W. (1980). *Pap., Annu. Meet. Am. Urol. Assoc.* Abst. 70, p. 87, 75th Annual Meeting of American Urological Association, San Francisco.

15. Tumor Markers in Prostatic Cancer

Fair, W. R., Heston, W. D. W., Kadmon, D., Crane, D. B., Catalona, W. J., Ladenson, J. H., McDonald, J. M., Noll, B. W., and Harvey, G. (1982). *J. Urol.* (in press).
Fishman, W. H., Inglis, N. L., Stolbach, L. L., and Krant, M. J. (1968). *Cancer Res.* **28,** 150.
Fishman, W. H., Inglis, N. L., Vaitukaitis, J., and Stolbach, L. L. (1975). *Natl. Cancer Inst. Monogr.* **42,** 63.
Fleisher, M., Grabstald, H., Whitmore, W. F., Jr., Pinsky, C. M., Oettgen, H. F., and Schwartz, M. K. (1977). *J. Urol.* **117,** 635.
Foti, A. G., Herschman, H., and Cooper, J. F. (1975). *Cancer Res.* **32,** 2446.
Foti, A. G., Cooper, J. F., Herschman, H., and Malvaez, R. (1977). *N. Engl. J. Med.* **297,** 1357.
Freiha, F. S., Pistenma, D. A., and Bagshaw, M. A. (1979). *J. Urol.* **122,** 176.
Galen, R. S., and Gambino, S. R. (1975). "Beyond Normality: The Predictive Value and Efficiency of Medical Diagnosis." Wiley, New York.
Ganem, E. J. (1956). *J. Urol.* **76,** 179.
Gold, P., and Freedman, S. O. (1965). *J. Exp. Med.* **121,** 439.
Goldberg, I. D., Levin, M. L., Gerhardt, P. R., Handy, V. H., and Cashman, R. E. (1956). *JNCI, J. Natl. Cancer Inst.* **17,** 155.
Grayhack, J. T., Wendell, E. F., Oliver, L., and Lee, C. (1979). *J. Urol.* **121,** 295.
Grayhack, J. T., Lee, C., Kolbusz, W., and Oliver, L. (1981). *Cancer* **45,** 1896.
Griffiths, J. C. (1980). *Clin. Chem. (Winston Salem, N.C.)* **26,** 433.
Guinan, P., Ablin, R. J., Barakat, H., John, T., Sadoughi, N., and Bush, I. M. (1973). *Urol. Res.* **1,** 101.
Guinan, P. D., Nourkayhan, S., and Bush, I. M. (1975). *Clin. Res.* **23,** 339A (abstr.).
Gutman, A. B., and Gutman, E. B. (1938). *J. Clin. Invest.* **17,** 673.
Henneberry, M., Engel, G., and Grayhack, J. T. (1979). *Urol. Clin. North Am.* **6,** 629.
Huggins, C., and Talalay, P. (1945). *J. Biol. Chem.* **159,** 399.
Juengst, D., Pickel, A., Elsaesser, E., Marx, F. J., and Karl, H. J. (1979). *Cancer* **43,** 353.
Kane, R. D., Mickey, D. D., and Paulson, D. F. (1976). *Urology* **8,** 559.
Kaneti, J., Thompson, E. M. P., and Reid, E. C. (1982). *J. Urol.* (in press).
Klein, L. A., and Shapiro, P. (1981). *Urology* **17,** 550.
Kuriyama, M., Wang, M. C., Papsidero, L. D., Killian, C. S., Shimano, T., Valenzuela, L. A., Nishiura, T., Murphy, G. P., and Chu, T. M. (1980). *Cancer Res.* **40,** 4658.
Kuriyama, M., Wang, M. C., Lee, C., Papsidero, L. D., Killian, C. S., Inagi, H., Slack, N. H., Nishiura, T., Murphy, G. P., and Chu, T. M. (1981). *Cancer Res.* **41,** 3874.
Loor, R., Wang, M. C., Valenzuela, L., and Chu, T. M. (1981). *Cancer Lett.* **14,** 63.
McGregor, R. F., and Johnson, D. E. (1976). *Urology* **8,** 127.
Mahan, D. E., and Doctor, B. P. (1979). *Clin. Biochem.* **12,** 10.
Martland, M., Hansman, F. S., and Robison, R. (1924). *Biochem. J.* **18,** 1152.
Menon, M., and Walsh, P. C. (1979). *In* "Prostate Cancer" (G. P. Murphy, ed.), pp. 175–200. P. S. G. Publishing Co., Littleton, Massachusetts.
Moore, R. A. (1935). *J. Urol.* **33,** 224.
Mundy, A. R. (1979). *Br. J. Urol.* **51,** 570.
Murphy, G. P. (1981). *Urology* **17,** Suppl. 3, 1.
Nadji, M., Tabei, S. Z., Castro, A., Chu, T. M., Murphy, G. P., Wang, M. C., and Morales, A. R. C. (1981). *Cancer* **48,** 1229.
Neufeld, L., Dubin, A., Guinan, P., Naborg, R., Ablin, R. J., and Bush, I. M. (1974). *Oncology* **29,** 376.
Papsidero, L. D., Wang, M. C., Valenzuela, L. A., Murphy, G. P., and Chu, T. M. (1980). *Cancer Res.* **40,** 2428.
Paulson, D. F., Piserchia, P. V., and Gardner, W. (1980). *J. Urol.* **123,** 697.
Pontes, J. E. (1980). *Urol. Clin. North Am.* **7,** 667.

Pontes, J. E. (1981). *Urology* **17,** Suppl. III, 38.
Quinones, G. R., Rohner, T. J., Drago, J. R., and Demers, L. M. (1981). *J. Urol.* **125,** 361.
Reynoso, G., Chu, T. M., Guinan, P., and Murphy, G. P. (1972). *Cancer* **30,** 1.
Rich, A. R. (1935). *J. Urol.* **33,** 215.
Roy, A. V., Brower, M. E., and Hayden, J. E. (1971). *Clin. Chem. (Winston-Salem, N.C.)* **17,** 1093.
Seligman, A. M., Chauncey, H. H., Nachlas, M. M., Manheimer, L. H., and Ravin, H. A. (1951). *J. Biol. Chem.* **190,** 7.
Silverberg, E. (1982). *Ca—Cancer. Clin.* **32,** 15.
Slack, N. H., Chu, T. M., Wajsman, L. Z., and Murphy. G. P. (1981). *Cancer* **47,** 146.
U.S. Bureau of the Census (1978). "Current Population Reports." U.S. Bureau of Census, Washington, D.C.
Unni Mooppan, M. M., Wax, S. H., Kim, H., Wang, J. C., and Tobin, M. S. (1980). *J. Urol.* **123,** 694.
Verhaegen, H., DeCock, W., DeCree, J., and Vergruggen, F. (1976). *Cancer* **38,** 1608.
Veterans Administration Cooperative Urological Research Group (VACURG) (1967). *Surg., Gynecol. Obstet.* **124,** 1011.
Vihko, P., Lukkarinen, O., Konttuni, M., and Vihko, R. (1981). *Cancer Res.* **41,** 1180.
Wajsman, Z., Chu, T. M., Saroff, J., Slack, N., and Murphy, G. P. (1979). *Urology* **13,** 8.
Wang, M. C., Valenzuela, L. A., Murphy, G. P., and Chu, T. M. (1979). *Invest. Urol.* **17,** 159.
Wang, M. C., Papsidero, L. D., Kariyama, M., Valenzuela, L. A., Murphy, G. P., and Chu, T. M. (1981). *Prostate* **2,** 89.
Watson, R. A., and Tang, D. B. (1980). *N. Engl. J. Med.* **303,** 497.
Webb, K. S., and Lin, G. H. (1980). *Invest. Urol.* **17,** 401.
Whitmore, W. F., Jr. (1956). *Am. J. Med.* **21,** 697.
Whitmore, W. F., Jr. (1973). *Cancer* **32,** 1104.
Woodard, H. Q. (1952). *Cancer* **5,** 236.
Wright, G. L., Jr., Schellhammer, P. F., Faulkner, R. J., Reid, J. W., Brassil, D. N., and Sieg, S. N. (1981). *Prostate (N.Y.)* **2,** 121.
Zweig, M. H., and Van Steirteghem, A. C. (1981). *JNCI, J. Natl. Cancer Inst.* **66,** 859.

CHAPTER 16

Markers of Gastrointestinal Cancer

YOUNG S. KIM   LAURENCE J. MCINTYRE

Gastrointestinal Research Laboratory
Veterans Administration Medical Center
San Francisco, California

Department of Medicine
University of California
San Francisco, California

I.	Introduction	299
II.	Oncodevelopmental Markers of Colorectal Cancer	300
	A. Immunological Markers	300
	B. Metabolic Markers	303
	C. Glycoconjugate Markers	303
III.	Oncodevelopmental Markers of Pancreatic Cancer	305
	A. Immunological Markers	306
	B. Glycoconjugate Markers	307
IV.	Oncodevelopmental Markers of Gastric Cancer	308
	A. Immunological Markers	308
	B. Metabolic Markers	309
	C. Glycoconjugate Markers	310
V.	Conclusions	310
	References	311

I. INTRODUCTION

It is estimated (Silverberg, 1982) that cancer of the gastrointestinal tract will have caused more deaths in the United States than any other form of cancer in 1982. The three regions of the gastrointestinal system in which cancer gives rise to the highest mortalities are the stomach, the pancreas, and the colon and rectum. This chapter will be restricted to a discussion of oncodevelopmental markers in cancers occurring at these sites. In the interest of clarity only the results obtained in human systems will be reported.

Many molecules have been proposed as oncodevelopmental markers, but, at present, almost all have been found to lack the necessary specificity. In some cases the same molecule can be found at low levels in normal tissues of the organ being studied, or the marker is found in other organs of the normal adult, or elevated levels of the molecule occur in nonmalignant conditions. Despite a lack of rigorous specificity, markers which had been once thought to be oncodevelopmental have proved useful in treating cancer. Moreover, such markers can often provide information about the cellular changes which occur during carcinogenesis. For these reasons, the definition of an "oncodevelopmental" marker has been applied quite loosely in this chapter.

The studies of oncodevelopmental markers have been divided into three main areas; immunological, metabolic, and glycoconjugate markers. There is considerable overlap, however, between these groups as, for example, many of the changes in glycoconjugates have been detected by immunological methods.

II. ONCODEVELOPMENTAL MARKERS OF COLORECTAL CANCER

Colorectal cancer has been a particularly fruitful area in the study of oncodevelopmental markers for human cancer. This is probably due to the high incidence of this type of cancer and to the comparative ease with which samples of normal and tumor tissue can be obtained from human subjects.

A. Immunological Markers

1. Carcinoembryonic Antigen

Certainly the most extensively studied of all oncodevelopmental markers for colorectal cancer has been carcinoembryonic antigen (CEA). CEA was first detected by Gold and Freedman (1965a) as an antigen present in extracts of human colonic tumors but absent in extracts of normal human colon. The same workers (Gold and Freedman, 1965b) also found that the antigens were present in other malignant tumors of the gastrointestinal tract epithelium and in fetal gut, liver, and pancreas. Subsequent studies (Slayter and Coligan, 1975) have shown CEA to be a glycoprotein with a molecular weight of about 180,000 consisting of 50–60% carbohydrate by weight. Although CEA was originally thought to be absent from normal colon, sensitive detection methods demonstrated that low levels of CEA could be measured in normal colonic mucosa (Khoo *et al.*, 1973). Based on the current evidence, it seems that in colorectal cancer an excess of CEA in the tumor tissue may be an oncodevelopmental marker.

When Thomson *et al.* (1969) developed a radioimmunoassay for CEA they demonstrated that patients with colorectal cancer had elevated serum levels of CEA compared to patients with cancer of other organs or with nonmalignant disease of the digestive organs. It was hoped that CEA would prove to be a sensitive and specific marker for colorectal cancer. Unfortunately, the extensive body of work which has accumulated since that time has not substantiated that hope (Conference Statement, 1981). It should be noted, however, that very high plasma levels of CEA (above 20 ng/ml) are highly suggestive of the presence of cancer.

Despite the limitations outlined above, CEA can be of use in the management of colorectal cancers. Sequential postoperative plasma CEA determinations have been used for the detection of tumor recurrence (Holyoke et al., 1972; Sorokin et al., 1974; Mach et al., 1974; MacKay et al., 1974). Sequential plasma CEA assays have also been shown to be useful in monitoring the clinical progression of patients undergoing radiotherapy or chemotherapy for unresectable colorectal cancer (Zamcheck, 1980).

As well as measuring the amount of CEA in plasma, the levels of CEA in colonic mucus and lavage samples and in fecal material have also been studied. When the CEA content of colonic lavage samples was examined (Winawer et al., 1977) the assay was not found to be suitable for the diagnosis of colorectal cancer. A recent study by Fujimoto et al. (1979) suggests that quantitative measurement of CEA in fecal samples may provide a sensitive method of detecting colorectal cancer, especially in the early stages. Antibodies to CEA are also being used in the radioimmunodetection of colorectal tumors (Goldenberg et al., 1980; Mach et al., 1980; Hine et al., 1980), and the results obtained so far are very promising. In one series of experiments colorectal tumors were detected with a sensitivity of 85% and a specificity of 98% (Goldenberg et al., 1980).

2. Pregnancy-Specific β_1-Glycoprotein

Pregnancy-specific β_1-glycoprotein (SP_1) is found in the serum of pregnant women (Bohn et al., 1981; Horne and Bremner, 1980). Using an immunoperoxidase technique it was possible to detect SP_1 in 60% of colorectal cancers (Horne and Bremner, 1980; Skinner and Whitehead, 1981). Studies have also been carried out to measure the amounts of SP_1 in the sera of cancer patients compared with normal controls (Horne and Bremner, 1980; Tatarinov and Sokolov, 1977; Engvall and Yonemoto, 1979). In most cases it now appears that very little of the SP_1 found in the tumor tissue can be detected in the serum of patients with colorectal cancer. This limits its usefulness as a marker for colorectal cancer.

3. Mucin Antigens

Ma et al. (1980), using immunofluorescent staining, have identified an antigen which is present in colonic carcinoma, absent in adult colon, and present in the fetus at 10 weeks although it later disappears. The mucin that the antiserum detects is of the nonsulfated type, as opposed to that of normal adult colon which is mainly sulfated. Bara and Burtin (1980) have also studied the changes in mucin-associated antigens which occur in colorectal carcinoma. One of the antigens they studied (M1) is associated with the mucus cells of surface gastric epithelium of the adult. M1 is distinct from the marker studied by Ma et al. (1980) which was not found in the normal adult stomach. Although M1 antigens were not found in normal human colon, when colonic tumor tissues were examined using immunofluorescence techniques, the mucin-associated antigen was detected in 29 of 53 tumors which contained mucins. Fetal human colon tissue also contains M1 antigen, demonstrating that it is an oncodevelopmental marker for colorectal cancer (Bara and Burtin, 1980). However the usefulness of both mucin-associated antigens will be limited by the fact that they are present in other organs of the normal adult.

4. β-Oncofetal Antigen

Another oncodevelopmental marker of human colorectal cancer which is also found in other normal adult organs is β-oncofetal antigen (BOFA). This antigen was identified using antisera raised against fractions of soluble extracts of human colon carcinoma tissue (Fritsché and Mach, 1975). BOFA can be detected in extracts of normal human colon. However, greatly elevated levels of the antigen are present in extracts of fetal colorectal tissue and of colorectal cancers. It is immunologically distinct from other oncodevelopmental markers such as CEA. Although BOFA appears to be an oncodevelopmental marker in colorectal cancer, it is also elevated in fetal and tumor tissue extracts of a variety of other organs. Goldenberg et al. (1978) have demonstrated that BOFA is abundant in normal adult kidney and liver.

5. Other Oncodevelopmental Antigens

Klavins et al. (1971) raised antiserum against whole fetal tissue and, after absorption, the antiserum was found to react with extracts of colon carcinoma as well as extracts of several other carcinomas. Extracts of a wide variety of normal adult organs did not react with the antiserum. At present, little more information is available concerning the identity of the antigens present in the carcinoma extracts. Edynak et al. (1972) found that the serum of a small number of cancer patients contained antibody against an antigen, subsequently called γ-fetoprotein, which was present in a wide range of tumors including 75% of colorectal carcinomas tested. This antigen is also found in fetal colon but not in normal adult colon and is distinct from CEA.

One other way in which oncodevelopmental markers of colorectal cancer can be detected by immunological methods is in the study of cell-mediated immune reactions. Hellström et al. (1970), using the inhibition of colony formation as a measure of reactivity, showed that colon carcinoma, fetal gut, and fetal liver cells have a common antigen which cannot be detected in adult colon or fetal kidney cells. Lymphocytes from patients with colonic carcinomas are immune to this antigen, whereas lymphocytes from control subjects are not. Similar results have been obtained using other methods of assaying cell-mediated immunity (Koldovsky and Weinstein, 1973; Zöller et al., 1979).

B. Metabolic Markers

1. Enzyme Markers

Herzfeld and Greengard (1980) studied the levels of enzymes in homogenates of tissue from normal adult colon, colon carcinoma, and normal fetal colon. They studied twelve enzymes involved in the metabolism of DNA, collagen, amino acids, and glucose. In fetal colon the levels of six of the enzymes were elevated relative to normal adult colon. Five of these enzymes also had elevated levels in colon carcinoma tissue. In addition, the amount of soluble aspartate aminotransferase is decreased in both carcinoma and fetal tissue. Munjal (1980) measured the activities of three enzymes in aqueous extracts of colonic carcinoma and fetal colon compared with normal adult colon. Two of the three were elevated in both fetal and cancer tissue relative to normal adult colon.

One enzyme which has been subjected to qualitative as well as quantitative analysis in the search for oncodevelopmental markers is thymidine kinase. Using both immunohistochemical (Balis et al., 1981) and electrophoretic techniques (Balis and Salser, 1980), it has been shown that the thymidine kinase of colonic tumors is related to a form present in placental tissue.

2. Glycogen Storage

The glycogen content of fetal colon tissue has been determined histologically and by quantitative assays (Rousset et al., 1979). The amount of glycogen in the cells decreased during gestation from high levels initially to a very low level by 36 weeks. Normal colorectal tissue contained a limited number of glycogen-positive cells, and the glycogen content was correspondingly low. On the other hand, glycogen-positive cells were readily visible in colon carcinomas varying from 10 to 100% of the cells, and the glycogen content was ten times higher than that of normal mucosa.

C. Glycoconjugate Markers

1. Blood Group Substances

One of the best-analyzed of all antigenic systems is that of blood group substances and in particular the ABH antigens. The specificities of these substances are determined by carbohydrate structures which can be found in both glycolipids and glycoproteins (Hakomori, 1981). The ABH blood group substances are not confined to erythrocytes and can be found on cells in many organs in the human. In the fetus, epithelial cells in all parts of the colon contain blood group substances. After birth, the ABH antigens disappear from the rectum and distal colon but remain in the proximal colon (Szulman, 1964; Wiley et al., 1981). Several investigators have studied the AB blood group substances in colorectal cancer with reference to the location of the tumor, and the results they obtained are shown in Table I. In general, it appears that tumors in the proximal colon often lose the blood group substance which is normally present while in the distal colon and rectum the opposite occurs, i.e., the cancer tissue expresses blood group substances. Both these changes may be oncodevelopmental in nature because Szulman (1964) showed that in the fetal colon there are alcohol-soluble blood group substances which appear early and then disappear to be followed by water-soluble blood group substances associated with secretion.

The deletion of blood group substances in some cases of human colorectal cancer has been studied in more detail at the cellular level. Watanabe and Hakomori (1976) have demonstrated that colonic neoplasia can result in the accumulation of a glycolipid precursor to the blood group antigens. It appears that the expression of ABH antigens in colonic carcinoma is often regulated by the levels of glycosyltransferases and may reflect a fetal cellular phenotype (Whitehead et al., 1979; Stellner et al., 1973).

TABLE I
Changes in AB Blood Group Antigens in Colorectal Cancer

Proximal colon	Change[a] Distal colon/rectum	Reference
100% (−)[b]		Davidsohn et al., 1966
variable[c]		Davidsohn et al., 1966
	61% (+)	Denk et al., 1974a
	47% (+)	Cooper and Haesler, 1978
	49% (+)	Abdelfattah-Gad and Denk, 1980
50% (−)	57% (+)	Wiley et al., 1981
	57% (+)	Cooper et al., 1980

[a] Changes are expressed as percentage of tumors examined in which AB blood group antigens were present (+) or absent (−).
[b] Nonmucinous.
[c] Mucinous.

2. Mucin

Two main types of mucin can be detected in human colorectal epithelial cells by histological techniques—sialomucin and sulfomucin (Filipe and Branfoot, 1976). When the colonic mucosa containing carcinoma was examined it was found that the histological staining of mucins was altered. In these regions the mucin was predominantly sialomucin (Filipe and Branfoot, 1976). This change in colorectal mucins can be classified as an oncodevelopmental marker since studies of fetal colon have shown that in the fetal colon the mucin is initially a sialomucin and sulfomucins only appear later in development (Lev, 1968; Lev and Orlic, 1974).

Recently, the carbohydrate structure of colorectal mucins in man has been probed using the binding of fluorescently labeled lectins to tissue sections (Boland et al., 1982). The lectin derived from peanuts did not bind to mucin in normal mucosa but did bind to the mucin in all cancer tissues examined (Boland et al., 1982). When specimens of fetal human colon were examined peanut lectin was found to bind to the mucin present at the base of fetal colonic crypt, suggesting that the carbohydrate structure recognized by the peanut lectin is an oncodevelopmental marker for human colorectal cancer (Boland and Kim, 1982).

3. Monosialoganglioside

Another promising marker for colorectal cancer that has been recently identified is the antigen detected by monoclonal antibodies raised against cell lines derived from human colorectal carcinomas (Herlyn et al., 1979). These antibodies were specific for colorectal carcinoma cells and did not bind to normal colonic mucosa or other normal and malignant human cells. Circulating antigen could be detected in 90% of patients with colorectal cancer and in cases of gastric or pancreatic cancer. None of the normal controls and only 2% of patients with other bowel diseases had antigen in their sera (Koprowski et al., 1981). The antigen responsible for the binding of the monoclonal antibodies to a human colorectal cancer cell line has been purified and was found to be a large monosialoganglioside (Magnani et al., 1981). The putative oncodevelopmental nature of this marker is suggested by the fact that it is present in human meconium which is a rich source of fetal glycolipids (Magnani et al., 1981).

III. ONCODEVELOPMENTAL MARKERS OF PANCREATIC CANCER

The number of oncodevelopmental markers which have been detected in human pancreatic cancer is much smaller than in colon cancer. Approximately

90% of pancreatic tumors are thought to originate in the ductal epithelial cells (Cubilla and Fitzgerald, 1975). For this reason tumors of the endocrine pancreas will not be discussed in this chapter.

A. Immunological Markers

1. Pancreatic Oncofetal Antigen

The first attempt to identify an oncodevelopmental marker specific for pancreas using immunological methods was made by Banwo et al. (1974). They raised antisera against stabilized homogenates of human fetal pancreas, and, after absorption with normal adult pancreas, the antisera could detect antigen in extracts of fetal pancreas and pancreatic carcinoma. The antigen was named pancreatic oncofetal antigen (POA). POA, purified by affinity chromatography, was found to have a molecular weight of 40,000 and to be a protein with a low carbohydrate content. POA is localized in the cytoplasm of the secretory cells of pancreatic adenocarcinomas and appears to be secreted by such cells. When serum POA levels were studied, 97% of patients with pancreatic carcinoma were positive (Hobbs et al., 1980). False positive results (16%) were only seen in patients who had chronic pancreatitis, and it is possible that some of these patients may have had adenocarcinoma. Preliminary experiments also indicate that serial POA levels can be used to monitor the course of pancreatic cancer during and after therapy (Hobbs et al., 1980). At least two other laboratories have produced antisera against POA. Gelder et al. (1978) found POA to be a single homogeneous glycoprotein with a molecular weight of between 800,000 and 900,000. Despite the differences in molecular characteristics, the anti-POA antisera of Gelder et al. and Hobbs et al. appear to be monospecific for the same antigenic moiety (Hobbs et al., 1980).

Gelder et al. (1978) used quantitative rocket immunoelectrophoresis to measure the levels of POA in the sera of over 700 patients. In this study 40% of patients with pancreatic cancer had elevated levels of POA in their sera. Fifteen percent of patients with pancreatitis also had elevated levels. The results obtained by Gelder et al. (1978) suggest that their assay system is less specific for the detection of pancreatic cancer than that of Hobbs et al. (1980). Further studies are necessary to determine the reason for this discordance.

Nishida et al. (1981) used double immunodiffusion assays to detect POA in the pancreatic juice of patients with pancreatic cancer. 72% of the cancer patients had a positive result compared with 10% of patients with pancreatitis or other disorders. Two other studies have also detected antigens which are oncodevelopmental markers for pancreatic cancer and which have very similar physical properties to POA (Mihas, 1978; Schultz and Yunis, 1979).

2. Carcinoembryonic Antigen

Plasma levels of CEA have been measured in a considerable number of studies on patients with pancreatic cancer (Zamcheck et al., 1972; LoGerfo et al., 1971; Hansen et al., 1974). In these and other studies, 74–100% of patients with pancreatic cancer had elevated levels of CEA. However, the problems of specificity and sensitivity discussed previously in relation to colorectal cancer also apply to pancreatic cancer. Chu et al. (1980) purified CEA from both pancreatic and colon cancer tissue and compared the immunological properties of the purified proteins. The results suggested that immunological differences exist between CEA from colon carcinoma and that from pancreatic cancer. This may lead to the development of more sensitive and specific assays for the detection of pancreatic cancer.

It has also been shown that CEA is present in pancreatic juice (McCabe et al., 1976), and the levels of CEA in the juice of control patients and patients with pancreatic carcinoma have been studied (Sharma et al., 1976; Carr-Locke, 1980; Nishida et al., 1980). Although there appear to be fewer false positives in this system than in the measurement of plasma CEA, there is still considerable overlap between the normal range and the values obtained with pancreatic cancer patients, limiting the usefulness of this assay as a diagnostic technique.

3. Other Oncodevelopmental Antigens

β-Oncofetal antigen (BOFA), described in the section on colorectal oncodevelopmental markers, has also been shown to be present in large amounts in human pancreatic tumor tissue and in fetal pancreas (Fritsché and Mach, 1975). As noted previously BOFA is not specific for one type of cancer and is found in normal adult kidney and liver. Another oncodevelopmental antigen which is found in pancreatic cancer tissue is oncofetal antigen-1 (OFA-1). Rees et al. (1981) found this antigen in two out of three pancreatic tumors that they examined. OFA-1 is found in a wide variety of human tumors, particularly those of ectodermal and mesodermal origin. It is absent in almost all normal tissues studied, and in fetal tissues OFA-1 is mainly found in the brain.

B. Glycoconjugate Markers

The only glycoconjugates which have been identified as oncodevelopmental markers for human pancreatic cancer are those responsible for blood group antigenicity. Although ductal cells are not distinguished from exocrine cells, it appears that normal human exocrine pancreatic tissue contains ABH antigens as determined by specific red cell adherence (Davidsohn, 1972). Eighty-four percent of pancreatic tumors are negative for ABH antigens in the same assay. When the blood group substances present in fetal pancreas were determined

(Szulman, 1964), cell wall blood group antigens were absent in most samples studied, and secretion-associated blood group substances appeared only at the later stages of development. At the present time the changes in blood group substances occurring in human pancreatic cancer have not been analyzed in any greater detail.

IV. ONCODEVELOPMENTAL MARKERS OF GASTRIC CANCER

A. Immunological Markers

1. Fetal Sulfoglycoprotein Antigen

One of the most extensively studied oncodevelopmental markers of gastric cancer has been fetal sulfoglycoprotein antigen (FSA). Häkkinen et al. (1968a) originally raised antisera against the glycoproteins present in the gastric juice from gastric cancer patients. After absorption, the antiserum obtained in this way was found to react with a sulfoglycoprotein present in fetal stomach and in gastric cancer tissue (Häkkinen et al., 1968a,b).

The absorbed anti-FSA has been used in a mass screening test for FSA in gastric juice in order to identify subjects requiring further investigation for gastric cancer (Häkkinen et al., 1980). In 39,706 persons tested, 3508 FSA secretors were identified, and 95% of these patients underwent endoscopy and biopsy. Of the FSA secretors, 36 gastric cancers and one gastric carcinoid were detected and, of these, 30 had been asymptomatic. In addition, a high proportion of the cancers were early cancers which have a much better prognosis, and, because of this, curative resection could be carried out in 78% of the cases. In the case of patients who were FSA positive but did not have cancer, more than one in four showed benign alterations of the gastric mucosa (Hakkinen, 1979).

2. Intestinal Antigens

Häkkinen et al. (1968a) also raised antisera against gastric juice glycoproteins from patients with gastric intestinal metaplasia. This antigen was present in normal colonic mucosa but was absent in the gastric mucosa of normal young subjects. When fetal stomach and gastric tumor tissue were examined a similar pattern was seen in that the intestinal antigen is present in the same location in both cases (Häkkinen et al., 1968a,b). DeBoer et al. (1969) have also been able to demonstrate an intestinal antigen which is present in fetal gastric epithelium and which disappears soon after birth. This antigen reappears in senescence, metaplasia, and neoplasia. A sulfated glycopeptidic antigen (SGA) was purified from gastric mucosa by papain digestion (Bara et al., 1978). Using immunofluorescence techniques this antigen was shown to be present in normal and fetal intestine but not in normal or fetal stomach. However, intestinal metaplasia in the vicinity of gastric cancer always contained SGA as did gastric cancers with colloid areas.

3. Carcinoembryonic Antigen

Carcinoembryonic antigen (CEA) can be detected immunohistochemically in many well-differentiated gastric cancers but to a lesser extent in undifferentiated tumors (Denk *et al.*, 1974b). In some studies, CEA has also been found in normal gastric mucosa (Khoo *et al.*, 1973; Ejeckam *et al.*, 1979). Because, as outlined in Section II on colorectal cancer, plasma CEA values are of little use in the detection of cancer, several laboratories have recently turned to the study of the levels of CEA in samples of gastric juice. Although the results are variable (Molnar *et al.*, 1976; Bunn *et al.*, 1979; Tatsuta *et al.*, 1980; Fujimoto *et al.*, 1979; Satake *et al.*, 1980), one important fact which emerges is that elevated CEA levels are always accompanied by at least benign changes in the gastric mucosa which may be premalignant.

4. Other Oncodevelopmental Antigens

The pregnancy-specific antigen SP1 described in Section II on colorectal cancer can also be detected, using immunoperoxidase techniques, in gastric cancer tissue (Bohn *et al.*, 1981). In addition to this, four other proteins which are present in placental extracts but not in fetal or adult human tissues can also be detected in gastric cancer tissue (Bohn *et al.*, 1981). However the same proteins are present in cancer tissue from a number of other organs, suggesting they are general oncodevelopmental markers and not specific for stomach.

As with colorectal cancer, oncodevelopmental markers present in gastric cancer can also be detected by measuring cell-mediated immune reactions (Zöller *et al.*, 1979). Another oncodevelopmental marker which has been detected in gastric cancer is α-fetoprotein (AFP). AFP is already a well-studied marker for liver cell carcinoma, but elevated serum levels of AFP have also been reported in gastric cancer. Since many of the reported cases were associated with hepatic metastases, Kodama *et al.* (1981) used immunohistochemical staining to study sections of gastric cancer tissue. They found that in 54% of the cases, cells positive for AFP could be seen. This marker can be considered to be oncodevelopmental since Gitlin *et al.* (1972) showed AFP to be present in human fetal gastrointestinal tract as well as in the liver and yolk sac.

B. Metabolic Markers

Hirsch-Marie *et al.* (1976) found that normal adult mucosa contained three forms of pepsinogen (II, III, and IV), but in fetal human stomach, type IV was the predominant form of enzyme with very small amounts of II and III. When gastric carcinoma tissue was compared with normal controls, pepsinogens II and III were often absent in the extracts, whereas type IV was always present.

C. Glycoconjugate Markers

As with the colon, human fetal stomach contains alcohol-soluble blood group substances early in gestation which later disappear and are replaced by water-soluble blood group substances associated with secretion (Szulman, 1964). ABH antigens are present in the adult stomach. These changes have led to conflicting results concerning the presence or absence of blood group substances in gastric cancer tissue since 1949 when the deletion of blood group activity in gastric cancer was first observed (Oh-uti, 1949). Denk *et al.* (1974b) explained the paradox by the fact that the diminution in blood group substances seen in most studies (Davidsohn *et al.*, 1966; Kapadia *et al.*, 1981) was of the water-soluble form and that when ABH antigens were found in gastric cancer tissue (Denk *et al.*, 1974b) they were of the alcohol-soluble type. Both these changes can be considered to be oncodevelopmental.

The deletion of blood group substances in gastric cancer is also accompanied by the appearance of precursor forms of the antigens such as I (Ma) (Kapadia *et al.*, 1981) and ceramide trisaccharide (Watanabe and Hakomori, 1976). The available evidence suggests that, as with colorectal cancer, the deletion of ABH antigens could be due to reductions in specific glycosyltransferases (Stellner *et al.*, 1973).

V. CONCLUSIONS

This chapter has demonstrated some evidence of the wide range and large number of oncodevelopmental markers which may be important in gastrointestinal cancer. At present the most significant aspect of these markers is in the early detection of cancer as, for all the malignancies discussed in this chapter, the earlier they are detected and treated, the better the prognosis. For this reason one of the future aims of the study of oncodevelopmental markers should be the improvement of the specificity and sensitivity of the markers. Monoclonal antibodies may well prove to be a very useful methodology in this area, as many of the markers are identified using immunological methods. A second method of improving specificity and sensitivity is the use of the simultaneous assay of several markers (McIntire, 1979). Preliminary evidence also suggests that the use of antibodies against two markers rather than one enhances the radioimmunodetection of tumors (Gaffar *et al.*, 1981).

The radioimmunodetection of tumors is another fruitful area for future research in oncodevelopmental markers. This will become even more important as assays for tumor markers become more sensitive and the malignant tissue becomes harder to localize.

A final aspect of oncodevelopmental markers in gastrointestinal cancer which deserves further study is the understanding of the cellular control processes involved in development and carcinogenesis and the relationship between the two.

ACKNOWLEDGMENTS

We would like to thank Drs. C. R. Boland and P. Lance for their critical review of this manuscript. We would also like to thank Ms. T. Harrington and Mr. R. Frautschi for their excellent typing. Our research was supported by USPHS Grants CA-14905 and CA-24321 from the National Cancer Institute through the National Large Bowel Cancer Project and the National Pancreatic Cancer Project, respectively, as well as by the Veterans Administration Medical Research Service.

REFERENCES

Abdelfattah-Gad, M., and Denk, H. (1980). *JNCI, J. Natl. Cancer Inst.* **64,** 1025–1028.
Balis, M. E., and Salser, J. S. (1980). *Prog. Cancer Res. Ther.* **13,** 235–241.
Balis, M. E., Higgins, P. J., and Salser, J. S. (1981). *Banbury Rep.* **7,** 129–130.
Banwo, O., Versey, J., and Hobbs, J. R. (1974). *Lancet* **1,** 643–645.
Bara, J., and Burtin, P. (1980). *Eur. J. Cancer* **16,** 1303–1310.
Bara, J., Paul-Gardais, A., Loisillier, F., and Burtin, P. (1978). *Int. J. Cancer* **21,** 133–139.
Bohn, H., Inaba, N., and Lüben, G. (1981). *Oncodev. Biol. Med.* **2,** 141–153.
Boland, C. R., and Kim, Y. S. (1982). *Clin. Res.* **30,** 34a.
Boland, C. R., Montgomery, C. K., and Kim, Y. S. (1982). *Proc. Natl. Acad. Sci. U.S.A.* **79,** 2051–2055.
Bunn, P. A., Jr., Cohen, M. I., Widerlite, L., Nugent, J. L., Matthews, M. J., and Minna, J. D. (1979). *Gastroenterology* **76,** 734–741.
Carr-Locke, D. L. (1980). *Gut* **21,** 656–661.
Chu, T. M., Holyoke, E. D., and Douglass, H. O. (1980). *J. Surg. Oncol.* **13,** 207–214.
Conference Statement (1980). *Cancer Res.* **41,** 2017–2018.
Cooper, H. S., and Haesler, W. E., Jr. (1978). *Am. J. Clin. Pathol.* **69,** 594–598.
Cooper, H. S., Cox, B. A., and Patchefsky, A. S. (1980). *Am. J. Clin. Pathol.* **73,** 345–350.
Cubilla, A. L., and Fitzgerald, P. J. (1975). *Cancer Res.* **35,** 2234–2248.
Davidsohn, I. (1972). *Am. J. Clin. Pathol.* **57,** 715–530.
Davidsohn, I., Kovarik, S., and Lee, C. L. (1966). *Arch. Pathol.* **81,** 381–390.
DeBoer, W. G. R. M., Forsyth, A., and Nairn, R. C. (1969). *Br. Med. J.* **3,** 93–94.
Denk, H., Tappeiner, G., and Holzner, J. H. (1974a). *Eur. J. Cancer* **10,** 487–490.
Denk, H., Tappeiner, G., Davidovits, A., Eckerstorfer, R., and Holzner, J. H. (1974b). *JNCI, J. Natl. Cancer Inst.* **53,** 933–942.
Edynak, E. M., Old, L. J., Vrana, M., and Lardis, M. P. (1972). *N. Engl. J. Med.* **286,** 1178–1183.
Ejeckam, G. C., Huang, S. N., McCaughey, W. T. E., and Gold, P. (1979). *Cancer* **44,** 1606–1614.

Engvall, E., and Yonemoto, R. H. (1979). *Int. J. Cancer* **23,** 759–761.
Filipe, M. I., and Branfoot, A. C. (1976). *In* "Pathology of Gastro-Intestinal Tract" (B. C. Morson, ed.), pp. 143–178. Springer-Verlag, Berlin and New York.
Fritsché, R., and Mach, J. P. (1975). *Nature (London)* **258,** 734–737.
Fujimoto, S., Kitsukawa, Y., and Itoh, K. (1979). *Ann. Surg.* **189,** 34–38.
Gaffar, S. A., Pant, K. D., Shochat, D., Bennett, S. J., and Goldenberg, D. M. (1981). *Int. J. Cancer* **27,** 101–105.
Gelder, F. B., Reese, C. J., Moossa, A. R., Hall, T., and Hunter, R. (1978). *Cancer Res.* **38,** 313–324.
Gitlin, D., Perricelli, A., and Gitlin, G. M. (1972). *Cancer Res.* **32,** 979–982.
Gold, P., and Freedman, S. O. (1965a). *J. Exp. Med.* **121,** 439–462.
Gold, P., and Freedman, S. O. (1965b). *J. Exp. Med.* **122,** 467–481.
Goldenberg, D. M., Garner, T. F., Pant, K. D., and van Nagell, J. R., Jr. (1978). *Cancer Res.* **38,** 1246–1249.
Goldenberg, D. M., Kim, E. E., Deland, F. H., Bennett, S., and Primus, F. J. (1980). *Cancer Res.* **40,** 2984–2992.
Häkkinen, I. (1979). *Immunol. Ser.* **9,** 342–357.
Häkkinen, I., Järvi, Ö., and Grönroos, J. (1968a). *Int. J. Cancer* **3,** 572–581.
Häkkinen, I., Korhonen, L. K., and Saxén, L. (1968b). *Int. J. Cancer* **3,** 582–592.
Häkkinen, I. P. T., Heinonen, R., Inberg, M. V., Järvi, Ö. H., Vaajalahti, P., and Vükari, S. (1980). *Cancer Res.* **40,** 4308–4312.
Hakomori, S.-I. (1981). *Semin. Hematol.* **18,** 39–62.
Hansen, H. J., Snyder, J. J., Miller, E., Vandervoorde, J. P., Miller, O. N., Hines, L. R., and Burns, J. J. (1974). *Hum. Pathol.* **5,** 139–147.
Hellström, I., Hellström, K. E., and Shepard, T. H. (1970). *Int. J. Cancer* **6,** 346–351.
Herlyn, M., Steplewski, Z., Herlyn, D., and Koprowski, H. (1979). *Proc. Natl. Acad. Sci. U.S.A.* **76,** 1438–1442.
Herzfeld, A., and Greengard, O. (1980). *Cancer* **46,** 2047–2054.
Hine, K. R., Bradwell, A. R., Reeder, T. A., Drolc, Z., and Dykes, P. W. (1980). *Cancer Res.* **40,** 2993–2996.
Hirsch-Marie, H., Loisillier, F., Touboul, J. P., and Burtin, P. (1976). *Lab. Invest.* **34,** 623–632.
Hobbs, J. R., Knapp, M. L., and Branfoot, A. C. (1980). *Oncodev. Biol. Med.* **1,** 37–48.
Holyoke, D., Reynoso, G., and Chu, T. M. (1972). *Ann. Surg.* **176,** 559–564.
Horne, C. H. W., and Bremner, R. D. (1980). *In* "Cancer Markers: Diagnostic and Developmental Significance" (S. Sell, ed.), pp. 225–247. Humana Press, Clifton, New Jersey.
Kapadia, A., Feizi, T., Jewell, D., Keeling, J., and Slavin, G. (1981). *J. Clin. Pathol.* **34,** 320–337.
Khoo, S. K., Warner, N. L., Lie, J. F., and Mackay, I. R. (1973). *Int. J. Cancer* **11,** 681–687.
Klavins, J. V., Mesa-Tejad, R., and Weiss, M. (1971). *Nature (London) New Biol.* **234,** 152–154.
Kodama, T., Kameya, T., Hirota, T., Shimosato, Y., Ohkura, H., Mukojima, T., and Kitaoka, H. (1981). *Cancer* **48,** 1647–1655.
Koldovsky, P., and Weinstein, J. (1973). *Natl. Cancer Inst. Monogr.* **37,** 33–35.
Koprowski, H., Herlyn, M., Steplewski, Z., and Sears, H. F. (1981). *Science* **212,** 53–55.
Lev, R. (1968). *Histochem. J.* **1,** 152–165.
Lev, R., and Orlic, D. (1974). *Histochemistry* **39,** 301–311.
LoGerfo, P., Krupey, J., and Hansen, H. J. (1971). *N. Engl. J. Med.* **285,** 138–141.
Ma, J., DeBoer, W. G. R. M., Ward, H. A., and Nairn, R. C. (1980). *Br. J. Cancer* **41,** 325–328.
McCabe, R. P., Kupchik, H. Z., Saravis, C. A., Broitman, S. A., Greeg, J. A., and Zamcheck, N. (1976). *JNCI, J. Natl. Cancer Inst.* **56,** 885–889.
Mach, J.-P., Jaeger, P., Bertholet, M. M., Ruegsegger, C. H., Loosli, R. M., and Pettavel, J. (1974). *Lancet* **2,** 535–540.

Mach, J.-P., Forni, M., Ritschard, J., Buchegger, F., Carrel, S., Widgren, S., Donath, A., and Alberto, P. (1980). *Oncodev. Biol. Med.* **1**, 49–69.
McIntire, K. R. (1979). *Immunol. Ser.* **9**, 521–539.
MacKay, A. M., Patel, S., Carter, S., Stevens, V., Laurence, D. J. R., Cooper, E. H., and Neville, A. M. (1974). *Br. Med. J.* **4**, 382–385.
Magnani, J. L., Brockhaus, M., Smith, D. F., Ginsburg, V., Blaszczhk, M., Mitchell, K. F., Steplewski, Z., and Koprowski, H. (1981). *Science* **212**, 55–56.
Medical Research Council Tumour Products Committee (1980). *Br. J. Cancer* **41**, 976–979.
Mihas, A. A. (1978). *JNCI, J. Natl. Cancer Inst.* **60**, 1439–1444.
Moertel, C. G., Schutt, A. J., and Go, V. L. W. (1978). *JAMA, J. Am. Med. Assoc.* **239**, 1065–1066.
Molnar, I. G., Vandevoorde, J. P., and Gitnick, G. L. (1976). *Gastroenterology* **70**, 513–515.
Munjal, D. D. (1980). *Clin. Chem. (Winston-Salem, N.C.)* **26**, 1809–1812.
Nishida, K., Yoshikawa, T., Kondo, M., and Thiele, H.-G. (1980). *Hepato-Gastroenterol.* **27**, 488–494.
Nishida, K., Kondo, M., Jessen, K., and Classen, M. (1981). *Hepato-Gastroenterol.* **28**, 102–105.
Oh-uti, K. (1949). *Tohoku J. Exp. Med.* **51**, 297–304.
Rees, W. V., Irie, R. F., and Morton, D. L. (1981). *JNCI, J. Natl. Cancer Inst.* **67**, 557–561.
Rousset, M., Robine-Leon, S., Dussaulx, E., Chevalier, G., and Zweibaum, A. (1979). *Front. Gastrointest. Res.* **4**, 80–85.
Satake, K., Yakashita, K., Kitamura, T., Tei, Y., and Umeyama, K. (1980). *Am. J. Surg.* **139**, 714–718.
Schultz, D. R., and Yunis, A. A. (1979). *JNCI, J. Natl. Cancer Inst.* **62**, 777–785.
Sharma, M. P., Gregg, J. A., Loewenstein, M. S., McCabe, R. P., and Zamcheck, M. (1976). *Cancer* **38**, 2457–2461.
Silverberg, E. (1982). *Ca—Cancer J. Clin.* **32**, 15–31.
Skinner, J. M., and Whitehead, R. (1981). *Cancer* **47**, 1241–1245.
Slayter, H. S., and Coligan, J. E. (1975). *Biochemistry* **14**, 2323–2330.
Sorokin, J. J., Sugarbaker, P. H., Zamcheck, N., Pisick, M., Kupchik, H. Z., and Moore, F. D. (1974). *JAMA, J. Am. Med. Assoc.* **228**, 49–53.
Stellner, K., Hakomori, S.-I., and Warner, G. A. (1973). *Biochem. Biophys. Res. Commun.* **55**, 439–445.
Szulman, E. A. (1964). *J. Exp. Med.* **119**, 503–516.
Tatarinov, Y. S., and Sokolov, A. V. (1977). *Int. J. Cancer* **19**, 161–166.
Tatsuta, M., Itoh, T., Okuda, S., Yamamura, H., Baba, M., and Taumura, H. (1980). *Cancer* **46**, 2686–2692.
Thomson, D. M. P., Krupey, J., Freedman, S. O., and Gold, P. (1969). *Proc. Natl. Acad. Sci. U.S.A.* **64**, 161–167.
Watanabe, K., and Hakomori, S.-I. (1976). *J. Exp. Med.* **144**, 644–653.
Whitehead, J. S., Fearney, F. J., and Kim, Y. S. (1979). *Cancer Res.* **39**, 1259–1263.
Wiley, E. L., Mendelsohn, G., and Eggleston, J. C. (1981). *Lab. Invest.* **44**, 507–513.
Winawer, S. J., Fleisher, M., Green, S., Bhargava, D., Leidner, S. D., Boyle, C., Sherlock, P., and Schwartz, M. K. (1977). *Gastroenterology* **73**, 719–722.
Zamcheck, N. (1980). *Prog. Cancer Res. Ther.* **13**, 219–234.
Zamcheck, N., Moore, T. L., Dhar, P., and Kupchik, H. (1972). *N. Engl. J. Med.* **286**, 83–86.
Zöller, M., Matzku, S., Schulz, V., and Price, M. R. (1979). *JNCI, J. Natl. Cancer Inst.* **63**, 285–293.

CHAPTER 17
The Carcinoembryonic Antigen and Its Cross-Reacting Antigens

P. BURTIN   M. J. ESCRIBANO
Laboratoire d'Immunochimie
Institut de Recherches Scientifiques sur le Cancer
94802 Villejuif, France

I.	Introduction	315
II.	Cell and Tissue Localization	316
III.	Biosynthesis and Catabolism of CEA	318
IV.	Physicochemical Characteristics	319
V.	CEA Evaluation as a Marker of Neoplasia	322
VI.	Monoclonal Antibodies against CEA	324
VII.	Utilization of Anti-CEA Antibodies in Radioimmunodetection	325
VIII.	Use of Anti-CEA Antibodies in Radioimmunotherapy	325
IX.	Antigens that Cross-React with CEA	326
	A. The Nonspecific Cross-Reacting Antigen (NCA)	326
	B. The Nonspecific Cross-Reacting Antigen 2 (NCA 2)	329
	C. Biliary Glycoprotein I (BGPI)	329
	D. Other Cross-Reacting Antigens	330
	E. Conclusions	330
	Reviews	331
	References	331

I. INTRODUCTION

The carcinoembryonic antigen (CEA) was first discovered by Gold and Freedman in 1965 in perchloric extracts of human colonic tumors and their metastases. CEA was also independently found by von Kleist and Burtin (1969a). Using immunoprecipitation in agar gels, Gold and Freedman (1965) detected CEA in extracts of all tumors arising from the endodermally derived digestive system epithelium as well as in embryonic gut, pancreas, and liver cells in the first two trimesters of gestation.

More sensitive immunological techniques, such as tissue localization by immunofluorescence (IF) and radioimmunoassay, demonstrated the presence of CEA in a wide variety of prenatal and adult tissues (Martin and Martin, 1971; Pusztaszeri and Mach, 1973). CEA radioimmunoassay has been of great use since the original work of Thomson *et al.* (1969).

Although CEA assay has only limited diagnostic value, this antigen is nonetheless a very interesting marker for the follow-up of cancer patients after surgery or other treatments.

A final topic of discussion in this paper is the discovery of several normal antigens that cross-react with CEA.

II. CELL AND TISSUE LOCALIZATION

The carcinoembryonic antigen is mainly present at the apical pole of malignant and normal epithelial cells.

In colorectal tumors CEA is shown at the luminal surface of tumor glands (Figs. 1 and 2) and produces a bright fluorescent line when stained with anti-CEA serum using immunofluorescence technique (von Kleist and Burtin,

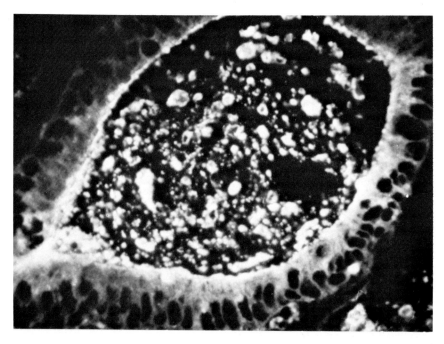

Fig. 1. Immunofluorescence pattern of a colonic tumor transverse section stained by anti-CEA serum. × 400.

17. CEA and Cross-Reacting Antigens

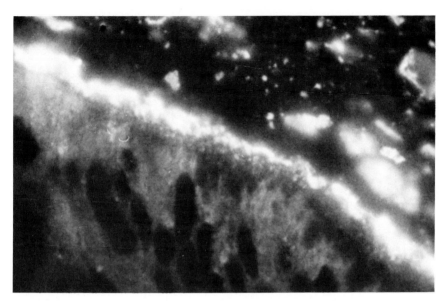

Fig. 2. Immunofluorescence pattern of a colonic tumor longitudinal section stained by anti-CEA serum. × 900.

1969b). By the same technique intraluminal deposits are also observed. The fluorescence patterns are related to tumor differentiation: the more anaplastic the tumor, the less CEA it contains. In very anaplastic tumors the staining can even be negative. Peritumoral areas containing hyperplastic glands are strongly positive as a rule even in samples taken more than 10 cm from the tumor site.

Staining of normal colonic mucosa with anti-CEA serum always produces weak results, but inflammatory mucosae, such as those present in hemorrhoids, give clearly positive reactions. The same is true for colorectal polyps either benign or premalignant. In general the staining is related to tissue differentiation: the more differentiated polyps produce the brightest images.

Ulcerative colitis mucosa also stains strongly, although not in premalignant stage. CEA is also present in other tumors and normal tissues (Table I).

In tumors of the stomach and ovary CEA is also localized at the luminal surface of tumor glands. Large amounts of CEA were characterized in the ovarian cysts as well, mainly in those exhibiting an intestinal differentiation (Van Nagell *et al.*, 1975). In these cases CEA is also found in the surface of the cyst epithelium.

Medullary carcinomas of the thyroid present CEA at both poles of tumor cells, i.e., apical and basal. Moreover, large positive CEA deposits were present in the connective peritumoral tissue (Ishikawa and Hamada, 1976; Burtin *et al.*, 1979).

TABLE I
Tissue Localization of CEA

Digestive tract
    Cancers of the colon and the rectum, the pancreas, the stomach
    Hepatomas
    Fetal organs: colonic and gastric mucosa, pancreas
    Normal and inflammatory colonic mucosa
    Polyps of the colon and rectum
Nondigestive organs
    Normal liver (traces)
    Normal bronchic mucosa (traces)
    Carcinomas of the lung, breast, cervix, etc.
    Carcinomas of the thyroid of medullary type (MCT)
    Ovarian mucoid cysts and carcinomas
Fluids and secretion products
    Plasma
    Saliva
    Feces and meconium
    Malignant pleural effusions and ascites

CEA is seldom seen in the cytoplasm. One exception is breast adenocarcinoma, which contains CEA in the cytoplasm of tumor cells as well as in premalignant tissue. Cytoplasmic staining is also observed in isolated tumor cells of gastrointestinal localization (Fig. 3) such as those designated "signet ring cells."

Altogether these results indicate that CEA detection by immunohistological technique does not provide an effective diagnostic tool for cancer.

Cultured tumors, mainly colonic ones, produce CEA, either *in vivo* after grafting in hamsters or nude mice, or *in vitro*. Established cell lines of colonic origin usually synthesize small amounts of CEA. Anti-CEA antibodies can induce polar redistribution (capping) of the CEA expressed on the membrane of cultured human intestinal cells.

Electron microscopy studies showed that CEA is present in the glycocalix (Gold *et al.*, 1970), thus it is not a structural component of the cell membrane, but rather a cell-associated antigen. Its fate is to be excreted, and relatively large amounts of CEA were found in stools.

III. BIOSYNTHESIS AND CATABOLISM OF CEA

Unlike other carcinoembryonic or oncofetal antigens, CEA is found exclusively in man. CEA is specifically synthesized by gastrointestinal tumor cells. This was demonstrated by the fact that human colonic tumors in organ culture

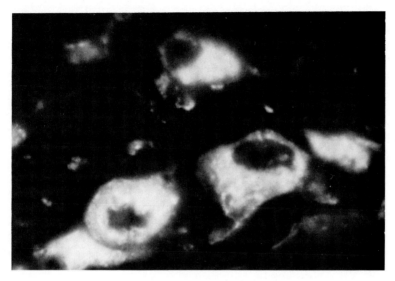

Fig. 3. Immunofluorescence pattern of isolated colonic tumor cells stained by anti-CEA serum. × 900.

contain CEA and that human colonic cancer cells transferred to golden hamsters or to nude mice continued to produce CEA.

Very little is known about the turnover of CEA. Circulating CEA in man is almost undetectable 2–14 days after surgery in patients who have undergone apparently curative resections. This suggests a rather rapid catabolism. Animals given CEA show different rates of clearance, probably reflecting heterogeneity within the population of CEA molecules.

IV. PHYSICOCHEMICAL CHARACTERISTICS

Carcinoembryonic antigen is a macromolecular glycoprotein which, in analytical ultracentrifugation, shows a single peak with a sedimentation coefficient of 6.8–7.2. The molecular weight estimated by polyacrylamide gel electrophoresis is approximately 180,000. Immunoelectrophoresis reveals a polydisperse line, having, most frequently, β mobility at pH 8.2–8.6. Extensive charge heterogeneity has been demonstrated by isoelectric focusing. Immunochemically, all CEA preparations are identical whatever the electric charge. This is true in immunoprecipitation in agar (double diffusion), but some degree of cross-reactivity among the samples has been observed in competitive inhibition techniques.

The CEA protein portion consists of a single polypeptide chain with six cystine disulfide bonds, whose integrity is essential for the CEA immunoreactivity. Various laboratories have reported similar CEA amino acid compositions (Table II). The major amino acid residues are aspartic and glutamic acid and/or their amine salts. There are low levels of basic and aromatic residues.

The carbohydrate content of CEA represents about 50% of its weight, although considerable variations on the carbohydrate composition have been observed in different tumors. The content of the major monosaccharides does not show great variation, but sialic acid can range between 0 and 20 mole/10^5 g. Changes in the carbohydrate content may explain the charge heterogeneity of CEA.

TABLE II
Chemical Composition of CEA

Amino acids	A (Burtin et al., unpublished data)	B (from R3)	C (from R4)
Asp (Asn)	13.1–14.5	15.7	14.4–15.8
Thr	8.3–8.5	8.7	8.7–9.1
Ser	10.7–10.8	11.4	10.9–11.3
Glu (Gln)	11.4–11.5	9.6	9.6–10.2
Pro	7.2–8	8.0	6.9–8.1
Gly	6.2–7	6.3	5.0–5.6
Ala	5.8–6.2	6.1	5.6–6.2
Val	6.2–6.6	6.6	6.8–7.2
Met	0–0	0.3	0 –0.8
Cys ½	0–1.33	1.6	1.7–2.2
Ileu	4.3–6.6	3.7	4.9–5.1
Leu	8.0–8.3	7.6	8.1–8.6
Tyr	4.3–4.35	4.0	4.0–4.4
Phe	2.4–2.7	2.2	1.8–2.5
His	1.9–1.95	1.7	1.5–1.8
Lys	2.7–3.1	2.9	2.2–2.6
Arg	3.6–4.1	3.1	3.4–3.8

Sugars	A (Burtin et al., unpublished data)	B (from R3)	C (from R4)
Fuc	13–14	19.5–21	16–23
Man	17–21	13.5–16	17–23
Gal	20–28	20 –26	19–23
Glc NAc	36–38	32 –43.5	35–42
Sialic acid	3–7	2.2– 6.5	ND
Percentage of carbohydrates	52–57	58	46–54

Despite the elevated carbohydrate content, most if not all, the antigenic determinants of the molecule appear to be located in the protein moiety. Thus, extensive destruction of the sugar residues has not influenced CEA reactivity in radioimmunoassay. On the contrary, breakage of the disulfide bonds largely destroys the immunoreactivity of CEA. However, some reports indicated a possible carbohydrate influence on immunoreactivity, as demonstrated by inhibition experiments using heterosaccharide fragments. The inhibitory activity of the carbohydrate compounds was, however, several thousand times lower than that of the entire CEA molecule. More specifically, Tomita *et al.* (1974) demonstrated that the precipitation of radiolabeled CEA by guinea pig anti-CEA serum was inhibited as much as 50% by acetylchitobiose. Similar results were never published concerning anti-CEA sera obtained in other species.

Several groups showed that at least some samples of CEA have patient-related blood group activity. Holburn *et al.* (1974) described CEA samples bearing A, B, or Lewis antigenic determinants, and demonstrated by radioimmunological methods that this determinant was really carried by the CEA molecule. This is important as blood group antigens may also contaminate CEA samples. Andreu *et al.* (1975) and Bali *et al.* (1976) presented evidence of H specificity on CEA molecules.

The A blood group reactivity was at first hardly understandable, as chemical analysis of CEA samples did not show the presence of *N*-acetylgalactosamine. *N*-Acetylglucosamine was the only amino sugar found in CEA molecules. Later, more accurate analyses detected small amounts of *N*-acetylgalactosamine. This sugar might be a component of some special carbohydrate chains of the mucin type that differ from the more common type found also in serum glycoproteins (N linkage between *N*-acetylglucosamine and asparagine). Blood group antigenic determinants are not related to the main epitopes of CEA, as proven by oxidation experiments made with periodic acid. Moderate oxidation of carbohydrate chains destroyed the blood group activity, whereas CEA epitopes remained unchanged.

As mentioned before, a substance similar to CEA was detected in normal colonic mucosa by immunohistological methods. Was this component identical to tumor CEA? It seems likely, as Martin and Martin (1971) were able to absorb an antiserum prepared against tumor CEA with perchloric extracts of normal colonic mucosa. Later, this very important question was answered by the isolation of normal CEA, either from normal mucosa (Fritsche and Mach, 1977) or from specimens of the intestinal washings of normal volunteers (Egan *et al.*, 1977). Both samples were found to be immunochemically identical to tumor CEA. Biochemical analysis showed that the amino acid composition (Egan *et al.*, 1977) and the N-terminal sequence (Shively *et al.*, 1978) of the CEA molecule were identical in both intestinal washings and tumor CEA. At present, there is still no evidence for a difference between normal and tumor CEA.

V. CEA EVALUATION AS A MARKER OF NEOPLASIA

Carcinoembryonic antigen was first found by Thomson et al. (1969) in the blood of patients with colonic carcinoma. Very sensitive techniques such as radio- or enzyme-immunoassay are required for CEA evaluation because its serum concentration stands at nanogram level. Different analytical methods show the upper limits of normal CEA concentrations as either 2.5 or 10 ng/ml. It is thus important to know which method was used before interpreting the results.

In the first study by Thomson et al. (1969), a strict specificity of CEA increase for digestive, mainly colonic, carcinomas was reported. Later reports from many laboratories confirmed the elevated levels in digestive tumors, but did not prove tumor and tissue specificity. Values of CEA above the normal threshold were found in carcinomas of the colon and rectum (70–80%), the pancreas (80–90%), and the stomach (70%), but also in 30–50% of the cases of various carcinomas, such as those of the lung, breast, and gynecological tract. Moreover, an increase in CEA was also observed in nonmalignant diseases of the digestive system, such as alcoholic cirrhoses and chronic pancreatitis, and in nondigestive diseases (chronic bronchitis). Furthermore, some apparently healthy people, especially heavy smokers, showed elevated levels of serum CEA.

A strong correlation exists in tumors between the number of CEA-positive cases and the extent of neoplasia. For example, only 20–40% of CEA-positive cases have been found in the first stages of colorectal tumors (Duke's stage A), but 50–60% in locally infiltrative carcinomas (Duke's stages B and C) and 80–90% in tumors with distant metastases.

The incidence of CEA increase in noncancerous diseases is variable (Table III). Two of these diseases, ulcerative colitis and Crohn's disease, have been extensively studied in the hope that the CEA assay might be of value in the early detection of malignant transformation. The results do not support any definite conclusion. From hundreds of sera examined, one can conclude that the concentration of CEA is often lower in benign diseases than in neoplasias, and that persistence of elevated levels can suggest a malignancy, but generally only in advanced phases of the disease. Moderately increased CEA levels, however, may be produced by certain diseases without malignancies.

All studies led to the same conclusions.

1. A normal CEA level does not exclude the presence of a malignant tumor, especially of small size.
2. Moderately increased CEA levels may be explained either by a malignancy or a noncancerous disease. Furthermore, the CEA level is not dependable for cancer screening surveys. It has some prognostic value, as colorectal tumors at a

TABLE III
Percentage of Serum or Plasma CEA Increases

I Digestive diseases	
Cancers of the colon and rectum (total)	65–90%
Cancers of the colon and rectum (Duke's stage A)	20–40%
Cancers of the stomach	50–70%
Cancers of the pancreas	80–90%
Hepatomas	50–60%
Noncancerous diseases	
Ulcerative colitis	10–50%
Crohn's disease	20–25%
Diverticulitis	20–25%
Colorectal polyps	15–40%
Gastroduodenal ulcers	0–30%
Chronic pancreatitis	40–60%
Liver cirrhosis	50–60%
Acute hepatitis	15–25%
II. Nondigestive diseases	
Cancers of the breast, lung, cervix, etc.	30–50%
Pulmonary emphysema	50–60%
Benign gynecological diseases	0–5%
III. Controls	
Blood donors	0–4%
Heavy smokers	10–20%

given stage have a worse prognosis if the CEA is elevated in comparison with cases exhibiting normal CEA levels.

The main interest of the CEA assay is patient surveillance after tumor removal. After surgery, the CEA level usually falls abruptly within the first 15 days. If it does not return to normal values within one month, an incomplete resection or occult metastasis is generally indicated.

Based on the results of several studies (see review by von Kleist and Burtin) one in 65 cases showed consistently elevated levels of CEA after complete surgical removal of tumors. Persistently elevated CEA levels were always noted when tumors could not be removed completely. Thus, a high level of CEA occurring after an apparently complete surgery, or decreasing only partially thereafter, is strong evidence for the existence of residual tumor or metastases that were not detected during the operation. However, as has been noted, it is important to wait at least 30 days after surgery before such a conclusion is drawn.

With repeated serum tests for CEA, several possibilities may occur. The first is the stability of CEA values in the normal zone. This generally indicates favorable evolution of the illness. In other cases, CEA reaches high levels only after several tests. A tumor recurrence, or more frequently, a metastasis, is the cause of this change and is irrevocably terminal. It is important to stress that the clinical signs of tumor evolution often begin several months after the CEA elevation is noted. In certain patients, CEA levels increase gradually and slowly; these cases are difficult to interpret. For some authors, especially Martin *et al.* (1977), when a slow increase of CEA levels is observed during three successive tests performed at 1-month intervals, this is by itself strong evidence for a malignant recurrence, strong enough to justify exploratory surgery. Others (NIH Consensus Development Statement, 1981, Steele *et al.*, 1980) are more hesitant and request only a thorough examination of the patient, for several reasons. First, the rise in CEA is sometimes explained by such ailments as hepatitis rather than cancer recurrence. Furthermore, the recurrence or metastases are not always resectable, especially in tumors of noncolonic origin. Finally, it is not certain that the successful resection of a metastasis means a better prognosis for the patient, as other metastases may develop later. Patients often undergo chemotherapy, radiotherapy, or both. It seems useful to follow the CEA level during these treatments. If they are effective, CEA returns to normal values. On the other hand, if CEA levels do not decrease, an alternate method of treatment should be considered.

VI. MONOCLONAL ANTIBODIES AGAINST CEA

The first monoclonal antibodies against CEA were obtained by Accola *et al.* (1980); many others have been produced in other laboratories. According to inhibition experiments, the antibodies belong to at least six groups, thus revealing at least six antigenic determinants on the part of the CEA molecule which is not shared by NCA (nonspecific cross-reacting antigen) (Miggiano *et al.*, 1980). Curiously, the antibodies of one of these six groups precipitated with CEA, showing that they react with an epitope that is repetitive on the CEA molecule.

Monoclonal antibodies to CEA were used in an enzyme immunoassay by Buchegger *et al.* (1982), producing results which correlated well with those obtained by conventional antisera. In particular, both methods showed an increase of CEA in noncancerous patients and in healthy people. Thus, "false positive" sera remained positive.

Kupchik *et al.* (1981) employed their anti-CEA monoclonal antibodies as an immunoabsorbent to purify CEA. The sample thus obtained had a better immunoreactivity than conventional CEA. The same authors raised a very interesting problem: do different monoclonal antibodies, able to precipitate only 30–40% of the labeled antigen, react with variants of this antigen? If this hypothesis was true, a better study of antigenic microheterogeneity would be possible. All the variants would be precipitated together by a conventional antiserum, thus providing a control for such a study.

VII. UTILIZATION OF ANTI-CEA ANTIBODIES IN RADIOIMMUNODETECTION

Animals bearing grafted human tumors were injected with radiolabeled anti-CEA antibodies by Mach *et al.* (1974) using nude mice and by Goldenberg *et al.* (1974) using hamsters. Later, the same trials were made in human patients. Goldenberg *et al.* (1980) simultaneously injected technetium-tagged serum albumin and heavily labeled ^{131}I affinity purified goat anti-CEA antibodies. The albumin allowed the calculation of background radioactivity due to the diffusion of the reagent and its computerized substraction. The specific localization of anti-CEA antibodies thus made possible the detection of tumors of colonic or ovarian origin not detected by other methods. These were either local tumors or distant metastases in both cases as small as 2 cm in diameter. By this method, Goldenberg *et al.* (1980) were able to detect almost all the tumors studied. Mach *et al.* (1979, 1981) were less enthusiastic as they had several "false negative" and a few "false positive" cases.

It must be stressed that only a small portion of the injected antibodies (less than 1%) reaches the tumor, hence the need for a very sensitive technique. Furthermore, the detection of liver metastasis was sometimes hindered by the nonspecific fixation of antibodies in the liver. This effect was greatly reduced by the use of Fab fragments instead of the entire antibody. Up to the present time, there have been no side effects reported due to the injection of such foreign proteins in human patients.

More recently, Mach *et al.* (1981) used labeled monoclonal anti-CEA antibodies for radioimmunodetection. Their results were still less positive than those reported by Goldenberg.

VIII. USE OF ANTI-CEA ANTIBODIES IN RADIOIMMUNOTHERAPY

The first trials in this field were made on patients by Order *et al.* (1980). This group claimed some partial remissions in hepatoma patients after injections of heavily labeled anti-CEA antibodies. Recently, Goldenberg *et al.* (1981) grafted a human colonic tumor, GW 39, into the cheek pouches of hamsters, then injected the animals with either purified anti-CEA goat antibody or normal globulin, both labeled with ^{131}I. The results varied according to the amount of radioactive label injected. A dose of 0.5 mCi, either of the anti-CEA antibody or normal globulin, did not produce any effect. One millicurie of anti-CEA antibody had a remarkable effect, almost stopping tumor growth. Eighty percent of the animals injected with 1 mCi survived after 60 days, whereas all the controls died after 30 days. The same amount of radioactive isotope bound to normal goat IgG delayed the growth of the tumor, but had no effect on survival rates. This difference between antibody and control globulin disappeared when 2 mCi of either reagent were injected: both had the same remarkable action on tumor growth and survival (80% in each category). The radioactive anti-CEA antibody had a preferential fixation in the tumor, and 2.5% of the injected antibody was found per gram of tumor 3 and 5 days after the injection. The fixation of the control goat globulin was not significantly higher in the tumor than in normal tissues, and less than 1% of the radioactivity could be detected per gram of tumor. Thus, it is not clear why this radioactive control globulin could delay tumor growth so efficiently.

IX. ANTIGENS THAT CROSS-REACT WITH CEA

When radioimmunoassays detected CEA in both normal and cancerous patients, it was postulated that "false positive" cases revealed the existence of a normal antigen able to cross-react with CEA. The hypothesis proved correct, as several antigens present in normal adult tissues were discovered. However, in competitive radioimmunoassay these antigens do not generally interfere with CEA evaluation and, hence, are not responsible for the "false positive" reactions.

A. The Nonspecific Cross-Reacting Antigen (NCA)

Among the various cross-reactive CEA antigens, the best known is a glycoprotein isolated by von Kleist *et al.* (1972) from colonic tumors. This antigen was named NCA (nonspecific cross-reacting antigen) because it was also found in relatively high levels in normal adult colon, spleen, and lung tissues, as well as in tumor extracts.

The first CEA cross-reacting antigen was, however, isolated by Mach and Pusztaszeri (1972) from normal lung perchloric extracts. According to its chemical nature, the antigen was called NGP (normal glycoprotein). NGP, as well as another antigen named CEX (CEA-associated protein) by Darcy *et al.* (1973) or still CCEA2 (colonic carcinoembryonic antigen 2) by the same group (Turberville *et al.*, 1973), and an antigen called CCA III by Newman *et al.* (1974) were actually seen to be identical with NCA.

17. CEA and Cross-Reacting Antigens

The physicochemical properties of NCA are very similar to those of CEA. In particular, it is soluble in perchloric acid and for this reason, often contaminates the conventional CEA preparations. The main physicochemical properties which distinguish NCA from CEA are the molecular weight (60,000 for NCA, 180,000 for CEA) and the lower carbohydrate content (30% in NCA, up to 60% in CEA). Otherwise, the amino acid composition of both proteins are quite similar, with the possible exception of methionine residues present in NCA and absent from CEA. The sequence of the first 26 amino acids at the N-terminal end for both molecules showed only one difference: position 21 contained valine in CEA and alanine in NCA (Engvall *et al.*, 1978).

Immunological techniques have clearly demonstrated that NCA is not a fragment of CEA, as NCA possesses specific antigenic determinants not found in CEA. Thus in double diffusion, NCA and CEA, when tested with anti-NCA sera, give a partial identity reaction showing that only some of the antibodies recognize CEA. Moreover, after absorption with CEA, the sera still precipitate with NCA. Conversely, after absorption with NCA, anti-CEA sera continue to react with CEA, but not with NCA.

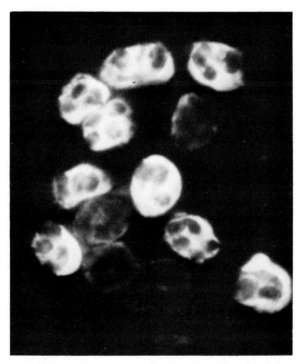

Fig. 4. Immunofluorescence pattern of peripheral blood polymorphs stained by anti-CEA serum. × 900.

Monoclonal antibodies against NCA-specific epitopes are now being prepared in our laboratory.

1. Tissue Localization

As shown by immunofluorescence, there are two main localizations of NCA: (1) The epithelial cells of the adult or fetal intestinal mucosa and gastrointestinal tumors. NCA, like CEA, is located at the apex of epithelial cells, but can be found more frequently in the cytoplasm. (2) The cells of the granulocytic and monocytic series (Burtin *et al.*, 1975; Bordes *et al.*, 1975). NCA was found in the granulocytes of peripheral blood (Fig. 4) as well as in immature bone marrow cells from the stage of promyelocytes (Fig. 5).

NCA was found in the azurophilic granulations of the polymorphs. Because of its resistance to proteases this antigen possibly assists in protecting these cells against their own hydrolases.

In addition, NCA is present in alveolar macrophages. NCA presence in monocytes of the peripheral blood (Burtin and Fondaneche, 1981) was shown only after adherence, probably because the antigen is inaccessible to the antibody when monocytes are in suspension. NCA was also detected in the cytoplasm of leukemic myeloblasts showing some degree of differentiation (type M2 of FAB classification). In chronic myeloid leukemias, many pathological cells are stained by anti-NCA serum, although the stain is weaker than for normal granulocytic cells (Heumann *et al.*, 1979; Burtin *et al.*, 1980).

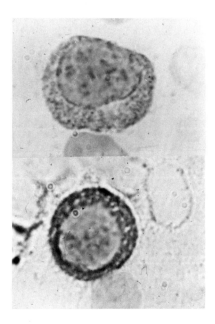

Fig. 5. Immunoperoxidase pattern of bone marrow immature granulocytic cells stained by anti-NCA serum. Top: promyelocyte. Bottom: myelocyte. × 1000.

2. NCA Assay

An NCA radioimmunoassay was set up in several laboratories (von Kleist et al., 1977; Wahren et al., 1980; Frenoy and Burtin, 1980). The normal value was found to be around 40 ng/ml by our group (Frenoy and Burtin, 1980). Important variations of this level were seen in several conditions: (1) In gastrointestinal carcinomas (von Kleist et al., 1977), a strong increase was occasionally observed, roughly parallel to that of CEA. (2) In inflammatory diseases (von Kleist et al., 1977), NCA elevations were found in lung ailments and hepatic cirrhoses. (3) In leukemias, a decrease of NCA was seen in acute myeloid leukemias and an increase was seen in chronic myeloid leukemias (Wahren et al., 1980; Frenoy and Burtin, 1980). In this disease, NCA levels were higher in the chronic phase than in blastic crisis, and somewhat correlated to the number of white blood cells and polymorphs in the blood.

The evolution of NCA levels was studied at different stages of chronic myeloid leukemias. After chemotherapy these levels fell to normal values, while pathological myeloid cells disappeared from the blood.

B. The Nonspecific Cross-Reacting Antigen 2 (NCA 2)

The nonspecific cross-reacting antigen 2 (NCA 2) was found in perchloric extracts of meconium and in stools by Burtin et al. (1973). This glycoprotein has an antigenic moiety in common with CEA, including that found on NCA, but containing another specific antigenic determinant, which is weak and difficult to detect.

NCA 2 has an approximate molecular weight of 120,000 and a β mobility. By immunofluorescence, NCA 2 was found in the cytoplasm of normal colonic cells, in gastrointestinal tumors, and in intestinal metaplasias of the gastric mucosa.

C. Biliary Glycoprotein I (BGPI)

Biliary glycoprotein I (BGPI) was discovered by Svenberg (1976) in human hepatic bile. It is a perchlorosoluble glycoprotein having specific antigenic determinants as well as common antigenic structure with both CEA and NCA. Thus absorption of anti-BGPI sera with CEA also abolished the reaction with NCA, while leaving a specific reaction with BGPI.

This antigen has a molecular weight of 80,000 and contains about 40% carbohydrate. The amino acid and carbohydrate composition are very similar to those of CEA. By immunofluorescence, BGPI was shown to be located at the surface of biliary canaliculi. Here again, BGPI covers the epithelium surface. Evaluation of the circulating BGPI by radioimmunoassay has been performed by Hammarström's group, which showed a level of 400 to 500 ng/ml in normal sera and a high increase in the sera of patients with hepatic disease of any type. At this time, BGPI has not proved to be of value as a diagnostic aid.

D. Other Cross-Reacting Antigens

Matsuoka *et al.* (1973) described several fecal antigens cross-reacting with CEA. One is NCA. Another was called NFCA, for normal fecal cross-reacting antigen. Recently, Kuroki *et al.* (1981) described another antigen, NFA 1, which had a small molecular weight (about 20,000) and bore a determinant common to CEA, but different from the determinants of NCA.

Another cross-reacting antigen, the CELIA (carcinoembryonic-like antigen), was found by Vuento *et al.* (1976), in the gastric juice of patients afflicted with gastric carcinomas.

A third component, the CEA-low, was found by Hammarström *et al.* (1978) in small amounts in the perchloric extracts of liver metastases of colonic tumors. It has a molecular weight around 120,000. It is less glycosylated than CEA and is antigenically incomplete in comparison to CEA, as the precipitin line of the latter shows a weak spur over that of CEA-low when reacting with an anti-CEA serum. It is not clear if CEA-low has a specific antigenic determinant; the antigen could be a degradation product of genuine CEA.

Finally, the TEX was isolated by Kessler *et al.* (1978) from extracts of colonic tumors. Its molecular weight is around 120,000. Its relationship with other antigens such as NCA or CEA-low is not yet known.

E. Conclusions

In conclusion, there are several components present in normal or tumor tissues that cross-react with CEA. Most of them are not fragments of CEA, as they have specific antigenic determinants. Furthermore, their tissue localization is different.

Their existence makes careful methodology necessary, especially in immunohistological studies. Obviously, data cannot be valid if they are obtained with an antiserum, the specificity of which is not clearly assessed (and obtained by careful absorption). On the contrary, radioimmunoassays are not disturbed by these cross-reactions, as long as competition methods are used. This can be explained by the low affinity of cross-reacting antigens when highly diluted antisera are employed (Primus *et al.*, 1977).

Quantitation of CEA is not influenced by the presence in the serum of molecules such as BGPI or NCA. This is fortunate because, as mentioned before, the levels of NCA and of BGPI are much higher than those of CEA. Cross-reactions

17. CEA and Cross-Reacting Antigens

did not likewise occur during assays of NCA and BGPI. However, when one uses sandwich techniques, either in enzyme- or radioimmunoassays, cross-reactions occur. It is thus necessary to prevent them by utilizing antisera, either truly specific, such as the anti-CEA sera obtained from baboons or guinea pigs, or rendered specific by absorption. More recently, the production of monoclonal antibodies has brought an elegant solution to this problem.

REVIEWS

Because of space limitations for this chapter, the reader is referred to some reviews indicated below where he or she can find many references concerning the biochemistry and tissue localization of CEA, as well as the clinical value of its radioimmunoassay. Other references are given either to mention some of the main articles that have been published in the field since 1965 or to quote more recent papers.

Burtin, P. (1978). *Ann. Immunol. (Paris)* **129C**, 185–198.
Fuks, A., Banjo, C., Schuster, J., Freedman, S. O., and Gold, P. (1974). *Biochim. Biophys. Acta* **417**, 123–152.
Hammarström, S., Svenberg, T., Hedin, A., and Sunblad, G. (1978). *Scand. J. Immunol.* **7**, Suppl. 6, 33–46.
Shively, J., and Todd, C. W. (1978). *Scand. J. Immunol.* **7**, Suppl. 6, 19–32.
von Kleist, S., and Burtin, P. (1977). *Prog. Gastroenterol.* **3**, 595–616.

REFERENCES

Accola, R. S., Carrel, S., and Mach, J. P. (1980). *Proc. Natl. Acad. Sci. U.S.A.* **77**, 563.
Andreu, G., Chavanel, G., and Burtin, P. (1975). *Immunochemistry* **12**, 921.
Bali, J. P., Magous, R., Lecou, C., and Mousseron-Canet, M. (1976). *Cancer Res.* **36**, 2124.
Bordes, M., Knobel, S., and Martin, F. (1975). *Eur. J. Cancer* **23**, 741.
Buchegger, F., Phan, M., Rivier, D., Carrel, S., Accolla, R., and Mach, J. P. (1982). *J. Immunol. Methods* **49**, 129.
Burtin, P., and Fondaneche, M. C. (1981). *Clin. Immunol. Immunopathol.* **20**, 146.
Burtin, P., Chavanel, G., and Hirsch-Marie, H. (1973). *J. Immunol.* **3**, 1926.
Burtin, P., Quan, P. C., and Sabine, M. C. (1975). *Nature (London)* **255**, 714.
Burtin, P., Calmettes, C., and Fondaneche, M. C. (1979). *Int. J. Cancer* **23**, 741.
Burtin, P., Flandrin, G., and Fondaneche, M. C. (1980). *Blood Cells* **6**, 263.
Darcy, D., Turberville, C., and James, R. (1973). *Br. J. Cancer* **28**, 147.
Egan, M., Pritchard, D. G., Todd, C. W., and Go, L. W. (1977). *Cancer Res.* **37**, 2638.
Engvall, E., Shively, J., and Wrann, M. (1978). *Proc. Natl. Acad. Sci. U.S.A.* **75**, 1670.
Frenoy, N., and Burtin, P. (1980). *Clin. Chim. Acta* **103**, 23.
Fritsche, R., and Mach, J. P. (1977). *Immunochemistry* **14**, 119.
Gold, P., and Freedman, S. O. (1965). *J. Exp. Med.* **122**, 467.
Gold, P., Krupey, J., and Ansari, H. (1970). *JNCI, J. Natl. Cancer Inst.* **45**, 219.
Goldenberg, D. M., Preston, D. F., Primus, F. J., and Hansen, H. J. (1974). *Cancer Res.* **34**, 1.
Goldenberg, D. M., Kim, E. E., De Land, F. H., Bennett, S., and Primus, J. E. (1980). *Cancer Res.* **40**, 2984.

Goldenberg, D. M., Gaffar, S. A., Bennett, S. J., and Beach, J. L. (1981). *Cancer Res.* **41,** 4354.
Hammarström, S., Svenberg, T., Hedin, A., and Sunblad, G. (1978). *Scand. J. Immunol., Suppl. 6,* 33.
Heumann, D., Candjardis, P., Carrel, S., and Mach, J. P. (1979). *In* "Carcino-Embryonic Proteins" (F. G. Lehmann, ed.). Elsevier-North Holland, Amsterdam, II, 3.
Holburn, A. M., Mach, J. P., MacDonald, D., and Newlands, M. (1974). *Immunology* **26,** 841.
Ishikawa, N., and Hamada, S. (1976). *Br. J. Cancer* **34,** 111.
Kessler, M.J., Shively, J., Pritchard, D. G., and Todd, C. W. (1978). *Cancer Res.* **38,** 1041.
Kupchik, H. Z., Zurawski, V. R., Hurrell, J. G., Zamcheck, N., and Black, P. H. (1981). *Cancer Res.* **41,** 3306.
Kuroki, M., Koga, Y., and Matsuoka, Y. (1981). *Cancer Res.* **41,** 713.
Mach, J. P., and Pusztaszeri, G. (1972). *Immunochemistry* **9,** 1031.
Mach, J. P., Carrel, S., Merenda, C., Sordat, B., and Cerottini, J. C. (1974). *Nature (London)* **248,** 704.
Mach, J. P., Forni, M., Ritschard, J., Carrel, S., Donath, A., and Alberto, P. (1979). *Protides Biol. Fluids* **27,** 205.
Mach, J. P., Buchegger, F., Forni, M., Ritschard, J., Berche, C., Lumbroso, T. D., Schreyer, M., Girardet, C., Accola, R., and Carrel, S. (1981). *Immunol. Today* **2,** 239.
Martin, E. W., James, K. K., Hurtubise, P. E., Catalano, P., and Minton, J. P. (1977). *Cancer* **39,** 440.
Martin, F., and Martin, M. (1971). *Int. J. Cancer* **6,** 352.
Matsuoka, Y., Hara, M., Takatsu, K., and Kitagawa, M. (1973). *Gann* **64,** 203.
Miggiano, V., Stähli, C., Häring, P., Schmidt, J., Le Dain, M., Glatthaar, B., and Staehelin, T. (1980). *Protides Biol. Fluids* **28,** 501.
NIH Consensus Development Statement (1981). *Cancer Res.* **41,** 2017.
Newman, E. S., Petras, S. E., Georgiadis, A., and Hansen, H. J. (1974). *Cancer Res.* **34,** 2125.
Order, S. E., Klein, J. L., Ettinger, D., Alderson, P., Siegelmann, S., and Leichner, P. (1980). *Cancer Res.* **40,** 3001.
Primus, J. F., Newman, E., and Hansen, H. J. (1977). *J. Immunol.* **118,** 55.
Pusztaszeri, G., and Mach, J. P. (1973). *Immunochemistry* **10,** 197.
Shively, J. E., Todd, C. W., Go, L. W., and Egan, M. L. (1978). *Cancer Res.* **38,** 503.
Steele, G., Zamcheck, N., Wilson, R., Mayer, R., Lokich, J., Ran, P., and Maltz, J. (1980). *Am. J. Surg.* **139,** 544.
Svenberg, T. (1976). *Int. J. Cancer* **17,** 588.
Thomson, D. M. P., Krupey, J., Freedman, S. O., and Gold, P. (1969). *Proc. Natl. Acad. Sci. U.S.A.* **64,** 161.
Tomita, J. T., Safford, J. W., and Hirata, A. A. (1974). *Immunology* **26,** 291.
Turberville, C., Darcy, D., Laurence, D., Johns, E., and Neville, A. (1973). *Immunochemistry* **12,** 841.
Van Nagell, J., Pletsch, Q., and Goldenberg, D. M. (1975). *Cancer Res.* **35,** 1433.
Vuento, M., Ruoslahti, E., Pihko, H., Svenberg, T., Ihamäki, T., and Siurala, M. (1976). *Immunochemistry* **13,** 313.
von Kleist, S., and Burtin, P. (1969a). *Cancer Res.* **29,** 1961.
von Kleist, S., and Burtin, P. (1969b). *Int. J. Cancer* **4,** 874.
von Kleist, S., Chavanel, G., and Burtin, P. (1972). *Proc. Natl. Acad. Sci. U.S.A.* **69,** 2492.
von Kleist, S., Troupel, S., King, M., and Burtin, P. (1977). *Br. J. Cancer* **35,** 875.
Wahren, B., Gahrton, G., and Hammarström, S. (1980). *Cancer Res.* **40,** 2039.

CHAPTER 18
Colorectal Cancer Markers: Clinical Value of CEA

NORMAN ZAMCHECK
Mallory Gastrointestinal Research Laboratory
Mallory Institute of Pathology Foundation
Boston City Hospital
Boston, Massachusetts

Harvard Medical School
Boston, Massachusetts

Department of Pathology
Boston University School of Medicine
Boston, Massachusetts

I.	CEA-Like Antigens	335
II.	Non-CEA-Like Antigens	336
III.	Clinical Use of CEA	336
IV.	Pathology	339
	A. Tissue Differentiation and Preoperative CEA Levels in Prognosis and Stratification	340
	B. CEA Activity in Other Body Fluids	341
V.	Radioimmunodiagnosis	342
	A. Radioimmunodetection with Antibodies to CEA	342
	B. The Use of Monoclonal Antibodies	342
VI.	CSA_p	343
VII.	Serum Galactosyltransferase Isoenzyme (GT-II)	343
VIII.	Colon Mucoprotein Antigens (CMA)	344
IX.	The Zinc Glycinate Marker	344
X.	Combinations of Markers	344
XI.	Urine Marker	345
XII.	Comment	346
	References	346

It is 18 years since Gold and Freedman (1965) published their first report of the presence of CEA in fetal tissues and human colonic cancer. A few of the milestones on the path bringing CEA to its present state are indicated in Table I.

TABLE I
Stages of CEA Evolution

1. Initial immunological definition of a protein related to colon cancer
2. Subcellular localization
3. Purification of CEA in quantity
4. Assays
    i. RIA for CEA
    ii. Improved purification and RIA
    iii. Double antibody RIA
    iv. Enzyme immunoassay
5. Identification of CEA-like (cross-reacting) molecules
6. Clinical-pathological correlations in benign and malignant disease states
7. Immunohistochemical staining of tissues
8. Pathophysiologic and metabolic studies
9. Improved clinical use of CEA in management
10. Radioimmunolocalization with CEA
11. Radioimmunotherapy
12. Monoclonal anti-CEA antibodies
13. Colonic cancer culture *in vitro*

It was almost five years after discovery of CEA before Thomson *et al.* (1969) developed a radioimmunoassay for CEA and reported their observations on the first series of patients with colorectal cancer. In subsequent years, improvements in purification of CEA and in assay methods were made (Egan *et al.*, 1972, 1974; Gold and Freedman, 1965; Hansen *et al.*, 1974), including an enzyme immunoassay (Rose *et al.*, 1978).

Beginning in late 1969, clinical and pathological studies of patients with benign and malignant diseases were initiated (Laurence *et al.*, 1972; LoGerfo *et al.*, 1971; Moore *et al.*, 1971), and these studies contributed to understanding of the pathophysiology and the clinical uses and limitations of the CEA assay. See reviews by Goldenberg (1978), Zamcheck *et al.* (1972), Zamcheck (1974), Zamcheck and Kupchik (1980), Krebs *et al.* (1978), Herberman (1979), Herberman and McIntire (1979), Lehmann (1979), Goldenberg *et al.* (1981), Sell (1980), and Goldenberg (1980).

Recently, the development of clinical radioimmunolocalization using radiolabeled anti-CEA antibody (Goldenberg *et al.*, 1978; Goldenberg, 1980) and preliminary studies of radioimmunotherapy have opened the door to new applications (Order *et al.*, 1980a). The developing technology for monoclonal anti-CEA antibodies promises the availability of large quantities of monospecific antibodies (Acolla *et al.*, 1980) for diagnostic and therapeutic purposes.

It is clear that we are still at the beginning of understanding the role of CEA in the pathobiology of human cancer.

I. CEA-LIKE ANTIGENS

Several molecules antigenically similar to CEA exist (Table II) and are detected in radioimmunoassays for CEA depending on the specificity of the antisera used. Studies of amino acid sequences suggest that the protein components of CEA and CEA-like molecules are closely related.

CEA-S appears to be similar in reactivity to CEA (Edgington et al., 1975). In contrast, nonspecific cross-reacting antigen (NCA) is not specifically elevated in colonic cancer (Von Kleist et al., 1972). NCA, normal glycoprotein (NGP), colonic carcinoembryonic antigen-2 (CCEA-2), breast carcinoma glycoprotein, and colon carcinoma antigen III (CCA-III) appear to be similar. NCA-2 appears not to be specifically related to colon cancer (Burtin et al., 1973) but may be similar to normal fecal antigen (NFA) (Matsuoka et al., 1973).

Todd and co-workers, suggest that TEX, a new CEA like marker, may be elevated in the serum of 29% of smokers whereas their CEA RIA values appear not to be elevated (Shively et al., 1979; Todd and Shively, 1978).

The structural chemistry, classification, and characterization of CEA, CEA-like, cross-reacting, and non-CEA molecules have been reviewed elsewhere (Hansen et al., 1971; Krupey et al., 1967; Plow and Edgington, 1979; Pritchard and Todd, 1979).

Whether improved diagnostic specificity lies in the development of assays which detect specific CEA species remains to be seen. Each such CEA species-

TABLE II
Some Colorectal Cancer Antigens and Candidates

Antigen	Reference
CEA-like antigens	
Carcinoembryonic antigen (CEA)	Gold and Freedman (1965)
Carcinoembryonic antigen-S (CEA-S)	Edgington et al. (1975)
Nonspecific cross-reacting antigen (NCA)	von Kleist et al. (1972)
Normal fecal antigen (NFA)	Matsuoka et al. (1973)
NCA-like (TEX)	Todd and Shively (1978)
Other non-CEA-like antigens	
Colonic mucoprotein antigen (CMA)	Gold and Miller (1978)
Zinc glycinate marker (ZGM)	Pusztaszeri et al. (1976)
Colon-specific antigen$_p$ (CSA$_p$)	Pant et al. (1978)
Membrane tissue antigen (MTA)	von Kleist et al. (1974)
Galactosyltransferase isoenzyme II (GT II)	Podolsky et al. (1978)
Basic fetoprotein (BFP)	Ishii (1979)
Monosialoganglioside detected by monoclonal antibody	Koprowski et al. (1981); Magnano et al. (1981)

specific assay will have to be systematically evaluated for its capacity to discriminate colonic from other neoplasms and from other nonneoplastic diseases.

II. NON-CEA-LIKE ANTIGENS

The non-CEA-like antigens listed (Table II) are increasing in number. How does one cope with so large a number?

The experiences gained in the study of CEA suggest a few guidelines. New candidate colorectal antigens and antibodies should be compared with known CEA, CEA-like, and non-CEA-like reagents. The ability to stain cancerous cells positively is a favorable but not sufficient indication of tissue specificity of a candidate marker.

Large amounts of standardized stable reagents are needed before initiating large-scale clinical testing. Then, with a reproducible assay technique in hand, preliminary screening of coded serum panels, such as those provided by the National Cancer Institute and the Mayo Clinic serum bank, may be performed.

Finally, close collaborations among laboratory and clinical investigators in the same institution and in other research centers are indispensable.

III. CLINICAL USE OF CEA

Serial CEA determinations serve as one indication of successful or incomplete resection of colorectal cancer (Sugarbaker *et al.*, 1976a); it gives evidence of recurrence; it is useful in monitoring the effect of chemotherapy (Mayer *et al.*, 1978) or radiation therapy (Sugarbaker *et al.*, 1976b); it aids in decision-making to alter therapy; and it provides a useful estimate of prognosis (Herrera *et al.*, 1976; Wanebo *et al.*, 1978). We have learned the prognostic importance of combining clinical and pathological staging (Dukes classification) with CEA levels (Zamcheck *et al.*, 1975).

Although rising CEA levels can predict recurrence of colorectal cancer and successful resection by second-look surgery is possible, few lives are saved (Martin *et al.*, 1979; Staab *et al.*, 1978; Steele *et al.*, 1980). One reason for this, we believe, is that plasma levels of CEA do not reflect the output of early or localized tumor. Also, since the circulating CEA level is regulated at normal levels by the normal liver, there is need to study the mechanisms by which CEA is cleared by the liver.

Plasma CEA levels provide a useful but not an absolutely quantitatable index of tumor mass and of spread.

Variability of plasma CEA levels is due to many factors (see Table III),

TABLE III
Factors Controlling Circulating CEA Levels[a]

A. Production (Total Tumor Load)
  1. The amount of CEA produced by cells
     Well differentiated cells: high CEA producers
     Poorly differentiated cells: low producers
     Undifferentiated cells: little or none
  2. Degree of invasion of tissues (Martin and Devant, 1973)
  3. Amount and location of spread
     (local, regional, and distant)
  4. Lymph nodes: barriers to tumor spread
     Barriers to CEA circulation?
B. Clearance and Excretion of CEA
  1. CEA cleared by liver. Liver metastases impair clearance by replacing normal liver cells
  2. Obstruction of extra- and intrahepatic bile ducts impede CEA excretion as well as clearance (probably). Examples: bile duct cancer, head of pancreas cancer, obstructing metastatic nodes

[a] From Zamcheck and Martin (1981).

including the marked variation of CEA production by colonic tumor cells as well as in amounts of primary colon cancer and of metastases therefrom.

The highest plasma CEA levels are seen in patients with liver metastases from colonic cancer, but even these may range widely from as little as 0 to over 100,000 ng/ml (O'Brien et al., 1980) (Table IV). This variability, seen in all patients with liver metastases from CEA-producing cancers, including pancreatic, breast, and small cell cancer of the lung as well as colonic cancer (O'Brien et al., 1980), may be due in part to the inability of the gland-forming cancers to discharge their stored CEA accumulations fully in the absence of normal glandular egress. Patients whose completely undifferentiated colonic tumors generate little or no CEA may, in fact, die of their advanced cancers without significant plasma elevations (Goslin et al., 1981).

The key role of the liver in regulating CEA levels is reviewed elsewhere (Loewenstein et al., 1980; O'Brien et al., 1980; Zamcheck, 1976). All varieties of liver disease, hepatitis, cirrhosis, cholestasis, (benign and malignant, intra- and extrahapatic) are associated with increased plasma levels of CEA.

Table IV summarizes the CEA levels, tumor primaries, differentiation, and metastatic sites from 19 autopsied patients with metastatic cancer whose CEA's were recorded within 8 weeks of death. All but one (patient No. 2) had liver metastases, and this patient had a 700 g signet ring cell pancreatic cancer which stained strongly for CEA and presumably produced large amounts of CEA.

The failure to detect elevated CEA levels in the plasma of patients with early colonic cancer may be due to the prevention of such rises by normal liver

TABLE IV
Plasma CEA, Tumor Primary and Differentiation and Metastatic Sites[a]

Patient	CEA (ng/ml)	Primary site	Differentiation	Metastases sites[b]
1	136,000	Colon	Well	LI., PL., LU.
2	48,000	Pancreas	Signet	AD., KI.
3	28,000	Stomach	Moderate	LI., PA., AD.
4	13,000	Colon	Moderate	LI., LU.
5	8,600	Colon	Moderate	LI., LU.
6	6,500	Colon	Well	LI., LU.
7	4,500	Colon	Well	LI., LU.
8	1,900	Colon	Moderate	LI., LU.
9	1,500	Cholan.ca+	Moderate	LU.
10	1,200	Lung	Moderate	LI., SP., BO.
11	1,000	Breast	Poor	LI., CO., BO.
12	340	Colon	Moderate	LI., AD., SP.
13	320	Colon	Moderate	LI., CE.
14	250	Colon	Moderate/well	LI.
15	240	Colon	Moderate	LI., LU., BR., KI., AD., JE.
16	210	Colon	Poor	LI., LU., KI., AD., PA.
17	67	Colon	Well	LI., LU., PL., ST., PE., SP.
18	39.1	Colon	Poor	LI., LU., AD., KI., BO., PA.
19	5.9	Colon	Well	LI.

[a] From O'Brien et al. (1980).
[b] LU., lung; PL., pleura; AD., adrenals; KI., kidney; LI., liver; PA., pancreas; SP., spleen; BO., bone; CE., cecum; BR., brain; JE., jejunum; ST., stomach; PE., peritoneum; +Cholangiocarcinoma.

function. We need to devise strategies for reversibly impeding this clearance so that the circulating level will more directly reflect the load of CEA produced by the tumor.

Thomas et al. (1977, 1981; Thomas and Summers, 1978) Toth et al. (1982) and others have studied the mechanisms of CEA clearance by the liver. Ashwell and Morell (1974) showed in animals that many desialylated glycoproteins may be extracted from the circulation by the liver. Hepatic uptake requires exposure of 2 galactose moieties on the glycoprotein molecule. This is accomplished by removing terminal sialic acid molecules. Sialylated CEA is initially taken up by Kupffer cells and transferred to the hepatocytes where desialylation may occur on the surface of the Kupffer cell and permit subsequent binding to the galactose-dependent binding site on the hepatocyte.

TABLE V
Asialoglycoprotein Binding to the Hepatic Binding Protein in Order of Binding Affinity

1. Orosomucoid (AGP)
2. Fetuin
3. Carcinoembryonic antigen
4. Ceruloplasmin
5. Haptoglobin
6. α-Macroglobulin
7. Thyroglobulin
8. Transferrin

There is competitive inhibition among circulating desialylated glycoproteins for the hepatic uptake site (Table V), those with greater affinity displacing those with lesser affinity. Inhibition of this mechanism or of the initial Kupffer cell uptake should be accompanied by a rise in plasma CEA levels reflecting true production by a tumor source.

Drivas et al. (1976) studied the phagocytic capacity of the reticuloendothelial system by measuring clearance from the plasma of micro-aggregated iodinated human serum albumen and showed that about one-half of the patients with obstructive jaundice, benign and malignant, had defective clearance, which improved after surgical decompression. The explanation for the effect of the cholestasis is not known, though damage to cell membranes by high concentrations of bile salts is one possibility. Whether cholestasis also impairs CEA clearance by Kupffer cells requires further study (Thomas et al., 1982).

Menard et al. (1980) pointed out that many antineoplastic agents may damage the liver by interfering with liver cell metabolism, by impairing biliary excretion, or by direct hepatotoxicity. Savrin and Martin (1981) reported a patient whose CEA levels increased with administration of 5-fluorouracil.

Another approach is to seek tumor markers which are not cleared by the liver, or to use markers which are freely excreted by the kidney, thereby enabling urine output to reflect circulating tumor burden. Studies of combined urine output and circulating marker levels are needed.

IV. PATHOLOGY

Histopathological evidence of tissue CEA production, indicated by positive tissue immunofluorescence or immunoperoxidase staining, is regularly found in well-differentiated colorectal adenocarcinomas, i.e., gland forming carcinomas, mucinous adenocarcinoma, and signet ring carcinoma, and plasma

TABLE VI

Plasma CEA and CEA Tissue Staining of Colorectal Cancers in Reference to Degree of Differentiation[a]

Plasma CEA (ng/ml)			Primary tissue differentiation (%)			CEA tissue staining
Pre-op	Relapse	Terminal	Well	Moderately	Poorly	
0.6	—	0.3	0	0	100[b]	—
1.9	2.0	2.0	0	0	100	—
1.0	1.0	1.0	0	0	100	—
2.3	2.0	2.0	0	0	100	—
2.0	1.7	6.4	0	25	75	—
5.0	D.F.[c]	D.F.	0	75	25	++
1.7	—	430	90	0	10	++
370	—	900	0	90	10	++
1.1	87	200	75	25	0	++
45	45	215	50	50	0	++

[a] Modified from Goslin et al. (1981).
[b] 100% "poorly differentiated" means undifferentiated.
[c] D.F., disease free.

CEA elevations are usually associated with all except early stages. Completely undifferentiated colonic carcinomas, showing complete absence of glands, mucus, or signet ring cells, fail to stain positively and to raise circulating CEA levels, even in advanced stages (Goslin et al., 1981).

Table VI compares the immunoperoxidase staining of two groups (5 each) of colon cancer tissues. One group, mostly undifferentiated, stained weakly or not at all for CEA. The second, moderately or well differentiated, stained strongly for CEA. The staining intensity corresponded with the plasma levels preoperatively, at relapse, and terminally. The variation in the preoperative CEA levels in the second group reflects in part the differing stages of disease when the patients were first studied.

Thus, the presence or absence of CEA staining of the primary colon cancer from a given patient is of value in helping to predict the usefulness of CEA as a monitor of disease progress in that patient. In the complete absence of staining, plasma CEA levels cannot be expected to provide useful monitoring.

A. Tissue Differentiation and Preoperative CEA Levels in Prognosis and Stratification

The clinical importance of distinguishing well from poorly differentiated tumors is increasingly apparent. Elevated plasma CEA values obtained prior to resection of Dukes/Kirklin C primary colorectal adenocarcinomas can predict patients at high risk of disease recurrence, while a preoperative CEA level less than 5 ng/ml in a well-differentiated primary adenocarcinoma is a good prognostic sign. CEA should, therefore, be included as an important prognostic factor in the stratification of patients for future adjuvant therapy. CEA is a poorer marker for poorly differentiated tumors, however, and thus a low preoperative plasma CEA level in a patient with a poorly differentiated colorectal cancer may be inconclusive. This is particularly so for undifferentiated cancers (Goslin et al., 1980).

Table VII summarizes some uses of tissue localization of tumor markers.

B. CEA Activity in Other Body Fluids

CEA-active substances are present in all extravascular body fluids, including saliva (Martin and DeVant, 1973), meconium (Rule, 1973), digestive fluids (Molner and Gitnick, 1976), gastric juice (Vuento et al., 1976), duodenal juice (Loewenstein et al., 1980), biliary juice (Svenberg, 1976), pancreatic juice (Rey et al., 1978; Sharma et al., 1976), small and large bowel washings (Go et al., 1975), pleural effusions (Rittgers et al., 1978), ascitic fluid (Loewenstein et al., 1978), urine (Wahren et al., 1975), spinal fluid (Hill et al., 1980), and cyst fluids (Fleisher et al., 1973).

These studies have shown general correlation between high levels of CEA and the presence of a CEA-producing malignancy, thus supporting their clinical usefulness, especially when combined with cytological or immunohistological testing. Nonmalignant causes of CEA-like activity, however, must also be taken into account.

Thus, the clinical physiology as well as the pathology can be expected to vary among new tumor markers as well as among individual patients.

TABLE VII

The Use of Tissue Antigen Localization in the Study of Tumor Markers[a]

1. To demonstrate presence of marker in tumor cells
2. To assess the differing specificity of candidate markers for malignant, normal, and benign tissues
3. To correlate amount and intensity of staining for marker with tumor cell type and differentiation
4. To compare tissue localization of candidate markers with that of other established markers
5. To screen antibodies (xenogeneic or monoclonal) for intensity and for selection and potential incorporation into clinical assays

[a] Uniform histopathological nomenclature should be used in correlating tumor marker production with histology in both clinical and basic studies of markers. The WHO classification of colorectal tumors is simple and useful (Morson and Sobin, 1976).

V. RADIOIMMUNODIAGNOSIS

A. Radioimmunodetection with Antibodies to CEA

Goldenberg and co-workers provided the watershed studies on radioimmunodetection of cancer (Goldenberg et al., 1978). They performed over 300 clinical studies using radiolabeled antibodies against CEA including 37 with colorectal cancer.

Although truly cancer-specific markers appeared not to be absolutely required, more specific antibodies, perhaps provided by hybridomas, may show improved findings. Because cancers are heterogeneous, combinations of different radiolabeled antibodies may prove to be better. Thus, in experimental studies in hamsters (GW-39 tumor model) the use of combined CEA and CSA_p antibodies gave better localization than did either alone (Gaffar et al., 1981), and Goldenberg suggested that mixtures of antibodies be considered for further studies in patients.

These studies are being assessed in several laboratories. Dykes et al. (1980) and Mach et al. (1980) confirmed the general findings (Table VIII). Mach expressed reservation with respect to widespread clinical application. It remains to be determined whether the use of monoclonal anti-CEA antibody results in improved detection.

B. The Use of Monoclonal Antibodies

1. Monoclonal Antibody to CEA

The development of a monoclonal antibody specific for CEA by somatic hybrids offers an important tool for producing large amounts of standardized monospecific anti-CEA antibodies (Acolla et al., 1980). It may also expedite the

TABLE VIII
Colorectal Cancer: Tumor Localization with Radiolabelled Anti-CEA Antibodies[a]

	Number of patients	Percentage detected
Goldenberg	37	85–90
Dykes (1980)	13	75
Mach (1980)	21 (sites)	57
Mach Extended Series	53	42
With Monoclonal AB	16	50

[a] Adapted from Letters to Editor, *New England Journal of Medicine* (Begent et al., 1980; Goldenberg et al., 1980; Mach et al., 1980; Order et al., 1980b), and the Radioimmuno-detection of Cancer (UICC Workshop) (Goldenberg, 1980).

further separation and identification of the several cross-reacting CEA-like molecules and the defining of their pathological, biological, and clinical distinctions.

2. Non-CEA Monoclonal Antibody to Colorectal Cancer

Using mouse monoclonal antibodies to colorectal cancer, Koprowski *et al.* (1981) developed an assay for the detection of non-CEA colorectal carcinoma-associated antigens in human serum. The antibody used reacted with colorectal, gastric, and pancreatic carcinoma cells. The assay was based on the inhibition of binding of one of the monoclonal antibodies to target preparations of colorectal carcinoma cells. This is described in more detail in Chapter 8.

3. Immunotherapy with Radiolabeled Anti-CEA and Monoclonal Antibodies to Colon Cancer Cells

The availability of monoclonal anti-CEA as well as non-CEA monoclonal antibodies to colon cancer cells has opened new avenues for targeting a variety of anti-cancer and cytotoxic agents to the cancer cell (Gilliland *et al.*, 1980; see also Order *et al.*, 1980b).

VI. CSA_p

Colon-specific antigen or CSA was originally detected and isolated by Goldenberg from a human colonic carcinoma system propagated in hamsters. Above 4×10^6 in molecular size, it could not be dissociated by detergents or by treatment with chaotropic reagents. Tryptic digestion of fractions, however, yielded smaller peptides. One such thermolabile polypeptide, CSA_p, MW 70,000–110,000, was isolated and purified. A radioimmunometric assay has been used to quantitate CSA_p in patients' sera (Pant *et al.*, 1978). There was no cross-reaction with CEA, α-fetoprotein, or ferritin. A radioimmunoassay has recently been prepared. Fifty-two percent of patients with colon carcinomas and 20% of those with pancreatic cancer had high CSA_p levels. Preliminary findings suggest that CSA_p is more specific for colorectal cancer than is CEA, although CEA is more sensitive than CSA_p. Discordance in positivities exists between the two markers in 42% of patients. Whether the use of the two combined will improve upon CEA assay results alone in the clinical diagnosis of colorectal cancer is under study.

VII. SERUM GALACTOSYLTRANSFERASE ISOENZYME (GT-II)

The enzyme GT-II (Podolsky *et al.*, 1978), an electrophoretically distinct form of galactosyltransferase is not a cancer- or tissue-specific marker. With the exception of 18/20 patients with celiac disease and 20% of patients with alcoholic hepatitis, none of 124 disease or normal controls possessed serum GT-II. Seventy-three percent of 117 patients with colorectal cancer were "positive," including 7 of 9 patients with Dukes B lesions, all of whom became negative after colectomy. Serum levels of GT-II correlated with extent of disease at surgery and with persistent disease. Highest levels were seen in patients with hepatic metastases. GT-II activity was detected in 28 of 32 cytologically positive effusions.

GT-II was positive in a broad spectrum of other malignancies, including pancreatic (83%), stomach (75%), breast (78%), bronchogenic (65%) as well as in esophageal, gallbladder, and prostatic cancers and hepatoma, Hodgkin's lymphoma, and chronic lymphatic leukemia.

The marker seems to be associated with mitotic activity. Whether it can be made more specific for colon cancer by combining its use with another marker remains to be seen. Its comparison with CEA is presently under study.

VIII. COLON MUCOPROTEIN ANTIGENS (CMA)

Colon mucoprotein antigens (CMA) extracted with phenol–water are present in normal and malignant colon tissue (Gold and Miller, 1978). The antigens (MW approximately 1.5×10^7) are not related to CEA. One type of CMA isolated from colon carcinomas, called TCMA, seems to differ from that of normal colonic mucosa. The concentration of TCMA was diminished on colon tumor cells that were made to differentiate *in vitro,* in contrast to CEA, which increased in more differentiated tumor cells. The tumor specificity of TCMA needs confirmation (Gold and Goldenberg, 1980).

IX. THE ZINC GLYCINATE MARKER

A glycoprotein 2×10^6 MW, or greater, is unrelated to CEA, AFP, BOFA, NCA, NCA-2, CSA_p, or acute phase reactant glycoproteins (Saravis *et al.,* 1978, 1979).

It is found in colonic and other malignancies and is detected by immunohistopathological staining in transitional and malignant epithelia of the gastrointestinal tract and in malignant effusions (O'Brien *et al.,* 1980), and in lesser amounts in normal and benign tissues (Doos *et al.,* 1978; O'Brien *et al.,* 1979).

X. COMBINATIONS OF MARKERS

McIntire (1979) emphasized the importance of the use of multiple immunoassays for circulating tumor markers. The primary basis for this recommendation was the lack of specificity of any of the markers for colorectal cancer. Hence, specific markers for non-CEA-producing undifferentiated cells are needed.

Another combination is that of CEA with an allegedly colon-specific marker, such as CSA_p, or with the monosialoganglioside antigen thought to be specific for colorectal cancer (Koprowski et al., 1981; Magnano et al., 1981).

Some data suggest that measurement of CEA and TPA (tissue polypeptide antigen) (Menendez-Botet et al., 1978) may enhance discrimination between "active" cancer and cancer without clinically overt signs (see also Holyoke and Chu, 1979).

In the effort to improve upon the CEA assay as a test for the earlier detection of recurrent colonic cancer, Steele et al. (1978) combined serial circulating immune complex (CIC) and CEA measurements, and noted that, in the presence of tumor recurrence, CEA levels and CIC trends rose in parallel whereas transient CEA rises in the absence of tumor recurrence were associated with inverse, falling CIC levels. Staab et al. (1980) also reported poorer prognoses in patients with rising immune complexes.

Numerous studies have purported to show that the combination of CEA plus one or more enzymes (alkaline phosphatase, acid phosphatase, glucosephosphate isomerase, γ-glutamyl transpeptidase, 5′-nucleotidase, and fructosebiphosphate aldolase) or ectopic hormones (adrenocorticotropin, human chorionic gonadotropin, gastrin, and parathormone among others) may detect liver metastases better or sooner than either alone (Neville and Cooper, 1976; Neville et al., 1978; Schwartz, 1971). Similarly, the measurement of acute phase proteins (haptoglobin, α_1-anti-trypsin, α_1-acid glycoprotein, and prealbumin) have been shown to aid in the monitoring of large bowel cancer by CEA and hepatic enzymes (Ward et al., 1977; Zamcheck and Kupchik, 1980).

The combination of CEA plus the leukocyte adherence index (LAI) text (Grosser et al., 1976) has been reported to improve preoperative identification in 85% of cases of Dukes B or C cancers. Sixty-five percent of these cases were LAI positive compared to 42% with CEAs greater than 2.5. The LAI was usually negative (13%) in metastatic colorectal cancer, whereas the CEA levels were consistently elevated (88%). The potential advantage of the LAI test in detecting early colonic cancers missed by CEA is supported by these findings (Ritts et al., 1980).

XI. URINE MARKER

Ideally, a small molecule excreted by the kidneys into the urine might permit better quantitation of tumor load than does serum or plasma measurement. Among such possible candidates are monosialoganglioside of Magnano et al. (1981), TPA, arylsulfatase B (Morgan et al., 1975), selected small molecular subunits of CEA and polyamines (Nishioka and Romsdahl, 1978), all of which remain to be studied.

XII. COMMENT

CEA is essential to our understanding of colon cancer from both fundamental and clinical viewpoints. It appears to be a hallmark molecule.

Promising new directions have emerged: monoclonal antibody production and the use of specific antibodies for immunodetection and as carriers of therapeutic and cytotoxic agents. The use of cultures of human colonic cancer cells to identify effective anti-cancer agents may make it possible to tailor anti-cancer therapy to the patient's own cancer. Whether CEA will monitor the *in vitro* response to therapy as it does the patient's is under study (Drewinko *et al.*, 1981; Calabresi *et al.*, 1979; Rutzky *et al.*, 1981 ; Boland and Kim, 1981).

ACKNOWLEDGMENTS

Supported by Public Health Service Grant No. CA-04486 from the National Cancer Institute. Modified from a presentation at the 1981 Workshop of the Colorectal Cancer Cadre, Dallas, January 8–10, 1981. *Cancer Bulletin* **33,** 141–151 (1981).

REFERENCES

Acolla, R. S., Carrel, S., and Mach, J. P. (1980). *Proc. Natl. Acad. Sci. U.S.A.* **77,** 563–566.
Ashwell, G., and Morrell, A. G. (1974). *Adv. Enzymol. Relat. Areas Mol. Biol.* **41,** 99–128.
Begent, R., Bagshawe, K.,Stanway, G., Keep, P., Searle, F., Newlands, E., Jewkes, R., Jones, B., and Vernon, P. (1980). *N. Engl. J. Med.* **303,** 1238.
Boland, C. R., and Kim, Y. S. (1981). *Workshop Large Bowel Cancer, 1981.*
Burtin, P., Chavanel, G., and Hirsch-Marie, H. (1973). *J. Immunol.* **3,** 1926.
Calabresi, P., Dexter, D. L., and Heppner, G. H. (1979). *Biochem. Pharmacol.* **28,** 1933–1941.
Doos, W. G., Saravis, C. A., Pusztaszeri, G., Burke, B., Oh, S. K., Zamcheck, N., and Gottlieb, L. S. (1978). *JNCI, J. Natl. Cancer Inst.* **60,** 1375–1382.
Drewinko, B., Yang, L. Y., Stragand, J. J., Barlogie, B., and Liebovitz, A. (1981). *Poster Session Workshop Colorectal Cancer, 19O1.*
Drivas, G., James, O., and Wardle, N. (1976). *Br. Med. J.* **1,** 1568–1569.
Dykes, P. W., Hine, K. R., Bradwell, A. R., *et al.* (1980). *Br. Med. J.* **1,** 220–222.
Edgington, T. S., Astarita, R. W., and Plow, E. F. (1975). *N. Engl. J. Med.* **293,** 103–107.
Egan, M. L., Lautenschleger, J. T., Coligan, J. E., and Todd, C. W. (1972). *Immunochemistry* **9,** 289–299.
Egan, M. L., Coligan, J. E., and Todd, C. W. (1974). *Cancer* **34,** 1504–1509.
Fleisher, M., Robbins, G. F., Breed, C. N., Fracchia, A. A., Urban, J. A., and Schwartz, M. K. (1973). *Clin. Bull.* **3,** 94–97.

Gaffar, S. A., Pant, K. D., Shochat, D., Bennett, S. J., and Goldenberg, D. M. (1981). *Int. J. Cancer* **27**, 101–105.
Gilliland, D. G., Steplewski, Z., Collier, R. J., Mitchell, K. F., Chang, T. H., and Koprowski, H. (1980). *Proc. Natl. Acad. Sci. U.S.A.* **77**, 4539–4543.
Go, V. L. W., Ammon, H. V., Holtermüller, K. H., Krag, E., and Phillips, S. F. (1975). *Cancer* **36**, 2346–2350.
Gold, D. V., and Goldenberg, D. M. (1980). In "Cancer Markers" (S. Sell, ed.), p. 329. Humana Press, Clifton, New Jersey.
Gold, D. V., and Miller, F. (1978). *Cancer Res.* **38**, 3204–3211.
Gold, P., and Freedman, S. O. (1965). *J. Exp. Med.* **121**, 439–462.
Goldenberg, D. W., ed. (1978). "Proceedings of the First International Conference on the Clinical Uses of Carcinoembryonic Antigen." J. B. Lippincott Co., Philadelphia.
Goldenberg, D. M., ed. (1980). *Cancer Res.* **40**, Part 2, 2953–3087.
Goldenberg, D. M., Deland, F. H., Kim, E., Bennett, S., Primus, F. J., van Nagell, J. R., Jr., Estes, N., DeSimone, P., and Rayburn, P. (1978). *N. Engl. J. Med.* **298**, 1394–1398.
Goldenberg, D. M., Deland, F. H., and Kim, E. E. (1980). *N. Engl. J. Med.* **303**, 1237–1238.
Goldenberg, D. M., Neville, M., Carter, A. C., et al. (1981). *Ann. Intern. Med.* **94**, 402–409.
Goslin, R. H., Steele, G., MacIntyre, J., Mayer, R., Sugarbaker, P., Cleghorn, K., Wilson, R., and Zamcheck, N. (1980). *Ann. Surg.* **192**, 747–751.
Goslin, R. H., O'Brien, M. J., Steele, G., Mayer, R., Wilson, R., Corson, J. M., and Zamcheck, N. (1981). *Am. J. Med.* **71**, 246–253.
Grosser, N., Martin, J. H., Proctor, J. W., and Thomson, D. M. P. (1976). *Int. J. Cancer* **18**, 39–47.
Hansen, J. J., Lance, K. P., and Krupey, J. (1971). *Clin. Res.* **19**, 143.
Hansen, J. J., Snyder, J. J., Miller, E., Vandevoorde, J. P., Miller, O. N., Hines, L. R., and Burns, J. J. (1974). *Hum. Pathol.* **5**, 139–147.
Herberman, R. B. (1979). *Dev. Cancer Res.* **1**.
Herberman, R. B., and McIntire, K. R. (1979). "Immunodiagnosis of Cancer," Parts 1 and 2. Dekker, New York.
Herrera, M. S., Chu, T. M., and Holyoke, E. D. (1976). *Ann. Surg.* **183**, 5–9.
Hill, S., Martin, E., Ellison, E. C., and Hunt, W. E. (1980). *J. Neurosurg.* **53**, 627–632.
Holyoke, D., and Chu, T. M. (1979). *Immunol. Ser.* **9**, 513–521.
Ishii, M. (1979). *Immunol. Ser.* **9**, 45–50.
Koprowski, H., Herlyn, M., Steplewski, A., and Sears, H. F. (1981). *Science* **212**, 53–55.
Krebs, B. P., LeLanne, C. M., and Schneider, M. (1978). *Int. Congr. Ser.—Excerpta Med.* **439**.
Krupey, J., Gold, P., and Freedman, S. O. (1967). *Nature (London)* **215**, 67–68.
Laurence, D. J. R., Stevens, U., Bettelheim, R., Darcy, D., Leese, C., Turberville, C., Alexander, P., Johns, E. W., and Neville, A. M. (1972). *Br. Med. J.* **3**, 605–609.
Lehmann, F. G. (1979). "Carcinoembryonic Proteins: Chemistry, Biology, Clinical Application," Vol. 1. Elsevier/North-Holland Biomedical Press, Amsterdam.
Loewenstein, M. S., Rittgers, R. A., Feinerman, A. E., Kupchik, H. Z., Marcel, R. B., Koff, R. S., and Zamcheck, N. (1978). *Ann. Intern. Med.* **88**, 635–638.
Loewenstein, M. S., Rau, P., Rittgers, R. A., Adhinarayanan, B. G., and Zamcheck, N. (1980). *JNCI, J. Natl. Cancer Inst.* **64**, 235–240.
LoGerfo, P. J., Krupey, J., and Hansen, H. J. (1971). *N. Engl. J. Med.* **285**, 138–141.
Mach, J. P., Carrel, S., Forni, M., Ritschard, J., Donath, A., and Alberto, P. (1980). *N. Engl. J. Med.* **303**, 1238–1239.
McIntire, K. R. (1979). *Immunol. Ser.* **9**, 521–539.
Magnano, J. L., Brockhaus, M., Smith, D. F., Ginsburg, V., Blaszcyk, M., Mitchell, K. F., Steplewski, Z., and Koprowski, H. (1981). *Science* **212**, 55–56.

Martin, E. W., James, K. J., Hurtubise, P. E., Catalano, P., and Minton, J. P. (1979). *Cancer* **39,** 440–446.
Martin, F., and Devant, J. (1973). *JNCI, J. Natl. Cancer Inst.* **50,** 1375–1379.
Matsuoka, U., Hara, M., Takatsu, K., and Kitagawa, M. (1973). *Gann* **64,** 203.
Mayer, R. J., Garnick, M. B., Steele, G. D., and Zamcheck, N. (1978). *Cancer* **42,** 1428–1433.
Menard, D. B., Gisselbrecht, C., Marty, M., Reyes, F., and Dhumeaux, D. (1980). *Gastroenterology* **78,** 142–164.
Menendez-Botet, C. J., Oettgen, H. G., Pinsky, C. M., and Schwartz, M. K. (1778). *Clin. Chem. (Winston-Salem, N.C.)* **24,** 868–872.
Molner, G. L., and Gitnick, I. G. (1976). *Gastroenterology* **70,** 513–515.
Moore, T. L., Kupchik, H. Z., Marcon, N., and Zamcheck, N. (1971). *Am. J. Dig. Dis.* **16,** 1–7.
Morgan, L. R., Jr., Samuels, M. S., Thomas, W., Krementz, E. T., and Meeker, W. (1975). *Cancer* **36,** 2337–2345.
Morson, B. C., and Sobin, L. H. (1976). "International Classification of Intestinal Tumors." World Health Organ., Geneva.
Neville, A. M., and Cooper, E. H. (1976). *Ann. Clin. Biochem.* **13,** 283–305.
Neville, A. M., Patel, S., Capp, M., Laurence, D. J. R., Cooper, E. H., Turberville, C., and Coombes, R. C. (1978). *Cancer* **42,** 1448–1451.
Nishioka, K., and Romsdahl, M. M.(1978). *Cancer Bull.* **30,** 205–209.
O'Brien, M. J., Zamcheck, N., Saravis, C. A., Burke, B., and Gottlieb, L. S. (1979). *Proc., Am. Soc. Clin. Oncol.* **20,** 401.
O'Brien, M. J., Bronstein, B., Zamcheck, N., Saravis, C. A., Burke, B., and Gottlieb, L. S. (1980). *JNCI, J. Natl. Cancer Inst.* **64,** 1291–1294.
Order, S. E., Leichner, P., Ettinger, D. S., Klein, J. L., and Strand, M. (1980a). *N. Engl. J. Med.* **303,** 1238.
Order, S. E., Klein, J. L., Ettinger, D., Alderson, P., Siegelman, S., and Leichner, P. (1980b). *Cancer Res.* **40,** 3001–3007.
Pant, K. D., Dahlman, H. L., and Goldenberg, D. M. (1978). *Cancer* **42,** 1626–1634.
Plow, E. G., and Edgington, T. S. (1979). *Immunol. Ser.* **9,** 181–239.
Podolsky, D. K., Weiser, M. M., Isselbacher, K. J., and Cohen, A. M. (1978). *N. Engl. J. Med.* **299,** 703–705.
Pritchard, D. G., and Todd, C. W. (1979). *Immunol. Ser.* **9,** 165–180.
Pusztaszeri, G., Saravis, C. A., and Zamcheck, N. (1976). *JNCI, J. Natl. Cancer Inst.* **56,** 275–278.
Rey, J. R., Krebs, B. P., and Delmont, J. (1978). *Excerpta Med.—Int. Congr. Ser.* **439,** 116–120.
Rittgers, R. A., Loewenstein, M. S., Feinerman, A. E., Kupchik, H. Z., Marcel, B. R., Koff, R. S., and Zamcheck, N. (1978). *Ann. Intern. Med.* **88,** 631–634.
Ritts, R. E., Jr., Shani, A., Weiland, L. H., Go, V. L. W., Thynne, G., and Moertel, C. G. (1980). *Proc., Am. Assoc. Cancer Res.* **21,** 216.
Rose, S. P., Jolley, M. E., Waindle, L. M., *et al.* (1978). *Abstr. 6th Meet., Int. Res. Group Carcinoembryonic Proteins* p. 206.
Rule, A. H. (1973). *Immunol. Commun.* **2,** 25–34.
Rutzky, L. P., Giovanella, B. C., Siciliano, M. J., *et al.* (1981). *Workshop Large Bowel Cancer, 1981.*
Saravis, C. A., Oh, S. K., Pusztaszeri, G., Doos, W., and Zamcheck, N. (1978). *Cancer* **42,** 1621–1625.
Saravis, C. A., Kupchik, H. Z., O'Brien, M., and Zamcheck, N. (1979). *Ser. Cancer Res.* **1,** 237–239.
Savrin, R. A., and Martin, E. W., Jr. (1981). *Cancer* **47,** 481–485.
Schwartz, M. K. (1971). *Cancer* **40,** 2620–2624.

Sell, S., ed. (1980). "Cancer Markers." Humana Press, Clifton, New Jersey.
Sharma, M. P., Gregg, J. A., Loewenstein, M. S., McCabe, M. P., and Zamcheck, N. (1976). *Cancer* **38,** 2457–2461.
Shively, J. E., Glassman, J. N. S., Engvall, E., and Todd, C. W. (1979). *Carcino-Embryonic Proteins Proc. Int. Soc. Oncodev. Biol. Med., 6th, 1978* Vol. 1, pp. 9–15.
Staab, H. J., Anderer, F. A., Stumpf, E., and Fischer, R. (1978). *J. Surg. Oncol.* **10,** 273–282.
Staab, J. H., Anderer, F. A., Stumpf, E., and Fischer, R. (1980). *Br. J. Cancer* **42,** 26–33.
Steele, G. D., Sonis, S., Stelos, P., Rittgers, R., Zamcheck, N., Finn, D., Maltz, J., Mayer, R., Lokich, J., and Wilson, R. E. (1978). *Surgery* **83,** 648–654.
Steele, G. D., Zamcheck, N., Wilson, R. E., Mayer, R., Lokich, J., Rau, P., and Maltz, J. (1980). *Am. J. Surg.* **139,** 544–546.
Sugarbaker, P. H., Zamcheck, N., and Moore, F. D. (1976a). *Cancer* **38,** 2310–2315.
Sugarbaker, P. H., Bloomer, W. D., Corbett, E. D., and Chaffey, J. T. (1976b). *AJR, Am. J. Roentgenol.* **127,** 641–644.
Svenberg, T. (1976). *Int. J. Cancer* **17,** 588–596.
Thomas, P., and Summers, J. W. (1978). *Biochem. Biophys. Res. Commun.* **80,** 335–339.
Thomas, P., Birbeck, M. S. C., and Cartwright, P. A. (1977). *Biochem. Soc. Trans.* **5,** 312–313.
Thomas, P., O'Neil, P. F., and Zamcheck, N. (1981). *Biochem. Soc. Trans.* **10,** 459–460.
Thomas, P., Toth, C. A., and Zamcheck, N. (1981). *Gastroenterology* **80,** 1302.
Thomson, D. M. P., Krupey, J., Freedman, S. O., and Gold, P. (1969). *Proc. Natl. Acad. Sci. U.S.A.* **64,** 161–167.
Todd, C. W., and Shively, J. E. (1978). *ACS Symp. Ser.* **80,** 342–346.
Toth, C. A., Thomas, P., Broitman, S. A., and Zamcheck, N. (1981). *Biochem. J.* **204,** 377–381.
von Kleist, S., Chavanel, G., and Burtin, P. (1972). *Proc. Natl. Acad. Sci. U.S.A.* **60,** 2492–2494.
von Kleist, S., King, M., and Burtin, P. (1974). *Immunochemistry* **11,** 249–253.
Vuento, M., Engvall, E., Seppala, M., and Ruoslahti, E. (1976). *Int. J. Cancer* **18,** 156–160.
Wahren, B., Edsmyr, F., and Zimmerman, R. (1975). *Cancer* **36,** 1490–1495.
Wanebo, H. J., Rao, B., Pinsky, C. M., Hoffman, R. G., Stearns, M., Schwartz, M. K., and Oettgen, H. F. (1978). *N. Engl. J. Med.* **299,** 448–451.
Ward, A. M., Cooper, E. H., Turner, R., Anderson, J. A., and Neville, A. M. (1977). *Br. J. Cancer* **35,** 170–178.
Zamcheck, N. (1974). *Adv. Intern. Med.* **19,** 413–433.
Zamcheck, N. (1976). *In* "Clinics in Gastroenterology" (P. Sherlock and N. Zamcheck, eds.), pp. 625–638. Saunders, Philadelphia, Pennsylvania.
Zamcheck, N., and Kupchik, H. Z. (1980). *In* "Manual of Clinical Immunology" (N. R. Rose and H. Friedman, eds.), 2nd ed., pp. 929–935. Am. Soc. Microbiol., Washington, D.C.
Zamcheck, N., and Martin, E. W. (1981). *Cancer* **47,** 1620–1627.
Zamcheck, N., Moore, T. L., Dhar, P., and Kupchik, H. (1972). *N. Engl. J. Med.* **286,** 83–86.
Zamcheck, N., Doos, W. G., Prudente, R., Lurie, B. B., and Gottlieb, L. S. (1975). *J. Hum. Pathol.* **6,** 31–45.

CHAPTER 19

hCG Expression in Trophoblastic and Nontrophoblastic Tumors

GLENN D. BRAUNSTEIN
Division of Endocrinology, Department of Medicine
Cedars-Sinai Medical Center
Los Angeles, California
UCLA School of Medicine
Los Angeles, California

I. Historical Overview	351
II. Chemistry, Biologic Activity, and Sites of Production	352
III. Methods of Measurement	354
IV. hCG in Gestational Trophoblastic Disease (GTD)	355
V. hCG in Germ Cell Tumors of the Testes	358
VI. hCG in Nontrophoblastic Neoplasms	360
References	365

I. HISTORICAL OVERVIEW

The first demonstration that the human placenta contained a gonadotropin was made by Hirose (1919, 1920) who studied the effects of implanted placental tissue on the uterus and ovaries of the rabbit. In 1927, Aschheim and Zondek discovered that the urine from pregnant women contained a unique gonadotropin that stimulated the ovaries of rats, mice, and rabbits. A few years later several investigators found that individuals with gestational and nongestational trophoblastic neoplasms excreted the gonadotropin in their urine (Zondek, 1929, 1930a,b; Heidrich et al., 1930; Brouha and Hinglais, 1931). The significance of these early observations were rapidly appreciated, and bioassays of human chorionic gonadotropin (hCG) were frequently used to diagnosis and monitor pregnancy and trophoblastic disease. These bioassays were also utilized to detect hCG production in the first reported cases of hCG production by nontrophoblastic tumors (McFadzean, 1946; Chambers, 1949). When the first effective chemotherapy for gestational trophoblastic disease was described by Li and co-workers (1956), the usefulness of serial hCG measurements to monitor the effects of therapy was also established.

During the 1960's several immunological methods for measuring hCG were developed. These had the advantages of being more specific, less costly, and less time consuming than the bioassays. The commercialization and dissemination of slide and tube pregnancy tests based on hemagglutination, latex particle agglutination, or agglutination-inhibition resulted in the widespread use of hCG measurements for pregnancy diagnosis, the diagnosis and monitoring of gestational and nongestational trophoblastic disease, and the diagnosis of some nontrophoblastic neoplasms. In the last decade, methods were developed to purify and structurally characterize the hCG molecule (Morgan et al., 1973, 1974, 1975; Bellisario et al., 1973; Carlsen et al., 1973). These studies resulted in the development of a new generation of techniques for specifically measuring hCG in biologic fluids (Vaitukaitis et al., 1972; Chen et al., 1976). The sensitivity and specificity of these new methods were such that they allowed the detection of hCG in the sera of many patients with nontrophoblastic neoplasms (Braunstein et al., 1973a). In addition, these tools enabled several groups of investigators to identify an hCG-like substance in normal human nontrophoblastic tissues (Braunstein et al., 1975, 1979; Chen et al., 1976; Yoshimoto et al., 1977, 1979a). To some extent this finding has blurred the distinction between ectopic and eutopic production of hCG by tumors, but has not diminished the usefulness of hCG measurements as an immunodiagnostic tool for the diagnosis and monitoring of certain neoplasms.

II. CHEMISTRY, BIOLOGIC ACTIVITY, AND SITES OF PRODUCTION

Human chorionic gonadotropin is a glycoprotein composed of two non-covalently linked subunits, designated α and β. The α subunit contains 92 amino acids and approximately 28% carbohydrate by weight, with an average total molecular weight of 14,500 (Birkin and Canfield, 1980). The hCG β subunit contains 145 amino acids, 28–36% carbohydrate and an average total molecular weight of 22,200 (Birkin and Canfield, 1980). The primary structure of the α subunit of hCG is nearly identical to the α subunits of the other human glycoprotein hormones—thyroid-stimulating hormone, follicle-stimulating hormone, and luteinizing hormone (Pierce et al., 1971). The β subunit of hCG is closely related to the β subunits of the other glycoprotein hormones, but shows much less homology than do the α subunits. Indeed, 80% of the first 115 amino acids in the hCG β subunit are in identical positions to those present in the β subunit of luteinizing hormone (Pierce et al., 1971; Closset et al., 1973; Shome and Parlow, 1973). In addition, the β subunit of hCG contains 30 amino acids at its carboxyl-terminal end which are not present in the β subunits of the other glycoprotein hormones (Carlsen *et al.*, 1973; Morgan *et al.*, 1975; Keutmann and Williams, 1977). Neither subunit has intrinsic biologic activity, but can be recombined *in vitro* to regenerate full biologic activity (Morgan *et al.*, 1974). There is a great deal of microheterogeneity in the hCG molecule, primarily related to differing amounts of carbohydrate (Vaitukaitis *et al.*, 1976a).

The major biologic activity of hCG is the stimulation of the granulosa, luteal and interstitial cells of the ovary, as well as the interstitial cells of the testes. This activity is qualitatively identical to that found with human luteinizing hormone, again attesting to their structural similarity (Catt and Dufau, 1973; Channing and Kammerman, 1974; Haour and Saxena, 1974). The hCG molecule also contains intrinsic follicle-stimulating hormone-like activity as well as thyroid-stimulating hormone-like activity (Nisula *et al.*, 1980).

During normal pregnancy, hCG appears in the maternal blood 6–9 days following conception and raises in a logarithmic fashion during the first trimester and reaches a peak at approximately 10 weeks after the last normal menstrual period (Braunstein *et al.*, 1973c, 1976). Following this, the maternal hCG concentrations decrease and reach a nadir at 17 weeks with relatively constant levels being maintained throughout the last half of pregnancy (Braunstein *et al.*, 1976). The source of hCG production during pregnancy is the syncytiotrophoblast. This has been confirmed by immunohistochemical methods (Midgley and Pierce, 1962; de Ikonicoff and Cédard, 1973), placental explant cultures (Gey *et al.*, 1938; Handwerger *et al.*, 1973; Golander *et al.*, 1978), and studies of cell-free synthesis of hCG subunits translated from placental messenger RNA (Chatterjee and Monro, 1977; Boime *et al.*, 1978; Fiddes and Goodman, 1979).

Several physiologic functions during pregnancy have been ascribed to hCG, although most of the evidence that has been accumulated has been indirect. These functions include the prolongation of the functional life of the corpus luteum during the cycle of conception (Strott *et al.*, 1969; Yoshimi *et al.*, 1969), stimulation of testosterone production by testicular Leydig cells during male fetal development which in turn results in appropriate sexual differentiation of male internal and external genitalia (Faiman *et al.*, 1981), and possibly participation in the relative suppression of the maternal immune system during pregnancy, which is presumably important in helping to prevent the rejection of the fetal and placental homografts (Adcock *et al.*, 1973; Carr *et al.*, 1973; Teasdale *et al.*, 1973; Beling and Weksler, 1974; Han, 1974; Hammarström *et al.*, 1979; Fuchs *et al.*, 1980). The putative immunosupressive properties of hCG have been an area of intense controversy.

Although the trophoblast is the undisputed physiologic source of hCG, an hCG-like substance has been demonstrated in a wide variety of normal human tissues (Braunstein *et al.*, 1975, 1979, 1980; Chen *et al.*, 1976; Yoshimoto *et al.*, 1977, 1979a). The substance resembles hCG immunologically and biologically and shares at least some physicochemical properties with hCG purified from pregnancy urine. Active secretion of this material has been demonstrated *in vitro* by normal human fibroblasts (Rosen *et al.*, (1980) and human fetal renal tissue (McGregor *et al.*, 1981). It is unknown whether this material has any physiologic function in the adult or if its production represents incomplete suppression of the fetal genome responsible for hCG production.

III. METHODS OF MEASUREMENT

As noted above, the first methods for measuring hCG in serum or urine were bioassays. The end point in most of these assays was based on a primary effect in the hCG target organ or measurement of a secondary effect of the sex steroids secreted by the gonadotropin-stimulated gonads. Examples of the former assays include the formation of hemorrhagic follicles and corpora lutea (Aschheim and Zondek, 1927), ovulation in rabbits (Friedman and Laphan, 1931), ovarian weight increase of mice and rats (Diczfalusy and Loraine, 1955), the depletion of ascorbic acid from the ovaries of pseudopregnant rats (Parlow, 1961), and the discharge of sperm from the testes of a toad (Galli-Mainini, 1947). The indirect assays include the estrogen-induced uterine weight increase in mice (Diczfalusy and Loraine, 1955) and the testosterone-induced increase in seminal vesicle and other accessary organ weight in either intact or hypophysectomized male rats (Van Hell *et al.*, 1964; Christiansen, 1967; McArthur, 1968). Because of the close homology between hCG and human luteinizing hormone, these bioassays are unable to distinguish between the two molecules. This lack of specificity combined with the expense of the maintaining of the animals and the relatively long length of time required for the assay limits the usefulness of *in vivo* bioassays. Recently *in vitro* bioassays for hCG have been developed. The two most popular are measurement of testosterone production by rat or mouse Leydig cells which have been exposed to hCG or luteinizing hormone (Dufau *et al.*, 1974; Van Damme *et al.*, 1974). These procedures are extremely sensitive, but again, lack specificity for hCG. since they also measure luteinizing hormone.

The early immunoassays for hCG utilized the intact hCG molecule as an immunogen. Antibodies raised against the intact molecule are unable to discriminate between hCG and luteinizing hormone (Paul and Ross, 1964). Since such antisera are generally used in the agglutination or agglutination-inhibition pregnancy tests, the sensitivity of such tests is set at a level high enough to eliminate interference with the levels of luteinizing hormone reached during the midcycle peak of this hormone in menstruating women or by the elevated levels found in postmenopausal women. They are generally able to detect between 1 and 2 IU of hCG/ml of urine or serum (Derman *et al.*, 1981). Since the majority of patients with hCG-producing nontrophoblastic neoplasms have serum hCG concentrations below 100 mIU/ml, the slide and tube pregnancy tests are not sensitive enough to detect hCG in the majority of such patients. Radioimmunoassays for hCG which use antisera raised against the intact molecule are quite sensitive, but are not specific for hCG because of the cross-reaction with luteinizing hormone (Bagshawe, 1973). Similarly, radioreceptor assays which use testicular or ovarian hCG receptors in place of anti-hCG serum to detect the molecule are not specific for hCG, since luteinizing hormone also binds to the receptors (Catt *et al.*, 1971; Landesman and Saxena, 1976).

In 1972 Vaitukaitis and co-workers developed a radioimmunoassay which circumvented many of the specificity problems of earlier immunoassays. They used the isolated subunit of the hCG molecule as the immunogen, and the antibody that was harvested was relatively specific for hCG and its β subunit. The immunoassay that these workers developed was capable of measuring 1 ng (5 mIU) of hCG/ml of serum in the presence of physiologic concentrations of luteinizing hormone, including those found in the blood of postmenopausal women and during midcycle surge of the hormone in cycling women. The assay can be performed within 24 hours or, if necessary, within 4 hours (Rasor and Braunstein, 1977). Because of its rapidity and wide availability, it is the assay of choice for measuring hCG in the blood or urine of patients with trophoblastic and nontrophoblastic tumors. Two other hCG-specific radioimmunoassays have been described recently. One assay uses an antiserum generated against the carboxyl-terminal portion of the hCG molecule (Chen *et al.*, 1976; Matsuura *et al.*, 1978). Since the luteinizing hormone molecule does not contain the carboxyl-terminal 30 amino acids of hCG, the antiserum used in this assay does not recognize luteinizing hormone at all. This assay, although quite specific for hCG, is presently too insensitive to be of clinical utility. Another recent development has been an hCG-specific radioimmunoassay that uses an antiserum raised against the reduced and S-carbamidomethylated hCG β subunit molecule (Pandian *et al.*, 1980). This radioimmunoassay does not measure luteinizing hormone and appears to be quite sensitive in detecting hCG.

IV. hCG IN GESTATIONAL TROPHOBLASTIC DISEASE (GTD)

Gestational trophoblastic neoplasms include hydatidiform mole, chorioadenoma destruens (invasive mole), and choriocarcinoma. Each of these tumors is characterized by the abnormal proliferation of hCG-secreting placental trophoblastic tissue. GTD may be present during an apparently normal or abnormal pregnancy, following a spontaneous or therapeutic abortion, or following delivery. Since hCG is normally found in maternal serum during each of these situations, the finding of hCG in the maternal serum or urine is not useful diagnostically. Indeed, at the time of diagnosis patients with GTD may have low, normal, or elevated concentrations of the hormone in relation to normal pregnancy levels (Hammond et al., 1967; Hammond and Parker, 1970; Bagshawe, 1976; Morrow et al., 1977; Gaspard et al., 1980). However, a urinary hCG level greater than 300,000 IU/24 hours or a serum hCG concentration greater than 200 IU/ml is suggestive of a molar pregnancy, although an occasional patient with multiple gestation may have hCG concentrations in this range at the time of the normal hCG peak during pregnancy (8–12 weeks) (Delfs, 1957; Borth et al., 1959; Braunstein et al., 1978). Serial measurements of hCG concentrations will distinguish between these groups of patients, since individuals with GTD will generally have increasing hCG concentrations over time, while hCG production during pregnancy declines rapidly following the 8–12 week peak (Braunstein et al., 1976). Approximately 20–30% of patients with GTD will have hCG concentrations that are less than 1 IU/ml and, therefore, may not be detected by the standard slide or tube pregnancy test (Hammond et al., 1967; Brewer and Eckman, 1974). Therefore, a negative urinary pregnancy test does not exclude the presence of GTD, and the more sensitive β subunit method should be used to confirm or negate the presence of GTD.

Following evacuation of a hydatidiform mole, the serum and urinary hCG levels show an early rapid decline followed by a slower disappearance (Fig. 1) (Bagshawe et al., 1973; Pastorfide et al., 1974; Goldstein et al., 1975; Morrow et al., 1977; Yuen and Cannon, 1981). Although the average time following evacuation for the hCG concentrations to reach the normal range is between 60 and 80 days (Morrow et al., 1977; Yuen and Cannon, 1981), hCG may be detected in the sera or urine of some women for 30–40 weeks (Bagshawe et al., 1973; Yuen and Cannon, 1981). It is of interest that the mean time required for hCG levels to become undetectable in the serum following first or second trimester therapeutic abortions or hysterectomies is between 27–40 days with a range of 16–45 days (Marrs et al., 1979). The longer time required for hCG concentrations to become undetectable in patients with uncomplicated molar pregnancies following evacuation of the mole undoubtably reflects the generally higher concentrations of hCG in such patients as well as the probable persistence of viable trophoblast in the uterus or elsewhere which is gradually destroyed by the hosts immune system. Since hCG concentrations in patients with molar pregnancies vary widely, the qualitative curve of the change in serum hCG concentrations in blood obtained at weekly intervals is more important than the actual quantitative level of the hCG or the duration of hCG detection following evacuation (Fig. 1). If the changes in hCG concentrations do not parallel the normal disappearance curve, or if there is a plateau in the serum concentration, chemotherapy should be administered, since such patients have persistent viable trophoblastic disease (Pastorfide et al., 1974; Hammond et al., 1980). hCG concentrations should be

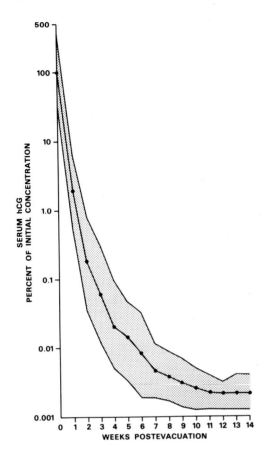

Fig. 1. Disappearance curve of maternal serum hCG following evacuation of hydatidiform mole. Circles represent mean values, and shaded area is 95% confidence limits. (Data adapted from Morrow *et al., Am. J. Obstet. Gynecol.* **128,** 1977, 424–430, by permission of the C. V. Mosby Company.)

measured at weekly intervals until they have been undetectable for 3 consecutive weeks. Following this, monthly determinations of serum hCG concentrations should be made for 6 months and at 2 month intervals for another 6 months (Ross *et al.,* 1965; Pastorfide *et al.,* 1974).

The level of hCG in the serum or urine of a patient with metastatic GTD does have some prognostic significance. Concentrations in the urine greater than 100,000 IU/24 hours or serum concentrations above 40 IU/ml in a patient with metastatic GTD is a poor prognostic sign, placing the patient in a high risk group (Hertz *et al.,* 1961; Ross *et al.,* 1965; Hammond *et al.,* 1967, 1973, 1980). The high hCG secretion in such patients undoubtable reflects a large trophoblastic tumor burden (Bagshawe, 1969). Indeed, there is a significant positive correlation between the initial hCG titer and the amount of chemotherapy required for remission of the GTD (Hammond *et al.,* 1967). Other features which indicate a poor prognosis include the presence of brain or liver metastases, unsuccessful prior chemotherapy, or an elapsed time of greater than 4 months between the antecedent pregnancy and the diagnosis of GTD (Hertz et al., 1961; Ross et al., 1965; Bagshawe, 1976; Hammond et al., 1980). Patients who have metastatic GTD who fall into this high risk group have a 20–40% remission rate with sequential single agent chemotherapy (Hammond and Parker, 1970; Goldstein, 1974), but a 66–80% survival rate with initial combination chemotherapy with three agents (Jones and Lewis, 1974; Hammond et al., 1980). In contrast, patients in the low risk group of metastatic GTD have a 95–100% cure rate with sequential single agent chemotherapy (Ross et al., 1965; Goldstein, 1974; Hammond et al., 1980).

Serial hCG measurements in the serum or urine are essential for monitoring the status of GTD during and following therapy. There is an excellent correlation between changes in the serum or urinary hCG titer and tumor burden measured by radiologic or radionucleotide examination of the size of the metastases (Bagshawe, 1969). Shortly after beginning chemotherapy in patients with metastatic GTD, there may be a transient rise in serum hCG concentrations presumably due to tumor necrosis and release of stored hormone (Tomoda et al., 1977). In patients suspected of having brain metastases, measurements of hCG in the cerebrospinal fluid may be helpful in establishing the diagnosis (Bagshawe and Harland, 1976; Soma et al., 1980). In patients without demonstrable brain metastases, the plasma hCG concentrations greatly exceeds the spinal fluid hCG level. When central nervous system metastases are present the plasma–spinal fluid ratio is decreased, usually with a ratio less than 45.

There has been recent interest in the measurement of the α subunit of the glycoprotein hormones in patients with GTD. Vaitukaitis and Ebersole (1976) suggested that the presence of free α subunit in the plasma urine of patients with GTD indicates a poor prognosis with resistance of the tumor to chemotherapy. Indeed, others have shown that elevated α subunit levels may reflect the presence of cerebral metastases (Dawood et al., 1977). Elevations of serum α subunit levels following apparently successful therapy for GTD, may indicate which patients are most likely to develop recurrence (Quigley et al., 1980a,b).

V. hCG IN GERM CELL TUMORS OF THE TESTES

Approximately 95% of all testicular tumors are of germ cell origin (Dixon and Moore, 1952; Braunstein et al., 1977). Such tumors exhibit five basic histologic patterns: seminoma (33–50%), embryonal carcinoma (20–33%), teratocarcinoma (10–33%), adult teratoma (10%), and choriocarcinoma (21%) (Kuhn and Johnson, 1972). The discovery that testicular germ cell tumors may secrete hCG was made Zondek (1929) who studied a patient with metastatic testicular tumor whose urine and tumor produced a typical Aschheim-Zondek reaction in the ovaries of mice. The sources of hCG in patients with such germ cell tumors are the trophoblastic giant cells, frequently seen in association with embryonal carcinoma and rarely with endodermal sinus tumors and seminomas and the syncytiotrophoblastic component of choriocarcinoma (Kurman et al., 1977).

Serum measurements of hCG are quite useful in the diagnosis and management of testicular tumors. Table I summarizes the results from multiple studies which have measured immunoreactive hCG by the β subunit method in the sera of patients with malignant germ cell neoplasms. As would be anticipated from the cellular source of hCG, patients with nonseminomatous testicular neoplasms had the highest frequency of hCG positivity in their sera. From a diagnostic standpoint, patients with a histologically pure seminoma, but who have detectable concentrations of immunoreactive hCG in their sera, pose a particularly difficult problem. Many of these patients when reevaluated are found to have nonseminomatous testicular tumors (Cochran, 1976; Lange et al., 1976, 1980). It is of interest that even when the testicle demonstrates only pure seminoma, nonseminomatous elements may be found at a later date in metastases (Maier et al., 1968). Thus, the presence of hCG in a patient with an apparent pure seminoma demands that a very careful search be made for nonseminomatous elements. The finding of an elevated AFP level in the serum of such a patient is unequivocal evidence that nonseminomatous elements are present in the patient's tumor (Lange et al., 1980). It should be emphasized that hCG measurements in patients with suspected or known testicular tumors should be performed by the β subunit radioimmunoassay method and not the less sensitive urine pregnancy test. Indeed, the urinary pregnancy test will be negative in one-third and equivo-

TABLE I
Frequency of hCG Detection in the Sera of Patients with Germ Cell Tumors of the Testes[a]

Tumor type	Number of patients	Number positive	Percentage positive
Seminoma	362	47	13.0
Nonseminoma	550	298	54.2
Embryonal carcinoma	82	42	51.2
Teratocarcinoma	20	7	35.0
Choriocarcinoma	12	12	100.0

[a] From Braunstein et al., 1973a; Waldmann and McIntire, 1974; Cochran et al., 1975; Lange et al., 1976; Newlands et al., 1976; Edsmyr et al., 1976; Vaitukaitis et al., 1976; Scardino et al., 1977; Schultz et al., 1978; Javadpour et al., 1978; Thompson and Haddow, 1979; Anderson et al., 1979; Scardino and Skinner, 1979; Narayana et al., 1979.

cal in another 10% of patients with germ cell tumors of the testes who have immunoreactive hCG in their circulation as measured by the β hCG radioimmunoassay (Braunstein et al., 1977).

Serial measurements of hCG (and AFP) are important in the management of patients with germ cell testicular tumors. Use of these markers have greatly reduced the staging error of patients following orchiectomy (Scardino et al., 1977). In addition, changes in the serum hCG concentrations parallel changes in the tumor burden in most patients. However, in some patients the hCG concentrations may decrease during or following therapy while the tumor progresses (Braunstein et al., 1973c; Perlin et al., 1976; Anderson et al., 1979). Such patients generally have complex tumors, such as teratocarcinomas or embryonal cell carcinomas, which contain trophoblastic, embryonal, and yolk sac elements. The reasons for this discordance are not clear, but are probably related to selective destruction of the trophoblastic elements by the chemotherapy. It is clear that a persistently elevated or raising hCG concentration in the serum of a patient with a germ cell neoplasm is indicative of residual neoplasm. Following effective therapy of an hCG-secreting testicular tumor, the hCG concentrations decrease with apparent serum half-life of 2–3 days (Willemse et al., 1981).

VI. hCG IN NONTROPHOBLASTIC NEOPLASMS

Prior to the development of the β hCG radioimmunoassay, several anecdotal reports noted that some nontrophoblastic tumors were capable of producing hCG ectopically. The tumors included bronchogenic carcinoma (Fusco and Rosen, 1966; Faiman et al., 1967; Becker et al., 1968; Rosen et al., 1968; Mochizuki et al., 1969; Dailey and Marcuse, 1969; Weintraub and Rosen, 1971), hepatoma or hepatoblastoma (Reeves et al., 1959; MacNab et al., 1952; Case records of the Massachusetts General Hospital, 1960; Behrle et al., 1963; Lipsett et al., 1964; Hung et al., 1963; Root et al., 1968), adrenal cortical carcinoma (McFadzean, 1946; Chambers, 1949; Higgins et al., 1956; Rose et al., 1968; McArthur, 1963), melanoma (Li, 1959), undifferentiated retroperitoneal carcinoma (Matteini, 1952), breast carcinoma (McArthur, 1963), and renal cell carcinoma (Case records of the Massachusetts General Hospital, 1972). The two clinical syndromes associated with ectopic hCG production are isosexual precocious puberty in boys with hepatoblastomas and gynecomastia in adult males.

Shortly after the development of the hCG β subunit radioimmunoassay, Braunstein and co-workers (1973a) studied the frequency of ectopic hCG production by examining the sera from 906 patients with nontrophoblastic cancers for the presence of hCG and found that 11.4% of the patients had detectable quantities of hCG in their sera. Since that initial study, many other investigators have retrospectively and prospectively studied the frequency of ectopic hCG production in patients with nontrophoblastic neoplasms using similar assay techniques (for references see Table II). The data from these studies are summarized in Table II and indicate that the incidence of ectopic hCG production is approx-

TABLE II
Frequency of hCG Detection in the Sera of Patients with Nontrophoblastic Neoplasms[a]

Tumor	Number of patients	Number positive for hcG	Percentage
Islet cell	83	37	44.6
Ovarian (epithelial)	260	101	38.8
Carcinoid	35	10	28.6
Gynecologic (nonovarian)	686	182	26.5
Breast	1311	336	25.6
Gastrointestinal	1211	251	20.7
Oropharynx	80	16	20
Esophagus	40	4	10
Gastric	159	28	17.6
Small intestine	27	3	11.1
Colorectal	118	23	19.5
Pancreatic	150	31	20.7
Liver	259	41	15.8
Biliary	10	2	20
Ears-Nose-Throat	83	17	20.5
Melanoma	194	38	19.6
Lung	1143	187	16.4
Sarcomas	76	11	14.5
Multiple myeloma	111	9	8.1
Genitourinary	268	19	7.1
Renal	118	7	5.9
Bladder	48	5	10.4
Prostate	102	7	6.9
Leukemia–lymphoma	814	29	3.6
Other	852	51	6.0
Totals	7127	1278	17.9

[a] Data compiled from Braunstein et al., 1973a,b; Goldstein et al., 1974; Sheth et al., 1974; Rosen et al., 1975; Tormey et al., 1975, 1977; Fishman et al., 1975, 1976; Lange et al., 1976; Vaitukaitis et al., 1976; Franchimont 1976, 1977, 1978; Stolbach et al., 1976; Samaan et al., 1976; Hagen et al., 1976; Gailani et al., 1976; Coombes et al., 1977; Broder et al., 1977; Kahn et al., 1977; Sufrin et al., 1977; Stone et al., 1977; Williams et al., 1977; Hattori et al., 1978; Waalkes et al., 1978; Dash et al., 1978, 1979a,b; Cowen et al., 1978; Rutanen and Seppälä, 1978; Dosogne-Guerin et al., 1978; Bender et al., 1979; Cove et al., 1979; Papapetrou et al., 1980; Gropp et al., 1980; Mackie et al., 1980; Menon and Stefani, 1980; Donaldson et al., 1980; Carenza et al., 1980; Blackman et al., 1980; Shinde et al., 1981; Öberg and Wide, 1981.

imately 18%. The highest frequency of ectopic hCG production has been found in patients with islet cell tumors of the pancreas, ovarian epithelial tumors, other gynecologic malignancies, gastrointestinal tract tumors, melanoma, and breast and lung carcinoma.

The prevalence of hCG production by nontrophoblastic neoplasms may actually be higher than the cumulative frequency noted in Table II. Papapetrou and co-workers (1980) compared the frequency of hCG presence in urine as compared to serum in 70 patients with nontrophoblastic malignancies. They found that 17.1% of these patients had immunoreactive hCG in the serum, while 44.3% had immunoreactive material in the urine. McManus and co-workers (1976) have demonstrated hCG in the cytoplasm or on the surface of 89% of tumors by the immunoperoxidase technique. None of the control tissues demonstrated the presence of hCG. Similar results for selected tumors have been reported by others (Acevedo et al., 1976; Horne et al., 1976; Castro et al., 1979, 1980), although some investigators have not noted as high a frequency (Hustin, 1978; Bellet et al., 1980). In addition, several groups have used the β hCG radioimmunoassay to measure hCG in tumor tissue extracts and have noted that 14–100% of cancers contain immunoreactive hCG (Castro et al., 1979, 1980; Hattori et al., 1978; Yoshimoto et al., 1979b). However, as noted above, most nonmalignant human tissues contain immunoreactive hCG. Therefore it is difficult to evaluate the significance of the mere presence of hCG in tumor tissues. Yoshimoto and co-workers (1979b) examined this question and found that 77% of a variety of cancers contain hCG in amounts indistinguishable from normal tissues, while 23% had concentrations that were above those seen in normal tissues.

Several problems concerning the use of hCG as a tumor marker have become apparent from analysis of the published information. The majority of studies have involved patients with relatively advanced neoplasms. There is little information regarding the utility of hCG as an immunodiagnostic agent for detecting cancer prior to the development of objective signs and symptoms of neoplasia. Williams and co-workers (1977) retrospectively examined sera from patients involved in the Framingham Heart Study who developed pancreatic or gastric carcinoma. Blood samples had been obtained from these patients 3–67 months prior to the diagnosis of the tumor. hCG was found in the serum of one patient in a sample obtained 26 months prior to the clinical diagnosis of pancreatic cancer. Since none of the other 16 patients with cancer had hCG present in their serum, it could not be determined if hCG production by hCG-secreting neoplasms consistently antedates the clinical manifestations of cancer.

The serum levels of hCG in patients with nontrophoblastic cancer are generally low, most often in the range 1–5 ng (5–25 mIU)/ml (Braunstein et al., 1973a; Sheth et al., 1974; Samaan et al., 1976; Gailani et al., 1976; Stolbach et al., 1976; Stone et al., 1977; Tormey et al., 1977; Williams et al., 1977; Coombes et al., 1977; Hattori et al., 1978; Bender et al., 1979; Dosegne-Guérin et al., 1978; Gropp et al., 1980; Mackie et al., 1980; Blackman et al., 1980).

These concentrations are the lower limit of detection in the β subunit radioimmunoassay, where the precision and specificity are the poorest. It is well documented that serum proteins can nonspecifically effect the β subunit radioimmunoassay at the lower end of the standard curve (Vaitukaitis *et al.*, 1972). Also, many of the commercially available β subunit radioimmunoassays do not have the same high degree of specificity for hCG that the NIH-distributed SB-6 antibody has. Therefore, some of the assays may detect physiologic concentrations of human luteinizing hormone.

Another important problem with the use of hCG as a tumor marker involves its reliability as an objective parameter of changes in tumor growth. Although in many instances the quantitative levels of hCG and changes in serum hCG concentrations in serial samples correlate with the clinical tumor burden (Tormey *et al.*, 1975; Stolbach *et al.*, 1976; Broder *et al.*, 1977; Waalkes *et al.*, 1978), this does not always occur, and discordance between hCG levels and tumor growth during therapy, as well as between hCG concentrations and other tumor markers, has been noted (Samaan *et al.*, 1976; Franchimont *et al.*, 1976, 1977; Stone *et al.*, 1977; Muggia *et al.*, 1975; Kahn *et al.*, 1977; Bender *et al.*, 1979; Cowen *et al.*, 1978; Carenza *et al.*, 1980; Stanhope *et al.*, 1979). When serial blood samples have been obtained in some patients with hCG-producing neoplasms, the hormone may be found intermittently and unrelated to therapy or changes in tumor burden (Braunstein *et al.*, 1973b; Stone *et al.*, 1977; Tormey *et al.*, 1977; Rutanen and Seppälä, 1978). These problems may be related to assay sensitivity and specificity, discontinuous tumor growth, or hCG synthesis occurring spontaneously or as a result of therapy, or intermittent release of hCG secondary to tumor necrosis. Spontaneous intermittent release of other hormones by tumors is well documented and has been termed "periodic hormonogenesis" (Bailey, 1971).

Perhaps the most serious problem with the utilization of hCG measurements in the serum of patients suspected of harboring a neoplasm is that of specificity. As previously noted, immunoreactive hCG has been detected in virtually all human tissues. Borkowski and Muquardt (1979) found that a material which resembled hCG immunologically and by gel filtration chromatography could be measured in plasma extracts from 12 of 16 male blood donors. These workers noted that the medium concentration was 19 pg/ml of blood. Although this is well below the limits of sensitivity of the β subunit radioimmunoassay, it is conceivable that some normal individuals and patients with benign disorders may be found to have concentrations of hCG in their sera that would be detected by the β hCG radioimmunoassay. Indeed, this has been found to be true. Table III summarizes the data on control patients from the various studies which measured hCG in the sera of cancer patients. The fact that 2.7% of patients without cancer have immunoreactive hCG in their circulation seriously limits the diagnostic utility of hCG measurements in nontrophoblastic tumors. The majority of patients without tumors have hCG levels less than 4 ng/ml, and we have found that patients with

TABLE III
Frequency of hCG Detection in the Sera of Normal Controls and Patients without Cancer

Type of control	Number of persons	Number positive for hCG	Percent positive
Normal	931	12	1.3
Benign disease	3144	100	3.2
Breast	121	7	5.9
Lung	154	5	3.2
Gastrointestinal tract	257	17	6.6
Gynecologic	513	35	6.8
Genitourinary tract	41	0	0
Central nervous system	10	0	0
Hypothyroid	28	0	0
Total	4075	112	2.7

a From Braunstein *et al.*, 1973a,b, 1981; Goldstein *et al.*, 1974; Vaitukaitis *et al.*, 1976; Lange *et al.*, 1976; Franchimont *et al.*, 1976, 1977, 1978; Hagen *et al.*, 1976; Gailani *et al.*, 1976; Hattori *et al.*, 1978; Dash *et al.*, 1978, 1979a,b; Williams *et al.*, 1977; Cowen *et al.*, 1978; Rutanen and Seppälä, 1978; Dosogne-Guérin *et al.*, 1978; Papapetrou *et al.*, 1980; Mackie *et al.*, 1980; Menon and Stefani, 1980; Carenza *et al.*, 1980; Blackman *et al.*, 1980; Donaldson *et al.*, 1980; Shinde *et al.*, 1981.

benign disorders uniformly have hCG levels less than 10 ng/ml, while 13% of patients with neoplastic hCG production have levels above this (Braunstein *et al.*, 1981). If one increases the normal level for hCG to 10 ng/ml, the false positive results would be eliminated, but the true positive rate for hCG production by nontrophoblastic neoplasms would decrease to 1.4% (Braunstein *et al.*, 1981). This indicates that serum hCG measurements are not useful in screening patients for nontrophoblastic malignancies.

Very few studies have examined whether the presence or absence of hCG has any prognostic implications in patients with nontrophoblastic tumors. Braunstein and associates (1973b) was unable to show any difference in survival between hCG-positive and hCG-negative patients with hepatocellular carcinoma. Similarly, Horne and co-workers (1976) found no survival differences between patients with breast cancer whose tumors contained immunoreactive hCG in comparison to those whose tumor did not. In contrast, Tormey and co-workers (1977) did find that patients with metastatic cancer who had immunoreactive hCG in their blood have a lower response rate to chemotherapy and a shorter duration of remission during therapy than did patients in whom hCG was undetectable in the blood. At the present time the data are just too sparse to draw any conclusion regarding the prognostic value of hCG measurements in patients with nontrophoblastic malignancies.

In summary, hCG measurements by the β subunit radioimmunoassay method are indispensible for the diagnosis of gestational and nongestational trophoblastic tumors as well as germ cell tumors of the testes. Serum hCG measurements are essential for monitoring the effects of therapy of gestational trophoblastic disease. Such measurements may also be useful in patients with germ cell tumors of the testes and hCG-producing nontrophoblastic tumors. Because of the occasional discordance between changes in hCG concentrations and tumor growth in such patients, a decrease in serum concentrations of hCG cannot be used as a definitive indication of a decrease in tumor burden. However, rising levels of the hormone in such patients is unequivocal evidence of persistent disease. With our present methodology, hCG measurements appear to be of no use in the screening of patients for nontrophoblastic cancers.

ACKNOWLEDGMENT

Supported in part by United States Public Health Service Grant HDCA 13042.

REFERENCES

Acevedo, H. F., Slifkin, M., Pouchet, G. R., and Rakhshan, M. (1976). *Program 58th Annu. Meet. Endocr. Soc.* Abstract No. 420.
Adcock, E. W., III, Teasdale, F., August, C. S., Cox, S., Meschia, G., Battaglia, F. C., and Naughton, M. A. (1973). *Science* **181**, 845–847.
Anderson, T., Waldmann, T. A., Javadpour, N., and Glatstein, E. (1979). *Ann. Intern. Med.* **90**, 373–385.
Aschheim, S., and Zondek, B. (1927). *Klin. Wochenschr.* **6**, 1322.
Bagshawe, K. D. (1969). "Choriocarcinoma. The Clinical Biology of the Trophoblast and Its Tumors." Williams & Wilkins, Baltimore, Maryland.
Bagshawe, K. D. (1973). *Methods Invest. Diagn. Endocrinol.* **2B**, 756–763.
Bagshawe, K. D. (1976). *Cancer* **38**, 1373–1385.
Bagshawe, K. D., and Harland, S. (1976). *Cancer* **38**, 112–118.
Bagshawe, K. D., Wilson, H., Dublon, P., Smith, A., Baldwin, M., and Kardana, A. (1973). *J. Obstet. Gynaecol. Br. Commonw.* **80**, 461–468.
Bailey, R. E. (1971). *J. Clin. Endocrinol. Metab.* **32**, 317–327.
Becker, K. L., Cottrell, J. C., Moore, C. F., Winnacker, J. L., Matthews, M. J., and Katz, S. (1968). *J. Clin. Endocrinol. Metab.* **28**, 809–818.
Behrle, F. C., Mantz, F. A., Jr., Olson, R. L., and Trombold, J. C. (1963). *Pediatrics* **32**, 265–271.
Beling, C. G., and Weksler, M. E. (1974). *Clin. Exp. Immunol.* **18**, 537–541.
Bellet, D., Arrang, J. M., Contesso, G., Caillaud, J. M., and Bohuon, C. (1980). *Eur. J. Cancer* **16**, 433–439.
Bellisario, R., Carlsen, R. B., and Bahl, O. P. (1973). *J. Biol. Chem.* **248**, 6797–6809.
Bender, R. A., Weintraub, B. D., and Rosen, S. W. (1979). *Cancer* **43**, 591–595.
Birkin, S., and Canfield, R. E. (1980). In "Chorionic Gonadotropin" (S. J. Segal, ed.), pp. 65–88. Plenum, New York.
Blackman, M. R., Weintraub, B. D., Rosen, S. W., Kourides, I. A., Steinwascher, K., and Gail, M. H. (1980). *JNCI, J. Natl. Cancer Inst.* **65**, 81–93.

Boime, I., Landefeld, T., McQueen, S., and McWilliams, D. (1978). In "Structure and Function of the Gonadotropins" (K. W. McKerns, ed.), pp. 235–257. Plenum, New York.
Borkowski, A., and Muquardt, C. (1979). *N. Engl. J. Med.* **301,** 298–302.
Borth, R., Luenfeld, B., Stamm, O., and DeWatteville, H. (1959). *Acta Obstet. Gynecol. Scand.* **38,** 417–423.
Braunstein, G. D., Vaitukaitis, J. L., Carbone, P. P., and Ross, G. T. (1973a). *Ann. Intern. Med.* **78,** 39–45.
Braunstein, G. D., McIntire, K. R., and Waldmann, T. A. (1973b). *Cancer* **31,** 1065–1068.
Braunstein, G. D., Grodin, J. M., Vaitukaitis, J., and Ross, G. T. (1973c). *Am. J. Obstet. Gynecol.* **115,** 447–450.
Braunstein, G. D., Rasor, J., and Wade, M. E. (1975). *N. Engl. J. Med.* **293,** 1339–1343.
Braunstein, G. D., Rasor, J., Adler, D., Danzer, H., and Wade, M. E. (1976). *Am. J. Obstet. Gynecol.* **126,** 678–681.
Braunstein, G. D., Friedman, M. B., Sacks, S. A., Thompson, R. W., and Skinner, D. G. (1977). *West. J. Med.* **126,** 362–377.
Braunstein, G. D., Karow, W. G., Gentry, W. C., Rasor, J., and Wade, M. E. (1978). *Am. J. Obstet. Gynecol.* **131,** 25–32.
Braunstein, G. D., Kamdar, V., Rasor, J., Swaminathan, N., and Wade, M. E. (1979). *J. Clin. Endocrinol. Metab.* **49,** 917–925.
Braunstein, G. D., Rasor, J., and Wade, M. E. (1980). In "Chorionic Gonadotropin" (S. J. Segal, ed.), pp. 383–410. Plenum, New York.
Braunstein, G. D., Rasor, J., Thompson, R., Van Scoy-Mosher, M., and Wade, M. E. (1981). *Clin. Res.* **29,** 98A.
Brewer, J. I., and Eckman, T. R. (1974). In "Controversy in Obstetrics and Gynecology II" (D. R. Reid and C. D. Christian, eds.), pp. 215–218. Saunders, Philadelphia, Pennsylvania.
Broder, L. E., Weintraub, B. D., Rosen, S. W., Cohen, M. H., and Tejada, F. (1977). *Cancer* **40,** 211–216.
Brouha, L., and Hinglais, H. (1931). *Gynecol. Obstet.* **24,** 42–56.
Carenza, L., DiGregorio, R., Mocci, C., Moro, M., and Pala, A. (1980). *Gynecol. Oncol.* **10,** 32–38.
Carlsen, R. B., Bahl, O. P., and Swaminathan, N. (1973). *J. Biol. Chem.* **248,** 6810–6827.
Carr, M. C., Sites, D. P., and Fudenberg, H. H. (1973). *Cell. Immunol.* **8,** 448–454.
Case Records of the Massachusetts General Hospital. Case 46451. (1960). *N. Engl. J. Med.* **263,** 965–971.
Case Records of the Massachusetts General Hospital. Case 13-1972. (1972). *N. Engl. J. Med.* **286,** 713–719.
Castro, A., Buschbaum, P., Nadji, M., Voigt, W., Tabei, S., and Morales, A. (1979). *Experientia* **35,** 1392–1393.
Castro, A., Buschbaum, P., Nadji, M., Voigt, W., Tabei, S., and Morales, A. (1980). *Acta Endocrinol. (Copenhagen)* **94,** 511–516.
Catt, K. J., and Dufau, M. L. (1973). *Nature (London) New Biol.* **244,** 219–221.
Catt, K. J., Dufau, M. L., and Tsuruhara, T. (1971). *J. Clin. Endocrinol. Metab.* **32,** 860–863.
Chambers, W. L. (1949). *J. Clin. Endocrinol.* **9,** 451–456.
Channing, C. P., and Kammerman, S. (1974). *Biol. Reprod.* **10,** 179–198.
Chatterjee, M., and Munro, H. N. (1977). *Biochem. Biophys. Res. Commun.* **77,** 426–433.
Chen, H. C., Hodgen, G. D., Matsuura, S., Lin, L. J., Gross, E., Reichert, L. E., Jr., Birken, S., Canfield, R. E., and Ross, G. T. (1976). *Proc. Natl. Acad. Sci. U.S.A.* **73,** 2885–2889.
Christiansen, P. (1967). *Acta Endocrinol. (Copenhagen)* **56,** 608–618.
Closset, J., Hennen, G., and Lequin, R. M. (1973). *FEBS Lett.* **29,** 97–100.
Cochran, J. S. (1976). *J. Urol.* **115,** 465–466.

Cochran, J. S., Walsh, P. C., Porter, J. C., Nicholson, T. C., Madden, J. D., and Peters, P. D. (1975). *J. Urol.* **114,** 549–555.
Coombes, R. C., Powles, T. J., Gazet, J. C., Ford, H. T., Nash, A. G., Sloane, J. P., Hillyard, C. J., Thomas, P., Keyser, J. W., Marcus, D., Zinberg, N., Stimson, W. H., and Neville, A. M. (1977). *Cancer* **40,** 937–944.
Cove, D. H., Woods, K. L., Smith, S. C. H., Burnett, D., Leonard, J., Grieve, R. J., and Howell, A. (1979). *Br. J. Cancer* **40,** 710–718.
Cowen, D. M., Searle, F., Ward, A. M., Benson, E. A., Smiddy, F. G., Eaves, G., and Cooper, E. H. (1978). *Eur. J. Cancer* **14,** 885–893.
Dailey, J. E., and Marcuse, P. M. (1969). *Cancer* **24,** 388–396.
Dash, R. J., Dutta, T. K., Purohit, O. P., Jayakumar, R. V., and Gupta, B. D. (1978). *Indian J. Cancer* **15,** 23–27.
Dash, R. J., Dutta, T. K., Gupta, S. K., Purohit, O. P., Jayakumar, R. V., and Gupta, B. D. (1979a). *Indian J. Med. Res.* **70,** 478–482.
Dash, R. J., Dutta, T. K., Gupta, S. K., Jayakumar, R. V., Purohit, O. P., Joshi, V. V., and Datta, B. N. (1979b). *Indian J. Pathol.* **22,** 97–102.
Dawood, M. Y., Saxena, B. B., and Landesman, R. (1977). *Obstet. Gynecol.* **50,** 172–181.
de Ikonicoff, L. K., and Cédard, L. (1973). *Am. J. Obstet. Gynecol.* **116,** 1124–1132.
Delfs, E. (1957). *Obstet. Gynecol.* **9,** 1–24.
Derman, R., Corson, S. L., Horwitz, C. A., Lau, H. L., and Soderström, R. (1981). *J. Reprod. Med.* **26,** 149–178.
Diczfalusy, E., and Loraine, J. A. (1955). *J. Clin. Endocrinol. Metab.* **15,** 424–434.
Dixon, F. J., and Moore, R. A., eds. (1952). "Atlas of Tumor Pathology." Armed Forces Inst. Pathol., Washington, D.C.
Donaldson, E. S., van Nagell, J. R., Pursell, S., Gay, E. C., Meeker, W. R., Kashmiri, R., and Van de Voorde, J. (1980). *Cancer* **45,** 948–953.
Dosogne-Guérin, M., Stolarczyk, A., and Borkowski, A. (1978). *Eur. J. Cancer* **14,** 525–532.
Dufau, M. L., Mendelson, C. R., and Catt, K. J. (1974). *J. Clin. Endocrinol. Metab.* **39,** 610–613.
Edsmyr, F., Wahren, B., and Silfversward, C. (1976). *Int. J. Radiat. Oncol., Biol. Phys.* **1,** 279–284.
Faiman, C., Colwell, J. A., Ryan, R. J., Hershman, J. M., and Shields, T. W. (1967). *N. Engl. J. Med.* **277,** 1395–1399.
Faiman, C., Winter, J. S. D., and Reyes, F. I. (1981). *In* "The Testis" (H. Burger and D. deKretser, ed.), pp. 81–105. Raven Press, New York.
Fiddes, J. C., and Goodman, H. M. (1979). *Nature (London)* **281,** 351–356.
Fishman, W. H., Raam, S., and Stolbach, L. L. (1975). *Semin. Oncol.* **2,** 211–216.
Fishman, W. H., Inglis, N. R., Vaitukaitis, J., and Stolbach, L. L. (1976). *Natl. Cancer Inst. Monogr.* **42,** 63–73.
Franchimont, P., Zangerle, P. F., Nogarede, J., Bury, J., Molter, F., Reuter, A., Hendrick, J. C., and Collette, J. (1976). *Cancer* **38,** 2287–2295.
Franchimont, P., Zangerle, P. F., Hendrick, J. C., Reuter, A., and Colin, C. (1977). *Cancer* **39,** 2806–2812.
Franchimont, P., Reuter, A., and Gaspard, U. (1978). *Curr. Top. Exp. Endocrinol.* **3,** 201–216.
Friedman, M. H., and Laphan, M. E. (1931). *Am. J. Obstet. Gynecol.* **21,** 405–410.
Fuchs, T., Hammarström, L., Smith, C. I. E., and Brundin, J. (1980). *Acta Obstet. Gynecol. Scand.* **59,** 355–359.
Fusco, F. D., and Rosen, S. W. (1966). *N. Engl. J. Med.* **275,** 507–515.
Gailani, S., Chu, T. M., Nussbaum, A., Ostrander, M., and Christoff, N. (1976). *Cancer* **38,** 1684–1686.
Galli-Mainini, C. (1947). *Sem. Med.* **1,** 337–340.

Gaspard, U. J., Reuter, A. M., Deville, J.-L., Vrindts-Gevaert, Y., Bagshawe, K. D., and Franchimont, P. (1980). *Clin. Endocrinol. (Oxford)* **13,** 319–329.
Gey, G. O., Jones, G. E. S., and Hellman, L. M. (1938). *Science* **88,** 306–307.
Golander, A., Barrett, J. R., Tyrey, L., Fletcher, W. H., and Handwerger, S. (1978). *Endocrinology* **102,** 597–605.
Goldstein, D. P. (1974). *In* "Controversy in Obstetrics and Gynecology II" (D. R. Reid and C. D. Christian, eds.), pp. 219–234. Saunders, Philadelphia, Pennsylvania.
Goldstein, D. P., Kosasa, T. S., and Skarin, A. T. (1974). *Surg., Gynecol. Obstet.* **138,** 747–751.
Goldstein, D. P., Pastorfide, G. B., Osathanondh, R., and Kosasa, T. S. (1975). *Obstet. Gynecol.* **45,** 527–530.
Gropp, C., Havermann, K., and Scheuer, A. (1980). *Cancer* **46,** 347–354.
Hagen, C., Gilby, E. D., McNeilly, A. S., Olgaard, K., Bondy, P. K., and Rees, L. H. (1976). *Acta Endocrinol. (Copenhagen)* **83,** 26–35.
Hammarström, L., Fuchs, T., and Smith, C. I. E. (1979). *Acta Obstet. Gynecol. Scand.* **58,** 417–422.
Hammond, C. B., and Parker, R. T. (1970). *Obstet. Gynecol.* **35,** 132–143.
Hammond, C. B., Hertz, R., Ross, G. T., Lipsett, M. B., and Odell, W. D. (1967). *Am. J. Obstet. Gynecol.* **98,** 71–78.
Hammond, C. B., Borchert, L. G., Tyrey, L., Creasman, W. T., and Parker, R. T. (1973). *Am. J. Obstet. Gynecol.* **115,** 451–457.
Hammond, C. B., Weed, J. C., and Currie, J. L. (1980). *Am. J. Obstet. Gynecol.* **136,** 844–858.
Han, T. (1974). *Clin. Exp. Immunol.* **18,** 529–535.
Handwerger, S., Barrett, J., Tyrey, L., and Schomberg, D. (1973). *J. Clin. Endocrinol. Metab.* **36,** 1268–1270.
Haour, F., and Saxena, B. B. (1974). *J. Biol. Chem.* **149,** 2195–2205.
Hattori, M., Fukase, M., Yoshimi, H., Matsukura, S., and Imura, H. (1978). *Cancer* **42,** 2328–2333.
Heidrich, L., Fels, E., and Mathias, E. (1930). *Beitr. Z. Klin. Chir.* **150,** 349–384.
Hertz, R., Lewis, J. L., Jr., and Lipsett, M. B. (1961). *Am. J. Obstet. Gynecol.* **82,** 631–637.
Higgins, G. A., Brownlee, W. E., and Mantz, F. A., Jr. (1956). *Am. Surg.* **22,** 56–79.
Hirose, T. (1919). *Mitt. Med. Fak. Tokio* **23,** 63.
Hirose, T. (1920). *J. Jpn. Gynecol. Soc.* **16,** 1055.
Horne, C. H. W., Reid, I. N., and Milne, G. D. (1976). *Lancet* **2,** 279–282.
Hung, W., Blizzard, R. M., Migeon, C. J., Camacho, A. M., and Nyhan, W. L. (1963). *Pediatrics* **63,** 895–903.
Hustin, J. (1978). *Gynecol. Obstet. Invest.* **9,** 3–15.
Javadpour, N., McIntire, K. R., and Waldmann, T. A. (1978). *Cancer* **42,** 2768–2772.
Jones, W. B., and Lewis, J. L., Jr. (1974). *Am. J. Obstet. Gynecol.* **120,** 14–20.
Kahn, C. R., Rosen, S. W., Weintraub, B. D., Fajans, S. S., and Gorden, P. (1977). *N. Engl. J. Med.* **297,** 565–569.
Keutmann, H., and Williams, R. (1977). *J. Biol. Chem.* **252,** 5393–5397.
Kuhn, C. R., and Johnson, D. E. (1972). *In* "Testicular Tumors" (D. E. Johnson, ed.), pp. 37–46. Medical Examination Publ., Co., New York.
Kurman, R. J., Scardino, P. T., McIntire, K. R., Waldmann, T. A., and Javadpour, N. (1977). *Cancer* **40,** 2136–2151.
Landesman, R., and Saxena, B. B. (1976). *Fertil. Steril.* **27,** 357–368.
Lange, P. H., McIntire, K. R., Waldmann, T. A., Hakala, T. R., and Fraley, E. E. (1976). *N. Engl. J. Med.* **295,** 1237–1240.
Lange, P. H., Nochomovitz, L. E., Rosai, J., Fraley, E. E., Kennedy, B. J., Bosl, G., Braisbane,

J., Catalona, W. J., Cochran, J. S., Comisarow, R. H., Cummings, K. B., deKernion, J. B., Einhorn, L. H., Hakala, T. R., Jewett, M., Moore, M. R., Scardino, P. T., and Streitz, J. M. (1980). *J. Urol.* **124,** 472–477.
Li, M. C. (1959). *Ann. N.Y. Acad. Sci.* **80,** 280–284.
Li, M. C., Hertz, R., and Spencer, D. B. (1956). *Proc. Soc. Exp. Biol. Med.* **93,** 361–366.
Lipsett, M. B., Odell, W. D., Rosenberg, L. E., and Waldmann, T. A. (1964). *Ann. Intern. Med.* **61,** 733–756.
McArthur, J. W. (1963). *Prog. Gynecol.* **4,** 146–172.
McArthur, J. W. (1968). *In* "Gonadotrophins" (E. Rosemberg, ed.), pp. 71–79. Geron-X, Los Altos, California.
McFadzean, A. J. S. (1946). *Lancet* **2,** 940–943.
McGregor, W. G., Raymoure, W. J., Kohn, R. W., and Jaffe, R. B. (1981). *J. Clin. Invest.* **68,** 306–309.
Mackie, C. R., Moosa, A. R., Go, V. L. W., Noble, G., Seizemore, G., Cooper, M. J., Wood, R. A. B., Hall, A. W., Waldmann, T., Gelder, F., and Rubenstein, A. H. (1980). *Dig. Dis. Syst.* **25,** 161–172.
McManus, L. M., Naughton, M. A., and Martinez-Hernandez, A. (1976). *Cancer Res.* **36,** 3476–3481.
MacNab, G. H., Moncrieff, S. A., and Budian, M. (1952). *In Annu. Rep.—Cancer Res. Campaign* **30,** 168.
Maier, J. G., Sulak, M. H., and Mittemeyer, B. T. (1968). *Am. J. Roentgenol., Radium Ther. Nucl. Med.* **102,** 596–602.
Marrs, R. P., Kletzky, O. A., Howard, W. F., and Mishell, D. R. (1979). *Am. J. Obstet. Gynecol.* **135,** 731–736.
Matsuura, S., Chen, H. C., and Hodgen, G. D. (1978). *Biochemistry* **17,** 575–580.
Matteini, M. (1952). *Rass. Neurol. Veg.* **9,** 252–271.
Menon, M., and Stefani, S. S. (1980). *Urol. Inst.* **35,** 291–293.
Midgley, A. R., and Pierce, G. B., Jr. (1962). *J. Exp. Med.* **115,** 289–294.
Mochizuki, M., Tokura, Y., and Tokura, T. (1969). *Acta Obstet. Gynaecol. Jpn.* **16,** 180–191.
Morgan, F. J., Birken, S., and Canfield, R. E. (1973). *Mol. Cell. Biochem.* **2,** 97099.
Morgan, F. J., Canfield, R. E., Vaitukaitis, J L., and Ross, G. T. (1974). *Endocrinology* **94,** 1601–1606.
Morgan, F. J., Birken, S., and Canfield, R. E. (1975). *J. Biol. Chem.* **250,** 5247–5258.
Morrow, C. P., Kletzky, O. A., Disaia, P. J., Townsend, D. E., Mishell, D. R., and Nakamura, R. M. (1977). *Am. J. Obstet. Gynecol.* **128,** 424–430.
Muggia, F. M., Rosen, S. W., Weintraub, B. D., and Hansen, H. H. (1975). *Cancer* **36,** 1327–1337.
Narayana, A. S., Loening, S., Weimer, G., and Culp, D. A. (1979). *J. Urol.* **121,** 51–53.
Newlands, E. S., Dent, J., Kardana, A., Searle, F., and Bagshawe, K. D. (1976). *Lancet* **2,** 744–745.
Nisula, B. C., Taliadouros, G. S., and Carayon, P. (1980). *In* "Chorionic Gonadotropin" (S. J. Segal, ed.), pp. 17–36. Plenum, New York.
Öberg, K., and Wide, L. (1981). *Acta Endocrinol. (Copenhagen)* **98,** 256–260.
Pandian, M. R., Mitra, R., and Bahl, O. P. (1980). *Endocrinology* **107,** 1564–1571.
Papapetrou, P. D., Sakarelou, N. P., Braouzi, H., and Fessas, P. H. (1980). *Cancer* **45,** 2583–2592.
Parlow, A. F. (1961). *In* "Human Pituitary Gonadotrophin" (A. Albert, ed.), pp. 300–310. Thomas, Springfield, Illinois.
Pastorfide, G. B., Goldstein, D. P., and Kosasa, T. S. (1974). *Am. J. Obstet. Gynecol.* **120,** 1025–1028.

Paul, W. E., and Ross, G. T. (1964). *Endocrinology* **75,** 352–358.
Perlin, E., Engeler, J. E., Jr., Edson, M., Karp, D., McIntire, K. R., and Waldmann, T. A. (1976). *Cancer* **37,** 215–219.
Pierce, J. G., Liao, T. H., Howard, S. M., Shome, B., and Cornell, J. S. (1971). *Recent Prog. Horm. Res.* **27,** 165–212.
Quigley, M. M., Tyrey, L., and Hammond, C. B. (1980a). *J. Clin. Endocrinol. Metab.* **50,** 98–102.
Quigley, M. M., Tyrey, L., and Hammond, C. B. (1980b). *Am. J. Obstet. Gynecol.* **138,** 545–549.
Rasor, J. L., and Braunstein, G. D. (1977). *Obstet. Gynecol.* **50,** 553–558.
Reeves, R. L., Resluk, H., and Harrison, C. E. (1959). *J. Clin. Endocrinol. Metab.* **19,** 1651–1660.
Root, A. W., Bogiovanni, A. M., and Eberlein, W. R. (1968). *J. Clin. Endocrinol. Metab.* **28,** 1317–1322.
Rose, L. I., Williams, G. H., Jagger, P. I., and Lauler, D. P. (1968). *J. Clin. Endocrinol. Metab.* **28,** 903–908.
Rosen, S. W., Becker, C. E., Schlaff, S., Easton, J., and Gluck, M. C. (1968). *N. Engl. J. Med.* **279,** 640–641.
Rosen, S. W., Weintraub, B. D., Vaitukaitis, J. S., Sussman, H. H., Hershman, J. M., and Muggia, F. M. (1975). *Ann. Int. Med.* **82,** 71–83.
Rosen, S. W., Weintraub, B. D., and Aaronson, S. A. (1980). *J. Clin. Endocrinol. Metab.* **50,** 834–841.
Ross, G. T., Goldstein, D. P., Hertz, R., Lipsett, M. B., and Odell, W. D. (1965). *Am. J. Obstet. Gynecol.* **93,** 223–229.
Rutanen, E.-M., and Seppälä, M. (1978). *Cancer* **41,** 692–696.
Samaan, N. A., Smith, J. P., Rutledge, F. N., and Schultz, P. N. (1976). *Am. J. Obstet. Gynecol.* **126,** 186–189.
Scardino, P. T., and Skinner, D. G. (1979). *Surgery* **86,** 86–93.
Scardino, P. T., Cox, H. D., Waldmann, T. A., McIntire, K. R., Mittemeyer, B., and Javadpour, N. (1977). *J. Urol.* **118,** 994–999.
Schultz, H., Sell, A., Norgaard-Pedersen, B., and Arends, J. (1978). *Cancer* **42,** 2182–2186.
Sheth, N. A., Saruiya, J. N., Ranadive, K. J., and Sheth, A. R. (1974). *Br. J. Cancer* **30,** 566–570.
Shinde, S. R., Adil, M. A., Sheth, A. R., Koppikar, M. G., and Sheth, N. A. (1981). *Oncology* **38,** 277–280.
Shome, B., and Parlow, A. F. (1973). *J. Clin. Endocrinol. Metab.* **36,** 618–621.
Soma, H., Takayama, M., Tokoro, K., Kikuchi, T., Kikuchi, K., and Saegusa, H. (1980). *Acta Obstet. Gynecol. Scand.* **59,** 445–448.
Stanhope, C. R., Smith, J. P., Britton, J. C., and Crosley, P. K. (1979). *Gynecol. Oncol.* **8,** 284–287.
Stolbach, L., Inglis, N., Lin, C., Turksoy, R. N., Fishman, W., Marchant, S., and Rule, A. (1976). *In* "Onco-Developmental Gene Expression" (W. H. Fishman and S. Sell, eds.), pp. 433–443. Academic Press, New York.
Stone, M., Bagshawe, K. D., Kardana, A., Searle, F., and Dent, J. (1977). *Br. J. Obstet. Gynaecol.* **84,** 375–379.
Strott, C. A., Yoshimi, T., Ross, G. T., and Lipsett, M. B. (1969). *J. Clin. Endocrinol. Metab.* **29,** 1157–1167.
Sufrin, G., Mirand, E. A., Moore, R. H., Chu, T. M., and Murphy, G. P. (1977). *J. Urol.* **117,** 433–438.
Teasdale, F., Adcock, E. W., III, August, C. S., Cox, S., Battaglia, F. C., and Naughton, M. A. (1973). *Gynecol. Invest.* **4,** 263–269.
Thompson, D. K., and Haddow, J. E. (1979). *Cancer* **43,** 1820–1829.
Tomoda, Y., Asai, Y., Arii, Y., Kaseki, S., Hideo, N., Miwa, T., Saiki, N., and Ishizuka, N. (1977). *Cancer* **40,** 1016–1025.

Tormey, D. C., Waalkes, T. P., Ahmann, D., Gehrke, C. W., Zumwatt, R. W., Snyder, J., and Hansen, H. (1975). *Cancer* **35,** 1095–1100.
Tormey, D. C., Waalkes, T. P., and Simon, R. M. (1977). *Cancer* **39,** 2391–2396.
Vaitukaitis, J. L., and Ebersole, E. R. (1976). *J. Clin. Endocrinol. Metab.* **32,** 1048–1055.
Vaitukaitis, J. L., Braunstein, G. D., and Ross, G. T. (1972). *Am. J. Obstet. Gynecol.* **113,** 751–758.
Vaitukaitis, J. L., Ross, G. T., Braunstein, G. D., and Rayford, P. L. (1976). *Recent Prog. Horm. Res.* **32,** 289–331.
VanDamme, M.-P., Robertson, D. M., and Diczfalusy, E. (1974). *Acta Endocrinol. (Copenhagen)* **77,** 655–671.
Van Hell, H. R., Matthijsen, R., and Overbeek, G. A. (1964). *Acta Endocrinol. (Copenhagen)* **47,** 409–418.
Waalkes, T. P., Gehrke, C. W., Tormey, D. C., Woo, K. B., Kuo, K. C., Snyder, J., and Hansen, H. (1978). *Cancer* **41,** 1871–1882.
Waldmann, T. A., and McIntire, K. R. (1974). *Cancer* **34,** 1510–1515.
Weintraub, B. D., and Rosen, S. W. (1971). *J. Clin. Endocrinol. Metab.* **32,** 94–101.
Willemse, P. H. B., Sleijfer, D. T., Koops, H. S., DeBruijn, H. W. A., Oosterhuis, J. W., Brouwers, T. M., Ookhuizen, T., and Marrink, J. (1981). *Oncodev. Biol. Med.* **2,** 129–134.
Williams, R. R., McIntire, K. R., Waldmann, T. A., Feinleib, M., Go, V. L. W., Kannel, W. B., Dawber, T. R., Castelli, W. P., and McNamara, P. M. (1977). *JNCI, J. Natl. Cancer Inst.* **58,** 1547–1551.
Yoshimi, T., Strott, C. A., Marshall, J. R., and Lipsett, M. B. (1969). *J. Clin. Endocrinol. Metab.* **29,** 225–230.
Yoshimoto, Y., Wolfsen, A. R., and Odell, W. D. (1977). *Science* **197,** 575–577.
Yoshimoto, Y., Wolfsen, A. R., Hirose, F., and Odell, W. D. (1979a). *Am. J. Obstet. Gynecol.* **134,** 729–733.
Yoshimoto, Y., Wolfsen, A. R., and Odell, W. D. (1979b). *Am. J. Med.* **67,** 414–420.
Yuen, B. H., and Cannon, W. (1981). *Am. J. Obstet. Gynecol.* **139,** 316–319.
Zondek, B. (1929). *Endokrinologie* **5,** 425–434.
Zondek, B. (1930a). *Chirurg* **2,** 1072–1073.
Zondek, B. (1930b). *Zentralbl. Gynaekol.* **54,** 2306–2308.

CHAPTER 20
Oncodevelopmental Antigens in Gynecologic Cancer

MARKKU SEPPÄLÄ
Department I of Obstetrics and Gynecology
University Central Hospital
Helsinki, Finland

I.	Introduction	373
II.	Placental Proteins	374
	A. Chorionic Gonadotropin	374
	B. Pregnancy-Specific β_1-Glycoprotein (SP_1)	376
	C. Other Placental Proteins	378
III.	α-Fetoprotein	381
IV.	Carcinoembryonic Antigen	383
V.	Other Antigens	387
	A. Ovarian Cancer-Associated Antigens OCAA and OCAA-1–OCAA-5	387
	B. Ovarian Cancer Antigens OvC-1–OvC-6	388
	C. Comparison of Various Ovarian Cancer-Associated Antigens	388
	D. Ovarian Cancer-Associated Urinary Oncofetal Peptide	389
	E. Oncodevelopmental Enzymes	389
VI.	Summary	389
	References	390

I. INTRODUCTION

The diagnosis of cancer depends on the ability to distinguish between normal and neoplastic tissues. In addition to elucidating the biology of cancer, research on oncodevelopmental antigens has yielded new sophisticated methods for more accurate histopathological diagnosis and improved the early detection and localization of some recurrent cancers. This chapter reviews the current state of the art relative to oncodevelopmental antigens in gynecologic cancer.

II. PLACENTAL PROTEINS

A. Chorionic Gonadotropin

1. hCG and Its Subunits in Trophoblastic Tumors

The measurement of chorionic gonadotropin (hCG) secretion is the best example of the use of a tumor marker for monitoring cancer patients. It has been estimated that a highly sensitive radioimmunoassay of hCG can detect secretion by as few as 10^4-10^5 cells, whereas 10^9-10^{12} cells are required before a tumor is detectable by other clinical methods (Bagshawe, 1975). Now that effective chemotherapy is available, the measurement of hCG by highly sensitive and specific radioimmunoassay (RIA) has led to detection of disease early enough to be completely cured.

Recent studies utilizing antiserum to the hCG β carboxyterminal peptide suggest that the urinary hCG assay is considerably more effective in detecting increased hCG production and, hence, persistent tumor burden, than the commonly used hCG β subunit RIA (Wehmann et al., 1981). In two women treated for choriocarcinoma, urinary hCG levels remained distinctly elevated, even though the serum hCG level was undetectable. In both cases the urinary assay reliably predicted the eventual reappearance of hCG in serum (Wehmann et al., 1981).

In addition to hCG, the normal placenta also contains subunits of hCG, the free α subunit being in excess of the β subunit (Vaitukaitis, 1974). The proportion of free α subunit is negligible in choriocarcinoma, and its secretion by the malignant trophoblast is suggested to be an unfavorable prognostic sign (Vaitukaitis and Ebersole, 1976). However, this has not been found in all studies (Rutanen, 1978). In Japan, Nishimura and co-workers (1981) have studied the biochemical properties of urinary hCG from normal pregnancy and trophoblastic disease. In vivo bioassay revealed that the bioactivity of hCG in choriocarcinoma was only about a half of hCG in normal pregnancy or hydatidiform mole. By contrast, the receptor binding activity in vitro of hCG in choriocarcinoma was about 3 times that of hCG in pregnancy. They observed a difference in the carbohydrate composition. While sialic acid was undetectable in hCG in choriocarcinoma, 8.5% sialic acid was found in hCG in normal pregnancy and hCG in hydatidiform mole. In Parlow rats, iodinated hCG in choriocarcinoma was taken up in large quantities by the liver in comparison to the ovary, whereas less of hCG in normal pregnancy was taken up by the liver. These differences may help to distinguish between normal pregnancy and choriocarcinoma in patients who have previously experienced a hydatidiform mole and who exhibit a resurge of hCG secretion.

The mechanisms controlling hCG secretion by the normal and malignant trophoblast are not known. The normal placenta contains gonadotropin-releasing hormone (GnRH) (Gibbons *et al.*, 1975; Khodr and Siler-Khodr, 1980) which is localized in the syncytiotrophoblast, and the same cells contain hCG (Seppälä *et al.*, 1980). GnRH extracted from the normal placenta is indistinguishable from synthetic GnRH by high pressure liquid chromatography and radioimmunoassay (Lee *et al.*, 1981b). GnRH has also been identified in choriocarcinoma cells (Seppälä *et al.*, 1980), and studies on cell columns have shown that both the normal placenta cells and BeWo cells (a choriocarcinoma cell line from the American Type Culture Collection, Rockville, Maryland) can release hCG in response to GnRH pulses. BeWo cells also contain GnRH-like immunoreactivity (Bützow *et al.*, 1982). These results indicate that hCG secretion by the trophoblast may be controlled by GnRH.

2. hCG in Nontrophoblastic Tumors

Early studies utilizing RIA with antiserum to the β subunit of hCG have demonstrated that patients with various types of nontrophoblastic cancers can have elevated serum hCG levels (Braunstein *et al.*, 1973). These include gynecologic cancer. However, when detectable the hCG levels are much lower than in trophoblastic disease. This was demonstrated by Rutanen and Seppälä (1978) who examined the circulating hCG levels in 380 patients with malignant and benign gynecologic tumors. They found hCG in 18% of patients with gynecologic cancer (Table I) but also in 14% of patients with nonmalignant gynecologic disease. The levels in cancer patients were higher, but always below 70 U/liter. The secretion of hCG was not related to clinical stage or histological differentiation. After radical surgery, hCG disappeared from serum in many patients suggesting that it was secreted by the tumor. However, there also were patients who transiently had immunoreactive hCG in serum, for the first time after radical surgery. The latter may result from increased secretion of pituitary gonadotropins following ovariectomy. Recent studies indicate that the human

TABLE I
Occurrence of hCG in Serum of 276 Patients with Nontrophoblastic Gynecologic Cancer[a]

Site of tumor	Elevated/total	Percentage elevated
Vulva	2/8	25%
Vagina	2/6	33%
Cervix	23/111	21%
Endometrium	17125	14%
Ovary	5/26	19%
All patients	49/276	18%

[a]From Rutanen and Seppälä (1978).

pituitary gland contains a substance which shares physicochemical and immunochemical properties with hCG (Chen *et al.*, 1976). This and results of clinical studies (Rutanen and Seppälä, 1978) raise legitimate questions regarding the usefulness of hCG measurement for monitoring nontrophoblastic gynecologic cancer. The same seems to apply to the measurement of the α subunit secretion. Although some HeLa cell strains release hCG and free α subunit in culture (Lieblich *et al.*, 1976), patients with carcinoma of the cervix or other types of nontrophoblastic gynecologic cancer do not have elevated levels of the alpha subunit (Rutanen and Seppälä, 1979).

B. Pregnancy-Specific β_1-Glycoprotein (SP$_1$)

1. SP$_1$ in Trophoblastic Tumors

SP$_1$ has been localized by immunohistochemical methods in the normal syncytiotrophoblast and in choriocarcinoma cells (Bohn and Sedlaceck, 1975; Tatarinov *et al.*, 1976). Elevated SP$_1$ levels have been found in the serum of patients with choriocarcinoma (Tatarinov *et al.*, 1974; Tatarinov and Sokolov, 1977), and these may occur in the absence of hCG (Searle *et al.*, 1978; Seppälä *et al.*, 1978). Since hCG is an excellent marker for choriocarcinoma, the measurement of SP$_1$ levels would be of clinical value only if "isolated SP$_1$ secre-

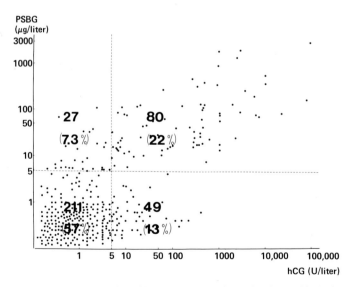

Fig. 1. Serum SP$_1$ and hCG levels in 367 serum samples from 17 patients with choriocarcinoma. Broken lines indicate the level which is not exceeded by any normal individual or patient with nonneoplastic disease. (Data from Rutanen and Seppälä, *J. Clin. Endocrinol. Metab.* **50**, 57–61, © 1980 by The Endocrine Society.)

tion" would reflect active disease. The significance of "isolated SP_1 secretion" was examined by Rutanen and Seppälä (1980) in 17 patients with choriocarcinoma, 69 patients with nontrophoblastic disease, and 85 healthy nonpregnant women. The highest level in normal individuals was 2.2 µg/liter, and small elevations up to 4.7 µg/liter were found in 1 out of 20 patients with liver disease, and 1 out of 20 patients with rheumatoid arthritis. In choriocarcinoma, SP_1 levels over 4.7 µg/liter were found in 62% of hCG positive samples (80/129) and 11% of hCG negative samples (27/238) (Fig. 1). The latter group is important, since an hCG negative result is considered to indicate remission. When such patients were followed up for 1 year without treatment, SP1 was still found in 6 out of 42 hCG negative samples (17%) from 3 patients. Choriocarcinoma patients who have been hCG negative for 1 year are known to have excellent prognosis with less than 2% recurrence rate (Hammond et al., 1973). Therefore, further chemotherapy has not been recommended on the basis of isolated SP_1 peaks (Rutanen and Seppälä, 1980).

The circulating SP_1 from choriocarcinoma patients is qualitatively different from SP_1 in pregnancy serum. Most of SP_1 from either source binds to concanavalin A (Con A). In choriocarcinoma, the Con A nonbinding fraction is greater (Fig. 2). This property is not caused by chemotherapy, as serum SP_1 from patients with untreated choriocarcinoma also has this characteristic, and it is not altered during chemotherapy (Koistinen et al., 1981). This difference may be clinically important. Patients who have experienced a hydatidiform mole are at an increased risk of developing choriocarcinoma and, therefore, are monitored by the hCG assay. When hCG is found, it can result from a new pregnancy or choriocarcinoma. If SP_1 can also be demonstrated, the difference in the amount

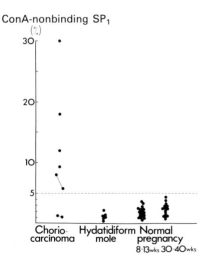

Fig. 2. The proportion of serum Con A-nonbinding SP_1 in normal pregnancy and patients with trophoblastic disease. (From Seppälä and Rutanen, 1982, by permission of Academic Press, Inc.)

of Con A nonbinding SP_1 may be helpful for making the differential diagnosis between normal pregnancy and malignant trophoblastic disease.

2. SP_1 in Nontrophoblastic Tumors

Patients with various types of nontrophoblastic malignant neoplasms may have elevated serum SP_1 levels (Tatarinov and Sokolov, 1977; Bagshawe et al., 1978), but this has not been confirmed in all studies (Engvall and Yonemoto, 1979). Small elevations of serum SP_1 level have been observed in patients with ovarian and other types of gynecological cancer (Crowther et al., 1979; Wurtz, 1979), but the origin and clinical significance of such peaks are not clear. The author has seen transient small SP_1 peaks in patients with cancer of the uterine cervix and endometrium, and these bear no relationship to clinical relapse (Fig. 3).

C. Other Placental Proteins

1. Human Placental Lactogen

Human placental lactogen (hPL) has been identified in the human placental tissue and in the serum of pregnant women (Ito and Higashi, 1961; Josimovich

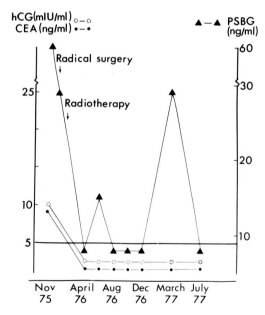

Fig. 3. Circulating levels of SP_1, CEA, and hCG, in a patient with invasive epidermoid cancer of the cervix. (Reprinted from Seppälä et al., 1982, by courtesy of Marcel Dekker, Inc.)

and McLaren, 1962). While a vast literature has accumulated on hPL in pregnancy, only few reports describe the measurement of hPL in trophoblastic and nontrophoblastic tumors. Circulating hPL levels in patients with untreated trophoblastic tumors are usually below the range for normal pregnancy (Ehnholm *et al.*, 1967; Saxena *et al.*, 1968). It has been suggested that low hPL levels in the presence of a normal circulating hCG concentration may distinguish between normal pregnancy and trophoblastic disease (Rosen *et al.*, 1975). Low levels of hPL have also been reported for some patients with nontrophoblastic cancer including gynecologic tumors (Crowther *et al.*, 1979; Weintraub and Rosen, 1971), but the measurement of hPL has not gained popularity in monitoring nontrophoblastic tumors.

2. Placental Protein 5 (PP_5)

PP_5 was isolated and characterized by Bohn and Winckler (1977). It is a glycoprotein with a molecular weight of 36,000–42,500. During pregnancy, PP_5 is secreted into the maternal circulation, where the levels rise from the twelfth week. In immunoperoxidase staining, PP_5 is localized in the syncytiotrophoblast (Seppälä *et al.*, 1979). The levels in pregnancy serum are higher than in pregnancy plasma (Obiekwe *et al.*, 1979), and in the presence of heparin or endogenous heparin-like substances PP_5 forms complexes and polymers (Salem *et al.*, 1980). PP_5 appears to be related both to the blood coagulation and fibrinolytic systems as, heparin binding apart, it has antiplasmin activity (Bohn and Winckler 1977; Siiteri *et al.*, 1982), and human placental urokinase inhibitor contains a substance which is immunologically and physicochemically related to PP_5 (Seppälä *et al.*, 1983).

a. Trophoblastic Tumors. In patients with untreated hydatidiform mole the PP_5 levels are elevated (Seppälä *et al.*, 1979; Grudzinskas *et al.*, 1980), but after evacuation the level rapidly decreases. The half-life of PP_5 in serum is short, approximately 15 minutes (Bohn and Winckler, 1977). In immunoperoxidase staining, PP_5 is localized in the syncytiotrophoblast of the mole, but it may be absent from invasive mole (Seppälä *et al.*, 1979).

Choriocarcinoma patients usually do not have detectable PP_5 in serum in spite of concomitant high hCG and SP_1 levels (Seppälä *et al.*, 1979). This was confirmed by Lee and co-workers (1981a) who studied serum levels of hCG, SP_1, and PP_5 in 14 patients with untreated hydatidiform mole and in 9 patients with untreated choriocarcinoma. There was no obvious difference in serum hCG levels between patients with hydatidiform mole and choriocarcinoma, but the SP_1 levels were lower in patients with choriocarcinoma, and PP_5 was not detectable at all.

The difference in PP_5 levels between hydatidiform mole and choriocarcinoma is not due to half-life, since these patients were all untreated. Immunoperoxidase studies have confirmed the difference in tissue PP_5 between benign and malignant trophoblastic disease: while PP_5 is present in hydatidoform mole, it can rarely, if at all, be found in choriocarcinoma (Seppälä et al., 1979; Lee et al., 1982). However, this difference was not seen in a recent study (Nisbet et al., 1982).

These results may shed light on the biology of trophoblastic disease. The distinct patterns of hCG, SP_1, and PP_5 seen in most studies suggest that either the synthetic mechanisms or the site of origin of these proteins are fundamentally different (Lee et al., 1981a). In light of the role of PP_5 as a protease inhibitor (Bohn and Winckler, 1977), the absence of PP_5 may contribute to the invasiveness of malignant trophoblast.

b. Nontrophoblastic Tumors. Using the enzyme-bridge immunoperoxidase (PAP) technique, Bohn and colleagues (1981) have reported finding of PP_5 in the cytoplasm of 3 out of 7 ovarian cancers and 2 out of 3 endometrial cancers. By contrast, using the same technique T. Wahlström (personal communication) did not find any PP_5 in 13 ovarian cancers (4 serous and 4 mucinous cystadenocarcinomas, 2 mesonephroid and 3 anaplastic carcinomas), or in 5 endometrial and 5 endocervical adenocarcinomas. Bohn et al., (1981) used higher antibody concentrations than Wahlström and the use of high antiserum concentrations is likely to give more positive reactions.

The author has studied the circulating PP_5 level in 29 patients with nontrophoblastic gynecologic cancers, in 37 samples from a patient with an α-fetoprotein-producing endodermal sinus tumor (EST) of the ovary, 19 patients with benign uterine myomas, and 80 apparently healthy women. PP_5 was found in 9 out of 37 samples of the patient with EST, but not in any other patients or healthy individuals. Even in the patient with EST the PP_5 levels were low ranging from 0.5 to 1.1 ng/ml, and they bore no relationship with AFP levels. Similar borderline PP_5 values may occur in some patients with nonmalignant liver disease. It is concluded that the measurement of serum PP_5 concentration cannot be used for monitoring patients with nontrophoblastic gynecologic cancer.

3. Placental Protein 10

Placental protein 10 (PP_{10}) is an α_1-glycoprotein which contains 6.6% carbohydrate and has a molecular weight of 48,000 (Bohn and Kraus, 1979). Immunohistochemical studies have localized PP10 in the syncytiotrophoblast of normal placenta and hydatidiform mole, but not in choriocarcinoma. PP10 has also been identified in the endometrial stroma cells and secretory epithelium (Wahlström et al., 1982). In addition, Bohn and co-workers (1981) have reported the occurrence of PP_{10} in the amnion epithelium and histiocytes of amnion, chorion villi, and decidua. Using the immunoperoxidase technique, the same authors found PP_{10} in 2 out of 7 ovarian cancers and 2 out of 3 endometrial cancers. Using radioimmunoassay, the same authors reported elevated serum PP_{10} levels in patients with a variety of malignant tumors (Bohn et al., 1981). Serum PP_{10} level is also elevated during pregnancy (Bohn et al., 1981). The role of PP_{10} as a tumor marker remains to be studied.

4. Placental Protein 11

Placental protein 11 (PP_{11}) is also an α_1-glycoprotein with a molecular weight of 44,000 and a carbohydrate content of 3.9% (Bohn and Winckler, 1981). PP11 has been identified in the chorion and chorion villi by immunoperoxidase technique, but not in normal fetal or adult tissues. However, PP_{11} was found in 3 out of 7 ovarian cancer tissues and 2 out of 3 endometrial cancer tissues (Bohn et al., 1981).

5. Placental Protein 12

Placental protein 12 (PP_{12}) is an α_1-glycoprotein which contains 4.3% carbohydrate and has a molecular weight of 25,000 (ultracentrifugation) or 51,000 (SDS–polyacrylamide gel electrophoresis) (Bohn and Kraus, 1980). PP_{12} may be identical with the chorionic α_1-microglobulin described by Petrunin et al. (1979). Using the immunoperoxidase technique, Wahlström et al. (1982) have identified PP_{12} in the syncytiotrophoblast of normal placenta and hydatidiform mole, but not in choriocarcinoma tissues. Studies by Bohn et al. (1981) indicate PP_{12} staining in the hysticocytes of normal amnion, villi, and decidua, but not in fetal or adult tissues. Wahlström et al. (1982) have found PP12 in secretory epithelial cells of the endometrium. Inaba et al. (1980) have reported finding by the immunoperoxidase method of PP_{12} in 1 out of 7 ovarian cancers and 1 out of 3 endometrial cancers.

Studies by PP_{12} radioimmunoassay (Bohn et al., 1981) have revealed that the levels in serum are higher in women than in men, and in some patients with malignant tumors the PP_{12} level is elevated. Significantly elevated serum PP_{12} levels are rare in trophoblastic and nontrophoblastic gynecologic tumors, and the measurement of serum PP_{12} level will hardly be useful for the diagnosis or monitoring of treatment of cancer patients (Rutanen et al., 1982).

III. α-FETOPROTEIN

α-Fetoprotein (AFP) is synthesized by the fetal liver, yolk sac, and to a small degree by the fetal gastrointestinal tract (Gitlin et al., 1972). In analogy with the embryonic sites of its production, AFP is secreted by most primary liver cancers and yolk sac tumors (Abelev, 1974). Serum AFP levels are usually normal in patients with choriocarcinoma (Seppälä et al., 1972), but some patients with a hydatidiform mole may have elevated levels (Seppälä and

Ruoslahti, 1974), and vesicular fluid from hydatidiform moles may contain high concentrations of AFP (Grudzinskas *et al.*, 1977). Based on these findings the measurement of AFP is not helpful in the management of patients with trophoblastic tumors.

Primary liver cancer and yolk sac tumors apart, AFP is rarely secreted by nontrophoblastic neoplasms. In a study of 92 patients with various gynecologic tumors, of which 53 were ovarian cancers, Seppälä and co-workers (1975) found normal AFP levels in all but one patient who had ovarian carcinoma with liver secondaries.

Germ cell tumors of the gonads may differentiate along either embryonic or extraembryonic pathways (Teilum, 1965). The latter may yield yolk sac or trophoblastic tissue and secrete AFP or hCG, respectively. Many germ cell tumors have mixed components (Fox, 1980). In pure yolk sac tumors, AFP is

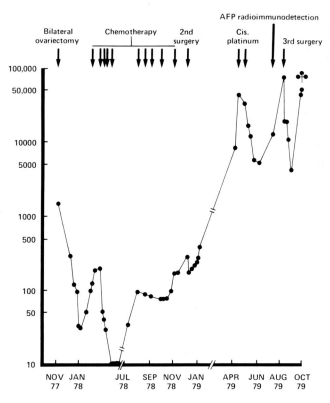

Fig. 4. Circulating AFP levels relative to treatment in a patient with bilateral ovarian yolk sac tumors. The half-life of AFP after surgery was 7 days indicating that removal of cancer had not been complete.

localized in cells lining the endodermal sinuses and in intra- and extracytoplasmic PAS-positive hyaline globules (Teilum *et al.*, 1974).

The normal half-life of human fetal AFP is 4–5 days (Seppälä and Ruoslahti, 1972). In patients with germ cell tumors postoperative estimation of half-life of AFP gives early information of the completeness of surgical treatment (Fig. 4).

Radiolabeled anti-AFP antibodies have been injected to patients with germ cell tumors, and attempts at localizing tumor deposits have been successful by subsequent external scanning of radioactive deposits (Goldenberg *et al.*, 1980). Although AFP is not a cell surface marker, the local antigen concentration has been sufficient for successful radiolocalization.

α-Fetoprotein from yolk sac tumors differs from fetal serum AFP by its lesser binding to Con-A (Ruoslahti *et al.*, 1978). While 5–10% of fetal serum AFP is of Con-A nonbinding type, we have seen a patient with a yolk sac tumor (Fig. 4) who had 40–60% of Con-A nonbinding AFP in serum.

IV. CARCINOEMBRYONIC ANTIGEN

Carcinoembryonic antigen (CEA) has been identified in premalignant and malignant lesions of the female reproductive system (Lindgren *et al.*, 1979; Goldenberg *et al.*, 1976; Rutanen *et al.*, 1978). CEA may also occur in the mucinous cyst fluid (van Nagell *et al.*, 1975b), ascitic fluid (Seppälä *et al.*, 1975), and serum of such patients (Barrelet and Mach, 1975; DiSaia *et al.*, 1975; Khoo and Mackay 1974; Seppälä *et al.*, 1975; van Nagell *et al.*, 1975a). CEA from the ascitic fluid from malignant serous or mucinous cystadenocarcinomas of the ovary appears to possess similar physical and immunologic properties to colon cancer CEA (Seppälä *et al.*, 1975).

Some premalignant lesions of the uterine cervix contain CEA (Lindgren *et al.*, 1979). Immunoperoxidase staining of normal cervical epithelium and dysplasias revealed that 25% of mild dysplasias were CEA-positive, whereas the normal epithelium was CEA negative (Lindgren *et al.*, 1979). The CEA positivity rate increased from premalignant lesions to invasive cervical cancer so that 60% at clinical stage I and 80% at stage IIb (FIGO classification) were CEA-positive. From these results it seems that CEA in tissue represents malignant potential.

The prognostic significance of CEA in cancer tissue was studied by Lindgren *et al.* (1979) on patients with epidermoid cancer of the uterine cervix. In a group of 60 patients who had been treated by radical surgery for stage I–IIa cancer, 37 (62%) were CEA-positive. No difference was observed in the survival rates between CEA-positive and CEA-negative patients at any time during the 10-year follow up. The 5- and 10-year survival rates were 65% in either group (Lindgren *et al.*, 1979). Thus, expression of CEA does not seem to make epidermoid cancer of the cervix more malignant. Elevated serum CEA levels are associated with poor prognosis (Levin et al., 1976), but this probably reflects a greater tumor burden rather than increased aggressiveness of CEA-positive lesions.

Routine immunoperoxidase staining of adenocarcinomas of the uterine cervix and endometrium can improve the histopathological diagnosis. In a study by Wahlström and co-workers (1979) tissue from 80% of patients with endocervical adenocarcinoma was CEA-positive, whereas this was true of 8% of endometrial adenocarcinomas. Mesonephric adenocarcinomas were CEA-negative irrespective of the site, and endometrial adenosquamous carcinomas were CEA-positive. After exclusion of these on morphological criteria, endometrial adenocarcinomas were all CEA-negative, while 80% of endocervical adenocarcinomas were CEA-positive (Table II). The histological variability of endocervical and endometrial adenocarcinomas may create problems of differential diagnosis, and specimens obtained by curettage of the endocervix may contain endometrial tissue. CEA staining may help distinction between the two types in those cases where the site of lesion is not clear by routine histopathologic examination.

Immunohistochemical staining of CEA may also be useful in ovarian tumors. Heald et al. (1979) have observed a higher CEA content in mucinous than serous tumors of the ovary, and a partial correspondence between the degree of malignancy of mucinous tumors, as assessed histologically, and their content of CEA. The authors suggest that examination of tissue CEA may allow more precise grading of the degree of malignancy.

Radioimmunodetection of ovarian tumors by anti-CEA antibodies has been attempted by van Nagell et al. (1980). They injected intravenously ^{131}I-labeled goat immunoglobulin G (IgG) against CEA to patients with ovarian cancer. The primary cancer was localized by subsequent external photoscanning in all 13 patients, and the metastases in 6 out of 9 patients. Computer-assisted tomography, ultrasonography, and angiography were less efficient. Lesions smaller than 2 cm diameter could not be detected. Thus, CEA in ovarian cancer can serve as a target for immunolocalization by radioactively labeled anti-CEA antibodies.

In various studies, elevated serum CEA levels have been observed in 7–57%

TABLE II

Tissue CEA in Endocervical and Endometrial Adenocarcinomas after Exclusion of Mesonephroid Adenocarcinomas and Those Containing Squamous Elements[a]

Origin of tissue	CEA positive/total	Percentage positive
Endocervical	131/152	(86%)
Endometrial	0/122	(0%)

[a]Wahlström et al. (1979).

TABLE III
Incidence of Raised Plasma CEA Levels (≥ 5 ng/ml) in Gynecologic Cancer

	Site		
Author	Cervix	Endometrium	Ovary
Barrelet and Mach, 1975	56%	45%	36%
Seppälä et al., 1975	50%	20%	21%
van Nagell et al., 1975a	40%	57%	45%
Kjörstad and Orjaseter, 1977		7%	
Rutanen et al., 1978	12%	7%	20%
van Nagell et al., 1979	24%	12%	22%

of patients with gynecologic cancer (Table III). Elevated levels are more common among patients with advanced disease (Table IV). The highest frequency of elevated CEA levels is seen in patients with endocervical adenocarcinoma and ovarian cancer, and the lowest in endometrial adenocarcinoma. Occasionally, high levels are encountered in patients with benign mucinous cystadenoma (Fig. 5). In spite of the great variation in the frequency of elevated CEA values, most workers agree that the raised levels are not so high as in colorectal cancer. After successful removal, the CEA level declines, and if it does not do so, residual tumor is likely to be left (Fig. 6).

In patients with CEA-positive carcinomas of the cervix, progressively rising plasma CEA levels have predicted recurrent disease in over 80% of patients, whereas serial plasma CEA values have correlated positively with clinical disease status in only 28% of patients whose tumors were devoid of CEA (van Nagell et al., 1979). In a follow-up study, the same group (van Nagell et al., 1978) found elevated preoperative CEA levels in 48% of 300 patients with

TABLE IV
Plasma CEA Levels (> 2.5 ng/ml) and Carcinoma of the Cervix Uteri

	Stage				
Author	O	I	II	III	IV
DiSaia et al., 1975	6%	26%	47%	71%	100%
van Nagell et al., 1975a	76%	63%	70%	92%	100%
DiSaia et al., 1977	5%	26%	47%	84%	85%
Rutanen et al., 1978[a]	14%	7%	8%	23%	50%
van Nagell et al., 1979[b]	—	38%	50%	62%	71%

[a]Values > 5 ng/ml.
[b]In patients whose tissue was CEA-positive.

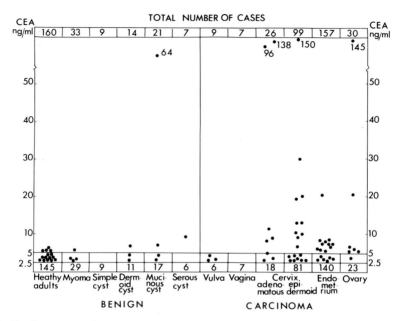

Fig. 5. Pretreatment circulating CEA levels in 160 apparently healthy adults and in 412 patients with malignant and nonmalignant gynecological tumors. (From Rutanen et al., 1978, by permission of *Cancer*.)

invasive carcinoma of the uterine cervix. Thirty patients had progressively increasing plasma CEA levels following therapy, of whom 29 developed recurrent cancer. A progressive rise of plasma CEA preceded the clinical diagnosis of recurrence by 1–23 months (mean 6 months) in 13 patients (van Nagell et al., 1978). In 16 patients the CEA level did not rise before the clinical diagnosis of recurrence. Khoo and Mackay (1974) have found a mean lead time of 10.7 weeks in 12 patients with gynecologic cancer, the range being 0–40 weeks. More frequent CEA assays may give longer lead times, and this is likely to become important once more effective treatment can be offered. However, not all authors are so optimistic about the value of the CEA assay in gynecologic cancer mainly because CEA levels become rarely elevated, and when they do so, it is a matter of advanced cancer. Moreover, effective additional treatment is not usually available. A Consensus Development Conference on CEA was held at the National Institutes of Health, September 29 to October 1, 1980. It was stated that the role of CEA in the postoperative and therapeutic monitoring of patients with gynecological neoplasms is less convincing than it is for colorectal cancer (NIH Consensus Development Conference Statement, 1981).

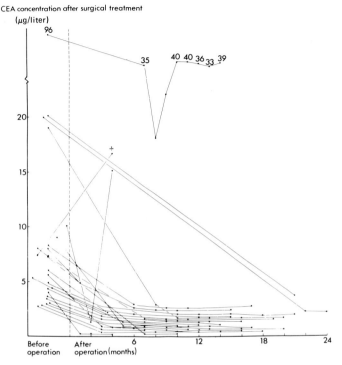

Fig. 6. Circulating CEA levels before and after surgery for gynecological cancer. Surgery was incomplete in a patient with stage IIb cervical epidermoid carcinoma in whom the CEA level did not decline. (From Rutanen et al., 1978, by permission of Cancer.)

V. OTHER ANTIGENS

A. Ovarian Cancer-Associated Antigens OCAA and OCAA-1–OCAA-5

Several antigens have been isolated from ovarian cancer tissue, some of them are specific and secreted in the serum or body fluids of cancer patients. Bhattacharya and Barlow (1973) isolated two tumor antigens, OCAA and OCAA-1, in ovarian cystadenocarcinomas that were not detectable in the normal ovary. OCAA was found in 70% of serous and all mucinous cystadenocarcinomas, but not in benign serous or mucinous cysts. OCAA-1 was present in 90% of malignant ovarian cancers, and it could also be identified in other malignant gynecologic, colon, breast, and pancreatic neoplasms. OCAA-1 may be considered as an oncodevelopmental antigen, since it is present in human umbilical cord serum and pregnancy serum in the third trimester (Bhattacharya and Barlow, 1979).

B. Ovarian Cancer Antigens OvC-1–OvC-6

Six antigens, OvC-1–6, were isolated from ovarian cancer tissue by Imamura et al. (1978). None of them was detected in normal ovary. OvC-1 and OvC-2 were unrelated to any well-defined tumor antigen. OvC-1 was found in 48 out of 93 epithelial cancers of the ovary, and also in malignant tumors of the cervix, breast, stomach, and pancreas. OvC-1 is not cancer specific, since it was present in normal kidney and lung, and in the placenta, amniotic fluid, cord serum, and pregnancy sera. OvC-2 was found in 8 out of 93 ovarian cancers, in the placenta, amniotic fluid, and fetal intestine, but not other types of tumors or normal adult tissues. OvC-3 was identified as pregnancy-associated macroglobulin (Stigbrand et al., 1976), and OvC-4 as CEA. OvC-5 was the normal cross-reacting antigen, NCA (von Kleist et al., 1972). OvC-6 was specific for the immunizing tumor only.

C. Comparison of Various Ovarian Cancer-Associated Antigens

A collaborative study on ovarian cancer associated antigens was organized by P. Sizaret of the International Agency for Research of Cancer and K. D. Bagshawe, J. N. Gennings, and F. Searle of the Charing Cross Hospital, London, in which various antigens and antisera were compared. The laboratory work was carried out by N. H. Axelsen in Copenhagen. Over 30 antigens and antisera were examined by fused rocket immunoelectrophoresis, crossed immunoelectrophoresis, and tandem crossed immunoelectrophoresis. CEA-like immunoreactivity was found in OCAA-5 of Bhattacharya and Barlow (1973), OCAA-2 of Lamerz et al. (1979), OvC-3 of Knauf and Urbach (1974), and OvC-4 of Imamura et al. (1978). OCAA-3 and OCAA-4 of Bhattacharya and Barlow (1973) corresponded to the M4 antigen of Bara et al. (1977). The following ovarian cancer-associated antigens were unrelated to any previously known protein and considered to be unique: OCAA of Bhattacharya and Barlow (1973), M1 antigen of Bara et al. (1977), embryonic prealbumin (Tatarinov and Kalashnikov, 1977), fucose-rich glycoprotein (Hamazaki and Hotta, 1975), and the ovarian tumor associated antigen of Bagshawe et al. (1980). Prospective clinical trials on these unique ovarian cancer antigens are required to assess their clinical utility.

D. Ovarian Cancer-Associated Urinary Oncofetal Peptide

Stenman and co-workers (1982) have isolated a peptide from the urine of a patient with ovarian cancer (OFP). OFP has an apparent molecular weight of 7000, and it contains no carbohydrate. A radioimmunoassay was developed, and the concentration of OFP in various body fluids was determined. In healthy subjects, the serum levels were 30–100 μg/liter and the urinary levels were 30–200 μg/liter. Urine from newborn infants contained 100–1500 μg/liter of OFP, and amniotic fluid at 12–18 weeks 350–1700 μg/liter. Urine of patients with ovarian cancer had OFP concentrations of 300-14,000μg/liter, and the serum levels of some patients were 50-1500μg/liter. These results suggest an oncodevelopmental nature of OFP. Studies on the clinical utility of measurement of OFP are required to assess its value in monitoring cancer patients.

E. Oncodevelopmental Enzymes

Fishman and co-workers (1975) have studied 833 cancer patients and found elevated levels of carcinoplacental alkaline phosphatase, Regan isoenzyme, in 43% of patients with ovarian adenocarcinoma. Increased levels of enzyme were correlated with advancing disease. Inglis and co-workers (1973) identified the D-phenotype of placental alkaline phosphatase, Nagao isoenzyme, in 82% of ascitic fluids from ovarian cancer patients. These observations add to our knowledge of the biology of cancer, but more studies are required to assess whether enzyme determinations can be clinically useful.

VI. SUMMARY

Oncodevelopmental antigens are expressed in a variety of tumors including trophoblastic and nontrophoblastic gynecologic cancer. They may elucidate tumor biology and help the clinician in the assessment of diagnosis and prognosis of certain tumors. The measurement of oncodevelopmental antigens in serum of a cancer patient may indicate early recurrent cancer. The measurement of hCG, together with effective treatment, has revolutionized the early treatment and survival of patients with malignant trophoblastic disease. AFP and hCG are secreted by ovarian germ cell tumors containing elements of the yolk sac or choriocarcinoma, and the markers can be used for histological classification as well as for monitoring treatment of such tumors. The development of more effective drugs, such as cis-platinum, and the concomitant use of supersensitive methods for the detection of early recurrence have improved the survival of patients with this highly malignant disease. CEA is expressed in most adenocarcinomas of the endocervix, but in only few endometrial adenocarcinomas, and we think the immunoperoxidase staining of CEA can help the histopathologist to decide on the origin of the lesion. In our experience, the circulating CEA level is elevated in 10–20% of patients with cervical squamous or adenocarcinoma or with ovarian mucinous adenocarcinoma, but this is usually a late phenomenon. In recurrent cancer the levels become rarely reelevated. In some patients, however, lead times of 1–23 months have been reported. This is likely to become important when more effective treatment modalities are available. Ovarian cancer has defied all attempts at early diagnosis and, hence, treatment of advanced cases remains unsatisfactory. New markers, such as ovarian cancer-associated antigens (OCAA and OCAA-1), ovarian cancer antigens (OvC-1 and OvC-2), or oncofetal ovarian cancer-associated urinary peptide (OFP), have been identified and purified. The development of more sensitive methods for their measurement and clinical studies will eventually show whether any of them can improve the early diagnosis of ovarian cancer.

ACKNOWLEDGMENTS

The original work referred to in this chapter has been supported by the Cancer Society of Finland, the Research Council for Medical Sciences, Academy of Finland, and the Sigrid Juselius Foundation, Finland. Antisera and purified antigens for the hCG assay and synthetic GnRH were generous gifts from the National Institute of Child Health and Human Development, and from the National Institute of Arthritis and Metabolic Diseases, National Institutes of Health, Bethesda, Maryland. The author wishes to thank Dr. Hans Bohn, Behringwerke AG, Marburg, W. Germany, for purified SP1, PP5, and PP12 and the antisera.

REFERENCES

Abelev, G. I. (1974). *Transplant. Rev.* **20**, 3–37.
Bagshawe, K. D. (1975). In "Medical Oncology: Medical Aspects of Malignant Disease" (K. D. Bagshawe, ed.), pp. 453–469. Blackwell, Oxford.
Bagshawe, K. D., Lequin, R. M., Sizaret, P., and Tatarinov, Y. S. (1978). *Eur. J. Cancer* **14**, 1331–1335.
Bagshawe, K. D., Wass, M., and Searle, F. (1980). *Arch. Gynecol.* **229**, 303–310.
Bara, J., Malarewicz, A., Loisillier, F., and Burtin, P. (1977). *Br. J. Cancer* **36**, 49–56.
Barrelet, V., and Mach, J. -P. (1975). *Am. J. Obstet. Gynecol.* **121**, 164–168.
Bhattacharya, M., and Barlow, J. J. (1973). *Cancer* **31**, 588–595.
Bhattacharya, M., and Barlow, J. J. (1979). *Int. Adv. Surg. Oncol.* **2**, 155–176.
Bohn, H., and Kraus, W. (1979). *Arch. Gynaekol.* **227**, 125–134.
Bohn, H., and Kraus, W. (1980). *Arch. Gynaekol.* **229**, 279–291.
Bohn, H., and Sedlacek, H. (1975). *Arch. Gynaekol.* **220**, 105–121.
Bohn, H., and Winckler, W. (1977). *Arch. Gynaekol.* **223**, 179–186.

Bohn, H., and Winckler, W. (1981). *Arch. Gynaekol.* **229**, 293–301.
Bohn, H., Inaba, N., and Lüben, G. (1981). *Oncodev. Biol. Med.* **2**, 141–153.
Braunstein, G. D., Vaitukaitis, J. L., Carbone, P. P., and Ross, G. T. (1973). *Ann. Intern. Med.* **78**, 39–45.
Bützow, R., Seppälä, M., and Wahlström, T. (1982). *In* "Pregnancy Proteins: Biology, Chemistry and Clinical Application" (J. G. Grudzinskas, M. Seppälä, and B. Teisner, eds.), pp. 77–81. Academic Press, Sydney.
Chen, H. -C., Hodgen, G. D., Matsuura, S., Lin, L. J., Gross, E., Reichert, L. E., Jr., Birken, S., Canfield, R. E., and Ross, G. T. (1976). *Proc. Natl. Acad. Sci. U.S.A.* **73**, 2885–2889.
Crowther, M. E., Grudzinskas, J. G., Poulton, T. A., and Gordon, Y. B. (1979). *Obstet. Gynecol.* **53**, 59–61.
DiSaia, P., Haverback, B. J., Dyce, B. J., and Morrow, C. P. (1975). *Am. J. Obstet. Gynecol.* **121**, 159–163.
DiSaia, P., Morrow, C. P., Haverback, B. J., and Dyce, B. J. (1977). *Cancer* **39**, 2365–2370.
Ehnholm, C., Seppälä, M., Tallberg, T., and Widholm, O. (1967). *Ann. Med. Exp. Biol. Fenn.* **45**, 318–319.
Engvall, E., and Yonemoto, R. H. (1979). *Int. J. Cancer* **24**, 759–761.
Fishman, W. H., Inglis, N. R., Vaitukaitis, J. L., and Stolbach, L. L. (1975). *Natl. Cancer Inst. Monogr.* **42**, 63–73.
Fox, H. (1980). *UICC Tech. Rep. Ser.* **50**, 22–33.
Gibbons, J. M., Jr., Mitnick, M., and Chieffo, V. (1975). *Am. J. Obstet. Gynecol.* **121**, 127–131.
Gitlin, D., Perricelli, A., and Gitlin, G. M. (1972). *Cancer* **32**, 979–982.
Goldenberg, D. M., Sharkey, R. M., and Primus, F. J. (1976). *JNCI, J. Natl. Cancer Inst.* **57**, 11–21.
Goldenberg, D. M., Kim, E. E., DeLand, F., Spremulli, E., Nelson, M. O., Gockerman, J. P., Primus, F. J., Corgan, R. L., and Alpert E. (1980). *Cancer* **45**, 2500–2505.
Grudzinskas, J. G., Kitau, M. J., and Clarke, P. G. (1977). *Lancet* **2**, 1088.
Grudzinskas, J. G., Gordon, Y. B., Obiekwe, B. C., Pendlebury, D., and Chard, T. (1980). *Am. J. Obstet. Gynecol.* **137**, 866–867.
Hamazaki, M. H., and Hotta, K. (1975). *Experientia* **31**, 241–243.
Hammond, C. B., Borchert, L. G., Tyrey, L., Creasman, W. T., and Parker, R. T. (1973). *Am. J. Obstet. Gynecol.* **115**, 451–457.
Heald, J., Buckley, C. H., and Fox, H. (1979). *J. Clin. Pathol.* **32**, 918–926.
Imamura, N., Takahashi, T., Lloyd, K. P., Lewis, J. L., Jr., and Old, L. J. (1978). *Int. J. Cancer* **21**, 570–577.
Inaba, N., Renk, T., Wurster, K., Rapp, W., and Bohn, H. (1980). *Klin. Wochenschr.* **58**, 789–791.
Inglis, N. R., Kirley, S., Stolbach, L. L., and Fishman, W. H. (1973). *Cancer Res.* **33**, 1657–1661.
Ito, Y., and Higashi, K. (1961). *Endocrinol. Jpn.* **8**, 279–287.
Josimovich, J. B., and McLaren, J. A. (1962). *Endocrinology* **71**, 209–220.
Khodr, G., and Siler-Khodr, T. M. (1980). *Science* **207**, 315–317.
Khoo, S. K., and Mackay, E. W. (1974). *Cancer* **34**, 542–548.
Kjörstad, K. E., and Orjaseter, H. (1977). *Cancer* **40**, 2953–2956.
Knauf, S., and Uhrbach, G. I. (1974). *Am. J. Obstet. Gynecol.* **119**, 966–970.
Koistinen, R., Heikinheimo, M., Rutanen, E. -M., Stenman, U. -H., Lee, J. -N., and Seppälä, M. (1981). *Oncodev. Biol. Med.* **2**, 179–182.
Lamerz, R., Schnabl, G., Stein, G., Kumper, H. J., and Brandt, A. (1979). *Carcino-Embryonic Proteins [Proc. Int. Soc. Oncodev. Biol. Med.], 6th, 1978* Vol. 2, pp. 509–514.
Lee, J. N., Salem, H. T., Al-Ani, A. T. M., Chard, T., Huang, S. C., Ouyang, P. C., Wei, P. Y., and Seppälä, M. (1981a). *Am. J. Obstet. Gynecol.* **139**, 702–704.

Lee, J. N., Seppälä, M., and Chard, T. (1981b). *Acta Endocrinol. (Copenhagen)* **96**, 394–397.
Lee, J. N., Wahlström, T., and Seppälä, M., Salem, H. T., Ouyang, P. C., and Chard, T., (1982). *Placenta* **3**, 67–70.
Levin, L., McHardy, J. E., Poulton, T. A., Curling, P. M., Kitau, M. J., Neville, A. M., and Hudson, C. N. (1976). *Br. J. Cancer* **33**, 363–369.
Lieblich, J. M., Weintraub, B. D., Rosen, S. W., Chou, J. Y., and Robinson, J. C. (1976). *Nature (London)* **260**, 530–532.
Lindgren, J., Wahlström, T., and Seppälä, M. (1979). *Int. J. Cancer* **23**, 448–453.
NIH Consensus Development Conference Statement (1981). *Cancer Res.* **41**, 2017–2018.
Nisbet, A. D., Bremner, R. D., Horne, C. H. W., Brooker, D. Twiggs, L. B., and Okagaki, T. (1982). *Am. J. Obstet. Gynecol.* **144**, 396–401.
Nishimura, R., Endo, Y., Tanabe, K., Ashikata, Y., and Tojo, S. (1981). *J. Endocrinol. Invest.* **4**, 349–358.
Obiekwe, B., Pendlebury, D. J., Gordon, Y. B., Grudzinskas, J. G., Chard, T., and Bohn, H. (1979). *Clin. Chim. Acta* **95**, 509–516.
Petrunin, D. D., Gryaznova, I. M., Petrunin, Y. A., and Tatarinov, Y. S. (1979). *Byull. Eksp. Biol. Med.* **35**, 600–602.
Rosen, S. W., Weintraub, B. D., Vaitukaitis, J. L., Sussman, H. H., Herschman, J. M., and Muggia, F. M. (1975). *Ann. Intern. Med.* **82**, 71–83.
Ruoslahti, E., Engvall, E., Pekkala, A., and Seppälä, M. (1978). *Int. J. Cancer* **22**, 515–520.
Rutanen, E. -M. (1978). *Int. J. Cancer* **22**, 413–421.
Rutanen, E. -M., and Seppälä, M. (1978). *Cancer* **41**, 692–696.
Rutanen, E. -M., and Seppälä, M. (1979). *Carcino-Embryonic Proteins [Proc. Int. Soc. Oncodev. Biol. Med.], 6th, 1978* Vol. 2, pp. 759–764.
Rutanen, E. -M., and Seppälä, M. (1980). *J. Clin. Endocrinol. Metab.* **50**, 57–61.
Rutanen, E. -M., Bohn, H., and Seppälä, M. (1982). *Am. J. Obstet. Gynecol.* **144**, 460–463.
Rutanen, E. -M., Lindgren, J., Sipponen, P., Stenman, U. -H., Saksela, E., and Seppälä, M. (1978). *Cancer* **42**, 581–590.
Salem, H. T., Obiekwe, B. C., Al-Ani, A. T. M., Seppälä, M., and Chard, T. (1980). *Clin. Chim. Acta* **107**, 211–215.
Saxena, B. M., Goldstein, D. P., Emerson, K., Jr., and Selenkow, H. A. (1968). *Am. J. Obstet. Gynecol.* **102**, 115–121.
Searle, F., Leake, B. A., Bagshawe, K. D., and Dent, J. (1978). *Lancet* **1**, 579–581.
Seppälä, M., and Ruoslahti, E. (1972). *Am. J. Obstet. Gynecol.* **112**, 208–212.
Seppälä, M., and Ruoslahti, E. (1974). *Int. Congr. Ser.—Excerpta Med.* **320**, 449–461.
Seppälä, M., and Rutanen, E. -M. (1982). In "Pregnancy Proteins: Biology, Chemistry and Clinical Application" (J. G. Grudzinskas, M. Seppälä, and B. Teisner, eds.), pp. 235–240. Academic Press, Sydney.
Seppälä, M., Bagshawe, K. D., and Ruoslahti, E. (1972). *Int. J. Cancer* **10**, 478–481.
Seppälä, M., Pihko, H., and Ruoslahti, E. (1975). *Cancer* **34**, 1377–1381.
Seppälä, M., Rutanen, E. -M., Heikinheimo, M., Jalanko, H., and Engvall, E. (1978). *Int. J. Cancer* **21**, 265–267.
Seppälä, M., Wahlström, T., and Bohn, H. (1979). *Int. J. Cancer* **24**, 6–10.
Seppälä, M., Wahlström, T., Lehtovirta, P., Lee, J. N., and Leppäluoto, J. (1980). *Clin. Endocrinol.* **12**, 441–451.
Seppälä, M., Rutanen, E. -M., Lindgren, J., and Wahlström, T. (1982). In "Biochemical Markers for Cancer " (T. M. Chu, ed.), pp. 321–350. Dekker, New York.
Seppälä, M., Rutanen, E. -M., Siiteri, J. E., Wahlström, T, Koistinen, R., Pietila, R., and Bohn, H. (1983) *Ann. N.Y. Acad. Sci.*, in press.

Siiteri, J. E., Koistinen, R., Salem, H. T., Bohn, H., and Seppälä, M. (1982). *Life Sci.* **30**, 1885-1891.
Stenman, U. -H., Huhtala, M. -L., Koistinen, R., and Seppälä, M. (1982). *Int. J. Cancer* **30**, 53-57.
Stigbrand, T., Damber, M. -G., and von Schoultz, B. (1976). *Protides Biol. Fluids* **24**, 181-188.
Tatarinov, Y. S., and Kalashnikov, V. V. (1977). *Nature (London)* **265**, 638-639.
Tatarinov, Y. S., and Sokolov, A. V. (1977). *Int. J. Cancer* **19**, 161-166.
Tatarinov, Y. S., Mesnyankina, N. V., Nikoulina, D. M., Novikova, L. A., Toloknov, B. O., and Falaleeva, D. M. (1974). *Int. J. Cancer* **14**, 548-554.
Tatarinov, Y. S., Falaleeva, D. M., Kalashnikov, V. V., and Toloknov, B. O. (1976). *Nature (London)* **260**, 263.
Teilum, G. (1965). *Acta Pathol. Microbiol. Scand.* **64**, 407-429.
Teilum, G., Albrechtsen, R., and Norgaard-Pedersen, B. (1974). *Acta Pathol. Microbiol. Scand., Ser. A* **82**, 586-588.
Vaitukaitis, J. L. (1974). *J. Clin. Endocrinol. Metab.* **38**, 755-760.
Vaitukaitis, J. L., and Ebersole, E. R. (1976). *J. Clin. Endocrinol. Metab.* **42**, 1048-1055.
van Nagell, J. R., Jr., Meeker, W. R., Parker, J. C., Jr., and Harralson, J. D. (1975a). *Cancer* **35**, 1372-1376.
van Nagell, J. R., Jr., Pletsch, Q. A., and Goldenberg, D. M. (1975b). *Cancer Res.* **35**, 1433-1437.
van Nagell, J. R., Jr., Donaldson, E. S., Gay, E. C., Rayburn, P., Powell, D. F., and Goldenberg, D. M. (1978). *Cancer* **42**, 2428-2434.
van Nagell, J. R., Jr., Donaldson, E. S., Gay, E. C., Hudson, S., Sharkay, R. M., Primus, F. J., Powell, D. F., and Goldenberg, D. M. (1979). *Cancer* **44**, 944-948.
van Nagell, J. R., Jr., Kim, E., Casper, S., Primus, F. J., Bennett, S., DeLand, F. H., and Goldenberg, D. H. (1980). *Cancer Res.* **40**, 502-506.
von Kleist, S., Chavanel, G., and Burtin, P. (1972). *Proc. Natl. Acad. Sci. U.S.A.* **69**, 2492-2494.
Wahlström, T., Lindgren, J., Korhonen, M., and Seppälä, M. (1979). *Lancet* **2**, 1159-1160.
Wahlström, T., Bohn, H., and Seppälä, M. (1982). *In* "Pregnancy Proteins: Biology, Chemistry and Clinical Application" (J. G. Grudzinskas, M. Seppälä, and B. Teisner, eds.), pp. 415-422. Academic Press, Sydney.
Wehmann, R. E., Ayala, A. R., Birken, S., Canfield, R. E., and Nisula, B. C. (1981). *Am. J. Obstet. Gynecol.* **140**, 753-757.
Weintraub, B. D., and Rosen, S. W. (1971). *J. Clin. Endocrinol. Metab.* **32**, 94-101.
Würz, H. (1979). *Arch. Gynaekol.* **227**, 1-6.

CHAPTER 21
Large-Scale AFP Screening for Hepatocellular Carcinoma in China

KAI-LI XU
Section of Biochemical and Immunological Diagnosis
Shanghai Cancer Institute
Shanghai, China

I. Introduction	395
II. A Project for Large-Scale Screening	396
A. Selection of a Screening Test	396
B. Organizational and Institutional Form	398
III. Results of Large-Scale AFP Screening	398
IV. Study on the Follow-Up of Low-Level AFP Subjects	401
A. Study on Selected Populations	401
B. Study of a Random Population	402
C. Study on the Interval of Testing	403
V. The Possibility of Prevention of Liver Cancer	405
VI. Conclusions	406
References	407

I. INTRODUCTION

Patients diagnosed as having localized tumors appear to survive longer than those with metastasized tumors. Therefore, several programs have been devised using appropriate screening tests to detect early cancer. Evidence of the effectiveness of cancer screening has been observed from certain screening tests, such as those for cervical and gastric cancer. The survival rate of cancer-screened patients improves tremendously after treatment, and their death rate even decreases (Okui and Tejima, 1980). However, the following conditions are necessary when screening large populations for a particular tumor.

1. To select an area with long-term exposure to carcinogens or a high-cancer-risk population

2. To choose a screening test with a false-positive rate under 1% that is easy to perform, sensitive, highly specific, and suitable for the field (Herberman, 1980)

3. To organize well the financing of the screening, the timely treatment of tumors in their early stages, and the elimination of social influence. It is particularly necessary to eliminate phobic psychology in the false-positive subjects

Hepatocellular carcinoma (HCC) is one of the more common cancers in Asia and Africa (UICC, 1980). Its virulence can be judged from its short course and the close relationship between incidence and mortality rates.

Since the association of liver cancer with increased α-fetoprotein (AFP) serum concentration was noted, the AFP test has been recommended for clinical use since the 1960s. In the 1970s, the determination of AFP concentration in human serum was used as a screening test for HCC in several regions. The AFP test has been widely used in some localities with a high incidence of HCC, although the value of using the AFP assay for large-population screening of HCC has been debated. Practice in the last 10 years has proved, however, that serum AFP determination is not only a comparatively ideal screening test for HCC, but also offers the following advantages.

1. Many asymptomatic HCC cases can be detected through screening, thus providing greater resectability and survival rates, and making early treatment possible.

2. A group of patients with low AFP levels was found among a highly sensitive group of HCC victims. This occurrence suggests a means of conducting further research on the mechanism of hepatocarcinogenesis, its prevention, and treatment.

II. A PROJECT FOR LARGE-SCALE SCREENING

A. Selection of a Screening Test

The key to the success of a screening project lies in the choice of a good detection technique, the efficiency of which depends on its sensitivity, specificity, reliability, and economic value, including also factors such as ease and convenience of operation, and perservation of samples.

1. Double Agar Gel Diffusion (AGD)

Being inexpensive, simple, convenient, and highly specific, AGD has been used for clinical diagnosis and mass surveys in the late 1960s and early 1970s in quite a few regions. A few asymptomatic patients with liver cancer were detected; however, the disease was overlooked in some patients because of the tests' low detection threshold.

2. Countercurrent Immunoelectrophoresis (CIEP)

Countercurrent immunoelectrophoresis (CIEP) is a rapid, inexpensive, highly specific, and sensitive detection method. Its threshold can be as low as 250 ng/ml. As AFP concentration is usually over 500 ng/ml in most cases of HCC, the test is useful for HCC screening. However, CIEP is now seldom used as a screening method in China because of its inability to detect some HCC cases exhibiting low AFP levels.

3. Reverse Hemagglutination Assay (RHA)

Reverse hemagglutination assay (RHA) is a simple and sensitive method. A skilled operator can test several hundred samples a day, and the screening threshold can be as high as 50 ng/ml. As horse serum and anti-AFP serum are added to this technique, the false-positive rate is decreased (Xu and Guan, 1979). Therefore, this detection method is most widely adopted for mass survey studies in China.

4. Radio Rocket Electrophoresis Autography (RREA)

Radio rocket electrophoresis autography (RREA) is an AFP quantitative method, being highly sensitive (up to 30 ng/ml) and remarkably specific. It is also suitable for mass surveys because expensive instruments are not needed (Sun *et al.*, 1979).

5. Radioimmunoassay (RIA)

Radioimmunoassay (RIA) is the most sensitive technique (up to 1 ng/ml). Because it is costly and complicated to operate and requires expensive equipment, it is not widely adopted as a screening test (Purves, 1973). RIA is used in many countries, however, as the best method for quantitative analyses and follow-up purposes.

6. Enzyme Immunoassay (ENZ)

Enzyme immunoassay (ENZ) is newer but less frequently used in China. Its sensitivity and reliability can be compared with those of the radioimmunoassay. As very complex instruments are not required, ENZ should be considered for mass surveys. Whether it can be used for large population screenings remains to be seen.

In summary, AFP is by no means the specific antigen of liver cancer. It is, therefore, necessary to take the following aspects into consideration when selecting a screening test: (a) the AFP background of a natural population, (b) the level of normal AFP elevation in nonmalignant conditions, (c) the distribution curve of AFP in the serum of patients with HCC, and (d) the AFP levels of people in high-incidence areas of HCC.

α-Fetoprotein concentrations from 20 to 500 ng/ml are found in benign liver diseases as well as early and later stages of HCC. Therefore, to prevent the occurrence of false positives without affecting the detection rate of early HCC, we recommend using the hemagglutination assay for primary screening and the radioimmunoassay for the follow-up of positive cases.

B. Organizational and Institutional Form

To guarantee the implementation of HCC screening, especially of large-scale screening, and to achieve economic and social objectives, it is necessary to have a powerful three-unit organization with (a) a screening unit using the most effective testing methods, (b) a detection laboratory, and (c) a hospital. Some of the suspected subjects should be admitted to the hospital for further analysis to eliminate false-positive cases and to ensure the timely treatment of early HCC.

III. RESULTS OF LARGE-SCALE AFP SCREENING

According to epidemiological reports, the incidence of HCC tends to rise slightly from year to year in Shanghai, China (Gao et al., 1981), and since HCC is readily progressive and essentially not treatable by the time it becomes symptomatic, it has been emphasized that considerable effort should be made toward its early detection and diagnosis, followed by the monitoring of prognosis.

In the early 1960s, Abelev reported that a hepatoma-specific antigen extracted from mouse hepatoma cells was immunologically identical to normal mouse AFP. The resynthesis of this fetal protein by the malignant cell represented a biochemical form of dedifferentiation and led to the detection of human AFP in the serum of an HCC patient (Tatarinov, 1964), suggesting that AFP may be a helpful diagnostic test for human HCC. Its diagnostic usefulness was rapidly established in clinical studies in many parts of the world. Although low levels of AFP also appear in nonmalignant cases, AFP is still the most highly specific tumor marker (Abelev, 1971).

In the last 10 years, many countries, such as Senegal (Leblanc et al., 1973), Japan (Nishi and Hirai, 1973), South Africa (Purves, 1973), and China (Coordinating Group for Research on Liver Cancer, China, 1974a,b) have been using the serum AFP detection method for large-scale screening in areas with a high incidence of HCC. Analysis of reports from the above-mentioned countries reveals that using AFP for screening studies can be divided into several stages mainly on the basis of the sensitivity of the AFP detection technique.

Leblanc *et al.* (1973) used a double diffusion method, screening at random 9864 residents of a region with a high incidence of HCC in Senegal. Through samples of finger blood taken over a 2-year period, three AFP-positive patients were without any clinical symptoms. Their postoperative condition was good, but they all died within 5 months. It was evident from these results that the double diffusion detection method alone was not a suitable screening test for the early discovery of HCC. However, it must be pointed out that the application of an AFP detection technique was at least a promising screening test.

From 1971 to 1974, the Coordinating Group of China (1974b) adopted the double diffusion and countercurrent immunoelectrophoresis methods for large-scale AFP screening, and a total of 417,664 persons were tested, 57 cases of HCC being found (the natural population detection rate being 13.6:100,000). Compared with the clinically treated group, the AFP-positive group was in the early clinical stage of HCC, as shown by smaller tumor nodules and limited lesions, as well as general symptoms, physical signs, isotope scanning, ultrasonic detection, and resection rate, which further illustrated the superiority of the double diffusion and countercurrent immunoelectrophoresis screening techniques in mass surveys. However, through a follow-up screening of a group of 118,138 selected from the original Shanghai study group, nine HCC cases were found successively in 8 months. These results indicated that applying a detection method of low or medium sensitivity for a single sample might give a detection–failure rate as high as 50%.

Both of the above studies employed relatively insensitive immunodiffusion assays for AFP. Purves (1973), using a more sensitive radioimmunoassay, failed to detect a single case of elevated AFP (defined by a cutoff of 100 ng/ml) among 5000 asymptomatic Bantu miners. This is surprising since the incidence of primary liver cancer among this population was 1 in 2700 overall. The investigators attempted to explain these unusual results on the basis of a very rapid doubling time (about 10 days) for this particular type of cancer.

With refinements in sensitivity, simplicity, convenience in methods, and improvement of blood collecting (finger blood preserved in plastic tubes) for large-scale AFP screening, at least two areas in China adopted the reverse hemagglutination assay for primary mass surveys, and radio rocket immunoelectrophoresis for quantitative follow-up of serum AFP.

The Shanghai Coordinating Group (1979) compiled statistics on some 1,967,511 staff members of factories and enterprises, screened from 1971 to 1976, with 300 HCC patients being detected in those years. The detection rate was 15.4:100,000, lower than the incidence in urban districts (see below). This might be due to the fact that a single blood sample was taken from each member

TABLE I
Results of AFP Screening in Random Populations

Reference	Areas	Incidence (per 100,000)	Methods[a]	Number screened	Total	Positive HCC 1/100,000	Asymptomatic
Leblanc et al. (1973)	Senegal	45	AGD	9864	6	60	
Coordinating Group (1974)	Shanghai (China)	23.4	AGD and CIEP	281,795 (Urban) 135,849 (Suburb)	32 25	11.36 18.40	3
Shanghai Coordinating Group (1979)	Shanghai (China)	23.7	RHA	1,967,511 (Urban)	300	15.4	134
Zhu (1981)	Qidong county (China)	60	RHA and RREA	1,223,912	425	38.5	167

[a] AGD, double agar gel diffusion; RHA, reverse hemagglutination assay; CIEP, counter immunoelectrophoresis; RREA, radio rocket electrophoresis autography.

of the mass survey. Among the AFP positives, only 20% were diagnosed as advanced patients by isotope scanning and ultrasonic detection, and 134 cases (44.7%) were in the early stages. The patient resection rate was 58.5%. The 1-, 2-, and 3-year survival rates after operation were 86.7%, 75.0%, and 57.1%, respectively. It is concluded that AFP screening is of significant diagnostic value in the early detection of HCC.

In addition, 1,223,912 individuals were screened in three consecutive years (1974–1977) in Qidong county, Jiangsu province, a high endemic area (Zhu, 1981). Four hundred seventy-five patients with liver cancer were detected. The detection rate was 38.5 : 100,000, being lower than the local incidence of HCC. Among the 475, 167 cases (35.2%) were in the early stages, and the survival rate for such cases 2 years after operation was up to 69%. Five years after operation, the survival rate was 26%.

The results mentioned above (Table I) indicate that the concentration of serum AFP increases as a function of the early stages of HCC growth. Therefore, by means of a high-sensitivity detection method for screening HCC, cases with smaller tumor nodules can usually be discovered. It is believed that HCC in its early stages has a more favorable prognosis than in the later stages of the disease.

IV. STUDY ON THE FOLLOW-UP OF LOW-LEVEL AFP SUBJECTS

Twenty to 25% of HCC patients reveal serum AFP levels similar to the levels found in patients with nonmalignant liver diseases (25–500 ng/ml). In the former cases, HCC patients in the early stage occupy a certain proportion. Alpert *et al.* (1971) reported that a few HCC patients having low concentrations of serum AFP experience slow tumor growth, and may survive more than 1 year without treatment. Therefore, one cannot assume that HCC is invariably accompanied by high serum concentrations of AFP.

A. Study on Selected Populations

Masseyeff *et al.* (1976) reported a determination of serum AFP concentration from 195 cirrhotic patients, 72.8% of them showing AFP levels less than 10 ng/ml, 22.1% showing levels of 10–2000 ng/ml, and 5.1% showing levels above 2000 ng/ml. Follow-up studies have revealed that the presence of HCC is somehow relevant to the level of AFP concentration; for instance, all of those in the group with high AFP levels were proved by histology to have HCC, while about 25% of the group with mid-levels of AFP developed the disease, but no one in the group with normal AFP levels acquired malignant growths.

Thomas *et al.* (1979) examined the serum AFP concentrations of 310 patients with chronic hepatitis. Among them, 37 cases were found to be associated with abnormal elevations of AFP (AFP greater than 10 ng/ml). After a follow-up in 2 years, about six HCC patients were found in the group with abnormal AFP elevations, but none was found in the group with normal AFP levels.

In addition, in a follow-up study of patients with persistent liver diseases, Okuda *et al.* (1975) detected four cases of HCC in individuals exhibiting rapidly rising AFP levels. However, in each of them, the tumor nodule was 4 to 6 cm in diameter. In one case, the tumor was resected and the patient survived 2 years following surgery.

These results reveal that patients with chronic hepatitis or cirrhosis with abnormally elevated AFP are most likely to develop cancer. Evidently it is of great significance to observe this sort of patient over a long period of time.

B. Study of a Random Population

Since 1974, high-sensitivity screening assays for HCC (RHA and RREA) have been adopted in China, and have revealed a most interesting phenomenon in mass surveys. In addition to subjects with high concentrations of AFP, many were also found with low levels, which prevented accurate diagnoses. Serial clinical studies on these subjects revealed a tendency for the early stages of HCC to exhibit low levels of AFP.

Xu (1978) reported on a 1974 Shanghai study in which a reverse hemagglutination assay (RHA) was performed on 115,000 random subjects in the city. Twenty cases of HCC were found at a rate of 17.1 per 100,000. These twenty patients showed high levels of AFP.

In addition, 270 subjects were also found to exhibit low levels of AFP. After a 2-year follow-up study, nine cases of HCC were found in the low level group. The 20 cases found in those who had high levels of AFP, and the nine cases found among those having persistently low levels of AFP gave a total detection rate of 25.2 per 100,000. The nine subjects with initially low AFP levels eventually exhibited higher concentrations. The shortest interval between the incidence of low and high levels was 4 months, and the longest was 17 months, with the average being 8 months.

Li *et al.* (1979) also found, after a 3-year follow-up study, two cases with low AFP levels who developed HCC. As described above, it can be predicted that about 20–30% of cancer cases whose AFP levels are lower than 500 ng/ml can be discovered through follow-up studies.

Another mass survey and follow-up of 10,417 persons was performed in Qidong county, using the RREA screening test (Sun *et al.*, 1979). Six cases of HCC were found with elevated concentrations of AFP. Three of these patients were still operable, and their preoperation AFP levels were 300, 800, and 1000 ng/ml. A 4-year postoperative follow-up study showed that all had recovered.

Recently Tang and Yang (1981) also observed resection rates as high as 83% (5/6) in asymptomatic patients with AFP concentrations rising over 500 ng/ml.

The results of these long-term studies indicate that the follow-up of low AFP subjects is useful in the detection of HCC in the early and subclinical detection of HCC. However, it must be noted that the values in the range of 20–500 ng/ml may be associated with benign liver diseases as well as with HCC; therefore, one must use other methods to avoid interference between nonmalignant and carcinoma-related abnormal high levels of AFP. For example, kinetic studies on both serum AFP concentration and SGPT (serum glutanic pyrivic transaminase) activity in those patients with simultaneously elevated AFP and SGPT may aid in the differential diagnosis of malignant versus benign diseases (Coordinating Group, 1979). Xu (1978) also suggested that among low AFP subjects, patients with long-term, simultaneous abnormalities in ZNTT (zinc turbid test) and AFP have a higher risk of carcinogenesis.

Recently, with the earlier demonstration of AFP concanavalin A binding variants (Ruoslahti et al., 1978; Mackiewicz and Breborowicz, 1980), we have devised a technique of high-sensitivity lectin affinity radioimmunoelectrophoresis, the sensitivity of which is 50 ng/ml. An examination of the sera of HCC and teratocarcinoma has revealed two kinds of AFP variants (Fig. 1). At present, this technique is being used for analysis of AFP variants with some subjects with abnormal elevations of AFP which are difficult to differentiate immediately between HCC or other malignant tumors (Xu and Chou, 1982).

C. Study on the Interval of Testing

Using high-sensitivity screening techniques for large-scale AFP screening and following up low AFP level subjects can increase the detection rate of HCC, lower the detection failure rate, and detect early HCC in time for treatment. However, as the relationship between AFP and carcinogenesis of human HCC is still unknown at present, it is difficult to describe in full the relationship between the growth of tumor cells and the kinetics of AFP based on the incidence in different populations and areas.

Purves (1973) proposed that the elevation of serum AFP level is associated with the doubling time of tumor cell growth. His hypothesis was that if tumor cells doubled in 10 days, 5 months would be required for the clinical symptoms to appear; if the doubling time was 50 days 16 months would be necessary.

Lehmann's hypothesis (1979) was that during the course of hepatocarcinogenesis, the variation of AFP concentration is in a nonlinear curve with cellular reproduction. The whole course of carcinogenesis can be divided into three periods: the latent period (tumor doubling time is unknown), the "fast" period (tumor doubling time is between 7 and 10 days), and the "slow" growing period (tumor doubling time is 50 days). AFP is in linear relation to malignant

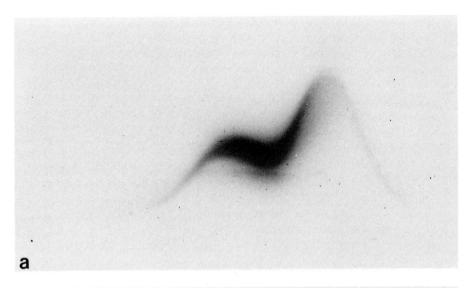

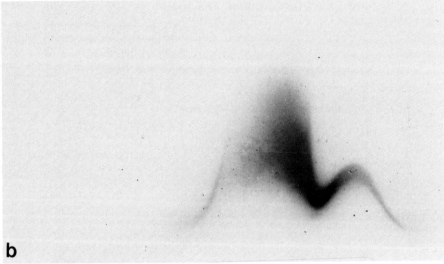

Fig. 1. (a) Autoradiogram of AFP variants in teratocarcinoma of the ovary by using lectin affinity radioimmunoelectrophoresis. (b) Autoradiogram of AFP variants in hepatocellular carcinoma by using lectin affinity radioimmunoelectrophoresis.

growth only in the "fast" growing period. It can be postulated, therefore, that AFP-positive cases with incipient HCC can be most readily traced during the "fast" growing period.

According to Xu's observation in 1977 (Fig. 2), the interval between the first

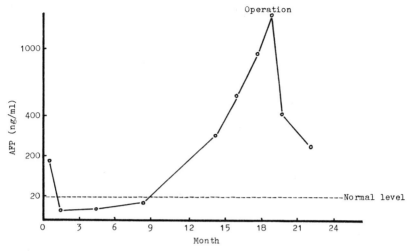

Fig. 2. Kinetics of serum AFP in a patient with hepatocellular carcinoma during follow-up.

and second peaks (from normal AFP levels to over 400 ng/ml) was about 6 months in one biphasic-type case, with an average of 8 months in nine HCC cases discovered through follow-up studies. Sun *et al.* (1979) proposed that when serum AFP concentration is under 1000 ng/ml, the tumor nodule is less than 3 cm and the growth of the tumor cell is parallel to AFP concentrations. It seems that in China the "fast" growing period of tumors in Lehmann's hypothesis may be about 6 to 10 months, and the variation of serum AFP concentration is probably limited to 1000 ng/ml. At the first Coordinating Conference of Liver Cancer, held in 1977 in China, Xu (1977) was the first to suggest screenings every 6 months in areas of high HCC incidence. It was subsequently proved that a screening twice a year can lower the detection failure rate of such screenings to 3.8% (Zhu, 1981), but annual screenings maintain a detection failure rate as high as 15% (Coordinating Group, 1979).

V. THE POSSIBILITY OF PREVENTION OF LIVER CANCER

Because of a lack of better therapeutic means, the study of HCC should not only emphasize the multiple screening techniques for the detection of early HCC, but should also stress the prevention of carcinogenesis.

It is interesting that at least in some patients with HCC, the elevation of AFP was biphasic—a transient increase followed by a sustained increase concomitant with the appearance of clinical symptoms of HCC, being similar to the dynamics of serum AFP observed in chemical carcinogenesis and in spontaneous liver carcinogenesis (Xu, 1977; Watabe, 1971; Jalanko *et al.*, 1978). Serial clinical and experimental studies showed that there is a long latent period before the onset and detection of liver carcinogenesis of humans. In this period, AFP may fluctuate or continue to remain at low or medium levels of abnormal elevation. As for the duration of this degree of the period of cellular reproduction, it may have something to do with bodily exposure to carcinogens or the degree of injury due to HBV (hepatitis B virus) infection. According to findings made in cirrhosis and chronic hepatitis follow-up studies in areas with low HCC occurrence (Lehmann, 1976; Thomas *et al.*, 1979), carcinogenesis was rarely found when AFP levels were normal. However, this indicates that an abnormal AFP elevation could really be taken as a dangerous signal of carcinogenesis of the human liver.

Therefore, to block or retard carcinogenesis, subjects with abnormally low concentrations of AFP should be removed from carcinogenic environments (Purves *et al.*, 1973) or treated with appropriate medication (Xu *et al.*, 1980; Oon *et al.*, 1980).

VI. CONCLUSIONS

The following statements are made preliminarily on the basis of the above-mentioned relationship (Section V) between clinical testing and research on AFP and the carcinogenesis of HCC (Fig. 3) (Xu, 1981).

Practice in recent years has encouraged efforts in the detection of incipient HCC in China, although the development of large-scale AFP screening for HCC

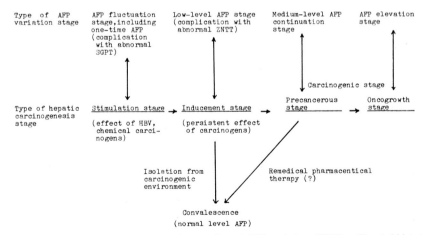

Fig. 3. Pattern of human hepatic carcinogenesis and AFP variation. ZNTT = Zinc turbid test; HBV = Hepatitis B virus.

would encounter restrictions by political, economic, medical, and therapeutic conditions as well as organizational limitations. The public would be justified in being pessimistic about the results of large-scale testing to detect early hepatoma if there were not at the same time increased research in the development of better therapy for established cases of hepatoma. However, through long-term AFP screenings for HCC in large populations, and continued follow-up studies on populations with high HCC incidence, some dynamics of serum AFP in the course of carcinogenesis have been observed. Among these results, attention should be paid mostly to the persistent low-level AFP stage of the preclinical cancerous stage. We, therefore, are of the view that if certain preventive measures, such as isolation from carcinogenic environments, and appropriate medical treatment are taken in this stage, the course of carcinogenesis of the human liver may be inhibited or retarded.

REFERENCES

Abelev, G. I. (1971). *Adv. Cancer Res.* **141,** 295–358.
Alpert, E., Hershberg, R., Schur, P. H., and Isselbacher, K. J. (1971). *Gastroenterology* **61,** 137–143.
Coordinating Group for Research on Liver Cancer, China (1974a). *In* "Proceedings of the Second International Symposium of Cancer Detection and Prevention" (C. Maltoni, ed.), pp. 655–658. Excerpta Medica, Amsterdam.
Coordinating Group for Research on Liver Cancer, China (1974b). *2nd. Congr. Cancer 11th, 1974.* (Florence, Italy).
Coordinating Group for Research on Liver Cancer, China (1979). *Adv. Med. Oncol., Res. Educ., Proc. Int. Cancer Congr. 12th, 1978* **9,** 291–295.
Gao, Y. T., Tu, J. T., Jin, F., and Gao, R. N. (1981). *Br. J.Cancer* **43,** 183–193.
Herberman, R. B. (1980). *In* "Cancer Control" (I. I. Kessler, ed.), pp. 121–142. Univ. Park Press, Baltimore, Maryland.
Jalanko, H., Virtanen, I., Engvall, E., and Ruoslahti, E. (1978). *Int. J. Cancer* **211,** 452–459.
Leblanc, L., Tuyns, A. I., and Masseyeff, R. (1973). *Digestion* **8,** 8–14.
Lehmann, F. G. (1976). *In* "Onco-Developmental Gene Expression" (W. H. Fishman and S. Sell, eds.), pp. 407–415. Academic Press, New York.
Lehmann, F. G. (1979). *In* "Alpha-Fetoprotein in Clinical Medicine" (H. K. Weitzel and J. Schneider, eds.), pp. 109–115. Thieme, Stuttgart.
Li, M.-L., Yu, E.-X., and Zhang, P.-L. (1979). *Res. Cancer Treat. Control* **5,** 9 (China).
Mackiewicz, A., and Breborowicz, J. (1980). *Oncodev. Biol. Med.* **1,** 251–261.
Masseyeff, R., Rey, J. F., Schenowitz, G., and Delmont, J. (1976). *Protides Biol. Fluids* **24,** 357–360.
Nishi, S., and Hirai, H. (1973). *Tumor Res.* **8,** 51.
Okuda, K., Kubo, Y., and Obata, H. (1975). *Ann. N.Y. Acad. Sci.* **259,** 248–252.
Okui, K., and Tejima, H. (1980). *Acta Chir. Scand.* **146,** 185–187.
Oon, C. J., Yosui, L., Chio, L. F., and Chan, S.-H. (1980). *Ann. Acad. Med. Singapore* **9,** 240.
Purves, L. R. (1973). *S. Afr. J. Sci.* **69,** 173–178.
Purves, L. R., Branch, W. R., Geddes, E. W., Manso, C., and Portugal, M. (1973). *Cancer* **31,** 578–587.

Ruoslahti, E., Engvall, E., Pekkala, A., and Seppälä, M. (1978). *Int. J. Cancer* **22**, 515–520.
Shanghai Coordinating Research Group for Research on Liver Cancer (1979). *Adv. Med. Oncol., Res. Educ., Proc. Int. Cancer Congr. 12th, 1978* **9**, 285–289.
Sun, T.-T., Wang, L.-C., Chang, Y.-L., Lu, C.-H., Hsia, C.-C., and Li, F.-M. (1979). *Chin. Med. J.* **92**, 17–25.
Tang, Z., and Yang, B. (1981). *Chin. J. Oncol.* **13**, 205.
Tatarinov, Y. S. (1964). *Vopr. Med. Khim.* **10**, 90–91.
Thomas, H. C., Lin, K. C., and Sherlock, S. (1979). *In* "Alpha-Fetoprotein in Clinical Medicine" (H. K. Weitzel and J. Schneider, eds.), pp. 135–137. Thieme, Stuttgart.
Union Internationale Le Cancer (UILC) (1980). *In* "Cancer Risk by Site" (T. Hirayama, J. A. H. Waterhouse, and J. F. Fraumeni, Jr., eds.), p. 66. UICC, Geneva.
Xu, K. L. (1977). *Natl. Conf. Liver Cancer, 1st,* 1977, p. 85 (Shanghai).
Xu, K.-L. (1978). *Nat. J. (Shanghai)* **5**, 148.
Xu, K.-L. (1981). *Natl. Conf. Liver Cancer, 2nd, 1981* p. 44.
Xu, K.-L., and Chou, J. (1982). *Biochemistry and Biophysics.* **6**, 30.
Xu, K.-L., Li, M.-L., Guan, S.-F., and Yu, E.-X. (1980). *Chin. J. Oncol.* **12**, 45.
Xu, S.-K., and Guan, S.-F. (1979). *Med. J. (Junior) China* **2**, 22–25.
Watabe, H. (1971). *Cancer Res.* **31**, 1192–1194.
Zhu, Y. R. (1981). *Chin. J. Oncol.* **3**, 35.

CHAPTER 22

Uses and Limitations of Tumor Markers

RONALD B. HERBERMAN
Biological Therapeutics Branch
Biological Response Modifiers Program, Division of Cancer Treatment
National Cancer Institute
Frederick, Maryland

I.	Introduction: Experience of the Past Ten Years	409
II.	The Promise of Current Research	411
III.	Future Directions	412
	A. Improved Classification of Disease	413
	B. Accurate Assessment of Primary Therapy	413
	C. Response to Continued Therapy and Early Detection of Recurrent Disease	414
IV.	Markers in Cancer Detection and Diagnosis	415
	References	417

I. INTRODUCTION: EXPERIENCE OF THE PAST TEN YEARS

For the past 5–10 years, there has been a strong feeling among tumor immunologists and oncologists that immunologic tests would provide a new generation of more sensitive and specific tests for detection, diagnosis, and monitoring of individuals with cancer. Recently, however, there has developed considerable pessimism and even skepticism as to whether immunodiagnostic procedures would have an important place in the diagnostic armamentarium of oncologists. The problem is, which attitude is more realistic and in line with the scientific evidence. First, particularly because it may reflect the current feelings of many oncologists, it seems worthwhile to consider some of the reasons for the feeling of pessimism or skepticism.

1. There seems to be a general inverse correlation between the amount of knowledge and the extent of investigation about a particular marker and the level of enthusiasm or optimism that is experienced. In many, probably most, cases, as we gain more insight into the actual value of a marker, we become increasingly aware of its limitations. There appear to be several phases which the work on tumor markers go through: (a) initial examination of its ability to discriminate between cancer and noncancer. This is most often, for the circulating markers, a limited examination of readily available serum specimens, involving obvious cancer, usually advanced, and healthy controls, usually considerably younger than the cancer patients. There are a large number of reports in this category. (b) Unfortunately there are many fewer studies that go on to a second phase, to determine more clearly the discriminatory value of a marker and to assess thoroughly the levels of markers in the whole spectrum of health and disease. For this, it is necessary to use age-matched normal controls, patients with relevant benign diseases, other cancers, and various other appropriate controls. Research on a substantial number of markers has reached this point, and although the results are not as promising in most cases as in the initial studies, a considerable portion of these assays have survived this stage. (c) The next stage relates to the determination of the actual value of a marker for a particular clinical application. This is usually dependent on time-consuming, expensive, and logistically difficult, well-designed clinical studies and therefore is approached only in a handful of cases. Even among the highly selected series of markers that have been reported to be promising for immunodiagnosis and have begun to be utilized for testing of large numbers of patients, only a portion have undergone this type of rigorous evaluation.

2. Because of some major problems in the adequate evaluation of markers, coupled with the growing availability of tests for various tumor markers, there is a rather widespread and uncontrolled utilization of tumor markers. Carcinoembryonic antigen (CEA) is the best known and most widely used example of tumor markers, and millions of assays are now being done each year, at an estimated cost of many millions of dollars, probably well in excess of 30–50 million dollars. A large portion of this testing is likely being performed for ill-defined reasons or with results that are not reliable because of technical problems in the testing laboratory or because their application for management is not adequately understood by the clinicians who ordered the tests.

3. Progress in this field appears quite slow. As a personal milestone, I have reflected back on a meeting at NIH in late 1978 that was designed to assess critically the status of tumor markers and the available documentation regarding their value for particular clinical applications (Herberman, 1979a). Relatively little additional solid information has come forth since then that was not known, or at least anticipated, 3 years ago. There are still very few immunodiagnostic tests that have become generally available to oncologists and that have demonstrated utility.

II. THE PROMISE OF CURRENT RESEARCH

With these points being mentioned, why do I still have feeling of sustained optimism?

1. Although the number of advances and new insights into tumor markers has been rather small, some have occurred and offer much promise. For example, the development and practical demonstration of radioimmunodetection at a clinical level, which had been predicted, from many earlier studies, to be almost hopeless or unfeasible. The successful demonstration of some value of this procedure (Goldenberg et al., 1978) has come despite the main use of an antibody to a marker, CEA, which is clearly not tumor specific or as selective as desired or as are some other, already available, markers. Another important success in this field relates to the use of the assays for α-fetoprotein (AFP) for screening for primary liver cancer among high risk populations. In a large-scale study in mainland China (Coordinating Group, 1974), the feasibility of mounting such a large-scale screening project has been demonstrated, and the results have shown an ability of the marker to detect at least a small number of cases of primary liver cancer at early, treatable, perhaps curable stages. Another example of an important practical application of tumor markers is the increased documentation of the value of AFP and human chorionic gonadotropin (hCG) for the management of patients with choriocarcinoma and for testicular cancer (Lange et al., 1977). Yet another impressive success has been the demonstration of the value of the radioimmunoassay for calcitonin, particularly after provocative stimulation, in the screening of families with medullary carcinoma of the thyroid (Wells et al., 1975). In a somewhat different area, the rapid application of the expanding body of information on markers for subpopulations of normal lymphoid cells to the classification of leukemias and lymphomas has already resulted in considerable revision of the classification of these diseases (Sallan et al., 1980). In several instances, this is being shown to be important for adequate management. For example, acute lymphocytic leukemia of the T cell type (T-ALL) has been shown to have a considerably poorer prognosis than common acute lymphocytic leukemia, and it seems likely that a different form of therapy will be needed for T-ALL.

There have also been some recent impressive advances in technology. For this field, monoclonal antibodies have been rapidly recognized to offer the potential for a whole new generation of tumor markers and perhaps even for new approaches to the use of tumor markers. However, I must introduce a note of caution at this point. It remains unclear as to just how much more powerful or selective the monoclonal antibodies will be. It is proving quite difficult to find monoclonal antibodies that detect specific antigens that are expressed only on tumor cells. Thus, the problems that are being encountered are similar to those which have been experienced for many years with the conventionally produced antisera. Although the monoclonal antibodies will clearly provide better defined reagents, for more consistent and standardized results, it is not clear that they will allow definitive discrimination between cancer and all normal tissues. There is a continuing need for experienced serologists to evaluate thoroughly the specificity of each new monoclonal antibody that appears promising. Several monoclonal antibodies which initially appeared to be tumor specific have been found, after very extensive evaluation of normal tissues, to be present in some normal tissues. It is possible that one of the main limitations with the current generation of monoclonal antibodies is that they are being produced in rodents, which may lack the ability to detect the small variations in specificities between tumor cells and normal tissues. More selective and hopefully tumor-specific reagents may come from the new attempts to make human monoclonal antibodies, by fusion of human B cells with human B cell lymphomas (Olsson and Kaplan, 1980). In any event, the monoclonal antibodies and other reagents for detection of tumor markers can now be used for more refined and careful examination of cells and tissues, because of the increasing availability of cell sorters, for objective and large-scale utilization of markers, and because of the development of sensitive and relatively simple enzyme-linked immunoassays. Certainly the widespread experience with immunohistochemical techniques for detection of markers in tissue sections also represents considerable progress with a promising approach (Kurman et al., 1978).

2. More generally, the potential ability of immunological procedures to detect very few molecules on a cell or materials released from a cell continues to provide a very strong set of tools for approaching the problem of identification of markers associated with a tumor.

3. Furthermore, the widespread experience, although largely disappointing, over the past few years with the current generation of markers, has provided an excellent learning experience for investigators in the field, who are now considerably more sophisticated and aware of the potential pitfalls and difficulties in going from the initial findings of a promising marker to the critical evaluation of its utility. It certainly should not take as long now to make this transition as it did for CEA, hCG, and AFP.

III. FUTURE DIRECTIONS

After this long introduction, let us now briefly consider where we stand and where we should be going in this field. [More detailed discussions of the tumor markers and thier potential clinical applications have been published elsewhere (Herberman and McIntire, 1979; Herberman, 1979b)].

Clearly most of the definite progress in this field has been with the use of markers for management of patients with malignant disease. This can be subdivided into several particular aspects.

A. Improved Classification of Disease

The improved classification of disease, particularly into subgroups which vary in their prognosis and responsiveness to certain treatments, includes the already widespread use of lymphoid cell markers for the classification of leukemias and lymphomas; the use of AFP and hCG for improved staging with testicular tumors; and the demonstrated potential for CEA and a few other markers to divide a particular stage of disease into prognostic categories. The information from the last example is still awaiting adoption by clinicians to help in the stratification of patients for clinical therapy trials.

The real question is whether the markers could add to the staging information that is obtained by currently available means, e.g., at surgery. It would be very helpful if it were possible to classify patients better in regard to responsiveness to therapy. This might be done preoperatively, to help to identify patients with unresectable disease. There have been some indications for the value of CEA in this regard for patients with lung cancer, with circulating levels above about 8 ng/ml, indicating the high likelihood for unresectable disease (Concannon et al., 1978). Immunologic approaches might also be helpful to identify cancer in metastatic sites. In addition to radioimmunodetection, one could envision the use of cytologic approaches for detection of small numbers of malignant cells in bone marrow or possibly even in the peripheral blood. For example, markers associated with metastatic lung or breast cancer might be expected to be most useful in this regard. At the time of surgery, the challenge would be to determine whether a tumor marker could subdivide patients within a particular TNM (standardized classification scheme indicating tumor size, regional lymph node status, and distant metastases) stage of disease. Examination of removed lymph nodes for small numbers of tumor cells might considerably improve the staging. After surgery, one would like to be able to utilize tumor markers to detect residual disease, which might lead to further surgery or to more aggressive chemotherapy or radiotherapy. For this, some of the circulating markers might be helpful. Identification of patients with continued elevation of markers might be expected to help in assigning patients to adjuvant chemotherapy studies.

B. Accurate Assessment of Primary Therapy

There are an increasing number of markers for more accurate assessment of the efficacy of primary therapy: for example, the persistence of marker elevation after operation or chemotherapy; the increased ability to search for occult metastases; the search for residual tumor cells by cytologic or immunohistochemical examination of lymph nodes, bone marrow, or even blood, particularly by the use of such selective markers as monoclonal antibodies, coupled with the possible use of a fluorescence activated cell sorter. For most of these applications, it is important to note the usual need for having tissues and/or pretherapy serum specimens to evaluate adequately the presence of a particular marker. Such specimens can be screened for the presence of a variety of markers, and one that is found can then be used for further follow-up of the patient. For this application, it is also important to consider the possible value of assays for immune responses; even relatively nonspecific immune response assays may reflect the tumor burden in the patient. For example, in recent studies performed in my laboratory it has been possible to demonstrate that depressed mixed lymphocyte culture reactivity in the early postoperative period in stage I lung cancer patients is associated with a particularly poor prognosis (Cannon *et al.*, 1980). Similarly, by performing mixed lymphocyte–tumor interactions with postoperative breast cancer patients, it has been possible to show that patients with strong reactivity to their autologus tumors have significantly better prognosis than do patients with the same pathologic stage of disease who have low or undetectable reactivity (Cannon *et al.*, 1981). Radioimmunodetection also appears to be quite promising for detection of occult metastases (Goldenberg *et al.*, 1978). However, this procedure has yet to be shown to be able to detect very small foci of tumor. The present ability seems to be limited to the localization of tumors at least 1–2 cm in diameter; yet, even this could be very helpful for management of various types of tumors. In a more experimental way, there is the potential to use markers to identify possible new types of effective chemotherapy or other treatments. One could sequentially screen a series of drugs in a particular patient, looking for early changes in the levels of circulating marker. Very soon after treatment, one might expect an increase in tumor marker levels related to necrosis of tumor cells if the tumor is responsive. A bit later, one would expect to find falling levels of the circulating marker associated with some response of the tumor to the agent. Such changes in marker levels should allow much more rapid identification of potentially useful drugs than can now be accomplished by following the usual clinical parameters. It is surprising that this approach has not been utilized more by chemotherapists. It has been well demonstrated to be a very potent approach for choriocarcinoma and testicular cancer. This approach should be particularly valuable in the search for new and more effective drugs for solid tumors, where currently only a limited number of useful agents are available. In fact it should be noted that methods for improved detection of recurrent disease will only be clinically useful to the extent that there are effective therapeutic procedures.

C. Response to Continued Therapy and Early Detection of Recurrent Disease

Assays for tumor markers can also be quite useful in following patients in regard to their response to continued treatment and regarding early detection of recurrent disease. Successes with this approach include the application of radioimmunoassays for hCG and AFP for choriocarcinoma and testicular cancer, radioimmunoassays for CEA in colorectal cancer, and some of the assays for ectopically produced hormones.

In regard to all of the above applications, it is important to note that the particular objective in this area of research is to detect recurrent disease reliably, prior to conventional means, in order that this lend to earlier or more aggressive therapy. The real determinant of successful utilization of the markers will be their leading to improved survival of the patients. It would be very helpful to obtain the cooperation of clinicians to couple such studies of tumor markers directly with the randomized clinical trials that are being performed for therapy. This approach would seem to be the most reasonable and economical way to obtain solid documentation of the value of markers in management in cancer patients.

It is likely that the markers that will be useful for the above management applications will mainly be those associated with metastatic tumor cells, since this is the main problem in this area, the early detection of occult metastases at the time of initial treatment or upon follow-up. We continue to face real problems of both sensitivity and specificity with the markers. In regard to the circulating markers, a critical question is how many tumor cells are needed to make enough molecules to enter the circulation and be detected, even by the highly sensitive radioimmunoassays or enzyme immunoassays. It has been estimated that the radioimmunoassays for hCG can detect about 10^4 tumor cells (K. D. Bagshawe, personal communication), and at this level a marker can be very helpful. However, if detection of the marker is only seen when 10^7 or 10^8 tumor cells are present, this may not provide sufficient lead time to be really valuable. There is clearly a need to develop more sensitive approaches to the detection of small numbers of tumor cells. The assays for circulating markers may not, in general, turn out to be sufficiently sensitive for this problem. In any situations where tumor cells themselves can be looked at, or where one can obtain secretions or blood from the region of tumor growth, one might anticipate considerably more sensitive detection of a marker.

IV. MARKERS IN CANCER DETECTION AND DIAGNOSIS

I would now like to turn to the more difficult, but the potentially more important, application of markers to the initial detection and diagnosis of cancer. In general, I suspect that markers which prove to be best for management of cancer patients will also be particularly useful for diagnosis. However, I would like to point out that there may well be markers for local tumors which will not be well expressed or applicable to metastases. This is of particular importance because of the clear desirability to detect tumors accurately at a local stage. For both detection and diagnosis, an important measure of success should be the following. Does the marker provide additional information beyond that of the available established techniques or does it help to identify those individuals who should be evaluated by difficult or potentially dangerous diagnostic procedures? The real criterion for success is, how do the markers make a difference in the ultimate survival of the patients or do they provide an increased chance for curative therapy? In this context, it is of interest to consider the possible use of markers for breast cancer relative to the established use and demonstrated value of mammography. Mammography clearly has some limitations, both in its inability to detect all small breast cancers and in its considerable rate of false-positive results. It should be noted that among women with positive mammograms, only one out of four to one out of ten who are biopsied can be shown to have breast cancer as opposed to benign breast disease. However, such a screening test can still be very helpful to identify subpopulations of individuals who are much more likely to have malignant disease. We can then accept the increased risk of surgery or other procedures to make a more definitive diagnosis. The radioimmunodetection procedure might be useful to complement or possibly even replace mammography, since it might be expected to accurately distinguish cancer lesions from various benign lesions in the breast.

In regard to initial diagnosis, the problem is to discriminate reliably between patients with cancer and those with benign lesions or disease of the same organ. In regard to circulating markers, one would need a marker that is elevated in a relatively high percentage of patients with localized cancer and which is low in most patients with benign disease.

A central question is, is that type of technique likely to have sufficient sensitivity and specificity? It may be unlikely that circulating markers in general will allow very early detection of small numbers of locally growing tumor cells. If it is possible to obtain cells, either by biopsy or by cytology, one would expect to have a much better chance to find small numbers of tumor cells in the specimens. For example, an approach to the application of tumor markers to screening lung cancer would be the examination for tumor markers on cells in the sputum. For the past several years, the National Cancer Institute in the United States has supported an early lung cancer detection project, based on careful examination of sputum cytology. Using standard cytologic techniques, it has been possible to detect a number of patients with *in situ* carcinoma of the lung, and those patients detected on the basis of positive cytology alone have had a high rate of apparently curative surgical removal of their tumors (Melamed *et al.*, 1977). Fluorescence or immunohistochemical staining of such cytology specimens by antibodies to selective lung tumor markers may aid in the early detection of cancer. One would require an antigen that is not on normal lung epithelial cells and is not a differentiation antigen but rather one which would be expressed on most or all lung cancer cells. The use of such a reagent with the fluorescence activated cell sorter might allow screening of large numbers of cells to detect positive cells at a low frequency.

However, an alternative approach is the potentially high sensitivity of assays for immune responses to tumor-associated antigens. It seems likely that immune responses to tumor antigens could occur very early in the course of disease, substantially earlier than one would expect to have release of suffucient quantities of tumor products into the circulation. However, most of these assays for immune responses are still in early stages of development. It will be necessary to develop assays which are considerably more sensitive and reproducible and suitable for large-scale application than is the current generation of assays.

It should be noted that the ideal would be to find positive results with a tumor marker without clear evidence of cancer by classical cytology, X-ray examination, or other accepted diagnostic techniques. This then would lead to a vigorous search for tumor. Such findings in a screening study would also require careful follow-up of the screened population to determine who subsequently develops cancer. It would be necessary to follow both the individuals who were either positive and negative for the particular marker to determine whether the incidence of subsequent cancer would be higher in the marker-positive group. It should be noted that in such studies, even if a marker was not sufficiently discriminatory to allow early diagnosis in a particular patient, if the marker would only be elevated or positive in the subpopulation of individuals at highest risk of developing cancer, such information could be quite useful for focusing the follow-up and for epidemiologic evaluation.

Clearly workers in this field still have much to do, but I am optimistic that continued real progress can be made and that markers will become established as parts of the armamentarium for diagnosis and management of cancer. Hopefully, subsequent publications on this topic will demonstrate the verification of my continued optimism for progress in this important field of research.

REFERENCES

Cannon, G. B., Dean, J. H., Herberman, R. B., Perlin, E., Reid, J., Miller, C., and Lang, N. P. (1980). *Int. J. Cancer* **25,** 9–17.
Cannon, G. B., Dean, J. H., Herberman, R. B., Keels, M., and Alford, C. (1981). *Int. J. Cancer* **27,** 131–138.
Concannon, J. P., Dalbow, M. H., Hodgson, S. E., Headings, J. J., Markopolulos, E., Mitchell, J., Cushing, W. J., and Liebler, G. A. (1978). *Cancer* **42,** 81–87.
Coordinating Group for the Research of Liver Cancer, China (1974). *In* "Proceedings of the Second International Symposium of Cancer Detection and Prevention" (C. Maltoni, ed.), pp. 655–658. Excerpta Medica, Amsterdam.
Goldenberg, D. M., Deland, F., Kim, E., Bennetts, S., Primus, F. J., Van Nagell, J. R., Jr., Estes, N., Desimone, P., and Rayburn, P. (1978). *N. Engl. J. Med.* **298,** 1384–1388.

Herberman, R. B. (1979a). *In* "Compendium of Assays for Immunodiagnosis of Human Cancer" (R. B. Herberman, ed.), Vol. I, pp. 629. Elsevier/North-Holland, Amsterdam.

Herberman, R. B. (1979b). *Proc. Arnold O. Beckman Conf. Cln. Chem., 2nd, 1978* pp. 347–374.

Herberman, R. B., and McIntire, K. R., eds. (1979). "Immunodiagnosis of Cancer." Dekker, New York.

Kurman, R. J., Scardino, P. T., McIntire, K. R., Waldmann, T. A., Javadpour, N., and Norris, H. J. (1978). *Scand. J. Immunol.* **8,** 127–130.

Lange, P. H., McIntire, K. R., Waldmann, T. A., Hakala, T. R., and Fraley, E. E. (1977). *J. Urol.* **118,** 593–596.

Melamed, M. R., Zaman, M. B., Flehinger, B. J., and Martini, N. (1977). *Am. J. Surg. Pathol.* **1,** 5–16.

Olsson, L., and Kaplan, H. (1980). *Proc. Natl. Acad. Sci. U.S.A.* **77,** 5429.

Sallan, S. E., Ritz, J., Pesanso, J., Gelber, R., O'Brien, C., Hitchcock, S., Coral, F., and Schlossman, S. F. (1980). *Blood* **55,** 395–402.

Wells, S. A., Ontjes, D. A., Cooper, G. W., Hennessy, J. F., Ellis, G. J., McPherson, H. T., and Sabiston, D. C., Jr. (1975). *Ann. Surg.* **182,** 362–368.

Subject Index

A

α-Acid glycoprotein, 345
α_1-Antitrypsin, 116, 121, 252, 345
α-Fetoprotein, 6, 7, 90, 91, 115, 243, 309, 381, 411, *see also* Alphafetoprotein
   in prostatic cancer, 290
   synthesis by hepatoma cell lines, 90
   from yolk sac tumors, 383
α-hCG, *see* Human chorionic gonadotropin
α-I-Antitrypsin, 6
α-L-Fucopyransoylceramide, 94
α-Lactalbumin, 6, 139
α-Macroglobulin, 339
α_2-Macroglobulin, 117, 121
α-Napthyl butyrate esterase, 209
AB blood group antigens, changes, in colorectal cancer, 304
ABH blood group substances, 304
Acetylchitobiose, 321
Acid phosphatase, 6, 211, 215, 283, 348
   collection of samples, 284
   isolation, 283
   localization, 283
   techniques for measuring, 284
      clinical relevance, 284
Actin, 7
Acute lymphocytic leukemia, 197, 198, 411
Adenocarcinoma
   colon, 193
   squamous mixed, tumor of lung, 270
Adenocorticotropic hormone, 6, 143, 145, 232, 261, 264, 345
   β-Endorphin, 232

β-Lipoprotein, 94
   ectopic syndrome, 95
   immunoreactive, 232
ADH, *see* Antidiuretic hormone
Adrenal chromaffin cells, 224
Adrenocortical stimulating hormone, 124
Adult progeria, Werner's, 191
AFP, *see* Alphafetoprotein
Alphafetoprotein, 5, 121, 137, 247, 359
   clinical application, and hCG, 243
   concanavalin A binding variants, 403
   countercurrent immunoelectrophoresis, 307
   double agar gel diffusion, 396
   enzyme immunoassay, 397
   gene, 90
   levels relative to treatment, 382
   metabolic half-life, 244
   molecular variants, 90
   oligosaccharide processing, 91
   production
      by germ cell tumor xenografts, 116
      by hepatic tumor xenografts, 117
   radioimmunoassay, 397
   radio rocket electrophoresis autography, 397
   random population, study, 402
   reverse hemagglutination assay, 397
   screening,
      hepatocellular carcinoma in China, 395
      in random populations, 400
      results of large scale, 398
   selected populations, study, 401
   testing interval, study, 403
Agammaglobulinemia, 196
Albinism, 191

Albumin, 116, 121
Alcian Blue, staining, 209
Alkaline phosphatase, 6, 116, 124, 139, 210, 211, 345
  activity, inducible, 99
  carcinoplacental, 389
  cryptic membrane, location, 9
  developmental, phase-specific, 98
  first trimester, 9
  heat-stable, 70
  immunologic determinants, 11
  intestinal, fetal, 9
  isoenzymes, 98
    genes, hypothetical genealogy, 5
    placenta specific, 114
    tumor associated, 10
  liver, human, 99
  regulation, in cell lines, 12
  role, 13
  three-gene loci, 10
  tissue unspecific, 9, 12
Amino acid sequence homology, between AFP and albumin, 90
Antibody localization
  in cancer patients, 169
  in xenografts of human tumors, in experimental animals, 168
Anti-CEA antibodies, see Monoclonal antibodies
Anti-colorectal carcinoma antibodies, 159
Anti-diuretic hormone, 124, 261
Anti-estradiol antiserum, 143
Antigenicity, common cell surface, 125
Antigenic microheterogeneity, 324
Antigens
  cell surface, 125
  cell wall blood group, 308
  depressed, 133
  developmental, 133
  in nucleolus, migrating, lage G_1 phase, 65
Anti-hepatoma nucleolar antisera, 39
Anti-lymphocyte globulin, 111
Anti-melanoma antibodies, 157
Anti-plasmin activity, 379
APUD
  cells, 8, 267, 229, 263
  tumors, 145
Arylsulfatase B, 345
Asialoglycoprotein, binding to hepatic binding protein, 339
Aspartate aminotransferase, 303
Astrocytomas, 157, 163, 198
Ataxia telangiectasia, 191, 224
Autoimmunity, 198

B

β-Endorphin, 232, 261, 264
  enkephalin, 145
β-Glucuronidase, 211
β-LPH, 145
β-Lipoprotein, see Adrenocorticotropin
β-Lipotropin, 232, 261, 264
β-melanocyte stimulating hormone, 124
β-oncofetal antigen, 302
β-oncofetal antigen, in pancreatic cancer, 307
B cell development, 135
B cell disorders, 135
B cell lymphomas, 7, 8
  human, 412
Basal cell cancer, 193
Basement membrane macromolecules, 14, 22
Benign fibrous tumors, 193
Bilateral malignant neurolemmoma, 198
Biliary glycoprotein I, 329
Bipedal lymphangiography, 247
Blastocyst, including ectoderm, endoderm, mesoderm, 5
Blood coagulation, 379
Blood group substances
  in colorectal carcinoma, 303
  in gastric cancer, 310
Bloom's syndrome, 191
Bombesin, 7, 261, 266
Brain tumors, 191
Breast cancer, 83, 139, 360, 361
  families, prone, 201
  nucleolar antigens, 56
Breast duct, normal, human, 149
Bright nucleolar fluorescence
  evaluation, in unknown species, 60
  in human malignant tumor specimens, 54
Bronchial mucosa, normal and neoplastic cell types, 262
Bronchogenic carcinoma, 360
5-bromodeoxyuridine, 13, 89, 99

C

C cell hyperplasia, 224, 233
C cells of the thyroid, 147
CCA III, see Nonspecific cross reacting antigen

Subject Index

CCEA2, *see* Carcinoembryonic antigen, colonic
Carcinoembryonic antigen, 6, 88, 116, 122, 139, 159, 228, 231, 234, 251, 254, 261, 289, 300, 339, 342, 383, 411
  acid phosphatase levels, correlation with, 289
  activity in non-blood body fluids, 341
  antibody concentrated in necrotic areas, 183
  antigens, like, 335
  biosynthesis and catabolism, 318
  carbohydrate content, 320
  carcinomas
    cervix, positive, 385
    colon cells, producing, 89
    lung, human, line, 90
  cell-associated antigen, 318
  chemical composition, 320
  clinical use, 336
  colonic, antigen 2, 326
  cross-reacting antigens, 315, 326
  cyst epithelium, 317
  degree of differentiation of tumors, 340
  evaluation as marker of neoplasia, 322
  evolution, stages, 334
  in gastric cancer, 309
  immunohistochemical staining, 384
  in ovarian cancer, 384
  in pancreatic cancer, 307
  in premalignant lesions of uterine cervix, 383
  levels
    before and after surgery, for gynecological cancer, 387
    factors controlling circulating, 337
    in gynecologic cancer ($\geq$5ng/ml), 385
    low in cells, 330
  localization, cell and tissue, 316
  monoclonal antibodies against, 324, 342
  patient-related blood group activity, 321
  physiochemical characteristics, 319
  production tumorigenicity relationship, 123
  prognostic significance, in cancer tissues, 383
  radioimmunodetection with antibodies to , 342
  synthesis, by colon tumor cells, 89
  tissue localization, 318
  xenograft GW-39, producing, 123
Carcinoembryonic antigen-S (CEA-S), 335
Carcinoembryonic-like antigen (CELIA), 330

Calcitonin, 7, 96, 221, 224, 228, 261, 264
  gene product, related coding regions, 235
  monomer, 96
  radioimmunoassay, 411
  secretion by normal C cells, 222
  tumor marker, 222
Cancer, *see also* Carcinoma, specific types
  colon, 201
  colorectal, 160
    urinary markers, 345
  endocrine aspects, 143
  epidermoid, of cervix, 143
  gastric, oncodevelopmental markers, 83, 160, 308
  gastrointestinal, 361
  gynecologic, oncodevelopmental antigens, 373
  liver, prevention possibility, 405
  lung, human, 361
    biochemistry, 266
    classification, 271
  prostate, 6
    incidence, relative, 281
    staging system, 280
    survival from, 281
    tumor markers in, 279
  radioimmunolocalization, 167
  thyroid, 191
Carcinogenesis
  course of, 403
  two-mutation model, 200
Carcinomas, *see also* specific types
  bony and soft tissue, 193
  breast, 122
    antigen expression, heterogeneity, 151
    human, 132, 141, 142, 150
    colon, human, in hamsters, 168
  embryonal, 6, 116, 118, 254, 358
    cell xenografts, 137
    human testis, 138
  gall bladder, 160
  human, cultured
    breast, 100
    pancreas, 100
    small cell lung carcinoma, response to irradiation, 269
    urinary bladder, 100
    thyroid, 100
  islet cell, 361
  lung, human, small cell, 95
  ovarian, 181, 191

Carcinomas (cont.)
  periampullary, 201
  rat yolk sac, 115
  renal cell, 200, 360
  skin, 99
Carotid body tumor, 193
Casein, 6
Cathepsin B, 211
  lysosomal protease, 102
  thiol proteinases, 215
CEA, see Carcinoembryonic antigen
CELIA, see Carcinoembryonic-like antigen
Cell matrix interactions
  in cancer, 30
  determinants of differentiation tumor invasion, 14
  in development, 28
Cell-mediated immune reactions, 302, 309
Cell membrane-associated
  glycolipids, 91
  glycoproteins, 91
Cells, cultured
  marker expression, 87
  mouse pituitary tumor, 95
Ceruloplasmin, 339
CEX, 326
Chemodactomas, 193
Chemotherapy, monitoring, 248
Cholecystokinin, 7
Chondroitin sulfate, 23
  proteoglycan, 23
Chondronectin, 28
Chondrosarcoma, 191
Choriocarcinoma, 99, 358, 376, 379, 411, 414
Chorionic α_1-microglobulin, 75
Chorioadenoma destruens, invasion mole, 355
Chorionic gonadotropin, 7, 374
  activity, like 112
Chromosome
  breakage disorders, 191
  deletion, constitutional, 200
  rearrangement cancer risk, in families, 199
Chronic lymphocytic leukemia, 193
Chronic myelogenous leukemia, 216
Circulating immune complex, 345
Classification of disease, with tumor markers, improved, 413
Collagen, 14, 22
  interstitial, 22
Collagenase, 31

Colon
  cancer, 201
  mucoprotein antigens, 344
Colon-specific antigen polypeptide, 342, 343
Colonic adenocarcinomas, 122
Colony stimulating factor, 124
Colorectal cancer, 160
  antigens, 335
  developmental, other, 302
  mucin, 301
  markers
    clinical value of CEA, 333
    immunological, 300
    metabolic, 303
  tumor localization with radiolabeled anti-CEA antibodies, 342
Concanavalin A, 97, 244, 377, 403
  affinity variants, 90
Corpus luteum, 353
Corticosteroids, 99
Corticotropin ACTH, 145
Corticoptopin-like intermediate peptide, 232
Cowden's disease, 191
Creatine kinase, (CK-BB)
  prostatic cancer, 290
Creatine phosphokinase, 261
Crohn's disease, 322
Cultured cancer cells, marker expression, 87
Cultured mouse pituitary tumor cells, 95
Cushing's syndrome, 232
Cyclic adenosine 3',5'-monophosphate, 89
Cystadenocarcinomas, 83, 380
Cytolysis, antibody-dependent, cell-mediated, 89
Cytotrophoblast, 114
  cells, 254
Cytoplasmic neurosecretory granules, 266
Cytosine arabinoside, 111

D

Defects, in cellular growth regulation, risk for familial colon cancer, 201
Developmental tissues, 4
Dexamethasone, 95
Dibutyrl-cAMP, 13, 97, 99
Difucosyl oligosaccharide, 163
Dihydrofolate reductase, 211
Dipeptidyl aminopeptidases, I, II, 211
Dopa decarboxylase, 228, 229

Down's syndrome, 191
DR antigens, 159
Duncan's disease, 191
1,2-dimethylhydrazine, carcinogen, 201

E

Ectodermal tumors, 6
Ectopic tumor syndromes, 4, 260
EGF, see Epidermal growth factor
Embryoblast tumors, 6
Embryonal carcinoma, 6, 116, 118, 254, 358
    cell xenografts, 137
    human testis, 138
Embryonal cell migration, 29
Embryonal epitheloid cells, 157
Embryonic liver adhesive glycoprotein, 14
Embryonic prealbumin, 388
Endocervical adenocarcinomas, 380
Endocrine cells, in nonpulmonary epithelial system, 265
Endoderm, cell lineage, 265
Endodermal tumors, 6
    sinus yolk sac, 7, 118, 359
Endometrial adenosquamous carcinomas, 384
Endometrial cancers, 380
Endorphin, 95
Endothelial cell tumors, 6, 7
Entactin, 23
Enzymes
    heterogeneity, 212
    markers in oncology, 214
        in colorectal cancer, 303
    specificity, 213
Ependymoma, 191
Epidermal growth factor, 97
    receptors, 102
Epidermoid cancer of cervix, 378
Epithelial cells, affinity to type IV collagen, 28
Epithelial mammary antigen, 143
Epithelial membrane antigen, 123, 140, 141
Epithelial-mesenchymal interactions, 14
Epstein-Barr virus
    B-cell proliferation, 196
    ectopic ACTH syndrome, 95
    lymphoproliferation, immune response, defective, 196
    membrane antigen, associated, 125
    mononucleosis, infectious, induced, 196

Erythrocytes, 162
Esophagus, 148
Estradiol receptors, location, 144
Estrogen receptor, 6, 143
    determinations, heterogeneity of distribution within cell tumor population, 144
Eutopic expression, oncodevelopmental proteins, 4
Extracellular matrices and components, structure, 22
Extraembryonic pathway, 138
Extramedullary hemopoiesis, 121

F

F9-like antigens, 253
Factor VIII related antigen, 6, 7
Familial adenocarcinoma syndrome, 193
Familial cancer
    mechanisms of neoplastic transformations, 189, 192, 193
    risk, non-X-linked genetic marker, 201
Familial dysplastic nevus syndrome, 193
Familial polyposis coli, 193, 201
Fanconi's syndrome, 191
Ferritin, 252, 254
Fetal genes, 51
Fetal glycolipids, 305
Fetal polygonal cells, 163
Fetal sulfoglycoprotein antigen, 308
Fetuin, 339
Fibromas, ovarian, 193
Fibronectin, 14, 23, 24, 25, 27, 28, 29, 92, 116, 137, 254
    bullous pemphigoid antigen, 23
    cell attachment site, 25
    cell motility, promotor, 28
    cell surface, in transformation, 30
    differentiation, directed, morphogenetic movements of certain cells, 28
    domain, structure, 24
    early embryonal development, 29
    glycopeptides, 92
    glycosaminoglycan-binding domain, 25
    human, 119
    involvement in differentiation, 29
    matrix incorporation, 27
    plasma, 27
    structural diversity, adhesive properties, 24
Fibrosarcoma, 191

Fibrosis, treatment-induced, 248
Flow cytoenzymology, advantages, 207, 208
Flow cytometry, cellular enzyme analysis, 209
Fluorescent methods, 210
Follicle-stimulating hormone, 6, 352
Fructosebiophosphate aldolase, 345
Fucogangliosides, 93
Fucolipids, 93

G

γ-glutamyltransferase, 99, 139, 211, 345
γ-LPH, 95
γ radiation, 198
Galactosyltransferase glycopeptide acceptor, 100
Gall bladder carcinomas, 160
Gamete neoplasms, 6
Gamma melanotrophin, 145
Gangliosides of transformed cells, 93
Gardner's syndrome, 193, 201
Gastin, 345
Gastric cancer
 immunological markers, 308
 metabolic markers, 309
 oncodevelopmental markers, 83, 160, 308
Gastric inhibitory peptide, 7
Gastric/pancreatic cancer phenotype, Lewis blood group type, 162
Gastrin, 7, 145
Gastrointestinal cancers, 160, 193, 361, see also Colorectal cancers
 cell lines, 159
 endocrine cell tumors, 7
 markers, 299
 monoclonal antibody-defined antigens, 161
Gene defect, 201
Genetic immunosuppression, 111
Genetic markers, linked to familial cancer risk, 200
Genetic syndromes and cancer, 190
Genitourinary cancers, 193, 361
Genodermatoses, 191
Germ cell neoplasms, 7, 8
Germ cell tumor xenografts, 115, 253
 human, 136
Germ cell tumors, of gonads, 382
Germinomas, 5
GGT, see γ-glutamyltransferase
Glicentin, 7
Glioblastoma line, CBT, 97

Glioma, 191
Glucocorticoids, 30
Glucosephosphate, 345
Glutamate-pyruvate transaminase, 201
Glycoconjugate markers
 colorectal carcinoma, 303
 pancreatic cancer, 307
Glycogen storage, in colon carcinoma, 303
Glycolipids, 159
 oncofetal type, 164
 precursor to blood group antigens, 304
Glycopeptides, fucose-containing, 92
 large, 92
 membrane-associated, 92
Glycoproteins, 97, 99, 157
 fucose-rich, 388
 hormones, 96
 normal, 326, 335
 sialosyl oligosaccharides, containing, 91
Glycosylated proteins, processing, 93
Glycosyltransferases, 100, 304
GnRH, see Gonadotropin-releasing hormone
Gonadotropin-releasing hormone, 375
Granulocytes, 162, 328
Growth hormone, 6

H

Hamartomatous disorders, 191
Hapten, human Le[a] blood group antigen, 162
Haptoglobin, 339, 345
hCG, see Human chorionic gonadotropin
Heavy chain immunoglobins, 7
Hematopoietic cell neoplasms, 7, 273
Hemopexin, 116
Heparan sulfate, 23
Heparin, 23
Hepatitis B
 surface antigen, 121
 virus DNA, 121
Hepatoblastoma, 117, 121
Hepatocarcinogenesis, 396
Hepatocellular carcinoma, 117, 191, 360, 364, 396
Hepatoma xenografts, 121
Hereditary cancer syndrome, 192
Heterogeneity
 antigen expression, 151
 breast carcinoma, 151
 tumor tissue, 120

Subject Index

Heterotransplanted tumor markers, oncodevelopmental markers, 109
Heterozygotes, glucose-6-phosphate dehydrogenase, 234
High-mannose type oligosaccharide chains, 93, 97
Histaminase, diamine oxidase, 7, 228, 230, 261
Histiocytic lymphoma, 198
  monocytic leukemia, 7, 8
HLA-associated, defective, immune response genes, 197
HLA Dr4, 197
Hodgkin's, non-Hodgkin's lymphomas, 193
Hormonal tumor markers, 231
Human B cell lymphomas, 412
Human blood group B determinant, 163
Human breast carcinoma, 132, 141, 142, 150
Human breast duct, normal, 140
Human cell surface components, 149
Human chorionic gonadotropin, 114, 245, 345, 411
  α-hCG, 117
  bioassays, 354
  biologic activity, 352, 353
  chemistry, production sites, 352
  choriocarcinoma, markers, 139
  detection, frequency in sera
    patients with nontrophoblastic neoplasms, 361, 364
    patients with testicular germ cell tumors, 359
  elevated plasma levels, 136
  expression in trophoblastic and nontrophoblastic tumors, 351
  germ cell tumors,
    producing, 183
    testes, 358
  gestational trophoblastic disease, 355
  HeLa cells, release, 376
  immunoassays, 354
  immunosuppressive properties, 353
  lectin-binding-defined variants, 251
  measurements
    methods, 354
    pregnancy diagnosis, 352
  metabolic half-life, 245
  microheterogeneity, 353
  nontrophoblastic cancer, patients, 362, 375
  nontrophoblastic neoplasms, 360
  occurrence in serum, patients with nontrophoblastic gynecologic cancer, 375
  placental trophoblastic tissue, secreting, 355
  production
    α-subunit, 96
    β-subunit, 96
    fetal genome responsible, 354
    xenografts of testicular, other tumors, 113
  secretion, control mechanisms, 374
  radioimmunoassay, specific, 355
  subunits
    secretion, from cultured cells, 97
    trophoblastic tumors, 374
  tumor marker, 362
Human colon carcinoma, in hamsters, 168
Human fibronectin, 119
Human germ cell tumor xenografts, 136
Human liver alkaline phosphatase, 99
Human lung cancer
  in culture, biochemistry, 266
  classification, 271
Human meconium, 305
Human melanoma, monoclonal antibody-defined antigens, 156
Human phaeochromocytoma, 146
hPL, see Human placental lactogen
Human placental lactogen, 71, 252, 254, 378
  in trophoblastic, nontrophoblastic tumors, 379
Human placental urokinase inhibitor, 379
Human testicular seminoma, 137
Human tumor nucleolar antigens, 37, 50
Hybridomas, 156
Hydatidiform mole, 355, 377, 379
5-hydroxytryptophan, 7, 230

I

Immune deficiency, 191
Immune deficiency, common variable, 191
Immune localization, nucleolar antigens, 61
Immunofluorescent techniques, 210
Immunolocalization, therapy of tumors, 14
Immunological markers, gastric cancer, 308
Immunoglobins
  heavy, 7
  light, 7
Immunoreactive ACTH, 232
Immunoreactive β-endorphin, in tumor extracts, 232
Immunosuppression, of recipient animals, 110

Immunotherapy, with radiolabeled anti-CEA, 343
Indium-III, 171
Inhibitors, DNA synthesis, 97
Insulin, 145
Interferon, 89
Intestinal antigens, in gastric cancer, 308
Intestine gene, 5
Islet cell
  adenomas, 193
  carcinoma, 361
Isomerase, 345
[^{131}I]-labeled antibody to CEA, 169
[^{125}I]-labeled normal immunoglobulin, 169

K

Kasahara isoenzyme, 9
KB nasopharyngeal carcinoma cells, 9
Keratins, 143
Kidney cancer, 191
Kultzchitzsky cells, 145, 224

L

Lactic dehydrogenase, 211, 252
  isoenzyme I, 116
  LDH-x, 252
  in prostatic cancer, 290
L-dopa, 230
  decarboxylase, 261, 263, 264, 266
Lactotransferrin, 157
Laminin, 14, 23, 26, 28, 30
  attachment of liver cells, promotor, 28
  in development of mammalian embryo, 30
  immunoperoxidase staining, 26
  in various teratocarcinoma-derived endodermal cell lines, 31
Large cell, undifferentiated cancer, 268, 274
Leiomyosarcoma, 198
Leukemia, 141
  acute, 191, 198
    lymphoblastic, 133, 193, 214, 411
    myelogenous, 193
  chronic
    lymphocytic, 193
    myelogenous, 216
Leukemia-lymphoma, 361
Leukemic myeloblasts, 328
Leukocyte adherence
  index, 345
  inhibition, 295
Leukocyte differential analyzer, 209
Leukocyte esterases, 209
Leukocyte migration inhibition, 295
Leukocyte peroxidase, 209
Li-Fraumeni syndrome, diverse neoplasms, 193
Light chain immunoglobins, 7
Liver cancer, prevention possibility, 405
Liver carcinogenesis, in humans, 406
Liver nucleolar antigens, 45
Lung cancer, 361
Luteinizing hormone, 352
Lymphocele, 248
Lymphocyte culture reactivity, depressed, mixed, 414
Lymphocyte-tumor interactions, mixed, 414
Lymphocytic leukemia, 7, 191
Lymphoma, 191, 411
  lymphoblastic, acute lymphocytic leukemia, 7
  malignant, 196
  prone family members, 196
Lymphoproliferative disorders, 191, 197
Lymphoreticular, 191
Lysosomal protease, cathepsin B, 102
Lysozyme, 7

M

M1 antigen, 388
  oncodevelopmental marker colorectal cancer, 302
Macrophage activation, 27
Malignant lymphoma, 196
Malignant melanoma, 193
Malignant schwannoma, 193
Mammography, 416
Marijuana smokers, 246
Marker enzymes, assayable by flow cytometry, 214
Marker expression, by cultured cancer cells, 87
Markers, combinations of, 344
Marrow prostatic acid phosphatase, radioimmunoassay, 289
Marrow replacement, 111
Maternal serum hCG, disappearance curve, 357
Meconium, 160
Medullary thyroid carcinoma, 7, 8, 96, 145, 193, 317, 411

Subject Index

control of gene expression, 235
hereditary
  clonal nature, 234
  two-mutational event, 234
histologic evolution, 227, 228
macroscopic, 224
microscopic, 233
rat model, 233
tumor markers, 221
Medulloblastoma, 193
Melanin, 7
Melanocyte stimulating hormone, α, β, λ, 232
Melanocytes of fetal origin, 159
Melanoma, 7, 191, 193, 360, 361
Melanotropin, 95
Membrane surface enzymes, 210
Meningioma, 191
Mesodermal tumors, 6
Mesonephric adenocarcinomas, 384
Metabolic markers of gastric cancer, 309
Methotrexate, 112
Methyl umbelliferone phosphate, 210
Microendocrine system, 8
Micrometastases
  detection, 139
  in lymph nodes, 32
Milk fat globule membrane, 139
Mitomycin D, effects on hepatoma xenograft line Li-7, 122
Mixed adeno-, squamous lung carcinomas, see Adenocarcinoma
Molar pregnancy, 356
Monoclonal antibodies, 149, 150, 155, 156
  anti-breast epithelial, 149
  anti-CEA
    necrotic areas, concentrated, 183
    use in radioimmunotherapy, 325
  human melanoma cells, specificity, 156
  immunohistology, 147
  use, to recognize tumor antigens, 155
Monclonal antibody-defined antigens
  gastrointestinal tract cancers, 161
  human gastrointestinal tumors, 159
  human melanoma, 158
  oncodevelopmental character, 163
Monosialoganglioside, 160, 345
  in colorectal cancer, 305
Mouse
  oviduct, normal, 148
  teratocarcinoma, 115
  thymectomized, irradiated, 122

Mucin
  antigens, associated, 302
  in colorectal cancer, 305
Multiple endocrine adenomatosis, 193
Multiple exostosis, 191
Multiple myeloma, 361
Murine carcinoma cells, embryonal, 125
Murine lymphosarcoma cells, 14
Murine teratocarcinoma cells, 148, 150, 252
Muscle cell tumors, 7
Mutation in single cell, 234
Myeloperoxidase, 209, 211
Myoepithelial cell tumors, 7
Myoglobin, 6
Myosin, 6
4-Methoxy-2-napthylamine, 210

N

N-acetylgalactosamine, 321
Nagao isoenzyme, 9, 389
Naphthol AS, derivatives, 210
NCA, see Nonspecific cross-reacting antigen
NCA-2, see Nonspecific cross-reacting antigen 2
Neonatal thymectomy, 111
Neoplasms, in family, diverse, 198
Neural crest, 145
  cells, 29
Neural development, 267
Neuroblastoma cells, 273
Neuroendocrine differentiation, 266
Neuroendocrine-related biomarkers, 264
Neurofibromas, 193
Neurofibromatosis, 191
Neuromas, 193
  acoustic, 191
Neuron-specific enolase, 261, 267
Neurotensin, 7, 233, 261
Neutral fucolipids, 93
Neutral glycolipids, 93
Nevoid basal cell carcinoma syndrome, 193
NFA-1, see Normal fecal antigen 1
NFCA, see Normal glycoprotein
Non-Regan isoenzyme, 9
Non-seminomatous germ cell tumors, 136, 254
  clinical monitoring
    α-fetoprotein, 248
    human chorionic gonadotropin, 248
  clinical staging, 247
  other tumor markers, 252

Nonspecific cross-reacting antigen, (NCA), 326, 335
  assay, 329
  physiochemical properties, 327
  tissue localization, 328
Nonspecific cross-reacting antigen 2, (NCA 2), 329
Nonspecific esterase, 211
Non-X-linked genetic marker, familial cancer risk, 201
Normal fecal antigen, (NFA), 335
Normal fecal antigen-1, 330
Normal fecal cross-reacting antigen, (NFCA), 330
Normal glycoprotein, (NGP), 326, 335
Novel radiation repair defect, 198
Novikoff hepatoma cells, 44
Nucleolar antigens, 38, 43, 44, 45, 48, 50, 56, 59, 61, 62
  in benign tumors, 59
  characterization, 61
  in normal breast tissue, 59
  mRNA, 48
  possible functions, 62
  rocket immunoelectrophoresis, 47
  tumor antigen, two-dimensional gel electrophoresis, 46
Nucleolar fluorescence, negative, in human tissues, 55
2-Naphthylamine, 210
5-Nitrosalicylaldehyde, 210
5-Nucleotidase, 210, 345

O

Oat cell carcinomas, 8
Occult focus of nonseminoma, 120
Occult metastases, early detection, 415
Oncodevelopmental antigens, gynecologic cancer, 373
Oncodevelopmental biology, 3, 4
Oncodevelopmental enzymes, 389
Oncodevelopmental markers, diagnostic value, 14
Oncodevelopmental proteins, 112
Oncofetal antigen-1, 307
Oncofetal type glycolipid, 164
Opsonin, 27
Oral potassium perchlorate, 172
Orosomucoid, 339
Osseus metastases, detection, 143

Osteogenic sarcoma, 6, 148, 199
Osteomas, 201
Ovarian cancer
  antigens, associated
    comparison, 388
    OCAA, OCAA-1 — OCAA-5, 387
    OvC-1 — OvC-6, 388
  epithelial, 361, 380
  tissues, 83
  urinary oncofetal peptide, associated, 389
Ovarian carcinoma, 181, 191
Ovarian fibromas, 193
Ovarian tumor-associated antigen, Bagshawe, 388
Ovarian tumors, 9

P

Pancreatic cancer, 160
  oncodevelopmental markers, 305
Pancreatic islet cell tumors, 145
  carcinoid tumors, 8
Pancreatic oncofetal antigen, 306
Parafollicular cell, thyroid, 222
Parathyroid adenoma, 193
Parathormone, 345
Parathyroid trophic hormone, 124
Peanut lectin, 305
Pentagastrin, 222
Pepsinogens in gastric cancer, 309
Periampullary carcinoma, 201
Periodic hormonogenesis, 363
Peripheral blood polymorphs, stained by anti-CEA serum, 327
Peroxidase, 211
Peutz-Jeghers syndrome, 191
Phagocytosis, 27
Pheochromocytoma, 191, 193
Phosphodiesterase inhibitors, 97
Phosphotyrosine, 13
  phosphatase, 13
Pituitary gland, 224
Pituitary glycoprotein hormones, FSH, LH, 97
Placental alkaline phosphatase, 5, 7, 9, 12, 250, 254
  in placenta, 11
  chimpanzee, 11
  orangutan, 11
Placental lactogen, 6
Placental proteins
  application as markers in oncology, 81

Subject Index

detection, classification, 70
localization, in normal tissues and cells, 78
occurrence in tumor patients, 81
quantitation in placental extracts and normal body fluids, 78
Placental protein-5, (PP-5), 6, 379
  inhibitor of proteolytic activity, plasmin and trypsin, 76
  in nontrophoblastic tumors, 380
  Radioimmunoassay, 80
  as tumor marker, 82
Placental protein-10, (PP-10), 380
  radioimmunoassay, 81
  as tumor marker, 83
Placental protein-11, (PP-11), 381
  as tumor marker, 83
Placental protein-12, (PP-12)
  in trophoblastic, nontrophoblastic tumors, 381
  radioimmunoassay, 81
  as tumor marker, 83
Placenta multigene family, 5
Placenta-specific α_1-microglobulin, 75
Placenta-specific proteins
  amino acid, carbohydrate compositions, 77
  in body fluids, 80
  in extracts from term placentas of Cynomolgus monkey, 79
  in extracts from term placentas of man, 79
  physical characteristics, 76
Plasma carcinoembryonic antigen
  increase as percentage of cancer and non-cancer cases, 323
  tumor primary, differentiation, metastatic sites, 338
Plasma fibronectin, 27
Plasminogen activator, 101
Polycythemia vera, 199
Polymorphism, allelic, 11
Polyoma-transformed BHK cells, 92
Polyoma virus, 92
Polypeptide, 7
PP-5, see Placental Protein 5
PP-10, see Placental protein 10
PP-11, see Placental protein 11
PP-12, see Placental protein 12
Pre-albumin, 116, 345
Prednisolone induction, placental-type phosphatase, 13
Pregnancy-associated α_2-glycoprotein, 71
Pregnancy-associated macroglobulin, 388

Pregnancy-associated plasma protein C, 72
Pregnancy-specific β_1-glycoprotein, 301
Pregnancy-specific glycoprotein (SP$_1$), 6, 309
Pregnancy zone protein, 71
Premalignant mole pattern, 194
Preneoplastic states, 190
Primary interaction of cells, with extracellular matrix, 27
Procollagen polypeptides, 22
Product diffusibility, 213
Progesterone receptors, 144
Prolactin, thyroid-stimulating hormone, 6
Proopimelanocortin precursor, 232
Prostate cancer, 6
  incidence, relative, 281
  staging systems, 280
  steroid hormones in, 291
  survival from, 281
  tumor markers in, 279
Prostate-specific antigen, 143
Prostatic acid phosphatase, 143, 212
  immunoassays, 286
  radioimmunoassay, 288
Proteases, 100
  from rat hepatoma cells, 102
  surface activity, 102
Protein kinases, 13
Proteoglycans, 22
Pulmonary endocrine cells, 265
Pulse cytofluorimeter method, 135
Pyruvate kinase, 210

R

Radiation resistance, *in vitro*, 271
Radioactive control globulin, 326
Radioimmunoassay, for calcitonin, 411
Radioimmunodetection, use of anti-CEA antibodies, 325
Radioimmunodiagnosis with CEA, 342
Radioimmunolocalization
  adverse reactions, 172
  alternative radionuclides, 185
  alternative to subtraction, 185
  antibody to hCG, 182
  carcinoembryonic antigen, 179
  clinical results, 179
  imaging, 172
  imaging characteristics, $^{99}\beta$-mTc and ^{131}I, 178
  monoclonal antibodies, 185

Radioimmunolocalization (cont.)
  nonspecific accumulation, radioisotope in normal organs, 176
  potential improvements, 184
  preparation, of patient, 172
  subtraction of ^{99m}Tc from ^{131}I image, 173
  testicular tumors, 253
  use of ^{131}I goat antibody to CEA, 178
  use of ^{131}I-labeled antibody to CEA, 170
    in patient with carcinoma of cecum, 182
  use of [^{131}I] rabbit antibody to hCG
    in drug-resistant choriocarcinoma, 184
    in patient with choriocarcinoma, 176
  variations in antibody distribution, normal tissues, 173
  variations in distribution, [^{99m}TC] albumin, ^{99m}TcO4, 177
Radioreceptor assays for hCG, 355
Radio resistance, 199
Rat yolk sac carcinoma, 115
Regan isoenzyme, 9, 98, 389
Release phenomenon, 249
Renal cell carcinoma, 200, 360
Retinoblastoma, 193, 200
Retroperitoneal lymphadenectomy, 247
Rhabdomyosarcoma, 198
Ribonuclease, and prostatic cancer, 291
Rosetting method, use of monoclonal anti-light chain
  antibody fixed to red blood cells, 136

S

Sarcoma, 191, 361
  growth factor, 102
  osteogenic, 6, 198, 199
SCC, see Small cell carcinoma
Schiller-Duval bodies, 136
Screening, organization and institutional form, 398
Screening tests, 281
  selection, 396
Secretin, 7
  associated blood group substances, 308
Seminomas, 5, 9, 116, 118, 136, 248, 385
  anaplastic, 120
  syncytial giant cells, 254
  tumor, like, 118
  tumor markers, 249

yolk sac tumors, ultrastructural and functional continuum, 137
Seromucoid, prostatic cancer, 291
Serum AFP
  detection method, 398
  kinetics, in hepatocellular carcinoma patient, 405
Serum αhCG subunit, elevations, 358
Serum CEA
  administration of intravenous [^{131}I]-labeled goat antibody to CEA, 174
  increase as percentage of cancer and non-cancer cases, 323
Serum Con-A nonbinding SP$_1$, 377
Serum CT concentration decline, after thyroidectomy, 226
Serum galactosyltransferase isoenzyme, 343
Serum markers, 283
Sex hormone binding globulin, 71
Sialic acid-Gal-GlcNAc chains, 92
Sialomucin, 305
Sialylated lacto-N-fucopentaose II, 160
Skeletal muscle tumors, 6
Skin carcinomas, 99
Small cell carcinoma
  biochemical markers, 259
  biochemistry, 270
  biology, 260
  clinical behavior, 263
  differentiation
    changes in status, 268
    lineage, 261
    non-SCC, 265
  lung cells, 95
  neural origin, 263
  undifferentiated, 124
SmIg, expression, by B cells and B cell tumors, 135
Smooth tumors, 6
Sodium butyrate, 13, 97, 99
Solubilized placental tissue proteins, 72
Somatostatin, 7, 145, 232, 261
SP$_1$, 79, 81, 82, 252, 254, 377
  heterogeneities, molecular and immunochemical, 76
  in hydatidiform mole, 82
  as inhibitor, mixed leukocyte reaction, 76
  in nontrophoblastic tumors, 378
  responsiveness of lymphocytes to phytohemagglutinin, depressed, 76

Subject Index

in trophoblastic tumors, 376
tumor marker experience, 81
Specialized migratory embryonic cell tumors, neural crest cell tumors, 7
Specific antibodies in cytopathology, immunohistology, 131
Specific oncoplacental proteins
 characterization, 75
 identification, systematic, 69
 purification, 73
Squamous cell, of skin, 191
src gene, 13
Staining artifact, 214
Steroid binding β-globulin, 71
Steroid hormones
 and metabolites, 7
 in prostatic cancer, 291
Substrate diffusion, 213
Sulfated glycopeptide antigen, 308
Sulfomucin, 305
Surface IgM, 125
Surface protease activity, 102
SV40-transformed 3T3 cells, 91
Syncytiotrophoblast, 114, 353, 379

T

T3 antigen, 133
T4 antigen, 134
T8 antigen, 134
T9 antigen, 133
T10 antigen, 133
T-ALL, see Leukemia, acute, lymphoblastic
T-cell antibodies, OKT series, 134
Technetum-99m, 171
Teratocarcinomas, 5, 92, 116, 159, 164, 358, 403
Teratomas, 5, 248, 358
Terminal deoxynucleotidyltransferase, 7, 133, 210, 211, 215
 positive cells, in bone marrow, 135
Testicular cancer, 411, 414
 clinical aspects, 242
 sensitivity and specificity
  α-fetoprotein in, 245
  human chorionic gonadotropin in, 245
Testicular malignant neoplasms, 83
Testicular tumors
 α-fetoprotein in, diagnosis, 247

human chorionic gonadotropin in, diagnosis, 247
 markers, immunohistochemistry, 254
 radioimmunolocalization, 253
Testosterone production, by testicular Leydig cells, 353
TEX, 330, 335
Theophylline, 89
Therapeutic target cells, 143
Three gem layer tumors, 5, 6
Thymectomized irradiated mice, 122
Thymidine kinase, 303
Thymus, 133, 224
Thyroid C cells, 145
Thyroid cancer, 191
Thyroglobulin, 339
Thyroid-stimulating hormone, 352
Tissue antigen localization, in study of tumor markers, 341
Tissue differentiation, preoperative CEA levels, 340
Tissue dispersal, 212
Tissue polypeptide antigen, 345
Tissue unspecific gene, 5
Tomographic imaging, 185
 computerized, 180, 247
Torre's syndrome, 193
Transferrin, 116, 121, 156, 339
 receptor, 133
Transition between small cell carcinoma and large cell, in undifferentiated lung cancer, 267
Translocation, balanced reciprocal, 200
Trophoblast proteins, developmental, phase-specific, 13
Trophoblast-specific β_1-glycoprotein, 72
Trophoblastic giant cells, 359
Trophoblastic neoplasms, 6
Trophoblastic tissue, 254
TSH β-subunit, 97
Tuberous sclerosis, 191
Tumor antigens, in xenograft-bearing animals, 125
Tumor-associated antigens, single epitopes, 156
Tumor cells, extravasation, 32
Tumor–host size inequality, 111
Tumor isoenzyme, 10
Tumor localization, [^{125}I]-labeled anti-CEA antibodies, in vivo, 123

Tumor markers
 accurate assessment of primary therapy, 413
 cancer detection and diagnosis, 415
 early detection of recurrent disease, 414
 response to continued therapy, 414
 uses and limitations, 409
Tumor nomenclature, embryologic basis, 4
Tumor nucleolar antigens, 44
 immunoelectrophoretic profile, 43
 in nontumor tissues, 51
Tumor promoter, 12-O-tetradecanoyl phorbol-13-acetate, 202
Tumors, *see also* specific types
 ampulla of Vater, 193
 ectodermal, 6
 embryoblast, 6
 lymphoid tissues, 133
 ovarian, 9
 testis, markers, 241
Turcot's syndrome, 193
Two-mutation model, carcinogenesis, 200
Tylosis, with esophageal carcinoma, 193
Type I, Werner's syndrome, 193
Type II, Sipple's syndrome, 193
Type III, mucosal neuromas and endocrine adenomatosis, 193
Type IV collagen, 22
Tyrosinemia, hereditary, 244
[^{99m}TC] sulfur colloid, liver scanning agent, 173

U

Ulcerative colitis mucosa, 317, 322
Ultrasonography, abdominal, 247
Undifferentiated retroperitoneal carcinoma, 360
Unitary cell of origin theory, for lung cancer, 270
Urinary markers, 292
 carcinoplacental isoenzyme (Regan N), 293
 cholesterol, total and non esterified, 292
 colorectal cancer, 345
 fibronectin, 294
 hydroxyproline, 293
 isoleucine, 294
 prostate fluid proteins, 295
 spermidine, 294
 urinary carcinoembryonic antigen, 293
Urokinase, 101

V

Vasoactive intestinal peptide, 7, 145, 233
Vasopressin, 95
 antidiuretic hormone, 145
Viral capsid antigen, 125
Viral transformation of cells, 92
Von Hippel-Lindau syndrome, 191

W

Wag/Rij rat medullary thyroid carcinoma, tumor model, 235
Wagner osteosarcoma, 168
Waldenstrom's macroglobulinemia, 193, 197
Walker carcinoma, 168
Wiscott-Aldrich syndrome, 191

X

X-linked agammaglobulinemia, 191
X-linked lymphoproliferative syndrome (XPL), 196
Xenografts, *see also* specific types
 Burkitt's lymphoma, 125
 choriocarcinoma, 168
  gastric, 114
  gestational, 114
  ovarian, 114
 embryonal cell carcinoma line, 112
 enzyme production, 124
 hormone production, 124
 tumors
  in hamster cheek pouches, 122
  human germ cell, 125
  marker-producing, 115
Xeroderma pigmentosum, 191

Y

Yolk sac carcinoma, 6, 122, 116, 254

Z

Zinc glycinate, marker in malignancy, 344